HEME PEROXIDASES

HEME PEROXIDASES

H. Brian Dunford
Department of Chemistry
University of Alberta
Edmonton, Alberta, Canada

WILEY-VCH

New York • Chichester • Weinheim • Brisbane • Singapore • Toronto

This book is printed on acid-free paper. ♾

Published simultaneously in Canada.

Library of Congress Cataloging-in-Publication Data:

Dunford, H. Brian.
Heme peroxidases / H. Brian Dunford.
p. cm.
Includes index.
ISBN 0–471–24244–6 (cloth : alk. paper)
1. Peroxidase. I. Title.
QP603.P4D86 1999
572′.791—dc21 98–28908

Printed in the United States of America.

10 9 8 7 6 5 4 3 2 1

CONTENTS

PREFACE

Why write a book on heme peroxidases? Because it is not there. There are review articles, edited volumes, and conference proceedings that touch upon, or are devoted to, various aspects of heme peroxidases. However, there has not been a systematic attempt to summarize the growth and current state of knowledge about heme peroxidases in a single volume since the 1964 book, *Peroxidase,* by B. C. Saunders and co-workers at Cambridge University. At that time not a single peroxidase primary sequence had been determined, nor had the first X-ray structure of an enzyme been produced. Many of the peroxidases described in this volume had not been discovered in 1964. A proper classification of heme peroxidases was not published until 1992.

Heme Peroxidases is organized in the following manner. Chapter 1 contains background and historical information, and other heme-containing enzymes are briefly discussed. Horseradish peroxidase and yeast cytochrome *c* peroxidase have been the subject of far more numerous investigations than any other peroxidases, and they are used to introduce peroxidase properties in Chapter 2. In Chapter 3 the family and superfamily classification of peroxidases and catalases devised by Karen Gjessing Welinder is introduced. Chapter 3 also contains an introduction to peroxidase cloning and crystal structures of representative members of the various superfamilies. A detailed discussion of work on the structural properties of yeast cytochrome *c* peroxidase, which for many years was the only peroxidase with a known crystal structure, forms the conclusion to Chapter 3.

Horseradish peroxidase provides the theme for Chapters 4–8. Its properties and mechanisms of its multitudinous reactions are described in Chapters 4–6. Because of the pioneering work on the horseradish enzyme, it is used as a method of introducing the battery of spectroscopic techniques that are now applied to peroxidases. In Chapter 7 the uses of electromagnetic radiation to obtain electronic, resonance Raman, X-ray absorption, and X-ray diffraction spectra are described. The determination of the unique saddle shape of peroxidase hemes by normal coordinate analysis is described. Also included in Chapter 7 is magnetic circular dichroism (MCD), which combines the use of optical and magnetic properties. Chapter 8 is devoted to the magnetic spectral techniques: nuclear magnetic resonance (NMR), electron paramagnetic resonance (EPR), and electron nuclear double resonance (ENDOR). Mössbauer spectroscopy, which combines the effects of

γ-radiation and a magnetic field, and complements EPR so well, is included. In each of the remaining chapters, I have integrated results from many techniques into the discussion of the various peroxidases. Chapters 9 and 10 are devoted to studies on yeast cytochrome *c* peroxidase: its reactions with small molecules in Chapter 9 and its unique reaction with the heme protein cytochrome *c* in Chapter 10.

The Welinder classification was used to organize the remaining chapters. The class I peroxidases, pea cytosolic ascorbate peroxidase and the bacterial catalase-peroxidases are described in Chapter 11. Chapter 12 is devoted to the fungal peroxidases, which are part of class II: lignin, manganese, and *Coprinus cinereus* peroxidases. Class III plant peroxidases, other than that from horseradish, are discussed in Chapter 13. Also included are results obtained with transgenic plants and the intriguing concepts of free radical chaperones in lignin biosynthesis and cancer in plants. Two peroxidases that do not fit into the Welinder scheme, chloroperoxidase, from *Caldariomyces fumago,* and the diheme peroxidase, from *Pseudomonas aeruginosa,* are described in Chapter 14.

The remaining chapters are devoted to animal peroxidases and catalases. Chapter 15 is devoted largely to myeloperoxidase from white cells and spleen. The much less studied eosinophil peroxidase is briefly described. The surprise peroxidase, prostaglandin H synthase, is discussed in Chapter 16; and two other important animal peroxidases, those from bovine milk and from thyroid glands are the topic of Chapter 17. Chapter 18 is devoted to the tetrameric catalases. Appendix A contains the amino acid sequences of many peroxidases.

Heme Peroxidases was designed to give the reader a feeling for the development of the field and the current state of knowledge as defined by results on the more intensively investigated peroxidases. Literature results to the end of 1997, with some from 1998, are included in the approximately 1700 references and 240 figures. However, the book was not intended to be a listing of all peroxidase references, a compilation that would have left no room for discussion. It became apparent as work on this volume progressed from its starting date in the fall of 1995 that heme peroxidases and heme catalases provided more than sufficient material for one volume. As a result, nonheme peroxidases and catalases, as well as model systems that include the microperoxidases derived from cytochrome *c,* were excluded.

I am indebted to the following friends and colleagues who reviewed various portions of the manuscript during its preparation: John Dawson, David Goodin, Lowell Hager, James Penner-Hahn, Yuchiong Hsuanyu, B. H. Vincent Huynh, Peter Jones, Vasilij Koshkin, R. E. D. (Ted) McClung, Francis Millett, Albin Otter, Giulietta Smulevich, Mario Sono, John Shelnutt, Robert Van Huystee, Nigel Veitch, and Karen Welinder. Also to the following who contributed original diagrams: Keiichi Fukuyama, Michael Gajhede, Michael

Garavito, David Goodin, Anette Henriksen, Thomas Poulos, Michael Rossmann, David Schuller, and Nigel Veitch.

Finally, I thank the many fine scientists from around the world with whom I have been associated. In my own laboratory I am especially grateful for the sustained, skillful work of Anne-Marie Lambeir, Leah Marquez, and Diana Metodiewa.

H. BRIAN DUNFORD

Department of Chemistry, University of Alberta
Edmonton, Canada

HEME PEROXIDASES

CHAPTER 1

Introduction. Historical Background

In this chapter background information on the peroxidases is presented, the relation of peroxidases to other heme proteins is discussed, and a brief overview of some important historical developments is given.

1.1. HEME PROSTHETIC GROUP

The common feature of all heme peroxidases is that their active sites contain very similar prosthetic groups. For nearly all known plant peroxidases the prosthetic group is *ferriprotoporphyrin IX* shown in Fig. 1.1.[1] The prosthetic groups in some mammalian peroxidases are different and have been a center of controversy. The different active site structures in lactoperoxidase and myeloperoxidase will be discussed later. Ferriprotoporphyrin IX, containing iron(III), is an example of a *hemin,* whereas ferroprotoporphyrin IX, containing iron(II), is a *heme.* However, there are at least five different oxidation states of the prosthetic group of a peroxidase, and as long as these are clearly designated, it is simpler to use the name heme, regardless of its oxidation state.

Ferriprotoporphyrin IX is made up of four *pyrrole* rings joined by *methene bridges* with iron(III) centering the molecule. Two hydrogen atoms attached to pyrrole nitrogens have been displaced by the iron. Therefore, the net charge on the heme with four-coordinate iron(III) is +1. Heme as purchased is usually the salt hemin chloride, with the chloride as a counterion. If the iron is absent, the molecule is a *porphyrin.* With no side chains and no iron the molecule is *porphine.* Some compounds based on the tetrapyrrole structure are shown in Fig. 1.2.

There are eight side chains in ferriprotoporphyrin IX as follows: four methyl groups (positions 1, 3, 5, and 8), two vinyl groups (2 and 4), and two propionate groups (6 and 7). The carbon atoms of the methene bridges are labeled α, β, γ and δ, with the α carbon being located between side-chain positions 2 and 3. The official name of the porphyrin is a mouthful: 1,3,5,8-tetramethyl-2,4-divinylporphine-6,7-dipropionic acid. Heme was first synthesized by Hans Fischer.

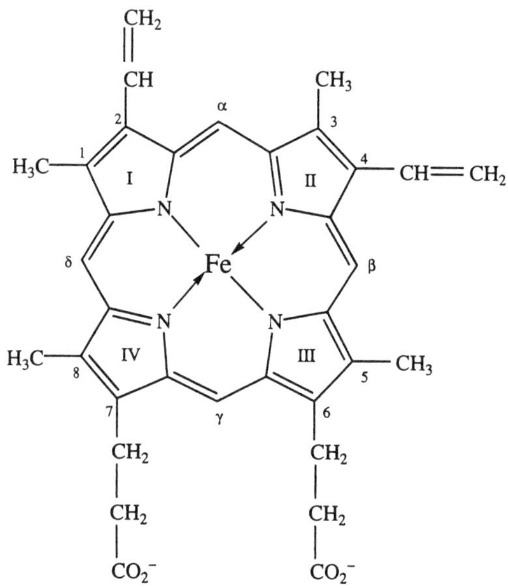

FIG. 1.1. Structure of ferriprotoporphyrin IX. The modified Fischer numbering system is shown. In the less widely used IUB/IUPAC system the 20 carbon atoms of the porphine skeleton are numbered clockwise, starting with the carbon atom between positions 1 and δ.

In its inorganic compounds iron is commonly six-coordinate, that is, it has six monodentate ligands attached. A long-standing mistake in peroxidase chemistry was the assumption that the heme iron is also six-coordinate. For the native resting enzyme, the iron(III) is most likely five-coordinate; and the evidence for the coordination number of five is documented later. Addition of common ligands such as cyanide or azide to the native enzyme results in six-coordinate iron(III).

The coordination positions of the iron are numbered. In peroxidases, positions 1–4 are occupied by the four pyrrole nitrogen atoms. Position 5 is located on the *proximal side* of the heme and is commonly occupied by the imidazole side chain of a histidine residue. Position 6 in the native resting enzyme is vacant and is located on the *distal side* of the heme (Fig. 1.3). The distal cavity is the region in which many of the peroxidase reactions occur. When cyanide is added to a peroxidase solution, it diffuses into the distal cavity and coordinates directly to the iron(III) so that the coordination number of the iron changes from 5 to 6. The methene bridges of a porphyrin are susceptible to attack. Cleavage of one methene bridge breaks the porphyrin ring and converts it into a linear structure (Fig. 1.2, B).[2,3]

In peroxidases, the heme is attached to the apoprotein by a linkage, in which an amino acid side chain is attached to the fifth coordination position

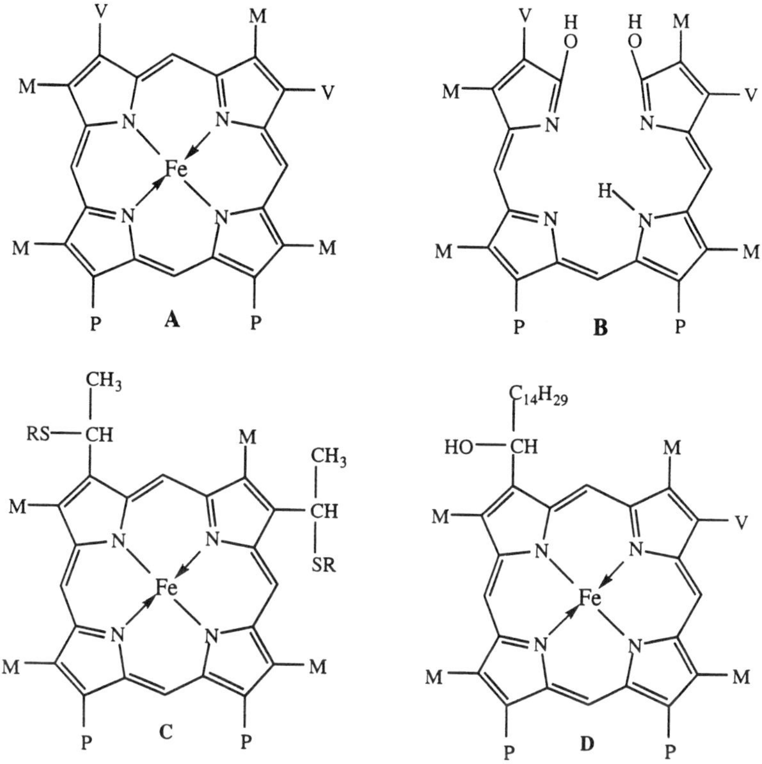

(A)

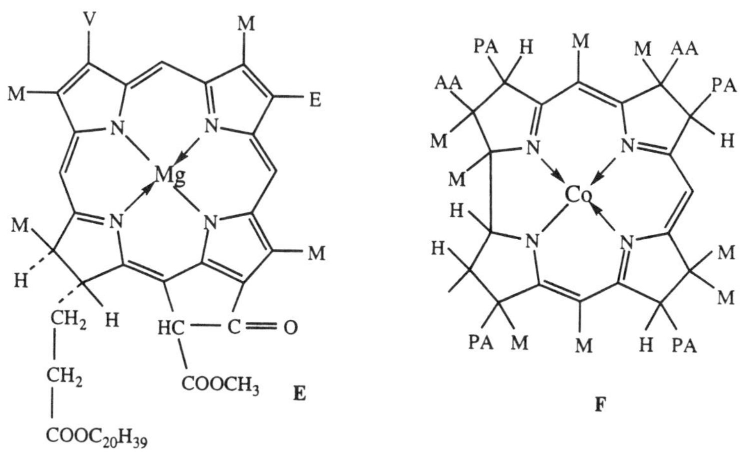

(B)

FIG. 1.2. Some biologically important molecules containing tetrapyrroles: (**A**) ferriprotoporphyrin IX, (**B**) the bile acid biliverdin, (**C**) heme *c,* the prosthetic group of cytochromes *c,* (**D**) heme *a,* prosthetic group of cytochromes of class *a,* (**E**) chlorophyll *a,* and (**F**) the tetrapyrrole portion of vitamin B_{12}. Side-chain designations: M, methyl; V, vinyl; P, propionate; E, ethyl; AA, acetamide group —CH_2CONH_2; PA, propionamide group —$CH_2CH_2CONH_2$; R, aliphatic group.

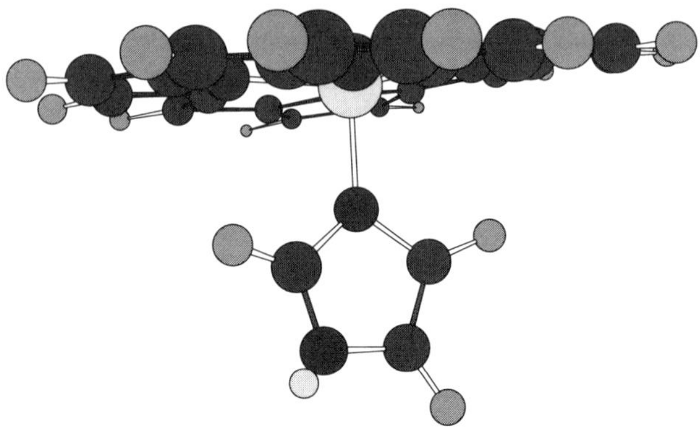

FIG. 1.3. Ball-and-stick model of a heme. The proximal side, below the heme, has an imidazole side chain coordinated to the iron. The distal side, above the heme, is the active site of peroxidases. Details of the heme side chains are not shown.

of the iron(III). This linkage is probably best described as being covalent. It can be broken in acid solution. The propionate side chains form hydrogen bonds with neighboring residues. Finally, weak hydrophobic forces, in which the organic porphyrin is in van der Waals contact with hydrophobic amino acids, complete the binding of ferriprotoporphyrin IX to the protein. The heme, which is usually not readily removed from the apoprotein, is called a prosthetic group. It remains attached to the protein throughout the enzyme cyclic reaction.

1.2. ROLE OF PEROXIDASES IN THE DEVELOPMENT OF MODERN ENZYMOLOGY

Peroxidases played a key role in the development of modern enzymology. A famous controversy occurred in the 1920s and 1930s between Richard Willstätter and James Sumner over the nature of enzymes. Willstätter had worked on the purification of horseradish peroxidase and was familiar with heme chemistry. His view was that enzymes are not catalysts and that trace inorganic contaminants were responsible for catalysis. Sumner worked on protein purification and was able to crystallize jack bean urease, the first crystallization of a protein. He was convinced that pure proteins were capable of catalysis. He actually visited Willstätter's laboratory but was unable to convince him of his views. The irony of Sumner's contribution is that it is now known that jack bean urease contains nickel ions. Partly from the work of these pioneers, the present-day knowledge evolved that many enzymes are pure proteins, made up solely of amino acid building blocks. Many others also contain metal ions, which may be structural elements or centers of catalysis. Still other enzymes, which are excreted via the endoplasmic reticulum, contain disulfide bridges, calcium ion, and have carbohydrate groups attached. The disulfide bridges and calcium ion increase protein stability, whereas the solubility in aqueous solution increases with the amount of glycosylation. It is also likely in the mammalian peroxidases that the sugar moieties play a role in immunological reactions.

Some of the most successful early studies on enzymes were on those that are secreted out of their parent cells. Purification of an enzyme that is outside of its cellular matrix is easier. Thus, the serine proteases initially were among the most widely studied enzymes. Horseradish peroxidase is located in the vacuole inside the cell. Nevertheless, it could be isolated and for a long time was the best known, most worked upon, and simplest example of a heme enzyme.

The amino acid sequence of a protein, its primary structure, was beyond the capabilities of scientists prior to World War II. Without primary structures an understanding of the three-dimensional structure was impossible. It was equally impossible to understand the mechanisms of enzyme action.

1.3. EARLY ENZYME KINETICS

Because of the confusion that has existed in the literature on peroxidase kinetics, a brief review of pertinent developments is justified. Some of the pre–World War II accomplishments in enzymology that have stood the test of time were in interpretation of rate measurements. The paper of Michaelis and Menten in 1913[4] established the basic equation for the initial rate of a single enzyme–single substrate reaction, as follows:

$$v = \frac{V_{max}}{1 + \dfrac{K_M}{[S]}} \tag{1.1}$$

In Eq. (1.1), V_{max} is the maximum attainable rate of product formation from the enzyme-catalyzed reaction (units Ms^{-1}), v is the rate for any given value of substrate concentration [S]. In the original formulation K_M was assumed to be the equilibrium constant for dissociation of substrate from enzyme.

$$ES \rightleftharpoons E + S \tag{1.2}$$

$$K_M = \frac{[E][S]}{[ES]} \tag{1.3}$$

Thus the concept of reversible binding of substrate to enzyme was established. If the concentration of enzyme is known, then Eq. (1.1) can be written in more fundamental form. The value of V_{max} can be defined in terms of total enzyme concentration $[E]_0$ as follows:

$$V_{max} = k_{cat}[E]_0 \tag{1.4}$$

where k_{cat} is the rate of product formation (units s^{-1}) for an enzyme totally saturated with substrate, also called the turnover number. By combination of Eqs. (1.1) and (1.4) one obtains

$$\frac{v}{[E]_0} = \frac{k_{cat}}{1 + \dfrac{K_M}{[S]}} \tag{1.5}$$

The steady-state approximation was introduced into chemistry by several workers in 1919 and into enzyme kinetics by Briggs and Haldane[5] in 1925. Briggs and Haldane more properly defined K_M in terms of rate constants, not an equilibrium constant, since the equilibrium between substrate and enzyme is shifted by the subsequent chemical reaction. For reversible substrate binding, it is best to regard K_M simply as one of the two parameters that can be determined experimentally from the shape of the rectangular hyperbola obtained by plotting $v/[E]_0$ versus [S]. Today K_M is commonly called the Michaelis constant, regardless of the theoretical interpretation placed upon it.

One can formulate a single-substrate reversible enzyme reaction as follows:

$$E + S \rightleftharpoons ES \rightleftharpoons E + P \tag{1.6}$$

The kinetic parameters obtained from studying the initial forward and reverse reactions can be related to the final equilibrium between initial substrate and final product. For forward and reverse k_{cat} values, use the subscripts F and R. Label the Michaelis constants for substrate (S) or product (P). The following equation can be derived from the equilibrium condition, where forward and reverse rates are equal.

$$K_{eq} = \frac{[P]_{eq}}{[S]_{eq}} = \frac{k_F K_P}{k_R K_S} \tag{1.7}$$

The above equation, commonly called the Haldane relation, is valid for any reversible single-substrate enzyme reaction, regardless of the actual mechanism of reaction.[6] The establishment of the basic equations for single–substrate single-product reactions paved the way for the development of theory for two–substrate two-product reactions, of which peroxidase–catalyzed reactions occupy a unique niche.

As we shall see, the concept of reversibility of substrate binding caused confusion in peroxidase chemistry, where irreversibility of the reaction is the usual occurrence, and enzyme–substrate complexes are difficult if not impossible to detect. Covalent compounds, not enzyme–substrate complexes, are the usual detectable intermediates in the peroxidase cycle.

1.4. SOME EARLY PEROXIDASE HIGHLIGHTS

Much of the work on enzymes in the nineteenth and early part of the twentieth centuries was not definitive because of inadequate theory, poorly developed techniques for studying proteins, and crude instrumentation. The early peroxidase literature is well reviewed by Paul[7] in 1963 and in the 1964 book by Saunders et al.[8]

Much of the early definitive work was that of Keilin and Theorell. Keilin initially studied cytochromes, which possess a similar heme group to that in peroxidases (Fig. 1.2). In a pioneering paper Keilin detected cytochromes spectroscopically in the wings of insects and showed how the spectra changed after an insect flapped its wings because the cytochrome became oxidized.[9] The techniques he developed were valuable in his studies on peroxidases.[10] Keilin spectroscopically detected what is now known as compound II of horseradish peroxidase in 1937.[11] This appears to be the first spectroscopic detection of an enzyme intermediate. Some hailed it as the elusive Michaelis–Menten complex, but it is now known to be a covalent compound, formed by binding of one oxygen atom from hydrogen peroxide to the iron of the heme.

Keilin thought he had the primary compound of horseradish peroxidase, but Theorell later showed that Keilin's compound was preceded by what is now known as compound I.[12] Keilin's compound was red in color; Theorell's was green. Theorell obtained the green compound, which rapidly and spontaneously converted to red. The late Karl-Gustav Paul, who simplified and improved Theorell's isolation procedure,[13] described to me at a peroxidase meeting in Geneva in 1986 some of his (Paul's) work on the highly reactive compound I. As Paul described it, upon addition of a peroxide to pure horseradish peroxidase, the solution changed from brown to bright green, and the color remained, much to everyone's amazement. Members of the laboratory stood contemplating the solution, and since it was lunchtime, withdrew to discuss what had happened. Upon returning to the laboratory the green solution remained unchanged! In this pioneering work, not only was compound I obtained, but the preparation was probably equal to that of highest purity and stability ever attained.

1.5. POST–WORLD WAR II DEVELOPMENTS

The explosion in discoveries about enzymes in the last half of the twentieth century dwarfs all previous knowledge. Research on peroxidases and heme enzymes has both contributed to and benefited from this explosion. Here is a capsule summary of some highlights.

1.5.1. Primary Structure and the Genetic Code

After years of work Frederick Sanger and co-workers completed the determination of the 51-amino acid residue sequence of insulin in 1953.[14–16] Some examples of the explosion of primary structure determinations are given in Appendix A. The accomplishments of Sanger are truly amazing. After showing how to obtain the amino acid sequence of a protein, he pioneered the sequencing of ribonucleic acid (RNA) and the rapid sequencing of deoxyribonucleic acid (DNA).

The brilliant work of Franklin, Wilkins, Crick, Watson, and others, in determining the double helical structure of DNA, well documented elsewhere, is regarded by many as the founding of modern molecular biology. One of the most exciting areas of current research is on the mechanism by which the genetic code is utilized in the ribosomes for protein synthesis.

Membrane-bound enzymes can be particularly difficult to isolate and purify. With the determination of the genetic code, it has now become easier to obtain protein primary sequence data from that of DNA, rather than by direct amino acid sequence determination of the protein.

1.5.2. X-Ray Determination of Secondary and Tertiary Structures

The common structural elements of proteins, the α-helix, β-pleated sheet, the turns, and loops constitute the secondary structure of proteins. The tertiary structure is the manner in which these elements are fit together. Because of the size of proteins, X-ray structural determinations cannot identify individual atoms or, in many cases, amino acid residues. However, structural determinations are possible once the primary sequence data is available and satisfactory crystals are grown.

The first breakthrough in X-ray crystallography of proteins involved the heme protein myoglobin, the structure of which was determined by John Kendrew et al. in 1960.[17] This was followed by the first enzyme structural determination, that of egg-white lysozyme, in 1965 by the group of David Phillips.[18,19] At an increasing rate, structures followed for oxyhemoglobin (Max Perutz et al.,[20] 1968), insulin (the group of Dorothy Hodgkin,[21] 1969), and ferric cytochrome *c* (R. E. Dickerson et al.,[22] 1971).

Pioneering work of a different sort on ribonuclease proved to be of fundamental importance. Christian Anfinsen showed that ribonuclease could be folded and unfolded reversibly. The molecule contained within itself sufficient information to restablish its proper tertiary structure after being denatured. Thus, three-dimensional structure is determined by the primary sequence. Recent work has defined the role of chaperone molecules, which can aid in attainment of the proper three-dimensional structure of larger proteins. Only an estimated 10% of endoplasmic-reticulum targeted proteins get help from chaperones.

The first crystal structure of a peroxidase was that of yeast cytochrome *c* peroxidase obtained by Joseph Kraut et al.[23] in 1980, followed rapidly by that of beef liver catalase by Michael Rossmann et al.[24] in 1981.

Questions arose as to whether crystal structures are the same as structures in solution. It would appear that they usually are. Structures of smaller proteins in solution can now be determined by nuclear magnetic resonance (NMR) techniques. Some differences between solution and crystal structures of cytochromes *c* have recently been determined in the NMR work of Ivano Bertini and co-workers.[25] As of 1995 the upper limit for the structure of a protein in solution is molecular weight of 26,000 daltons, determined using a 600-MHz NMR machine.

1.5.3. Site-Directed Mutagenesis

Once methods of sequencing of DNA became established and the genetic code was understood, the way was paved to manipulate nucleic acids to modify protein structures at specific sites. The development of restriction enzymes made it possible to cleave nucleic acids at specific sites, so that deletions and insertions could be controlled. The pioneering work of Michael Smith on methods of oligonucleotide synthesis and the later polymerase chain reaction, developed by Kary B. Mullis, made it possible to synthesize custom-designed proteins. Michael Smith was involved in the first site-directed mutagenesis of a peroxidase, that from yeast.

1.5.4. Kinetics of Two-Substrate Reactions

The basic behavior of enzymes that catalyze reactions involving two different substrates was not well understood until 1953. Robert Alberty pointed out clearly the kinetic differences in behavior in two-substrate, two-product reactions.[26] In particular, he delineated the unique kinetic behavior of an enzyme, which has the following mechanism. One substrate binds, followed by first product release. A modified enzyme is the result, to which the second substrate binds, followed by second product release.

The latter was popularized as a ping-pong mechanism by W. W. Cleland.[27,28] Important contributions to steady-state enzyme kinetics were also made by Keith Dalziel.[29] Using the notation of Cleland, the ping-pong mechanism can be represented as shown in Fig. 1.4.

The conventional ping-pong mechanism is reversible. A consequence of the reversibility is that there is a finite upper limit in rate. For lower substrate

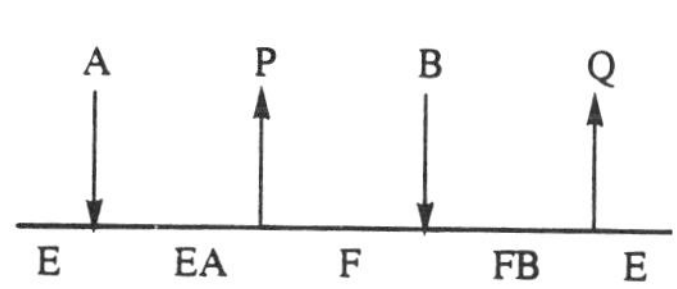

FIG. 1.4. Ping-pong mechanism. The substrate (or coenzyme) A reacts with the enzyme E to modify it to a new form F. Coenzymes such as reduced nicotinamide adenine dinucleotide (NADH) often behave kinetically in the same way as substrates, forming a reversible reactive complex with the enzyme.

concentrations, substrate binding is rate-limiting. For an enzyme saturated with substrate, product release is rate-limiting. The rate equation for the first few percent of a reversible ping-pong reaction is the following:

$$\frac{v}{[E]_0} = \frac{k_{cat}}{1 + \dfrac{K_A}{[A]} + \dfrac{K_B}{[B]}} \tag{1.8}$$

K_A is the Michaelis constant for substrate A, and K_B for substrate B. This equation for a ping-pong mechanism contains one term less in the denominator than the equations for other reversible two-substrate mechanisms. There are three unknown quantities, k_{cat}, K_A, and K_B, which can be determined by kinetic analysis.

The Haldane relation for the above equation is:

$$K_{eq} = \frac{[P]_{eq}[Q]_{eq}}{[A]_{eq}[B]_{eq}} = \frac{k_F{}^2 K_P K_Q}{k_R{}^2 K_A K_B} \tag{1.9}$$

The derivation of appropriate steady-state kinetic equations for reversible enzyme-catalyzed reactions was greatly facilitated by the King–Altman rules.[30] The ping-pong mechanism exhibits strikingly different kinetic behavior from other two-substrate–two-product reactions, leading to so-called parallel line kinetics. As we shall see, peroxidase kinetics have similarities to ping-pong kinetics, with the major difference that the reactions are usually irreversible. As a consequence of the irreversibility, there is no upper limit in rate. When more of both substrates is added, the reaction continues to increase in rate.

The establishment of the mechanism of a peroxidase reaction had a long and tortuous path, complicated by the dominance of concepts about enzyme–substrate complexes. It is now known that there are several mechanisms.[31] The most common mechanism that appears to be universally accepted today is the following:

$$\text{Native peroxidase} + H_2O_2 \xrightarrow{k_1} \text{Compound I} + H_2O \tag{1.10}$$

$$\text{Compound I} + AH \xrightarrow{k_2} \text{Compound II} + \bullet A \tag{1.11}$$

$$\text{Compound II} + AH \xrightarrow{k_3} \text{Native peroxidase} + \bullet A + H_2O \tag{1.12}$$

$$H_2O_2 + 2AH = 2H_2O + 2\bullet A \tag{1.13}$$

AH is a reducing substrate, such as a phenol, and •A is the corresponding free radical. The establishment of this mechanism, the nature of the oxidized intermediates, compounds I and II, and subsequent reactions of the free radicals generated in the peroxidase cycle are discussed in detail in Chapters 4–6.

The peroxidase reaction cycle is an irreversible ping-pong mechanism as illustrated in Fig. 1.5. The concept of irreversibility in an enzyme reaction, sometimes misunderstood, has a clear meaning. For a reaction to occur, the enzyme and its substrate must collide and form some type of encounter complex. For an irreversible reaction, the rate at which the complex reacts is fast compared to the rate at which it dissociates and hence reverts to initial reactants, which is the case in most peroxidase reactions. The consequence is that the rate of the reaction appears to have no upper limit. The larger the concentrations of reactants the faster the reaction. Thus, for example, a plot of the pseudo-first-order rate of compound I formation versus hydrogen peroxide concentration is linear and does not display saturation. The observed irreversibility does not mean the back reaction to initial reactants has zero rate, but that its rate is not measurable. As substrate concentration increases, eventually either saturation will occur or some other process such as enzyme inactivation will intervene.

Because of the 2 : 1 stoichiometry between hydrogen peroxide and phenol, it is necessary to define the rate clearly, either in terms of rate of H_2O_2 disappearance or of phenol disappearance. The rate of disappearance of phenol is twice that of H_2O_2. For H_2O_2 the following equations apply:

$$v = -\frac{d[H_2O_2]}{dt} \tag{1.14}$$

$$\frac{v}{[E]_0} = \frac{1}{\dfrac{1}{k_1[H_2O_2]} + \dfrac{k_2 + k_3}{k_2 k_3[AH]}} \tag{1.15}$$

Since the rate constant for compound I disappearance is typically an order of magnitude larger than that for compound II disappearance, $k_2 \gg k_3$, and the latter equation simplifies to

$$\frac{v}{[E]_0} = \frac{1}{\dfrac{1}{k_1[H_2O_2]} + \dfrac{1}{k_3[AH]}} \tag{1.16}$$

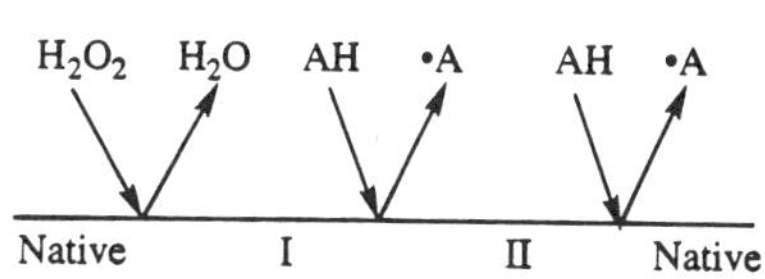

FIG. 1.5. Peroxidase cycle shown in the fashion of a ping-pong mechanism. Because complexes between substrates and enzyme are very short lived, and difficult if not impossible to detect, product is shown as departing as fast as substrate comes in contact with enzyme. I is compound I and II compound II.

There is no k_{cat} or K_M term to be seen in the latter equation. It is a grievous error to try to analyze peroxidase kinetics in terms of k_{cat} and K_M. There are two unknowns, k_1 and k_3, the second-order rate constants for the formation of compound I and reduction of compound II (units $M^{-1}s^{-1}$). Thus, peroxidases have a simpler steady-state rate equation than exists for any reversible two-substrate mechanism. Because of the irreversibility of peroxidase reactions, equilibria between reactants and products are not established, and there is no Haldane relation.

1.5.5. Discovery of Prostaglandins

The discovery of the prostaglandins was reported in 1964. Surprises in science continually occur, and one of the bigger ones occurred when the enzyme that initiates the arachidonic acid cascade was isolated and characterized. It was found that prostaglandin H endoperoxide synthase, which converts arachidonic acid into prostaglandins G_2 and H_2, has an intact peroxidase incorporated into its structure. Thus, the biochemistry of the prostaglandins, and the related leukotrienes and thromboxanes, is intimately connected with peroxidase chemistry. The properties of prostaglandin H synthase are discussed in detail in Chapter 16.

1.6. RELATION TO CYTOCHROME *c*, MYOGLOBIN, HEMOGLOBIN, AND CYTOCHROME P450

The cytochromes, the globins, and the cytochrome P450 enzymes all contain heme, as do most peroxidases. The differences in the chemistry that they perform is striking and must be attributed to the interaction of the protein with the heme.

1.6.1. Cytochrome *c*

Cytochrome *c* contains heme *c* (Fig. 1.2). It has six-coordinate iron with a thiolate and imidazole group in the fifth and sixth coordination positions, and a protein structure that holds these ligands firmly in place. The iron is thus inaccessible to potentially reactive ligands. Reduced cytochrome *c* contains iron(II) and the oxidized form iron(III). The mechanism of electron transport in and out of redox proteins is a research topic of great current interest. A heme group, which is highly aromatic, is the miniature equivalent of a copper bar in its ability to transport electrons. However, to penetrate the protein matrix, does an electron prefer to follow a pathway along covalent chemical bonds or does it prefer to jump through free space? The final answer is not yet established.

1.6.2. Myoglobin and Hemoglobin

The oxygen transport and storage proteins hemoglobin and myoglobin contain ferroprotoporphyrin IX. Iron(II), not iron(III), is at the center of the globin heme. Peroxidase with its iron(III) will not bind molecular oxygen, but when the iron is reduced to iron(II), peroxidases bind oxygen in similar fashion to the globins.

Molecular oxygen, formally called dioxygen, is best represented by its molecular orbital description as containing a triple bond with two unpaired electrons:

$$\bullet \mathrm{O}\equiv\mathrm{O}\bullet$$

There is a pair of nonbonding electrons on each atom. How can there be two unpaired electrons? Because they occupy different antibonding orbitals. The accurate description of the electronic structure of molecular oxygen was a great triumph of molecular orbital theory, since other theories of chemical bonding failed to predict the properties of molecular oxygen.

Why can iron(II) bind molecular oxygen when iron(III) cannot? One might naively expect the reverse. Since molecular oxygen is rich in electrons, one might think it would be more strongly attracted to iron(III), which has one electron less than iron(II). The answer is that the extra electron in an iron(II) porphyrin, compared to a porphyrin containing iron(III), completes an electron pair appropriately positioned to bond to the oxygen. The process is called π-back bonding and it stabilizes the iron–dioxygen complex.

The different protein environment in myoglobin, compared to a peroxidase, leads to a lower reduction potential for the iron, so that an iron(II) porphyrin is the stable form.

1.6.3. Cytochrome P450

Cytochrome P450 in its resting state contains iron(III) protoporphyrin IX as do peroxidases. However, it is activated by reduction to iron(II) followed by reaction with molecular oxygen. Electron transfer reactions are believed to reduce the iron(II)-oxygen complex from a formal oxidation state of +6 to +4 so that a species equivalent to compound I of a peroxidase, containing a single oxygen atom attached to the iron, is obtained. This putative compound I can insert an oxygen atom into a carbon–hydrogen bond and form an epoxide from a carbon–carbon bond. Cytochrome P450 has a thiolate fifth ligand attached to the iron, as has chloroperoxidase. The accepted mechanism for oxygen atom insertion into a C—H bond is described as an oxygen rebound. The oxygen atom of the oxidized iron abstracts the hydrogen atom to form a hydroxyl radical, which cannot diffuse away because of the immediate presence of the substrate. The hydroxyl radical reacts with the newly formed carbon radical so an alcohol group is generated.

1.7. CYTOCHROME *c* OXIDASE

The terminal receptor for electrons in mammalian catabolic reactions is molecular oxygen, and the enzyme that catalyzes the reaction is cytochrome *c* oxidase. Cytochrome *c* oxidase was discovered by Otto Warburg and initially was referred to as the respiratory enzyme. The reaction of the enzyme, which is highly exothermic, is the following:

$$O_2 + 4e^- + 4H^+ \rightarrow 2H_2O \tag{1.17}$$

Electrons are supplied by reduced cytochrome *c*. The enzyme pumps extra protons across a membrane, so that energy is stored as in a battery. This energy is used in adenosine 5′-triphosphate (ATP) synthesis. Thus, cytochrome *c* oxidase is a machine as well as an enzyme.

Cytochrome *c* oxidase contains heme *a*, heme a_3, and two copper ions. Heme *a* and heme a_3 are identical (Fig. 1.2); the designation is to indicate location. There are 13 subunits involved in the structure of the mammalian membrane-bound enzyme. The structure of cytochrome *c* oxidase defied the skill of X-ray crystallographers until 1995.[32,33]

Thus, we see that there are heme proteins that shuttle between iron(II) and iron(III) states, and heme proteins that shuttle between free iron(II) and iron(II)–molecular oxygen complexes. There are heme enzymes that utilize peroxides as their oxidizing substrates, and others that utilize molecular oxygen.

1.8. NOBEL PRIZES

Some indication of the importance of the research mentioned in this chapter is the following list of winners of Nobel prizes and the reason for the awards. The contributions, mainly of physicists, to the development of an array of modern spectrophotometric techniques, several of whom were rewarded with Nobel prizes, are discussed in Chapters 7 and 8.

Chemistry

1915	Richard Willstätter	Chlorophyll structures
1930	Hans Fischer	Hemin synthesis
1958	Frederick Sanger	Insulin primary structure
1962	John Kendrew/Max Perutz	Crystal structures of myoglobin and hemoglobin
1964	Dorothy Hodgkin	Crystal structures of vitamin B_{12} and insulin
1972	Christian Anfinsen (shared)	Reversible unfolding and folding of ribonuclease
1980	Frederick Sanger (shared)	Sequencing of nucleic acid
1993	Kary Mullis/Michael Smith	Polymerase chain reaction/site-directed mutagenesis

Physiology or Medicine

1931	Otto Warburg	Respiratory enzyme (cytochrome *c* oxidase)
1946	James Sumner (shared)	Protein purification and crystallization
1955	Hugo Theorell	Redox enzymes, horseradish peroxidase
1962	Francis Crick/James Watson/ Maurice Wilkins	Double helical structure of DNA
1982	Bengt Samuelsson	Discovery of prostaglandins

REFERENCES

Porphyrins

1. J. E. Falk, *Porphyrins and Metalloporphyrins,* Elsevier, Amsterdam, 1964.
2. K. M. Smith, ed., *Porphyrins and Metalloporphyrins,* Elsevier, Amsterdam, 1975 (a new edition based on the original volume by J. E. Falk).
3. D. Dolphin, ed., *The Porphyrins,* Vols. I–VII, Academic, New York, 1978 (the first five volumes deal with porphyrin chemistry, the last two with heme-containing enzymes).

Early Kinetics

4. L. Michaelis and M. L. Menten, *Biochem. Z.* **49,** 333–369 (1913).
5. G. E. Briggs and J. B. S. Haldane, *Biochem. J.* **19,** 338–339 (1925).
6. J. B. S. Haldane, *Enzymes,* MIT Press, Cambridge, 1965 (republication of the original book published by Longmans, Green in 1930).

Early Peroxidase Research

7. K. G. Paul in *The Enzymes*, 2nd ed., P. D. Boyer, H. Lardy, and K. Myrback, eds., Academic, New York, 1963, Vol. 8, Part B, pp. 227–274.
8. B. C. Saunders, A. G. Holmes-Siedle, and B. P. Stark, *Peroxidase,* Butterworths, London, 1964.
9. D. Keilin, *Proc. Roy. Soc.* (*London*), **98B,** 312–339 (1925).
10 K. A. C. Elliott and D. Keilin, *Proc. Roy. Soc.* (*London*) **114B,** 210–222 (1934).
11. D. Keilin and T. Mann, *Proc. Roy. Soc.* (*London*) **122B,** 119–133 (1937).
12. H. Theorell, *Enzymologia* **10,** 250–252 (1941).
13. K. G. Paul, *Acta Chem. Scand.* **12,** 1312–1318 (1958).

Protein Sequencing

14. F. Sanger and H. Tuppy, *Biochem. J.* **49,** 463–490 (1951).
15. F. Sanger and E. O. P. Thompson, *Biochem. J.* **53,** 353–374 (1953).
16. H. Brown, F. Sanger, and R. Kitai, *Biochem. J.* **60,** 556–565 (1955).

Three-Dimensional Structures

17. J. C. Kendrew, R. E. Dickerson, B. E. Strandberg, R. G. Hart, D. R. Davies, D. C. Phillips, and V. C. Shore, *Nature* **185,** 422–427 (1960).
18. C. C. F. Blake, D. F. Koenig, G. A. Mair, A. C. T. North, and D. C. Phillips, *Nature* **206,** 757–761 (1965).
19. C. C. F. Blake, G. A. Mair, A. C. T. North, D. C. Phillips, and V. R. *Proc. Roy. Soc.* (*London*) **B167,** 365–377 (1967).
20. M. F. Perutz, M. H. Muirhead, J. M. Cox, and L. C. G. Goaman, *Nature* **219,** 131–139 (1968).
21. M. J. Adams, T. L. Blundell, E. J. Dodson, G. G. Dodson, M. Vijayan, E. N. Baker, M. M. Harding, D. C. Hodgkin, B. Rimmer, and S. Sheats, *Nature* **224,** 491–495 (1969).
22. R. E. Dickerson, T. Takano, D. Eisenberg, O. B. Kallai, L. Samson, A. Cooper, and E. Margoliash, *J. Biol. Chem.* **246,** 1511 (1971).
23. T. L. Poulos, S. T. Freer, R. A. Alden, S. L. Edwards, U. Skogland, K. Takio, B. Eriksson, N. Xuong, T. Yonetani, and J. Kraut, *J. Biol. Chem.* **255,** 575–580 (1980).
24. M. R. N. Murthy, T. J. Reid III, A. Sicignano, N. Tanaka, and M. G. Rossmann, *J. Mol. Biol.* **152,** 465–499 (1981).
25. L. Banci, I. Bertini, H. B. Gray, C. Luchinat, T. Reddig, A. Rosato, and P. Turano, *Biochemistry* **36,** 9867-9877 (1997).

Modern Enzyme Kinetics

26. R. A. Alberty, *J. Am. Chem. Soc.* **75,** 1928–1932 (1953).
27. W. W. Cleland, in *The Enzymes,* 3rd ed., P. D. Boyer, ed., Academic, New York, 1970, Vol. II, pp. 1–65.
28. W. W. Cleland, in *The Enzymes,* 3rd ed., D. D. Hackney, ed., Academic, New York, 1990, Vol. XIX, pp. 99–158.

29. K. Dalziel, *Acta Chem. Scand.* **11,** 1706–1723 (1957).
30. E. L. King and C. Altman, *J. Phys. Chem.* **60,** 1375–1378 (1956).
31. H. B. Dunford, in *Peroxidases in Chemistry and Biology,* J. Everse, K. E. Everse, and M. B. Grisham, eds., CRC Press, Boca Raton, FL, 1991, Vol. II, Chapter 1, pp. 1–24.

Cytochrome c Oxidase Structure

32. S. Iwata, C. Ostermeier, B. Ludwig, and H. Michel, *Nature* **376,** 660–669 (1995).
33. T. Tsukihara, H. Aoyama, E. Yamashita, T. Tomizaki, H. Yamaguchi, K. Shinzawa-Itoh, R. Nakashima, R. Yaono, and S. Yoshikawa, *Science* **269,** 1069–1074 (1995).

CHAPTER 2

Model Peroxidases from Yeast and Horseradish, Cloned Enzymes, and Comparison to Metmyoglobin

2.1. INTRODUCTION

There are many reasons for choosing both yeast cytochrome *c* peroxidase and horseradish peroxidase as models for other peroxidases. Some of the reasons are listed in chronological order:

Horseradish Peroxidase	
1937	First identification of a compound II (Keilin and Hartree)[1]
1941	First identification of a compound I (Theorell)[2]
1976	First complete peroxidase primary sequence (Welinder)[3,4]

Yeast Cytochrome *c* Peroxidase	
1940	Discovery of the enzyme (Altschul, Abrams and Hogness)[5]
1980	Primary sequence published (Yonetani et al)[6]
1980 and 1984	X-ray crystal structure published and refined, the first for any peroxidase (Kraut et al.)[7,8]
1986 and 1987	Recombinant enzyme expressed in yeast[9] and in *Escherichia coli*[10]
1992	Structure used as the reference for the plant peroxidase superfamily (Welinder)[11]

Thus, many of the important discoveries and achievements in peroxidase research were first made by investigations of these two enzymes and comparisons to them. Neither enzyme is an ideal model for all peroxidases. For example, horseradish peroxidase contains disulfide bonds, calcium ions, and has carbohydrate attachments; yeast cytochrome *c* peroxidase does not. Yeast cytochrome *c* peroxidase has both an atypical compound I structure and an atypical reducing substrate, the very large molecule cytochrome *c*. However, between the two of them, they provide an excellent starting point to discuss all of the known peroxidases.

2.2. HORSERADISH PEROXIDASE

2.2.1. Reaction Cycle of Horseradish Peroxidase

Theorell's original preparation of compound I yielded a green solution that rapidly changed to red, the compound II first observed by Keilin and Mann. Both the green and red compounds were reduced by hydroquinone back to the native enzyme. Theorell felt that Keilin's compound was not physiologically active since it was not the primary compound. On the time scale of Theorell's experiments, compound II would not have been observed as an intermediate in compound I reduction by hydroquinone because compound II would immediately be reduced by excess of the reducing substrate. Thus, although the components of the normal peroxidase cycle had been detected, their significance was not yet appreciated. A further complication was the preconception that the intermediates were enzyme–substrate complexes with finite dissociation constants, rather than covalent compounds. There are several reviews that describe the early complicated peroxidase history.[12–14]

The roles of compounds I and II were largely sorted out by the work of Britton Chance and Phillip George. Chance measured the rate of reaction of horseradish peroxidase (HRP) and hydrogen peroxide.[15] He obtained a rate constant of $(9 \pm 2) \times 10^6\ M^{-1}s^{-1}$ at pH 4.7 in 0.01 M actetate buffer. He also measured spectra of compounds I and II. Chance indicated spontaneous conversion of compound I to compound II with a unimolecular rate constant (units s^{-1}). The spontaneous conversion occurs in the absence of an added reducing substrate, and its rate is highly dependent on the purity of the preparation of compound I (Section 1.4). George showed that the reduction of compound II to native enzyme is a one-electron process,[16,17] and utilizing the fact that the conversion of compound I to compound II is accelerated by reducing agents, he correctly formulated the following reaction scheme in which compound I contains two oxidizing equivalents compared to the native enzyme and compound II one equivalent:

$$\text{HRP} + H_2O_2 \rightarrow \text{Compound I}$$

$$\text{Compound I} + AH_2 \rightarrow \text{Compound II} + \bullet AH$$

$$\text{Compound II} + AH_2 \rightarrow \text{HRP} + \bullet AH$$

George showed that the iron in compound II is in a +4 oxidation state [iron (IV)], and he postulated an iron(V) state for compound I.

There is no doubt that the spontaneous conversion of compound I to compound II does occur in the absence of added (exogenous) reducing substrate. The nature of this conversion has plagued peroxidase chemistry and biochemistry to the present day.

The above reaction scheme is incomplete. Addition of the three reactions shows that two product water molecules are missing. As documented below, it is now known that both compounds I and II contain a ferryl group ($Fe^{IV}{=}O$). Therefore one oxygen atom from hydrogen peroxide is incorporated into compound I and is the cause of the two additional oxidizing equivalents. The remainder of the hydrogen peroxide becomes a water molecule. If the iron in native horseradish peroxidase is six-coordinate with water in the sixth position, then compound I formation should be shown as

$$\mathrm{HRP} + H_2O_2 \rightarrow \text{Compound I} + 2H_2O$$

It is now known that the iron(III) in the resting enzyme is five-coordinate with no water covalently bound. Therefore, only one water molecule should be shown as a product of compound I formation, and the second water molecule must be formed in a subsequent reaction. The correct balanced three-step reaction sequence follows:

$$\mathrm{HRP} + H_2O_2 \rightarrow \text{Compound I} + H_2O \tag{2.1}$$

$$\text{Compound I} + AH_2 \rightarrow \text{Compound II} + \bullet AH \tag{2.2}$$

$$\text{Compound II} + AH_2 \rightarrow \mathrm{HRP} + \bullet AH + H_2O \tag{2.3}$$

Note that water is a product of compound II reduction. It is formed from the ferryl oxygen atom. The sum of the correct three reactions is:

$$H_2O_2 + 2AH_2 = 2H_2O + 2\,\bullet AH \tag{2.4}$$

A detailed discussion of the subsequent reactions of the free radicals •AH is given in Chapters 4–6.

2.2.2. Native Horseradish Peroxidase

The evidence that the iron(III) of peroxidases is five-coordinate is as follows:

1. Nuclear magnetic resonance (NMR),[18,19] electron paramagnetic resonance (EPR),[20, 21] and resonance Raman[22] data indicate five-coordinate iron(III) in peroxidases.
2. Originally from the X-ray structural data for yeast cytochrome *c* peroxidase, the iron(III) was described as six-coordinate, the same as in metmyoglobin. However, the electron density for the water molecule is

located too far from the iron to form a covalent bond. There is also a question whether the first crystallized enzyme was fully active.

3. Native horseradish peroxidase and metmyoglobin both contain iron(III). When their nitric oxide complexes, contained in glasses frozen at 4 K, are flash photolyzed, the nitric oxide dissociates, leaving the uncomplexed protein. For horseradish peroxidase the spectrum of the native enzyme containing five-coordinate iron(III) is obtained. However, the metmyoglobin spectrum is different from that of the normal six-coordinate metmyoglobin. The obvious explanation is that the unnatural five-coordinate metmyoglobin is obtained.
4. When fluoride is complexed with the iron(III) of globins and peroxidases, the iron is six-coordinate for all species. This is reflected in the molar absorptivities of the Soret region absorption peak for a series of fluoride-complexed globins and peroxidases, which are all the same: $(138 \pm 7)\ mM^{-1}cm^{-1}$. For native iron(III) peroxidases with no added ligand the molar absorptivities are $(105 \pm 6)\ mM^{-1}cm^{-1}$ and for native iron(III) globins they are $(177 \pm 9)\ mM^{-1}cm^{-1}$. The obvious explanation for this difference: the iron(III) in the native peroxidases is five-coordinate and in the globins it is six-coordinate with water known to be the sixth ligand.[23]

Although there are continuing references to six-coordinate iron(III) in native heme peroxidases, these appear to be either quotes of erroneous literature or descriptions of enzymes that are not fully activated. The optical spectrum of native horseradish peroxidase is shown in Fig. 2.1. Note the maximum of the Soret region band at 403 nm and the shoulder at 380 nm. The shoulder at 380 nm is characteristic of five-cordinate iron(III) in peroxidases. For metmyoglobin, which has water in the sixth coordination position of the iron(III), the shoulder is merged into the main peak, which has a higher molar absorptivity as discussed above.

The ratio of the absorbance at 403 nm to that at 280 nm is called the reinheitzahl (RZ), or purity number of horseradish peroxidase. The absorbance at 280 nm is due to aromatic amino acid residues with a contribution from the heme. A rough rule-of-thumb is that the higher the purity number, the purer the enzyme preparation, but this does not necessarily indicate greater activity. Purity numbers in the range 3.0–3.4 are indicative of pure horseradish peroxidase preparations.

2.2.3. Compound I of Horseradish Peroxidase

The optical spectrum of compound I is shown in Fig. 2.1. The molar absorptivity of the peak in the Soret region (near ultraviolet) decreases by a factor of approximately two when the native enzyme is converted to compound I, and

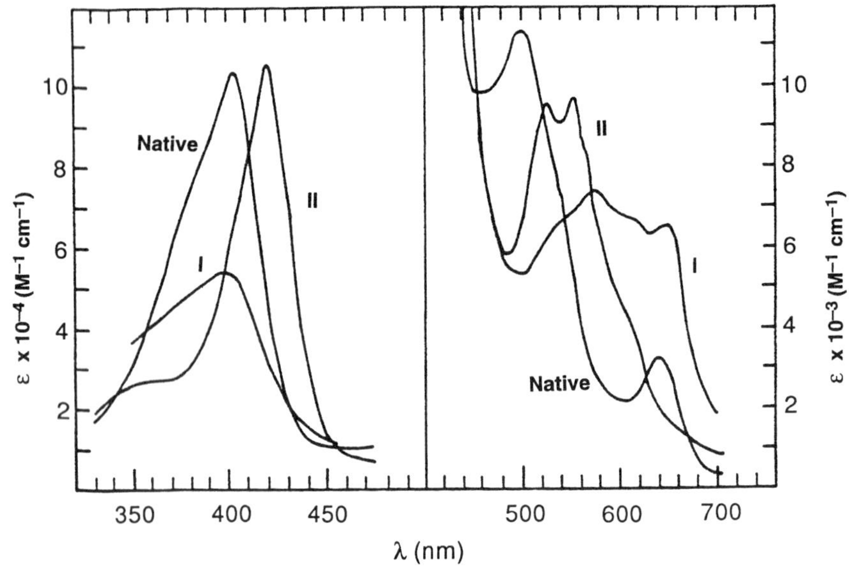

FIG. 2.1. Soret and visible region spectra of horseradish peroxidase and its compounds I and II. [Reproduced from H. B. Dunford, *Adv. Inorg. Biochem.* **4,** 41–68 (1982) with permission from Elsevier Science.]

this is typical of most peroxidases. The basic structure of compound I was not elucidated until 28–30 years after its discovery.

The first breakthrough occurred by use of Mössbauer spectrocopy.[24] (See Chapter 8 for a detailed discussion). Large Mössbauer spectral changes occurred between native enzyme and compound I but spectra of compounds I and II were very similar. Since all available evidence indicated that compound II contained iron(IV), it followed that compound I also contained iron(IV).

The second important evidence came from studies on a model compound.[25] Two-electron oxidation of a porphyrin containing cobalt(II) yielded a stable π-cation radical readily detected by EPR spectroscopy. The optical absorption spectrum of the oxidized compound, compared to the starting material, showed the characteristic drop in molar absorptivity of a native enzyme being converted to compound I. The conclusions from this work have stood the test of time. The combined results of the Mössbauer spectra and model compound work indicate compound I of horseradish peroxidase contains iron(IV) and a porphyrin π-cation radical.

The question then arose: Why had the EPR spectrum of horseradish peroxidase compound I never been detected? The answer: It does have an EPR spectrum but it appears to be weak.[26,27] The signal is perturbed by interactions of the iron(IV) with the porphyrin free radical. Presence of the porphyrin cation radical was confirmed by electron nuclear double resonance

(ENDOR),[28] NMR,[29] and resonance Raman[30] spectral measurements, self-consistent-field theoretical calculations,[31] and extended Hückel calculations combined with optical and EPR spectral data on the enzyme and model compounds.[32] Spectroscopic techniques are discussed in detail in Chapters 7 and 8.

Following is a discussion of the ferryl ($Fe^{IV}{=}O$) group. George postulated its existence in compound II in 1953.[16,17] Evidence for its existence in compound I has accumulated from a variety of sources. Similar Mössbauer spectra for compounds I and II, discussed above, indicate that $Fe^{IV}{=}O$ is present in both compounds. Steady-state kinetics on a peroxidase reaction cycle in the presence of cyanide showed that the iron of compounds I and II was blocked to cyanide, whereas the normal binding of cyanide to the iron(III) of the native enzyme occurred and inhibited the cycle. These results rule against a complex I, a reversible complex of intact hydrogen peroxide and iron, and are compatible with the presence of a covalently bound ligand in the sixth coordination position of the iron(IV) of both compounds I and II.[33] Titration of native enzyme with a peroxy acid indicated oxygen atom incorporation.[34] Transient-state kinetic experiments of compound I formation, also using a peroxy acid, showed that the rate of formation of both compound I and normal acid were identical, again indicating oxygen atom incorporation as part of compound I formation.[35]

2.2.4. Compound II of Horseradish Peroxidase

The optical spectrum of horseradish peroxidase compound II is shown in Fig. 2.1. A comparison of native enzyme with compound II shows that the maximum of the peak in the Soret region is shifted from 403 to 420 nm and the peak intensity is increased slightly. Many papers in the literature show a compound II Soret peak less intense than that of native enzyme, indicating that less than 100% of the enzyme has been converted to compound II. The double peak in the visible region of the spectrum, with maxima at 527 and 554 nm, is a characteristic feature of compound II.

Compound II is more stable than compound I and hence easier to characterize. Much of the more difficult work in identifying the ferryl $Fe^{IV}{=}O$ group of compound I, described in the previous section, was readily applied to compound II.[36] There is now universal acceptance that compound II contains an $Fe^{IV}{=}O$ group.

2.3. YEAST CYTOCHROME *c* PEROXIDASE

Yeast cytochrome *c* peroxidase is in some ways simpler and easier to work with than horseradish peroxidase. It does not contain heterogeneous carbohydrate attachments. For many years it was the only peroxidase with a known crystal structure. The crystal structure of its complex with cytochrome *c* is also known.[37] The reactions of oxidized enzyme with reduced cytochrome *c* would appear to be an ideal system to study long-range electron transfer, work described in Chapter 10. The discussion here is limited to the pertinence of knowledge about yeast cytochrome *c* peroxidase in constructing a model for all peroxidases, particularly for those in the plant kingdom. Work on yeast cytochrome *c* peroxidase has been reviewed.[38–42]

2.3.1. Native Yeast Cytochrome *c* Peroxidase

The development of the current state of knowledge about yeast cytochrome *c* peroxidase occurred in similar fashion to that for horseradish peroxidase, with at times considerable confusion in the literature. There were conflicting reports in the earlier literature over the coordination number of iron(III) in native yeast cytochrome *c* peroxidase. The cause of the confusion was that much work was performed on a partially inactivated form of the enzyme, obtained by improper storage or by freeze–thaw cycles.[37,43] Electron paramagnetic resonance[44] and resonance Raman[45–48] spectra show that the active form of the enzyme contains mainly five-coordinate iron(III). The structure of the inactive form is unknown.[44] The spectrum of native yeast cytochrome *c* peroxidase with five-coordinate iron(III) is shown in Fig. 2.2. The RZ, or purity, number of pure yeast cytochrome *c* peroxidase is in the range of 1.25–1.30. The maximum of the Soret band is at 408 nm and the shoulder is at 360–380 nm. This shoulder, characteristic of five-coordinate iron(III) in

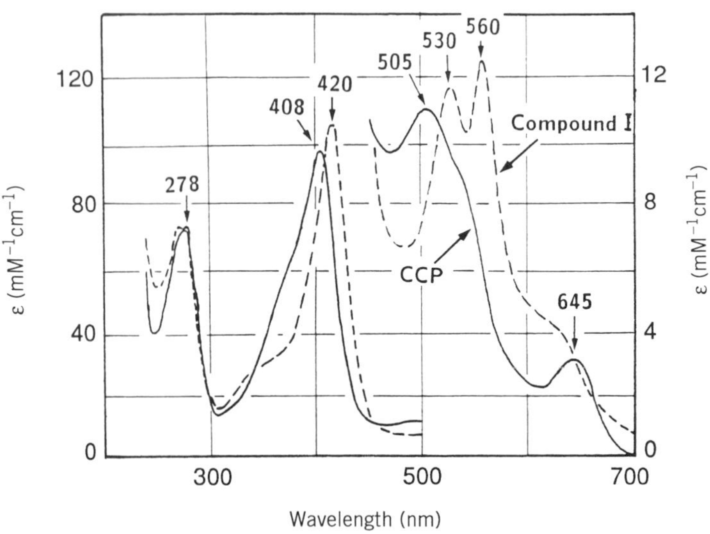

FIG. 2.2. Soret and visible region spectra of yeast cytochrome *c* peroxidase and its compound I. [Reproduced with permission from T. Yonetani and H. Anni, *J. Biol. Chem.* **262,** 9547–9554 (1987).]

peroxidases, tends to disappear with the presence of six-coordinate iron(III) and merge with the major peak in the Soret region. Thus, the RZ value, or purity number, of yeast cytochrome *c* peroxidase increases with increasing amounts of six-coordinate iron(III). The higher the percentage of the hexa-coordinate species, the lower the activity.[44]

2.3.2. Compound I of Yeast Cytochrome *c* Peroxidase

The ghost of the Michaelis–Menten complex has haunted the literature on yeast cytochrome *c* peroxidase. The product of the reaction of native enzyme with hydrogen peroxide was originally called complex ES, which underwent an evolution to compound ES, and finally to the widely accepted name of compound I.

The EPR spectrum of compound I of yeast cytochrome *c* peroxidase was published in 1966.[49] An intense signal was observed, indicative of a free radical electron largely unperturbed by the ferryl iron. The location of the free radical electron in compound I led to a merry chase involving many groups. It is now accepted that the unpaired electron is located on Trp191 on the proximal side of the heme, based upon site-directed mutagenesis[50] and ENDOR studies.[51, 52]

Compound I contains the $Fe^{IV}{=}O$ group typical of other peroxidase compounds I. The difference in the yeast peroxidase compound I is the location of the extra oxidizing equivalent on Trp191 rather than on the heme group. Because the free radical electron is undergoing little interaction with the $Fe^{IV}{=}O$ group, the optical spectrum of yeast cytochrome *c* compound I resembles that of compound II of horseradish peroxidase. The peak in the Soret region is shifted to 420 nm, compared to 408 nm for the native enzyme peak, and is not decreased in intensity (Fig. 2.2).

2.3.3. Compound II of Yeast Cytochrome *c* Peroxidase

Reduction of compound I to II has been believed to occur via pairing of the cation free radical electron on Trp191. The optical spectrum changes little during the electron addition, since the porphyrin $Fe^{IV}{=}O$ group does not change. If the reduction of compound I occurred at the $Fe^{IV}{=}O$ site, then a hybrid compound II would be produced, which contained iron(III) and the unchanged free radical on Trp191, in which case the compound II spectrum would resemble that of the native enzyme. Reduction of compound II is accomplished by reduction of the $Fe^{IV}{=}O$ group to iron(III), or possibly by reduction of the Trp191 radical site, if a reverse form of compound II exists.

Thus, the conventional wisdom has been that the reaction cycle of yeast cytochrome *c* peroxidase is the same as that of horseradish peroxidase, with the difference being that the free radical site of compound I is located on Trp191 instead of on the porphyrin ring. However, see the discussions in Chapters 9 and 10, particularly about the nature of compound II.

2.4. CLONED ENZYMES

Yeast cytochrome *c* peroxidase was one of the first enzymes to be cloned and was expressed in yeast itself.[9] The yeast enzyme, with no carbohydrate attachments, offered advantages over glycoproteins. Expression and mutagenesis in *E. coli* was found to be simpler and better, with easier subsequent purification (Section 9.5).[10]

The *E. coli* system was adopted for the expression of cloned horseradish peroxidase. Because the host cells do not possess the machinery for glycosylation, the overexpressed pure protein formed inclusion bodies. Peroxidase molecules had to be extracted, refolded, and activated by heme addition. These difficult experiments were successfully performed, albeit with a very low yield of active enzyme.[53] Folding was found to be critically dependent upon concentrations of urea, calcium ion, heme, and oxidized glutathione. The latter was required for successful formation of disulfide bonds. Thus, it was demonstrated that glycosylation was not required for heme activity. The road was opened, but not smoothly paved, for successful crystallization and X-ray structural determination. An alternative route to fully active deglycosylated horseradish peroxidase is via reaction of anhydrous trifluoromethanesulfonic acid with holoenzyme under mild conditions.[54]

Help in elucidation of the horseradish peroxidase crystal structure came from an unexpected source. Peanut peroxidase belongs to the same family as horseradish peroxidase, and crystals of the intact wild-type enzyme were successfully grown and the structure elucidated. The peanut peroxidase structure (Section 13.1.2) was used to help unravel the structure of the horseradish enzyme.

An alternative expression system for horseradish peroxidase was developed.[55] A baculovirus transfer vector was used in combination with insect tissue culture. This system has the advantage that disulfide linkages are readily formed because of an oxidizing environment. A fully glycosylated active enzyme is produced.

2.5. COMPARISONS OF HORSERADISH AND YEAST ENZYMES AND RELATION TO THE GLOBINS

Sequences of horseradish peroxidase and cytochrome *c* peroxidase on the distal and proximal sides of the heme are compared in Table 2.1. Key residues on both the distal and proximal sides are conserved.

Myoglobin and the peroxidases from yeast and horseradish all possess the same prosthetic group (ferriprotoporphyrin IX) and the same proximal ligand (the imidazole side chain of a histidine residue), and they all possess a distal histidine group. Why are they adapted to different functions? Most important the different protein folds and different packing of the heme. Furthermore,

TABLE 2.1. Sequence Homologies for Yeast Cytochrome *c* Peroxidase and Horseradish Peroxidase Showing the Histidines in the Distal Cavity and the Proximal Histidines That Have Imidazole Side Chains Bonded to Iron(III)

Distal Side of the Heme

*Cc*P	Arg48	Leu49	Ala50	Trp51	**His52**	Thr53	Ser54
HRP	Arg38	Leu39	His40	Phe41	**His42**	Asp43	Cys44

Proximal Side of the Heme

*Cc*P	Val168	Ala169	Leu170	Met171	Gly172	Ala173	**His174**	Ala175	Leu176	Gly177	Lys178
HRP	Val164	Ala165	Leu166	Ser167	Gly168	Gly169	**His170**	Thr171	Phe172	Gly173	Lys174

the distal arginine residue, present in the peroxidases but not in myoglobin, has been identified as a key residue in the formation of compound I. An amino acid backbone carbonyl is hydrogen bonded to the proximal imidazole ring in myoglobin and aspartate side chain in the peroxidases. The different bonding results in a longer proximal Fe—Nε bond length in the globins.[56] The globins are clearly from a different family than the peroxidases, and there is no meaningful sequence homology. The more basic group in the peroxidases contribute a higher electron density to the iron, which is believed to help stabilize the iron(III). The reduction potentials of the iron(III)/iron(II) couple for the peroxidases are −194 mV (yeast enzyme),[57] −220 mV (horseradish)[58] compared to +50 mV for myoglobin.[59] Comparisons of magnetic circular dichroism spectra of horseradish peroxidase and metmyoglobin are made in Section 7.3.2.

When excess hydrogen peroxide is reacted with metmyoglobin, a second-order rate constant of $3.5 \times 10^2\ M^{-1}s^{-1}$ is obtained, almost five orders of magnitude slower than the rate of formation of compound I from peroxidases.[60] In addition a compound-II-like optical spectrum is observed with isosbestic points with the spectrum of the original metmyoglobin. An attractive hypothesis is that instead of compound I formation, compound II is formed directly with concomitant formation of the hydroxyl radical.[61]

$$Fe^{III} + H_2O_2 \rightarrow Fe^{IV}{=}O + \bullet OH + H^+ \qquad (2.5)$$

A hydrogen atom is released into unbuffered solution,[62] which is not the case in peroxidase compound I formation.[63,64] Upon contact of hydrogen peroxide with the iron(III), the O—O bond is much less polarized than in the peroxidases. Hence, homolytic cleavage of the O—O bond is much more likely with resultant liberation of a hydroxyl radical. The heme group of metmyoglobin is much more susceptible to attack in the presence of excess hydrogen peroxide than is the heme of peroxidases, which is consistent with the postulate that hydroxyl radicals are generated.[65] An EPR signal[66] could be due to a free radical generated by hydroxyl radical attack on the protein. As a caution to the above simplistic explanation, the reaction kinetics become more complex when the hydrogen peroxide is not in large excess,[66] and oxygen evolution occurs. The rate of oxygen evolution needs to be measured to determine whether it can be correlated with the rate of the spectral changes observed when myoglobin reacts with hydrogen peroxide.

An alternative explanation is that a compound I species is formed, with molecular oxygen combining with the free radical site on the protein to form a peroxyl radical, identified by EPR.[67] Attack on the protein by the hydroxyl radical followed by combination with oxygen also provides a rationale for the observed results. Some reactions were shown to involve the heme iron of myoglobin, others a protein radical.[68] Linoleic acid was shown to react with heme $Fe^{IV}{=}O$, not the protein radical.[69]

Cyanide binding to metmyoglobin is much slower than to peroxidases, and mutation experiments support the conclusion that bond formation and disruptions within the distal pocket are rate-limiting steps.[70] Recent work with recombinant metmyoglobin showed that, upon reaction with hydrogen peroxide, a tyrosyl peroxyl radical is formed. Site-directed mutagenesis experiments demonstrated that the site of peroxyl radical formation was Trp14.[71] Intensive studies with mutant myoglobins have shown that its peroxidatic activity can be greatly increased.[72,73] The properties of a double mutant, where His64 and Val68 were exchanged, have been studied.[74] The pH and driving force dependence of intramolecular oxyferryl heme reduction in myoglobin has been investigated using ruthenium(II) labels covalently attached to the protein.[75]

Whatever the details of the reaction of wild-type metmyoglobin with hydrogen peroxide, it is clear that metmyoglobin is a very poor peroxidase in terms of its ability to form compound I. The absence of a distal arginine residue in metmyoglobin deserves emphasis.

REFERENCES

1. D. Keilin and T. Mann, *Proc. Roy. Soc.* (*London*) **122B,** 119–133 (1937).
2. H. Theorell, *Enzymologia* **10,** 250–252 (1941).
3. K. G. Welinder, *FEBS Lett.* **72,** 19–23 (1976).
4. K. G. Welinder, *Eur. J. Biochem.* **96,** 483–502 (1979).
5. A. M. Altschul, R. Abrams, and T. R. Hogness, *J. Biol. Chem.* **136,** 777–79 940).
6. K. Takio, K. Titani, L. H. Ericsson, and T. Yonetani, *Arch. Biochem* *ophys.* **203,** 615–629 (1980).
7. T. L. Poulos, S. T. Freer, R. A. Alden, S. L. Edwards, U. Skogland Takio, B. Eriksson, N. Xuong, T. Yonetani, and J. Kraut, *J. Biol. Chem.* **255,** 80 (1980).
8. B. C. Finzel, T. L. Poulos, and J. Kraut, *J. Biol. Chem.* **259,** 13036 (1984).
9. D. B. Goodin, A. G. Mauk, and M. Smith, *Proc. Natl. Ac* *U.S.A.* **83,** 1295–1299 (1986).
10. L. A. Fishel, J. E. Villafranca, J. M. Mauro, and J. Kraut, *mistry* **26,** 351–360 (1987).
11. K. G. Welinder, *Curr. Opin. Struct. Biol.* **2,** 388–393 (1
12. K.-G. Paul, *The Enzymes, Part B,* 2nd ed., in P. D H. Lardy, and K. Myrback, eds., Academic, New York, 1963, Vol. 8, 7.
13. B. C. Saunders, A. G. Holmes-Siedle, and B. P. S *oxidase,* Butterworths, London, 1964.
14. A. S. Brill in *Comprehensive Biochemistry,* n and E. H. Stotz, eds., Elsevier, Amsterdam, 1966, Vol. 14, Chapte 49).
15. B. Chance, *Arch. Biochem. Biophys.* **22,** 2
16. P. George, *Biochem. J.* **54,** 267–276 (195
17. P. George, *Biochem. J.* **54,** 220–230 (1

18. A. Lanir and A. Schejter, *Biochem. Biophys. Res. Commun.* **62,** 199–203 (1975).
19. G. N. La Mar, M. J. Chatfield, D. H. Peyton, J. S. de Ropp, W. S. Smith, R. Krishnamoorthi, J. D. Satterlee, and J. E. Erman, *Biochim. Biophys. Acta* **956,** 267–276 (1988).
20. S. Vuk-Pavlovic and Y. Siderer, *Biochem. Biophys. Res. Commun.* **79,** 885–889 (1977).
21. K. Kobayashi, M. Tamura, K. Hayashi, H. Hori, and H. Morimoto, *J. Biol. Chem.* **255,** 2239–2242 (1980).
22. T. G. Spiro, J. D. Stong, and P. Stein, *J. Am. Chem. Soc.* **101,** 2648–2655 (1979).
23. H. B. Dunford, *Adv. Inorg. Biochem.* **4,** 41–68 (1982).
24. T. H. Moss, A. Ehrenberg, and A. J. Bearden, *Biochemistry* **8,** 4159–4162 (1969).
25. D. Dolphin, A. Forman, D. C. Borg, J. Fajer, and R. H. Felton, *Proc. Natl. Acad. Sci. U.S.A.* **68,** 614–618 (1971).
26. R. Aasa, T. Vänngård, and H. B. Dunford, *Biochim. Biophys. Acta* **391,** 259–264 (1975).
27. C. E. Schultz, P. W. Devaney, H. Winkler, P. G. Debrunner, N. Doan, R. Chiang, R. Rutter, and L. P. Hager, *FEBS Lett.* **103,** 102–105 (1979).
28. J. E. Roberts, B. M. Hoffman, R. Rutter, and L. P. Hager, *J. Biol. Chem.* **256,** 2118–2121 (1981).
29. G. N. La Mar, J. S. de Ropp, K. M. Smith, and K. C. Langry, *J. Biol. Chem.* **256,** 237–243 (1981).
30. R. H. Felton, A. Y. Romans, N.-T. Yu, and G. R. Schonbaum, *Biochim. Biophys.* ...ta **434,** 82–89 (1976).
31. G... Loew and Z. S. Herman, *J. Am. Chem. Soc.* **102,** 6173–6174 (1980).
32. L. ... Hanson, C. K. Chang, M. S. Davis, and J. Fajer, *J. Am. Chem. Soc.* **103,** 663– (1981).
33. M. L. ...n, H. B. Dunford, and J. M. T. Raycheba, *Can. J. Biochem.* **51,** 627–631 (19...
34. G. R. Sch... ...um and S. Lo, *J. Biol. Chem.* **247,** 3353–3360 (1972).
35. S. A. Ade... ...nd H. B. Dunford, *Eur. J. Biochem.* **132,** 147–150 (1983).
36. G. Rakhit, ...iro, and M. Uyeda, M. *Biochem. Biophys. Res. Commun.* **71,** 803–808 (197...
37. H. Pelletier and ... ...ut, *Science* **258,** 1748–1755 (1992).
38. T. Yonetani, *Adv...* ...*iol.* **33,** 309–335 (1970).
39. T. Yonetani in *The...* ...*es,* P. D. Boyer, ed., Academic, New York, 1976, Vol. 13, pp. 345–361.
40. H. R. Bosshard, H. An... ...Yonetani in *Peroxidases in Chemistry and Biology,* J. Everse, K. E. Everse ... B. Grisham, eds., CRC Press, Boca Raton, FL, 1991, Vol. II, pp. 51–84.
41. G. Smulevich in *Biomolecu...* ...*scopy,* R. J. H. Clark and R. E. Hester, eds., Wiley, New York, *Part A,* 1... ...3–193.
42. A. M. English and G. Tsapra... ...*org. Chem.* **43,** 79–125 (1995).
43. R. A. Mathews and J. B. Witte... ...*ol. Chem.* **254,** 5991–5996 (1979).
44. T. Yonetani and H. Anni, *J. Biol...* ... 9547–9554 (1987).

45. G. Sievers, K. Österlund, and N. Ellfolk, *Biochim. Biophys. Acta* **581,** 1–14 (1979).
46. G. Smulevich, A. R. Mantini, A. English, and J. M. Mauro, *Biochemistry* **28,** 5058–5064 (1989).
47. S. Dasgupta, D. L. Rousseau, H. Anni, and T. Yonetani, *J. Biol. Chem.* **264,** 654–662 (1989).
48. G. Smulevich, Y. Wang, S. L. Edwards, T. L. Poulos, A. M. English, and T. G. Spiro, *Biochemistry* **29,** 2586–2592 (1990).
49. T. Yonetani, H. Schleyer, and A. Ehrenberg, *J. Biol. Chem.* **241,** 3240–3243 (1966).
50. J. E. Erman, L. B. Vitello, J. M. Mauro, and J. Kraut, *Biochemistry* **28,** 7992–7995 (1989).
51. A. L. P. Houseman, P. E. Doan, D. B. Goodin, and B. M. Hoffman, *Biochemistry* **32,** 4430–4443 (1993).
52. J. E. Huyett, P. E. Doan, R. Gurbiel, A. L. P. Houseman, M. Sivaraja, D. B. Goodin, and B. M. Hoffman, *J. Am. Chem. Soc.* **117,** 9033–9041 (1995).
53. A. T. Smith, N. Santama, S. Dacey, M. Edwards, R. C. Bray, R. N. F. Thorneley, and J. F. Burke, *J. Biol. Chem.* **265,** 13335–13343 (1990).
54. J. W. Tams and K. G. Welinder, *Anal. Biochem.* **228,** 48–55 (1995).
55. C. Hartmann and P. R. Ortiz de Montellano, *Arch. Biochem. Biophys.* **297,** 61–72 (1992).
56. R. Sinclair, S. Hallam, M. Chen, B. Chance, and L. Powers, *Biochemistry* **35,** 15120–15128 (1996).
57. C. W. Conroy, P. Tyma, P. H. Daum, and J. E. Erman, *Biochim. Biophys. Acta* **537,** 62–69 (1978).
58. H. A. Harbury, *J. Biol. Chem.* **225,** 1009–1024 (1957).
59. J. C. Cassatt, C. P. Marini, and J. Bender, *Biochemistry* **14,** 5470–5475 (1975).
60. G. M. Clore, A. N. Lane, and M. R. Hollaway, *Inorg. Chim. Acta* **46,** 139–146 (1980).
61. P. George and D. H. Irvine, *Biochem. J.* **52,** 511–517 (1952).
62. P. George and D. H. Irvine, *Biochem. J.* **60,** 596–604 (1955).
63. H. Yamada and I. Yamazaki, *Arch. Biochem. Biophys.* **165,** 728–738 (1974).
64. C. W. Conroy and J. E. Erman, *Biochim. Biophys. Acta* **527,** 370–378 (1978).
65. K. Sugiyama, R. J. Highet, A. Woods, R. J. Cotter, and Y. Osawa, *Proc. Natl. Acad. Sci. U.S.A.* **94,** 796–801 (1997).
66. T. Yonetani and H. Schleyer, *J. Biol. Chem.* **242,** 1974–1979 (1967).
67. D. J. Kelman, J. A. DeGray, and R. P. Mason, *J. Biol. Chem.* **269,** 7458–7463 (1994).
68. S. I. Rao, A. Wilks, and P. R. Ortiz de Montellano, *J. Biol. Chem.* **268,** 803–809 (1993).
69. S. I. Rao, A. Wilks, M. Hamberg, and P. R. Ortiz de Montellano, *J. Biol. Chem.* **269,** 7210–7216 (1994).
70. Y. Dou, J. S. Olson, A. J. Wilkinson, and M. Ikeda-Saito, *Biochemistry* **35,** 7017–7113 (1996).
71. J. A. DeGray, M. R. Gunther, R. Tschirret-Guth, P. R. Ortiz de Montellano, and R. P. Mason, J. Biol. Chem. **272,** 2359–2362 (1997).
72. S. Okazaki, T. Matsui, and Y. Watanabe, *J. Am. Chem. Soc.* **118,** 9784–9785 (1996).

73. S. Okazaki, T. Matsui, and Y. Watanabe, *J. Am. Chem. Soc.* **119,** 6666–6667 (1997).
74. Y. Dou, J. Admiraal, M. Ikeda-Saito, S. Krzywda, A. J. Wilkinson, T. Li, J. S. Olson, R. C. Prince, I. J. Pickering, and G. N. George, *J. Biol. Chem.* **270,** 15993–16001 (1995).
75. C. W. Fenwick, A. M. English, and J. F. Wishart, *J. Am. Chem. Soc.* **119,** 4758–4764 (1997).

CHAPTER 3

Heme Peroxidase and Catalase Families and Superfamilies: Crystal Structures

3.1. INTRODUCTION. STRUCTURAL BASIS FOR FAMILIES

Following a discussion of the peroxidase–catalase superfamilies in this chapter, a brief introduction will be given to three-dimensional structures of representative peroxidases and catalases. This will be followed by a more detailed discussion of the structure of yeast cytochrome *c* peroxidase, for many years the only known peroxidase crystallographic structure. However, the catalase structure was determined shortly after that of the yeast peroxidase. Detailed discussions of the many other X-ray structures that have now been determined, including that of catalase, are located elsewhere in appropriate chapters.

Development of a satisfactory classification system brings order out of chaos. Such a classification, based upon structure and function, has been developed for peroxidases and catalases.

X-ray crystallography has been the most successful method of obtaining three-dimensional structures. Despite the explosion in new three-dimensional structures, it has been outpaced by the rate of new primary structure determinations. For unknown structures, *a priori* prediction of structure from primary sequence has long been a goal of many researchers. However, it has had limited success.

A practical alternative to *a priori* prediction of structure has been to use X-ray data for one or two enzymes and to determine whether possibly related enzymes could be fit into the same three-dimensional structures. This was the approach used by Karen G. Welinder, who is undoubtedly the world's authority for plant peroxidases on correlation of primary sequence data of one peroxidase with tertiary structures, known from crystallographic data, for related family members. She published the complete amino acid sequence of horseradish peroxidase in 1976, the first for any peroxidase.[1] This was followed by sequence data for yeast cytochrome *c* peroxidase obtained by Takashi Yonetani et al.[2,3] and the crystal structure of C*c*P refined to 1.7 Å resolution obtained in the laboratory of Joseph Kraut.[4,5] Several peroxidase primary sequences are listed in Appendix A.

3.2. PEROXIDASE SUPERFAMILIES

3.2.1. Plant Peroxidase Superfamily

Welinder demonstrated that yeast cytochrome *c* peroxidase, despite the fact that it acts upon a much larger reducing substrate than many peroxidases, was a satisfactory model for the classification of all plant peroxidases.[6] This conclusion was supported when partial data on the crystal structure of horseradish peroxidase isoenzyme E5 became available[7,8] and could be compared to those for cytochrome *c* peroxidase.[9] Unfortunately, complete details of the horseradish peroxidase E5 crystal structure have never been published; the original data appears to be wrong in the orientation of one helix. Since then, many more plant peroxidase crystal structures have become available and have shown that the Welinder classification is correct.

Yeast cytochrome *c* peroxidase has a structure of 10 α-helices, labeled A to J, connected by surface amino acid chains of various lengths (Fig. 3.1). Helix B is close to the heme on the distal side and helix F on the proximal side. The reason for Welinder's success in classifying the peroxidases lies in her ability to recognize the various helical segments of structure, separated by surface amino acid chains that vary in length from one peroxidase to the next. The helical structures are preserved within a family, the connecting chains are not.

There are 9 invariant residues and 22 nearly invariant residues for all of the plant peroxidases.[10] Using the cytochrome *c* peroxidase amino acid numbering, these residues are as follows: Arg48 and His52 of helix B and

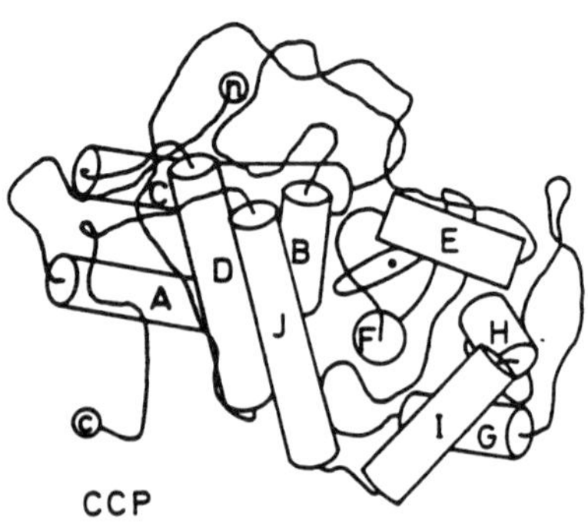

FIG. 3.1. Helical structure of yeast cytochrome *c* peroxidase. The heme is located between helices B and F. (Reproduced with permission from K. G. Welinder in *Plant Peroxidases 1980–1990,* C. Penel, Th. Gaspar and H. Greppin, eds., University of Geneva Press, Geneva, 1992, pp. 1–24.)

Asn82 are on the distal side of the heme. Val169 andHis175 of helix F and Asp235 are on the proximal side. Asp106 of helix D, Gly129 and Arg130 form a buried salt bridge. His175 is the fifth ligand of the heme iron and is hydrogen-bonded to Asp235.

On the basis of sequence homologies, Welinder has proposed three families of peroxidases which belong to the plant peroxidase superfamily.[9,11]

Class I. *Peroxidases of prokaryotic origin.* These include yeast cytochrome *c* peroxidase, chloroplast and cytosolic ascorbate peroxidases, and the gene-duplicated bacterial catalase-peroxidases.[12]

The bacterial catalase-peroxidases are typically dimers or tetramers. Each subunit is approximately double the molecular weight of the yeast enzyme. Their N-terminal end has bound heme and high sequence homology to yeast cytochrome *c* peroxidase.[13] The C-terminal end appears to have lost its heme and has much lower homology. No crystal structure of a catalase-peroxidase had been reported at the time of submission for publication.

The two published class I crystal structures are those of the yeast and pea cytosolic enzymes. The pea cytosolic enzyme is dimeric but the chloroplast enzymes are monomeric.[9] Class I peroxidases have no carbohydrate, no cystine bridges, no calcium ions, and no signal peptide for secretion. The Trp191 residue of the yeast enzyme appears to be conserved in other class I peroxidases. A striking difference between ascorbate peroxidase and yeast cytochrome *c* peroxidase is that the former has a metal binding site near the proximal tryptophan residue, apparently occupied by a potassium ion.[14]

Class II. *Secreted fungal peroxidases.* Examples are the lignin and manganese peroxidases of *Phanerochaete chrysosporium* and the ink cap mushroom peroxidase from *Coprinus cinereus* (alternative name *Arthromyces ramosus*). These class II peroxidases are discussed in detail in Chapter 11. Chloroperoxidase from *Caldariomyces fumago* and the cytochrome *c* peroxidase from *Pseudomonas aeruginosa* are different and do not fit into any of the three plant peroxidase families. The latter two enzymes are discussed in Chapter 14.

The fungal secretory peroxidases have an amino terminal signal peptide sequence for secretion through the endoplasmic reticulum. They possess about 5% carbohydrate, two calcium ions, and four conserved disulfide bridges.[9] The proximal tryptophan residue, conserved in the class I peroxidases, is replaced by phenylalanine or leucine in class II. One of the calcium binding sites, close to the proximal phenylalanine of lignin peroxidase, corresponds approximately to the potassium site in ascorbate peroxidase. X-ray structures have been published for three class II peroxidases: those of lignin, manganese, and *C. cinereus* (*A. ramosus*).

Class III. *Classical secretory plant peroxidases.* The most intensively studied example is horseradish peroxidase isoenzyme C, but the first published X-ray structure is that of the very similar peanut peroxidase. The class III peroxidases possess two calcium ions, an N-terminal signal peptide for excretion, four conserved disulfide bridges in different locations from those in class II, and extra helices that play a role in access to the heme edge. Glycosylation ranges from 0 to 25%. Some have a C-terminal propeptide that may target them for vacuoles.[9] Known class III structures are those for peanut, horseradish, and barley peroxidases.

The Welinder classification systematizes existing knowledge about peroxidases, leads to predictions about unknown properties, and provides a framework for future work. One of Welinder's predictions is particularly spectacular. It is the following: plant peroxidase enzymes consist of two halves that have a common origin.[15] Thus, gene duplication is predicted to have occurred. The 10 conserved helices in plant peroxidases can be considered as two five-helical structures, one for the distal domain and the other for the proximal domain. The proposed correlation is shown in Fig. 3.2. There are conserved introns in plant peroxidase genes to support this idea, so that if it is correct, gene fusion must have occurred. Gene duplication is supported by the large number of correlations now available.

3.2.2. Animal Peroxidase Superfamily

Recent crystal structures for myeloperoxidase[16] and prostaglandin H synthase[17] as well correlations based on sequence data[18] show that the animal peroxidases form a distinct family. The helices that make up the heme binding pocket of plant and animal peroxidases are similar, however.[17]

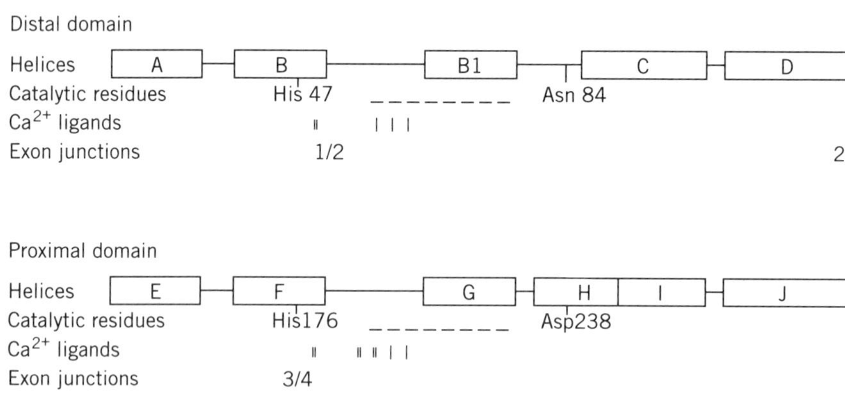

FIG. 3.2. Proposed correlation between the five helices of the distal domain with the five helices of the proximal domain for peroxidases, showing the correlation for those peroxidases that bind calcium ions. (Reproduced with permission from K. G. Welinder and M. Gajhede in *Plant Peroxidases. Biochemistry and Physiology,* K. G. Welinder, S. K. Rasmussen, C. Penel, and H. Greppin, eds., University of Geneva Press, Geneva, 1993, pp. 35–42.)

3.2.3. Catalase Superfamily

Catalases are efficient catalysts of the reaction

$$2H_2O_2 \rightarrow 2H_2O + O_2 \tag{3.1}$$

which is called the *catalatic* reaction. There are ample reasons why catalases and peroxidases should be grouped together. Some peroxidases can catalyze the catalatic reaction and catalases can also catalyze some of the same reactions catalyzed by peroxidases. Catalases have similar heme prosthetic groups to those found in peroxidases.

Catalases occur in both the plant and animal kingdoms. The crystal structures of two tetrameric catalases are known.[19,20] On the basis of these structures and primary sequence correlations, it is clear that catalases should form a separate superfamily.

3.3. CRYSTAL STRUCTURES OF PEROXIDASES AND CATALASES

After the first crystal structure of a peroxidase was published in 1980, a long time interval elapsed before a second peroxidase structure was obtained (Table 3.1). In the meantime catalase structures were determined. Yeast cytochrome *c* peroxidase and the catalases have no carbohydrate attachments, which greatly facilitated crystal growth. The problem of carbohydrate groups has been overcome in some instances by using cloned enzymes expressed in *E. coli*, which prevents glycosylation. Programs for analysis have been continually improved,

TABLE 3.1. Dates of Publication of Peroxidase and Catalase Crystal Structures

1980, 1984	Yeast cytochrome *c* peroxidase
1981, 1985	Beef liver catalase
1986	*Penicillium vitale* catalase
1991, 1993	Horseradish peroxidase E5 (preliminary data)
1992	Canine myeloperoxidase
1993	Lignin peroxidase
1993	*P. aeruginosa* cytochrome *c* peroxidase
1994	*C. cinereus* (*A. ramosus*) peroxidase
1994	Manganese peroxidase
1994	Prostaglandin H synthase
1995	Pea cytosolic ascorbate peroxidase
1995	Chloroperoxidase
1995, 1996	Human myeloperoxidase
1996	Peanut peroxidase
1997	Horseradish peroxidase C
1998	Barley peroxidase

and the use of synchroton radiation has greatly reduced the time required to accumulate diffraction data. As a unique example of the current speed at which X-ray diffraction data can now be accumulated and analyzed, I cite the work of David Goodin, discussed below. Not trained as a crystallographer, he learned the techniques, and he has been obtaining X-ray results on site-directed mutants at an amazing rate.

3.3.1. Structures of Plant Peroxidases

The gross structural features of representatives of the three plant peroxidase families are compared in Fig. 3.3 (see below and color insert). In all cases the heme group is located between helices B and F. Different additional helical groups occur for the different families. Thus, class I and II peroxidases possess an additional short helix B′ between helices B and C. The class III peanut peroxidase lacks the B′ helix but has three additional helices, D′, F′, and F″.[21]

3.3.2. Structures of Animal Peroxidases

Canine Myeloperoxidase A schematic representation of the structure of canine myeloperoxidase is shown in Fig. 3.4. Myeloperoxidase consists of two

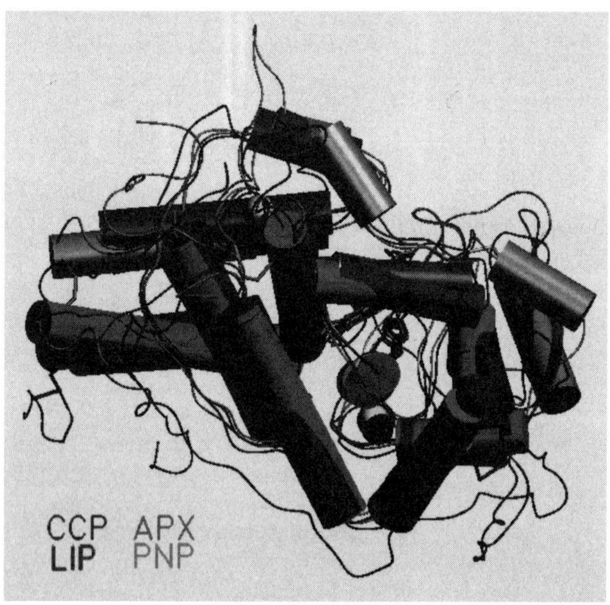

FIG. 3.3. Ribbon and helical structures for representative peroxidases. See color insert. Class I, yeast cytochrome *c* peroxidase (CCP) and ascorbate peroxidase (APX). Class II, lignin peroxidase (LIP). Class III, peanut peroxidase (PNP). [Reproduced with permission from D. J. Schuller, N. Ban, R. B. van Huystee, A. McPherson, and T. L. Poulos, *Structure* **4,** 311–321 (1996).]

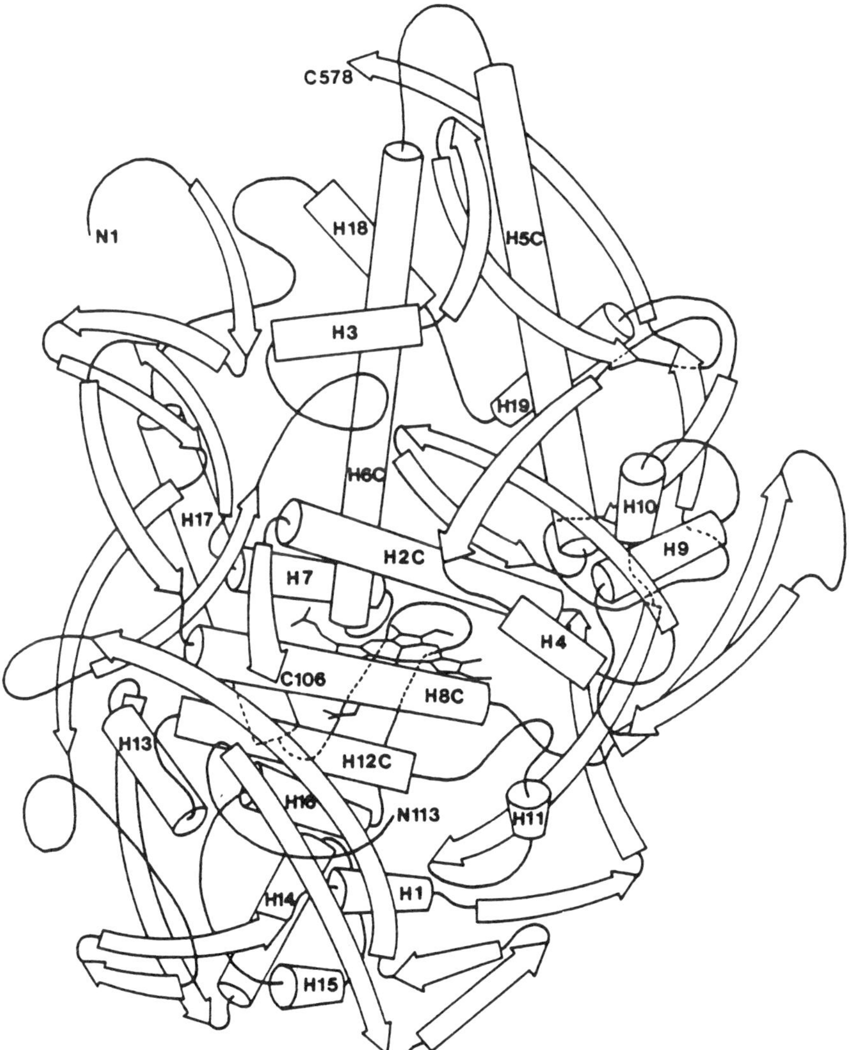

FIG. 3.4. Schematic drawing of one of the identical halves of myeloperoxidase. Helices are shown as cylinders, labeled from H1 to H19, and extended chains as ribbons. The small polypeptide starts at N1 and ends at C106; the large starts at N113 and ends at C578. Core helices have the suffix C. [Reproduced with permission from J. Zeng and R. E. Fenna, *J. Mol. Biol.* **226,** 185–207 (1992).]

identical halves linked covalently by a single disulfide bridge. Each half contains two polypeptides consisting of 108 and 466 amino acid residues, a heme prosthetic group, a calcium ion, and at least three sites of asparagine-linked glycosylation.[16] One helix, H2, from the smaller polypeptide and four helices of the larger polypeptide, H5, H6, H8, and H12, form a core unit containing the heme group. Histidine 336 of helix 8 is the proximal heme ligand. Histidine 95 of helix 2 appears to be a catalytic residue on the distal side. The other catalytic residue was suggested to be arginine 239.[16] Sequence homologies indicate that myeloperoxidase, lactoperoxidase, eosinophil peroxidase, and thyroid peroxidase are closely related.

Prostaglandin H_2 Synthase-1. Prostaglandin H_2 synthase-1 is also related to the other mammalian peroxidases.[17] The enzyme is a dimer consisting of two identical subunits of molecular weight approximately 70,000, containing a heme and three sites of N-linked glycosylation. Remarkably, the same five helices as found in myeloperoxidase, H2, H5, H6, H8, and H12, provide the prostaglandin H synthase core that contains the heme. Helices 8 and 12 are on the proximal side of the heme. His388 is the proximal ligand. On the distal side His207 and Gln203 are close to but not bound to the heme iron of prostaglandin H synthase. In addition to the 19 helices, numbered in the same manner as the myeloperoxidase helices, prostaglandin H synthase has additional structure beginning at the N-terminal end of the mature enzyme. There is a short compact domain with a structure very similar to that of the epidermal growth factor. This is followed by four helices, labeled A–D, with hydrophobic exteriors which link the enzyme to the membrane surface. The hydrophobic helices A– C, which would lie roughly parallel to the membrane plane, also provide a channel into the center of the monomer, which is the arachidonic acid binding site.

Structures of representatives of class I (yeast cytochrome *c* peroxidase), class II (fungal lignin peroxidase), and two animal peroxidases (ovine prostaglandin H synthase-1 and human myeloperoxidase) are displayed in Fig. 3.5 (see p. 41 and color insert).

3.3.3. Structure of Catalases

A preliminary report on the structure of the tetrameric catalase from *Penicillium vitale* was published in 1980.[22] However, complete primary sequence data were not available, so it was not possible to determine an accurate three-dimensional structure. The structure of beef liver catalase was published in 1981.[23]

The beef liver catalase tetramer forms a dumbell-like structure with the four heme groups embedded within the interior, 20 Å below the molecular surface and 23 Å from the tetrameric center (Fig. 3.6). The hemes are accessible through hydrophobic channels. Thirteen helices account for 26% and sheet residues for 12% of the molecule. There are nine β-sheet strands. A schematic

FIG. 3.5. Representations of four different peroxidase structures obtained in four different laboratories: those of lignin, yeast, and human myelo- peroxidases, and bottom left, that of ram prostaglandin H synthase isoenzyme 1. See color insert. The arrows point to calcium ion binding sites and the asterisks locate corresponding vestigal, but nonfunctional, sites. (Figure kindly supplied by Dr. Michael Garavito, Department of Biochemistry, Michigan State University.)

representation of one of the identical subunits is shown in Fig. 3.7, which emphasizes the sheet portion of the molecule. An alternative view of a *P. vitale* subunit, shown in Fig. 3.8, emphasizes the helical structures. Each subunit of the beef enzyme has four structural domains: an extended amino terminal arm that stabilizes the quaternary structure, an antiparallel eight-stranded β-

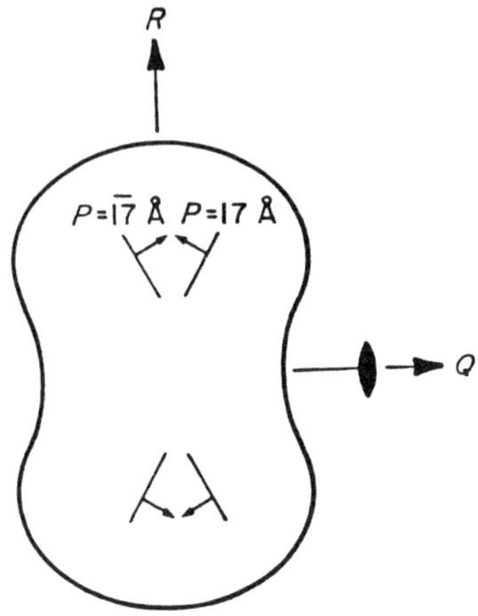

FIG. 3.6. Overall shape and symmetry of the tetrameric catalase molecule showing the approximate location of the buried heme groups. [Reproduced with permission from M. R. N. Murthy, T. J. Reid III, A. Sicignano, N. Tanaka, and M. G. Rossmann, *J. Mol. Biol.* **152,** 465–499 (1981).]

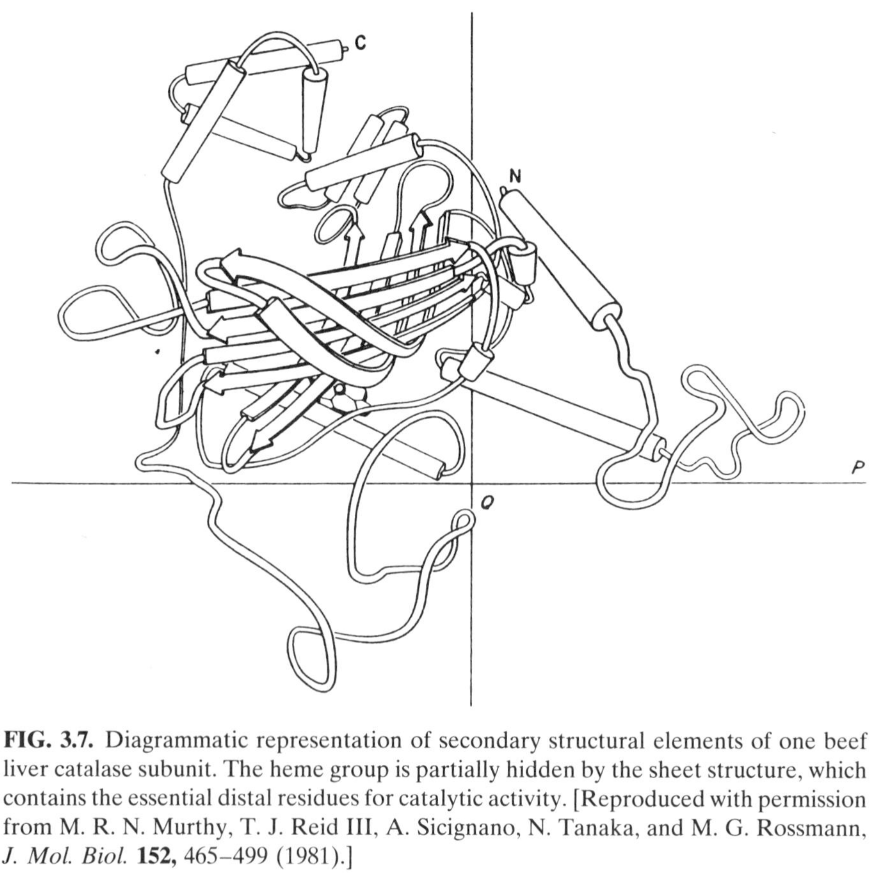

FIG. 3.7. Diagrammatic representation of secondary structural elements of one beef liver catalase subunit. The heme group is partially hidden by the sheet structure, which contains the essential distal residues for catalytic activity. [Reproduced with permission from M. R. N. Murthy, T. J. Reid III, A. Sicignano, N. Tanaka, and M. G. Rossmann, *J. Mol. Biol.* **152,** 465–499 (1981).]

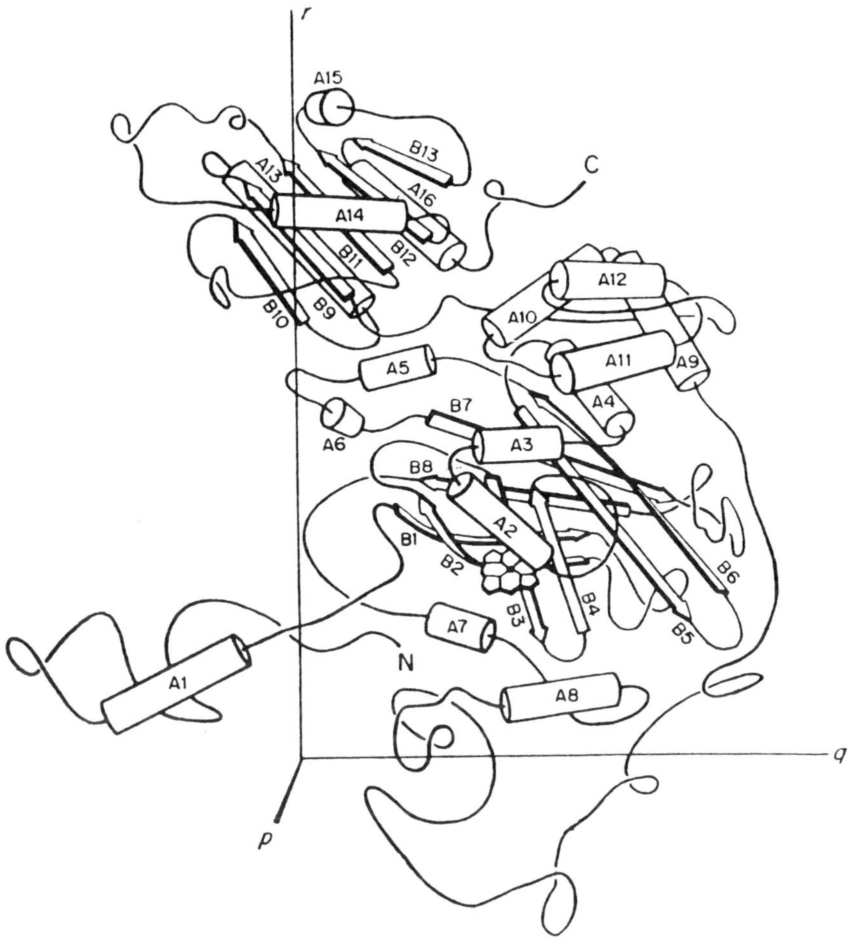

FIG. 3.8. Diagrammatic representation of secondary structural elements of one *Penicillium vitale* catalase subunit. [Reproduced with permission from B. K. Vainshtein, V. R. Melik-Adaman, V. V. Barynin, A. A. Vagin, A. I. Grebenko, V. V. Borisov, K. S. Bartels, I. Fita, and M. Rossmann, *J. Mol. Biol.* **188,** 49–61 (1986).]

barrel that provides the distal residues, a random "wrapping domain" around the exterior that also provides the proximal tyrosine ligand, and a final α-helical structure resembling the E–H helices of the globins. The beef catalase three-dimensional structure was used to help solve the similar *P. vitale* catalase structure.[24]

It was demonstrated that each subunit of beef liver catalase tightly binds one reduced nicotinamide adenine dinucleotide (NADH) molecule.[25] With this knowledge, the corresponding electron density was readily identified in the X-ray diffraction results.[26]

It is clear that the catalases are very different structurally from the peroxidases. The tyrosyl proximal ligand is unique to catalases, as is the β-sheet structure on the distal side of the heme. The catalases are related to the peroxidases through some common heme chemistry involving hydrogen peroxide.

3.4. DETAILS OF THE YEAST CYTOCHROME *c* PEROXIDASE STRUCTURE

3.4.1. Native and Mutant Enzyme Structures—Changes in Steady-State Rates

There are now more entries in the Brookhaven protein structure data bank for cytochrome *c* peroxidase obtained from bakers' yeast than the total for all other peroxidases. The original crystal structure of the native enzyme,[4] published in 1980, was obtained at 2.5 Å resolution by use of two isomorphous mercury derivatives. Only partial amino acid sequence data was available so that there were some errors in assignment of residue locations. However, the distal tryptophan, histidine, and arginine residues as well as the proximal imidazole ligand from a histidine residue were correctly identified. Three antiparallel β pairs were detected (Fig. 3.3). The iron(III) was thought to be 0.3–0.5 Å out of the heme plane in the direction of the imidazole ring in the fifth coordination position. The dish shape of the heme was not identified, and water was incorrectly reported to occupy the sixth coordination position, now known to be too far to coordinate to the iron in wild-type cytochrome *c* peroxidase.

The native enzyme crystal structure was refined to 1.7 Å resolution in 1984.[5] Through a combination of the X-ray results and sequence determination from the corresponding gene,[27] a single residue, missed in the amino acid sequence determination,[28] was identified. The correct sequence is listed in Appendix A. It was deduced that the proximal histidine 175 is hydrogen bonded to aspartate 235. The backbone structure of the enzyme is shown in Fig. 3.9.

The crystal structure of the recombinant enzyme, expressed in *E. coli,* was determined at 2.2 Å resolution, and the iron was shown to be a five-coordinate species. In the same study the structural effects of the Trp51Phe, Trp191Phe, and Asp235Asn substitutions were determined.[29]

A limitation of mechanistic deductions from crystal structures is shown by the suggestion that either Trp51 or Met172 could provide the protein free radical site of compound I.[5] Neither site is the correct location.[30] Studies of six mutations of the distal tryptophan 51 site showed that the mutations converted the iron(III) from a five-coordinate to six-coordinate species.[31,32]

From a combination of crystallographic, kinetic, electron paramagnetic resonance (EPR), and optical spectral studies of mutants in which the proximal histidine was replaced, it was concluded that the proximal His175

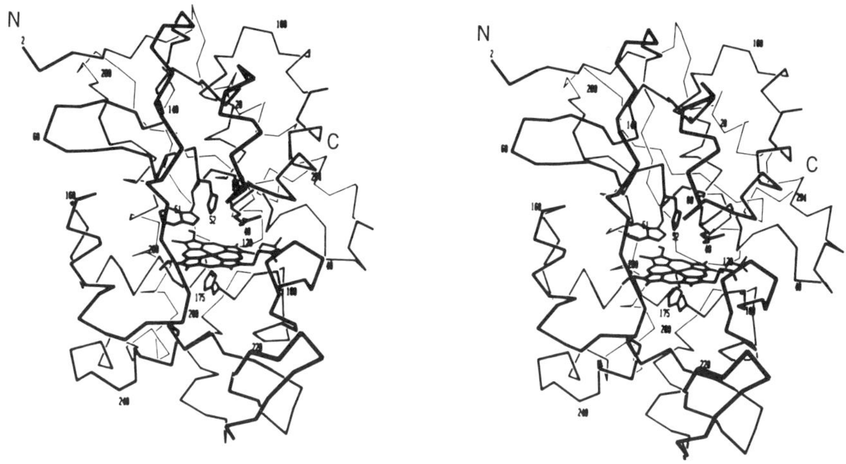

FIG. 3.9. Stereoscopic view of the α-carbon backbone of cytochrome *c* peroxidase. [Reproduced with permission from B. C. Finzel, T. L. Poulos, and J. Kraut, *J. Biol. Chem.* **259,** 13027–13036 (1984).]

can be replaced by other coordinating residues without destroying the ability of the enzyme to function.[33] The effect on compound I reactions was not examined. On the other hand the distal histidine plays a vital role, shown in a combined crystallographic and kinetic study of the His52Leu mutant of the cloned enzyme.[34] The X-ray structure was also determined for the Arg48Leu mutant.[35]

A detailed crystallographic, optical, and EPR spectroscopic, steady-state kinetic, and potentiometric study was made on the effect of mutations on the proximal Asp-His-Fe triad of recombinant cytochrome *c* peroxidase.[36] The triad was likened to the catalytic triad of the serine proteases. The structure, shown in Fig. 3.10 (see p. 46 and color insert), contains hydrogen bonds from the carboxylate of Asp235 to both Trp191 and His175. Three aspartate mutants were studied: Asp235Asn, Asp235Ala, and Asp235Glu. The first two mutations caused significant structural changes. The mutated enzymes had a very low turnover rate for the oxidation of horse cytochrome *c,* and the iron was converted from the high-spin, five-coordinate state of the recombinant enzyme to a six-coordinate species. The replacement of aspartate by glutamate led to much more subtle but equally interesting results. Little perturbation of the active site structure occurred. However, the small change in hydrogen bond geometry was sufficient to increase the reduction potential by 70 mV, alter the anisotropy of the Trp191 free radical EPR signal, reduce the turnover rate to 41% of that for the recombinant enzyme, and reduce the iron ligand field as measured by the zero field splitting. The hydrogen bond to His175 appears to be weakened. Major influences of the catalytic triad may be to

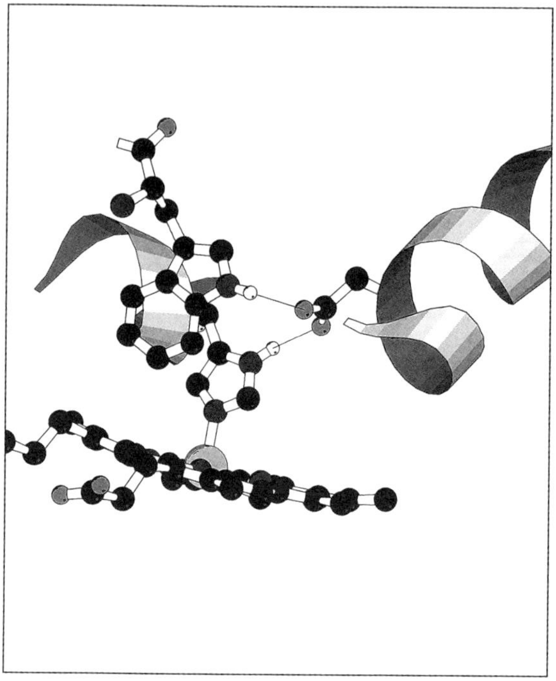

FIG. 3.10. View of the Asp-His-Fe triad of yeast cytochrome *c* peroxidase. See color insert. The thin lines indicate hydrogen bonds between Asp235 and both His175 and the Trp191 free radical center. [Reproduced with permission from D. B. Goodin and D. E. McRee, *Biochemistry* **32,** 3313–3324 (1993).]

correctly orient Trp191 for efficient coupling of the free radical to the heme and to maintain a five-coordinate high-spin iron center in the resting enzyme.

Construction and structural determination of the His175Gly mutant revealed that the heme is converted into a bisaquo complex. Exogenous histidine replaces the water proximal ligand. The untethered imidazole complex does not restore steady-state enzyme turnover.[37] Structures of the bisaquo complex and the bound imidazole complex of the His175Gly mutant are shown in Fig. 3.11 (see p. 47 and color insert). The His175Gly mutant has been examined in solution by resonance Raman spectroscopy in a study that defines the coordination number and spin state of the iron.[38] The mutant exists exclusively as a five-coordinate species in solution.

The effect of mutations near the δ-meso edge of the heme on reactivity was investigated.[39] The side chains of the Ala147Met and Ala147Tyr mutants are positioned over the δ-meso edge. Phenol oxidation rate was reduced by a factor of 5 in the Ala147Tyr mutant. The results were interpreted in terms

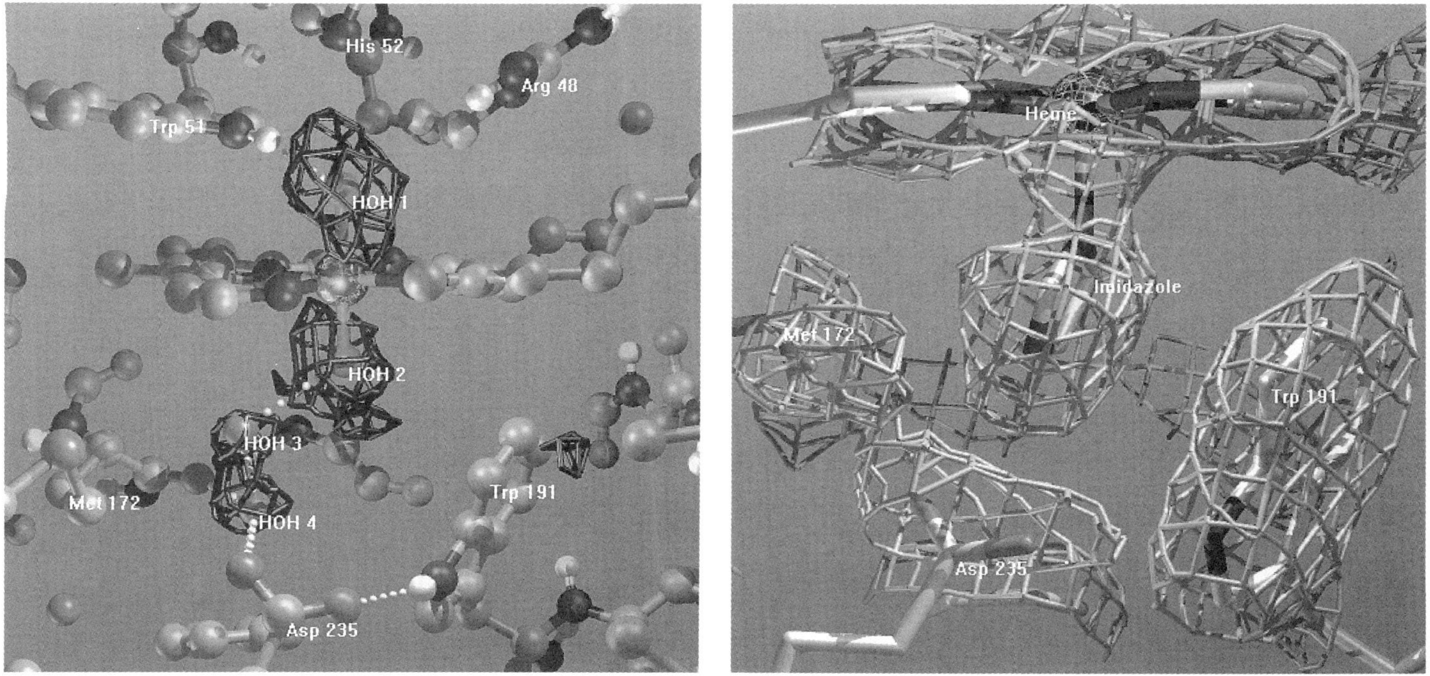

FIG. 3.11. Crystal structures of the His175Gly mutant and its imidazole complex in the vicinity of the heme. See color insert. Left, bisaquo complex of the mutant. Right, the imidazole complex. [Reproduced with permission from D. E. McRee, G. M. Jensen, M. M. Fitzgerald, H. A. Siegel, and D. B. Goodin, *Proc. Natl. Acad. Sci. U.S.A.* **91,** 12847–12851 (1994).]

of phenol reaction at the δ-meso edge, a result anticipated in the studies of Ortiz de Montellano and co-workers.[40,41] A dramatic change in the aniline oxidation by the Ala147Met was observed. The latter may be caused by a specific binding interaction of aniline with the mutant enzyme. A major conclusion was that cytochrome *c* and small reducing substrates react at different sites. An unanticipated and unexplained enhancement of cytochrome *c* oxidation rate occurred with the Ala147Tyr mutant.

3.4.2. Structures of Complexes

The crystal structures of the fluoride, cyanide, and nitric oxide complexes of the ferric enzyme and the carbon monoxide complex of the ferrous enzyme were determined, and the results on the heme environment are summarized in Fig. 3.12.[42] The results are of interest partly because ligand binding is often taken to mimic part of various enzymatic reactions. The Arg48 guanidinium group exhibits the largest total movement among the four complexes, at least 3.0 Å. It either hydrogen bonds to, or forms an ion pair with, the fluoride anion. Recent spectroscopic analyses have demonstrated the unique properties of fluoride as a probe of the hydrogen bonding in the distal cavity.[43]

3.4.3. Creation and Properties of an Artificial Cavity

Further information about the tryptophan 191 site was obtained from the crystal structures of the Trp191Gly and Trp191Gln mutants.[44,45] Replacement of the large tryptophan by much smaller residues created a cavity that was a natural potassium ion binding site, as deduced by comparison with other known cation binding enzymes. The site is surrounded by partial negative charges that would stabilize the Trp191 cation radical of compound I.

The artificial cavity, shown in Fig. 3.13 (see p. 50 and color insert), was studied further. It was found that a series of substituted imidazoles would bind in the cavity in their protonated cationic forms. The work opens the possibility of binding heterocycles into the cavity that could then be oxidized by the enzyme. It was shown that potassium ions compete weakly with the imidazoles for the cavity site. It was also shown that the Trp191Gly Asp235Asn double mutant displays little binding affinity, although the cavity

FIG. 3.12. Comparison of the structures of native cytochrome *c* peroxidase with the structures when the sixth coordination position of the ferric enzyme is occupied by cyanide, nitric oxide, and fluoride, and with its ferrous form containing bound carbon monoxide. [Reproduced with permission from S. L. Edwards and T. L. Poulos, *J. Biol. Chem.* **265,** 2588–2595 (1990).]

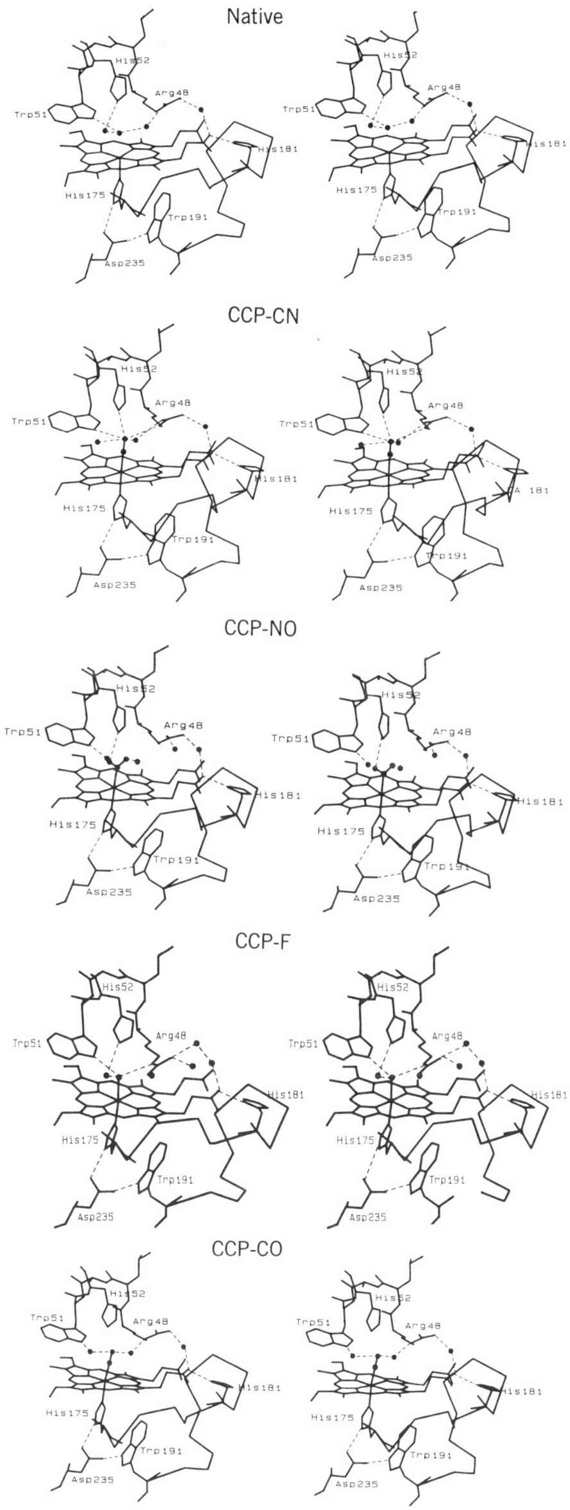
Native
His52
Arg48
Trp51
His181
His175
Trp191
Asp235
CCP-CN
His52
Arg48
Trp51
His181
His175
Trp191
Asp235
CCP-NO
His52
Arg48
Trp51
His181
His175
Trp191
Asp235
CCP-F
His52
Arg48
Trp51
His181
His175
Trp191
Asp235
CCP-CO
His52
Arg48
Trp51
His181
His175
Trp191
Asp235

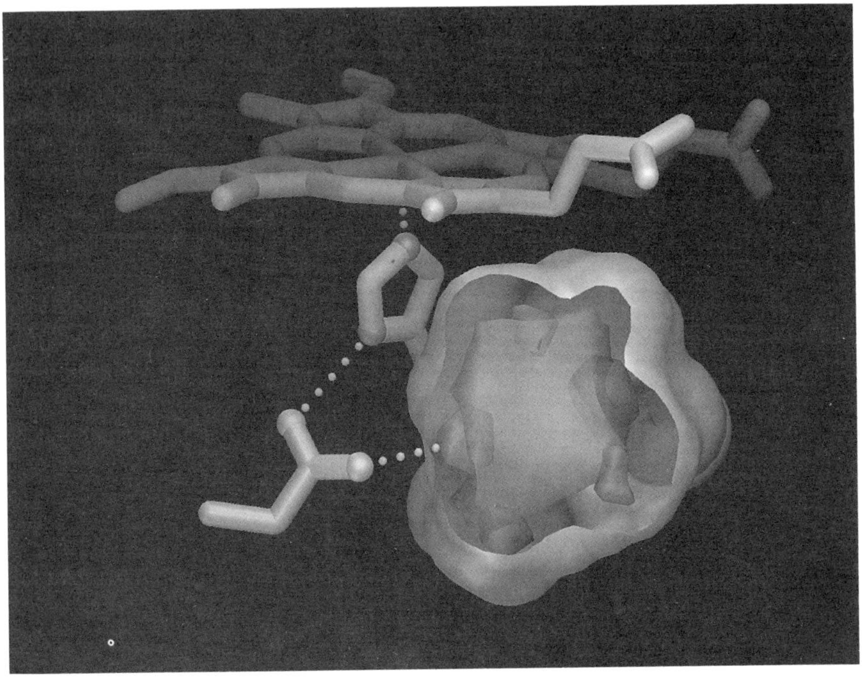

FIG. 3.13. Cavity created in the Trp191Gly mutant. See color insert. The surface of the cavity, in gray, is cut away to show the interior, colored yellow. Locations of water molecules bound in the cavity are shown in blue. [Reproduced with permission from M. M. Fitzgerald, M. J. Churchill, D. E. McRee, and D. B. Goodin, *Biochemistry* **33,** 3807–3818, (1994).]

remains largely unchanged. Thus the role of negative charges in stabilizing the binding process was demonstrated.[46] The mechanism of entry of molecules into the cavity was a mystery because the cavity is completely buried from solvent access. The binding of a larger molecule, protonated benzimidazole, induced permanent structural changes.[47] A hinged main-chain rotation at two residues (Pro190 and Asn195) opens a large solvent-accessible channel into the cavity. For bound benzimidazole the rotation is permanent, but for smaller molecules transient occurrence of the rotation could explain their access to the cavity. The result provides an example of the fact that an X-ray structure is an average of all atomic positions, whereas a reaction mechanism may involve transient, nonaverage structures.[48] Oxidation of 2-aminothiazole in the artificial cavity was studied,[49] and the ligand binding properties of the cavity were utilized to analyze unconventional CH ⋯ O hydrogen bond strengths.[50]

3.4.4. Compound I Structure

The normal peroxidase cycle involves native enzyme and two compounds that participate in the reaction sequence native enzyme → compound I → compound II → native enzyme. For the three reactions there are also three transition states, for a total of six species. Most X-ray results give the structure of one of these species, the native enzyme. In the case of the yeast enzyme a second X-ray structure was obtained, that of compound I.[51] Crystals of the native enzyme at 2°C were soaked for 20 min in a solution of 50 mM potassium phosphate, pH 6.0, 10 mM hydrogen peroxide, and 35% (v/v) 2-methyl-2,4-pentanediol. The crystals were then transferred into a hydrogen-peroxide-free solution, otherwise identical, and cooled to −15°C. Little overall change in structure resulted from compound I formation. A major change from the structure of the native enzyme was a shift of about 1.5 Å of the Arg48 side chain toward the ligand binding site, suggesting that it becomes hydrogen bonded to the ferryl oxygen atom. This compound I structure was confirmed and refined.[52] The side chain of Thr180 moved 0.15 Å in the direction of Met230, leading to the suggestion that the free radical of compound I is located on a cluster of Met230, Met231, and Trp191. The Trp191 free radical site was clearly defined between 1988 and 1995 in a series of studies involving site-directed mutagenesis, and EPR, kinetic, and electron nuclear double resonance (ENDOR) measurements, as discussed in Section 9.2.2.

3.4.5. Compound III Structure

The X-ray structure of a plausible model for a further intermediate or a transition state was obtained.[53] Cytochrome *c* oxyperoxidase or compound III, an analog of oxymyoglobin, can be represented as a $Fe^{III}-OO^{-}$ species. It was used as a model for the unstable $Fe^{III}-OOH$, a plausible intermediate in the formation of compound I from native or resting enzyme and hydrogen peroxide, since the two species differ by a single proton and a single electron. The structural determination was carried out on the Trp191Phe mutant. A stereo view of the oxyperoxidase heme pocket is shown in Fig. 3.14. The results indicate that the dioxygen can form a hydrogen bond with the distal histidine but not the distal arginine residue. On the basis of the structural results, a model for compound I formation was created, as shown in Fig. 3.15. As the hydrogen peroxide approaches the iron(III) site, it transfers a proton to the Nε atom of His52 and the $Fe^{III}-OOH$ species is formed. The proton on the Nε atom is transferred to the O_B atom forming water as a leaving group. Formation of further hydrogen

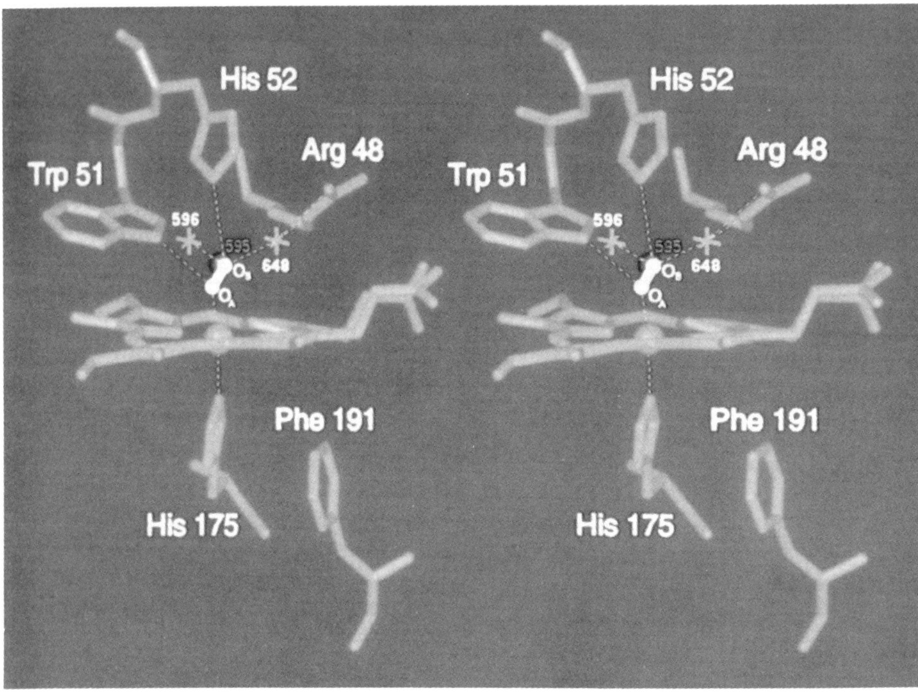

FIG. 3.14. Stereosopic view of the oxyperoxidase structure of the Trp191Phe mutant of yeast cytochrome *c* peroxidase, showing three water molecules in the vicinity of the ferryl oxygen atom. [Reproduced with permission from M. A. Miller, A. Shaw, and J. Kraut, *Nature Struct. Biol.* **1,** 524–531 (1994).]

bonds, distal arginine and tryptophan to ferryl oxygen, accompany the ferryl oxygen formation of compound I.

3.5. STRUCTURES OF THE CYTOCHROME *c* PEROXIDASE–CYTOCHROME *c* COMPLEXES

The structure of the 1 : 1 complex of the yeast enzyme with yeast iso-1 cytochrome *c* was determined at 2.3 Å resolution from crystals grown in 150–

FIG. 3.15. Modeling of the formation of compound I. (*A*), Transfer of proton from hydrogen peroxide to His51. (*B*), Formation of FeIII—OOH. (*C*), the O—O bond is broken. (*D*), Water departs. [Reproduced with permission from M. A. Miller, A. Shaw and J. Kraut, *Nature Struct. Biol.* **1,** 524–531 (1994).]

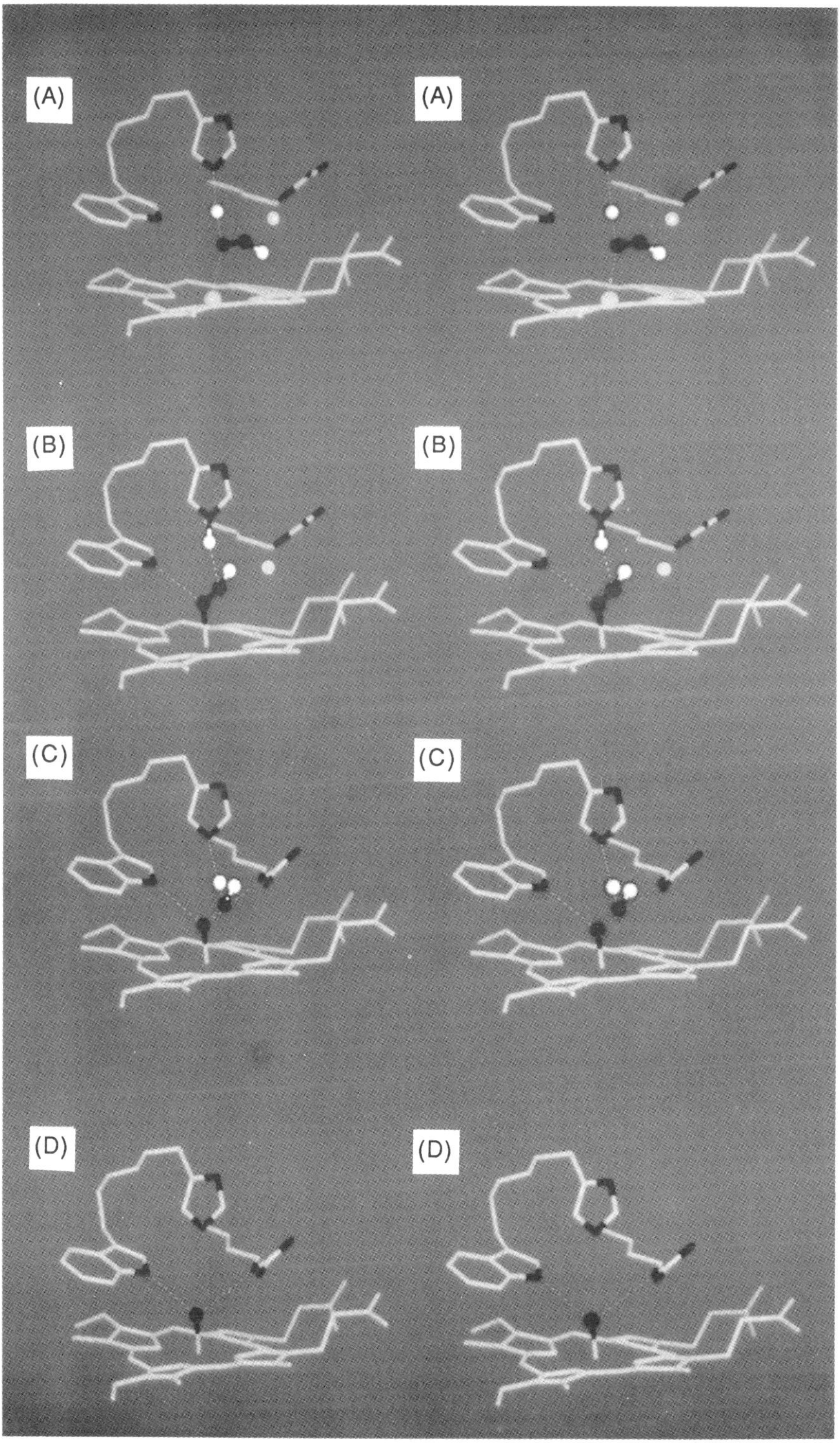
(A)
(A)
(B)
(B)
(C)
(C)
(D)
(D)

200 mM sodium chloride.[54] The structure suggested an electron-transfer pathway from the heme *c* of the cytochrome to the enzyme heme unlike any other that had been proposed. It involves following a single backbone chain of the enzyme, consisting of Ala194, Ala193, Gly192, and finally Trp191. The corresponding structure of the complex of the yeast enzyme with horse cytochrome *c*, determined with crystals grown at low ionic strength, was similar to the complex containing yeast iso-1 cytochrome *c*. Crystals of the horse complex could only be grown at low ionic strength, 0–5 mM phosphate buffer. Both types of crystals were grown at pH 7. The α-carbon structures of the

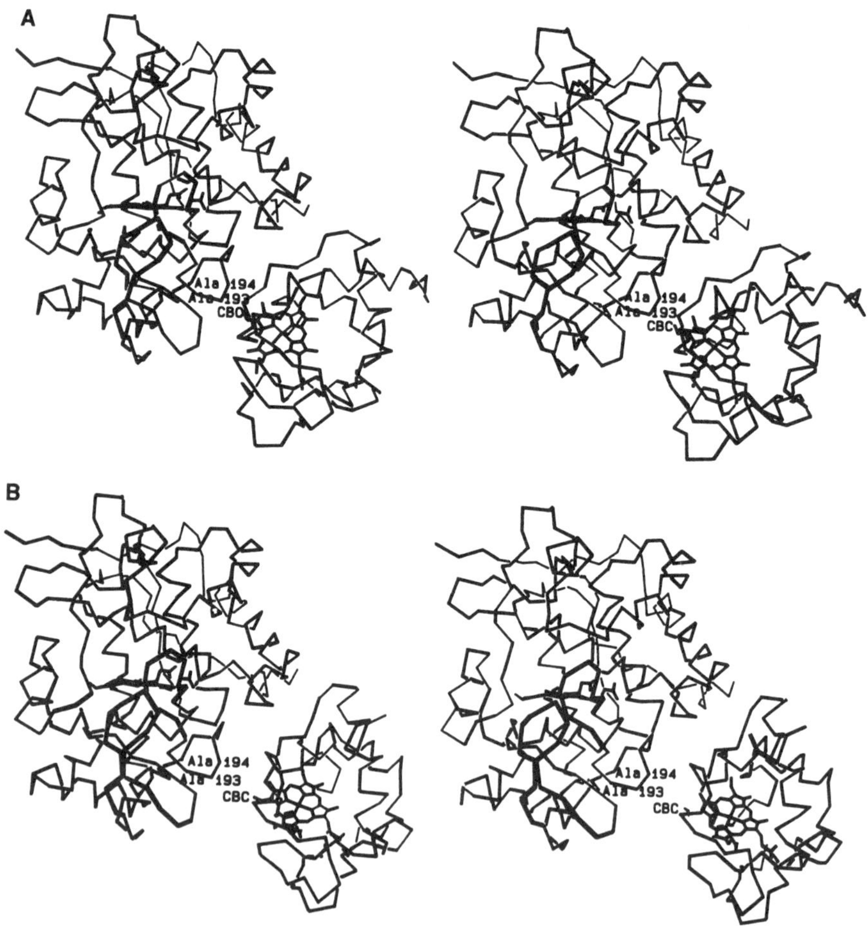

FIG. 3.16. α-Carbon representations of the crystal structures of yeast cytochrome *c* peroxidase, (*A*) with yeast iso-1 cytochrome *c* and (*B*) with horse cytochrome *c*. [Reproduced with permission from H. Pelletier and J. Kraut, *Science* **258,** 1748–1755 (1992).]

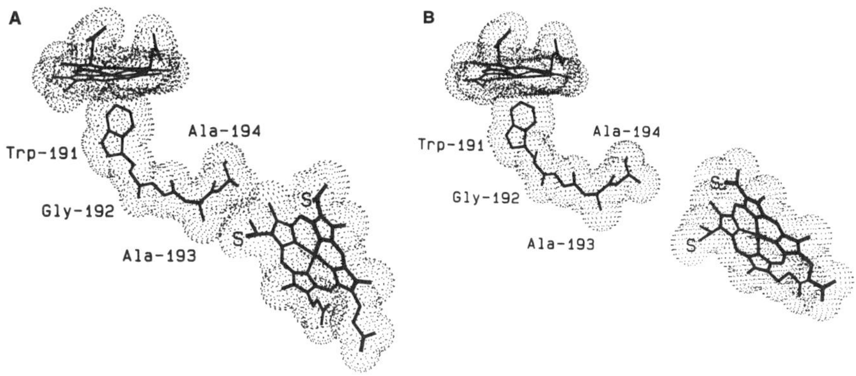

FIG. 3.17. Proposed electron-transfer pathways from the heme *c* to the heme of yeast cytochrome *c* peroxidase, (*A*) for the yeast cytochrome and (*B*) for the horse cytochrome. [Reproduced with permission from H. Pelletier and J. Kraut, *Science* **258,** 1748–1755 (1992).]

complexes are shown in Fig. 3.16 and the proposed electron-transfer pathways in Fig. 3.17.

REFERENCES

1. K. G. Welinder, *FEBS Lett.* **72,** 19–23 (1976).
2. K. Takio, K. Titani, L. H. Ericsson, and T. Yonetani, *Arch. Biochem. Biophys.* **203,** 615–629 (1980).
3. H. R. Bosshard, H. Anni, and T. Yonetani in *Peroxidases in Chemistry and Biology,* J. Everse, K. E. Everse, and M. B. Grisham, eds., CRC Press, Boca Raton, FL, 1991, Vol. II, pp. 51–84.
4. T. L. Poulos, S. T. Freer, R. A. Alden, S. L. Edwards, U. Skogland, K. Takio, B. Eriksson, N. Xuong, T. Yonetani, and J. Kraut, *J. Biol. Chem.* **255,** 575–580 (1980).
5. B. C. Finzel, T. L. Poulos, and J. Kraut, *J. Biol. Chem.* **259,** 13027–13036 (1984).
6. K. G. Welinder and L. Norskov-Lauritsen in *Molecular and Physiological Aspects of Plant Peroxidases,* H. Greppin, C. Penel and Th. Gaspar, eds., University of Geneva, Geneva, 1986, pp. 61–70.
7. Y. Morita, B. Mikami, H. Yamashita, J. Y. Lee, S. Aibara, M. Sato, Y. Katsube, and N. Tanaka in *Biochemical, Molecular and Physiological Aspects of Plant Peroxidases,* J. Lobarzewski, H. Greppin, C. Penel, and Th. Gaspar, eds., University of Geneva Press, Geneva, 1991, pp. 81–88.
8. Y. Morita, J. Funatsu, and B. Mikami in *Plant Peroxidases. Biochemistry and Physiology,* K. G. Welinder, S. K. Rasmussen, C. Penel, and H. Greppin, eds., University of Geneva, Geneva, 1993, pp. 1–4.
9. K. G. Welinder, *Curr. Opin. Struct. Biol.* **2,** 388–393 (1992).

10. K. G. Welinder, J. M. Mauro, and L. Norskov-Lauritsen, *Biochem. Soc. Trans.* **20,** 337–340 (1992).
11. K. G. Welinder, in Ref. 7, pp. 3–13.
12. A. Hochman in *Plant Peroxidases. Biochemistry and Physiology,* K. G. Welinder, S. K. Rasmussen, C. Penel, and H. Greppin, eds., University of Geneva, Geneva, 1993, pp. 103–112, 1993.
13. K. G. Welinder, *Biochim. Biophys. Acta* **1080,** 215–220 (1991).
14. W. R. Patterson and T. L. Poulos, *Biochemistry* **34,** 4331–4341 (1995).
15. K. G. Welinder and M. Gajhede, in Ref. 8, pp. 35–42.
16. J. Zeng and R. E. Fenna, *J. Mol. Biol.* **226,** 185–207 (1992).
17. D. Picot, P. J. Loll, and R. M. Garavito, *Nature* **367,** 243–249 (1994).
18. T. L. Poulos and R. E. Fenna in *Metal Ions in Biological Systems,* H. Siegel, ed., Marcel Dekker, New York, 1994, Vol. 30, pp. 25–75.
19. M. R. N. Murthy, T. J. Reid III, A. Sicignano, N. Tanaka, and M. G. Rossmann, *J. Mol. Biol.* **152,** 465–499 (1981).
20. B. K. Vainshtein, W. R. Melik-Adamyan, V. V. Barynin, A. A. Vagin, A. I. Grebenko, A. I., V. V. Borisov, K. S. Bartels, I. Fita, and M. G. Rossmann, *J. Mol. Biol.* **188,** 49–61 (1986).
21. D. J. Schuller, N. Ban, R. B. van Huystee, A. McPherson, and T. L. Poulos, *Structure* **4,** 311–321 (1996).
22. B. K. Vainshtein, V. R. Melik-Adaman, V. V. Barynin, and A. A. Vagin, *Doklady Akad. Nauk U.S.S.R.* **250,** 242–246 (1980).
23. M. R. N. Murthy, T. J. Reid III, A. Sicignano, N. Tanaka, and M. G. Rossmann, *J. Mol. Biol.* **152,** 465–499 (1981).
24. B. K. Vainshtein, V. R. Melik–Adaman, V. V. Barynin, A. A. Vagin, A. I. Grebenko, V. V. Borisov, K. S. Bartels, I. Fita, and M. Rossmann, *J. Mol. Biol.* **188,** 49–61 (1986).
25. H. N. Kirkman and G. F. Gaetani, *Proc. Natl. Acad. Sci. U.S.A.* **81,** 4343–4348 (1984).
26. I. Fita and M. G. Rossmann, *Proc. Natl. Acad. Sci. U.S.A.* **82,** 1604–1608 (1985).
27. J. Kaput, S. Glotz, and G. Blobel, *J. Biol. Chem.* **257,** 15054–15058 (1982).
28. K. Takio, K. Titani, L. H. Ericsson, and T. Yonetani, *Arch. Biochem. Biophys.* **203,** 615–629 (1980).
29. J. Wang, J. M. Mauro, S. L. Edwards, S. J. Oatley, L. A. Fishel, V. A. Ashford, N.-H. Xuong, and J. Kraut, *Biochemistry* **29,** 7160–7173 (1990).
30. D. B. Goodin, A. G. Mauk, and M. Smith, *J. Biol. Chem.* **262,** 7719–7724 (1987).
31. G. Smulevich, Y. Wang, J. M. Mauro, J. Wang, L. A. Fischel, J. Kraut, and T. G. Spiro, *Biochemistry* **29,** 7174–7180 (1990).
32. D. B. Goodin, M. G. Davidson, J. A. Roe, A. G. Mauk, and M. Smith, *Biochemistry* **30,** 4953–4962 (1991).
33. K. Choudhury, M. Sundaramoorthy, A. Hickman, T. Yonetani, E. Woehl, M. F. Dunn, and T. L. Poulos, *J. Biol. Chem.* **269,** 20239–20249 (1994).
34. J. E. Erman, L. B. Vitello, M. A. Miller, A. Shaw, K. A. Brown, and J. Kraut, *Biochemistry* **32,** 9798–9806 (1993).

35. L. B. Vitello, J. E. Erman, M. A. Miller, J. Wang, and J. Kraut, *Biochemistry* **32,** 9807–9818 (1993).
36. D. B. Goodin and D. E. McRee, *Biochemistry* **32,** 3313–3324 (1993).
37. D. E. McRee, G. M. Jensen, M. M. Fitzgerald, H. A. Siegel, and D. B. Goodin, *Proc. Natl. Acad. Sci. U.S.A.* **91,** 12847–12851 (1994).
38. J. Sun, M. M. Fitzgerald, D. B. Goodin, and T. M. Loehr, *J. Am. Chem. Soc.* **119,** 2064–2065 (1997).
39. S. K. Wilcox, G. M. Jensen, M. M. Fitzgerald, D. E. McRee, and D. B. Goodin, *Biochemistry* **35,** 4858–4866 (1996).
40. G. D. DePillis, B. P. Shista, A. G. Mauk, and P. R. Ortiz de Montellano, *J. Biol. Chem.* **266,** 19334–19341 (1991).
41. P. R. Ortiz de Montellano, *Annu. Rev. Pharmacol. Toxicol.* **32,** 89–107 (1992).
42. S. L. Edwards and T. L. Poulos, *J. Biol. Chem.* **265,** 2588–2595 (1990).
43. F. Neri, D. Kok, M. A. Miller, and G. Smulevich, *Biochemistry* **36,** 8947–8953 (1997).
44. M. M. Fitzgerald, M. J. Churchill, D. E. McRee, and D. B. Goodin, *Biochemistry* **33,** 3807–3818 (1994).
45. M. A. Miller, G. W. Han, and J. Kraut, *Proc. Natl. Acad. Sci. U.S.A.* **91,** 11118–11122 (1994).
46. M. M. Fitzgerald, M. L. Trester, G. M. Jensen, D. E. McRee, and D. B. Goodin, *Protein Sci.* **4,** 1844–1850 (1995).
47. M. M. Fitzgerald, R. A. Musah, D. E. McRee, and D. B. Goodin, *Nature Struct. Biol.* **3,** 626–631 (1996).
48. G. A. Petsko, *Nature Struct. Biol.* **3,** 565–566 (1996).
49. R. A. Musah and D. B. Goodin, *Biochemistry* **36,** 11665–11674 (1997).
50. R. A. Musah, G. M. Jensen, R. J. Rosenfeld, D. E. McRee, and D. B. Goodin, *J. Am. Chem. Soc.* **119,** 9083–9084 (1997).
51. S. L. Edwards, N. H. Xuong, R. C. Hamlin, and J. Kraut, *Biochemistry* **26,** 1503–1511 (1987).
52. V. Fülöp, R. P. Phizackerley, S. M. Soltis, I. J. Clifton, S. Wakatsuki, J. Erman, J. Hajdu, and S. L. Edwards, *Structure* **2,** 201–208 (1994).
53. M. A. Miller, A. Shaw, and J. Kraut, *Nature Struct. Biol.* **1,** 524–531 (1994).
54. H. Pelletier and J. Kraut, *Science* **258,** 1748–1755 (1992).

CHAPTER 4

Horseradish Peroxidase. I: Ligand Binding, Redox Potentials, Formation of Its Compounds, and Some of Their Reactions

Horseradish peroxidase might be regarded as the grandparent of all peroxidases. Not in terms of evolution, but in scientific terms, it precedes all other peroxidases. It had been intensively studied in many different laboratories before most other peroxidases had been identified. Because of the wealth of information available on horseradish peroxidase, and because of its historic importance, it deserves a special place in the detailed descriptions of peroxidases that follow in this and subsequent chapters. An introduction to the properties of horseradish peroxidase was given in Chapter 2.

4.1. ISOENZYME C OF HORSERADISH PEROXIDASE

There are many isoenzymes of horseradish peroxidase (see Chapter 5). The dominant isoenzyme is a neutral form designated as C. Commercial preparations consist mainly of C with perhaps some B isoenzymes also present, but the physical and chemical properties of B and C are identical for all practical purposes.[1] The only difference appears to be in electrophoretic mobility. The discussion in this chapter is restricted to isoenzyme C (or to commercial preparations that may be regarded as C).

4.2. LIGAND BINDING TO HORSERADISH PEROXIDASE

4.2.1. Introduction to Kinetics of Ligand Binding

The simplest of peroxidase reactions is the reversible binding of ligands. The interpretation of experimental results is usually straightforward and can lead to valuable insights into the nature of the active site. Results obtained on inorganic model systems can aid in the interpretation of results.

The binding of a ligand to horseradish peroxidase may be represented in the following manner:

$$\mathrm{Fe^{III}} + \mathrm{L_{tot}} \underset{k_{\mathrm{app}}}{\overset{k_{\mathrm{app}}}{\rightleftharpoons}} \mathrm{Fe^{III}{-}L_{tot}} \tag{4.1}$$

$$\frac{d[\mathrm{Fe^{III}{-}L_{tot}}]}{dt} = k_{\mathrm{app}}[\mathrm{Fe^{III}}][\mathrm{L_{tot}}] - k_{-\mathrm{app}}[\mathrm{Fe^{III}{-}L_{tot}}] \tag{4.2}$$

where Fe^{III} is the heme iron and L is the ligand. It is simpler to describe the heme iron of the native enzyme in terms of its oxidation state rather than by its actual or formal charge, which varies with changes in its environment. The ligand has the subscript *tot* for *total concentration,* since it may exist in more than one form, most commonly acidic and basic. The subscript *app* for *apparent* on the second-order rate constant k for the binding of ligand indicates that no *a priori* assumptions have been made concerning which form of the ligand is binding; the value of k_{app} is calculated on the basis of the total amount of ligand present. Similarly, no assumptions are made about the form of the ligand released in the dissociation process, except that it must be the same form that is bound. The dissociation of the complex has rate constant k_{-app}.

In a comprehensive study, the pH of the solution will be varied over the widest possible range. Ionic strength must be kept constant, and, when a change in buffers is made, it is preferable to obtain overlapping data to make sure that no specific buffer effects are occurring.

Once an accurate pH rate profile is obtained, then application of kinetic theory leads to conclusions about the reaction mechanism. I illustrate the procedure with one example. The plots in Fig. 4.1 show that the slope of the plot of log k_{app} versus pH goes from 0 to -1 and the inflection point corresponds to the pK_a of the ligand. The simplest possible explanation of the binding process is that the acid form of the ligand binds to a neutral form of the enzyme. There is no pH dependence for the rate constant for dissociation of the enzyme–ligand complex k_{-app}, so there is no proton uptake or release upon dissociation. The principle of microscopic reversibility states that the path of the forward reaction must also be the path of the reverse reaction. Therefore the acid form of the ligand dissociates from the complex. The quantitative treatment is as follows for the particular case that the acid form of the ligand is binding:

$$\mathrm{Fe^{III}} + \mathrm{HL} \underset{k_{-\mathrm{app}}}{\overset{k_{\mathrm{app}}}{\rightleftharpoons}} \mathrm{H^{+}{-}Fe^{III}{-}L^{-}} \tag{4.3}$$

$$\frac{d[\mathrm{H^{+}{-}Fe^{III}{-}L^{-}}]}{dt} = k_1[\mathrm{Fe^{III}}][\mathrm{HL}] - k_{-1}[\mathrm{H^{+}{-}Fe^{III}{-}L^{-}}] \tag{4.4}$$

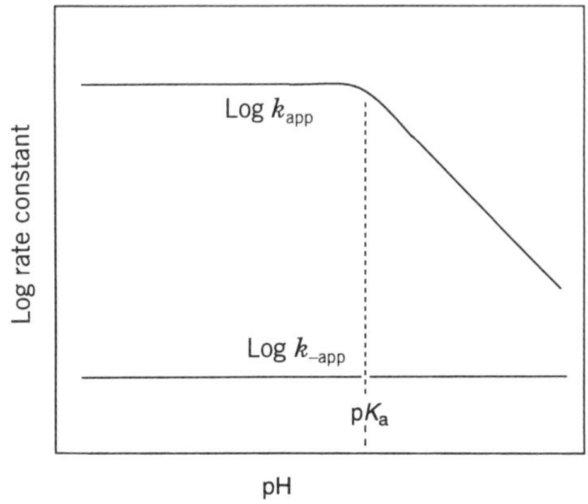

FIG. 4.1. Schematic representation of reversible ligand binding to a peroxidase in which the acid form of the ligand HL binds to the enzyme with a rate constant k_{app} and dissociates with rate constant k_{-app}. The slope of the plot of log k_{app} versus pH changes from 0 to -1 at the inflection point, which yields the pK_A value for the acid dissociation of the ligand.

In this example the proton of HL is binding at a different site on the enzyme from the anion L^-.

Equations (4.1) and (4.2) are operational equations in which no mechanistic assumptions have been made. Equations (4.3) and (4.4) are based upon a specific mechanism. The mechanism can be tested by equating the forward and reverse rates in Eqs. (4.2) and (4.4).

$$k_{app}[Fe^{III}][L_{tot}] = k_1[Fe^{III}][HL] \tag{4.5}$$

which leads to

$$k_{app} = \frac{k_1}{1 + \dfrac{K_A}{[H^+]}} \tag{4.6}$$

where K_A is the dissociation constant of the ligand HL. In similar fashion it can be shown that

$$k_{-app} = k_{-1} \tag{4.7}$$

Equations (4.6) and (4.7) fit the experimental data in Fig. 4.1 so a plausible mechanism has been obtained. The value of K_A can be determined from both kinetic and titration data.

An alternative, more complex, and hence perhaps less likely, explanation is that the basic form of the ligand L^- binds to an acidic form of the enzyme, in which case the total number of protons involved in the reaction is the same. If the rate constant k_1 for the latter mechanism exceeds the diffusion-controlled limit, then the mechanism is proven to be incorrect.[2]

The real power of the kinetic approach is that mechanisms can be excluded that do not fit the experimental data, thus narrowing the number of possibilities. When properly used, enzyme kinetics is a very powerful tool, a comment equally applicable to a battery of spectroscopic techniques, theoretical calculations, and site-directed mutagenesis.

4.2.2. Fluoride Binding to Horseradish Peroxidase

The kinetics of fluoride binding and dissociation from horseradish peroxidase (HRP) was studied using a joule-heating temperature jump apparatus.[3,4]

$$\mathrm{HRP_{tot}} + \mathrm{F_{tot}} \underset{k_{-\mathrm{app}}}{\overset{k_{\mathrm{app}}}{\rightleftharpoons}} \mathrm{HRP{-}F_{tot}} \tag{4.8}$$

$$\frac{1}{\tau} = k_{\mathrm{app}}[\mathrm{F}]_{\mathrm{tot}} + k_{-\mathrm{app}} \tag{4.9}$$

where τ is the relaxation time of the reaction. Fluoride binds to iron(III) of the peroxidase to convert it from a five-coordinate to a six-coordinate species. The relaxation time τ has units of seconds (s); its reciprocal is the first-order rate constant measured for a particular concentration of fluoride. The forward, or on, rate constant k_{app} showed a marked pH dependence, whereas the reverse, or off, rate constant $k_{-\mathrm{app}}$ showed little change as a function of pH. All of the main features of the reaction can be accounted for by the following mechanism:

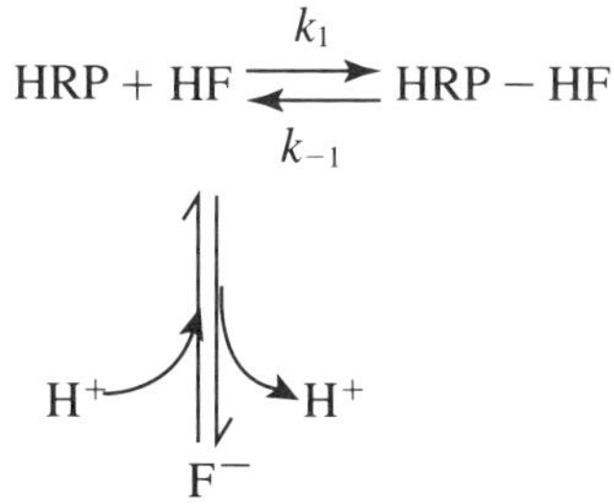

(4.10)

$$K_{\mathrm{A}} = \frac{[\mathrm{H^+}][\mathrm{F^-}]}{[\mathrm{HF}]} \tag{4.11}$$

The mechanism predicts that when the pH is greater than the value of the pK_A for HF, the on rate will decrease with increasing pH because the concentration of HF decreases. The mechanism also predicts that the off rate is pH-independent because there is no involvement of a proton from the bulk solution. There is a profound implication in this mechanism. The enzyme reacts with the protonated form of fluoride, HF, and the proton is incorporated into the enzyme structure. Available evidence from inorganic chemistry indicates that the HF molecule would not remain intact upon binding. The fluoride ion binds to iron(III) and the proton is accepted elsewhere. Upon dissociation from the enzyme the proton is returned to the departing fluoride ion.

The value of k_{app} decreases logarithmically with increasing pH; k_{-app} is essentially pH-independent. The rate constants k_1 and k_{-1} are pH-independent constants. The value of k_1 is $4.2 \times 10^6\ M^{-1}s^{-1}$, and of k_{-1} is $3.9 \times 10^2\ s^{-1}$ as determined from experiments over the pH range 5.0–7.5 . The treatment given here is simpler than in the original papers, and I believe better. By excluding data below pH 5.0, potential complications caused by the presence of the bifluoride ion $F{-}H{-}F^-$, and the potential effect of a high-voltage current on a protein with a large net positive charge are eliminated. Also, effects at low pH of actetate buffer and nitrate anion used to maintain constant ionic strength, which were not appreciated at the time, are eliminated. A subsequent study of azide binding to chloroperoxidase, using an iodine laser temperature jump apparatus, eliminated possible complications caused by heating with an electric current, and is described in Section 14.1.4.

4.2.3. Cyanide Binding to Horseradish Peroxidase

The mechanism of cyanide binding to horseradish peroxidase is similar to that for fluoride.[4,5] The pH dependence of the on rate constant for cyanide is different from that for fluoride because most of the accessible pHs have values less than the pK_A value for HCN (9.14) whereas for fluoride the reverse is true. Rate constants for cyanide binding in both water and deuterium oxide as a function of pH/pD are shown in Fig. 4.2.

The following mechanism fits the experimental data:

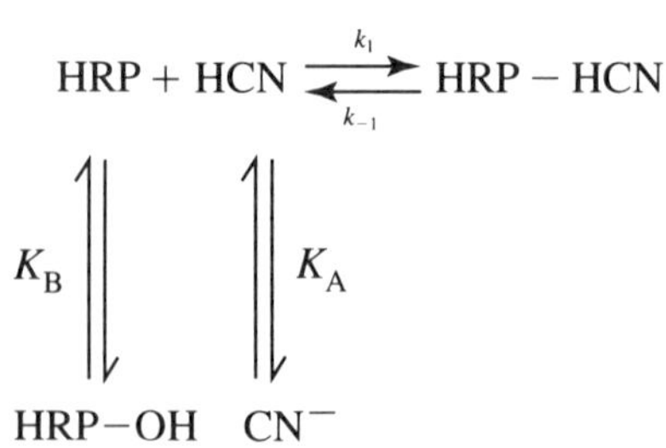

(4.12)

where K_A is the Brønsted acid dissociation constant for HCN (proton exchange is not shown) and K_B is the equilibrium constant for formation of the alkaline

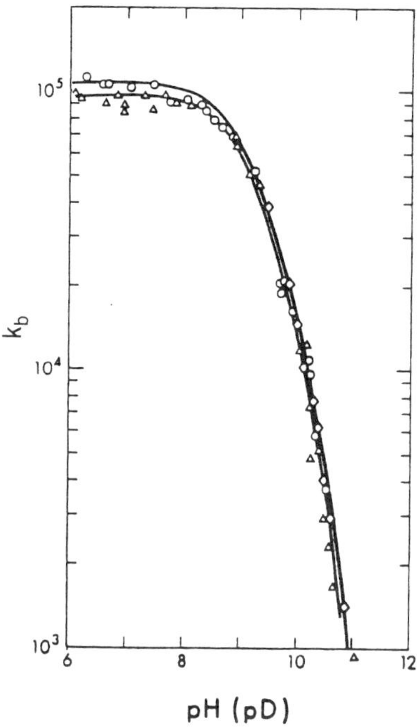

FIG. 4.2. Plot of the apparent second-order rate constant for the binding of cyanide to native horseradish peroxidase as a function of pH in water (upper curve) and as a function of pD in deuterium oxide (lower curve). [Reproduced with permission from H. B. Dunford, W. D. Hewson, and H. Steiner, *Can. J. Chem.* **56,** 2844–2852 (1978).]

form of horseradish peroxidase. The cyanide ion binds to iron(III) and the proton from HCN is accepted into the protein structure. The corresponding rate equations for the pH dependence of the reactions are:

$$k_{app} = \frac{k_1}{\left[1 + \frac{K_A}{[H^+]}\right]\left[1 + \frac{K_B}{[H^+]}\right]} \tag{4.13}$$

$$k_{-app} = k_{-1} \tag{4.14}$$

Three terms determine the value of k_{app}: k_1, K_A, and K_B. Their values in water, determined from the fit to the experimental data, are $k_1 = 1.06 \times 10^5$ $M^{-1}s^{-1}$, $pK_A = 9.14$ and $pK_B = 10.9$. Thus thermodynamic constants can be obtained from kinetic studies. The converse is not true.

Data at low pH for cyanide binding have been excluded from the above discussion, again because of the complicating effect of acetate buffer and potassium nitrate.[6,7] As can be seen in Section 4.5, cyanide binding appears to mimic compound I formation. When the ionization of hydrocyanic acid is taken into account, the pH rate profiles for cyanide binding and compound I formation are identical in shape. The main differences are that cyanide binding occurs at a rate lower by a factor of about 200 than that for compound I formation, and there is no k_H/k_D isotope effect, within the experimental error. A sharp contrast exists between the pH rate profiles for cyanide binding and compound I formation for lactoperoxidase (Chapter 17).

4.3. PROTON BALANCE TITRATIONS FOR INTERCONVERSIONS OF HORSERADISH PEROXIDASE AND ITS COMPOUNDS

Important clues to the structural and catalytic behavior of horseradish peroxidase were obtained by the determination of whether protons are taken up or released when one enzyme species is converted to another. By use of a pH stat in unbuffered solutions, proton uptake or release can be detected. Thus it was determined that compound I formation from native horesradish peroxidase does not involve proton uptake or release.[8] This result was confirmed in independent experiments, which were extended to all known interconversions.[9] Compound I formation from hydrogen peroxide and native enzyme is known to involve water release, so the reaction is:

$$\text{Native peroxidase} + H_2O_2 \longrightarrow \text{Compound I} + H_2O \qquad (4.15)$$

The ligand binding experiments, described in Section 4.2, indicate that a proton is taken up by the native enzyme when HF or HCN binds. The ligand binding reactions appear to mimic the first part of compound I formation. Therefore, it would appear in compound I formation that a proton is first taken up by the enzyme and later released, so that the net reaction neither consumes nor releases protons. The enzyme might be capable of a more efficient reaction for compound I formation, in which the proton is only partially transferred to the enzyme along a hydrogen bond from one oxygen atom of H_2O_2, and then is transferred back to the second oxygen atom of H_2O_2, forming the H_2O leaving group.

Titration of compound I with ferrocyanide showed that the net reduction reaction is as follows:[9]

$$\text{Compound I} + e^- + H^+ \longrightarrow \text{Compound II} \qquad (4.16)$$

Similarly, compound II reduction involves addition of a proton to the enzyme.[9] It is now known a proton is released as part of the H_2O leaving group, so that the overall reaction is:

$$\text{Compound II} + e^- + H^+ \longrightarrow \text{Native peroxidase} + H_2O \qquad (4.17)$$

These results are consistent with reducing substrates such as phenols and aromatic amines being hydrogen atom donors to the enzyme, as they have long been described in the biochemistry literature.[10]

The following additional stoichiometries were also reported, involving the interconversions ferrous enzyme → native ferric enzyme, ferrous enzyme → compound III, compound II → compound III, and compound III → native enzyme.[9] I have added water release where appropriate so that oxygen atom balance is maintained.

$$\text{Ferrous enzyme} \longrightarrow \text{Ferric enzyme} + e^- + H^+ \qquad (4.18)$$

The ferrous enzyme has a pK_a value of 7.17, so that its additional proton, compared to the ferric enzyme, is lost in alkaline solution. Conversion of the ferrous enzyme to compound III in acid solution does not involve proton transfer.

$$\text{Ferrous enzyme} + O_2 \longrightarrow \text{Compound III} \qquad (4.19)$$

Since the ferrous enzyme has an additional proton compared to that found in the native ferric enzyme in acid solution, compound III also contains an additional proton.

In the presence of a large excess of H_2O_2, compound II is converted into compound III,[11,12] a reaction that is unlikely to be important *in vivo*.

$$\text{Compound II} + H_2O_2 \longrightarrow \text{Compound III} + H_2O \qquad (4.20)$$

The important results of all these reactions has been summarized in a pentagonal diagram, which shows the correct stoichiometry for all reactions. Conversion of compound III to native enzyme can be simply shown as:

$$\text{Compound III} \longrightarrow \text{Native peroxidase} + HO_2\bullet \qquad (4.21)$$

which may be a reversible reaction. A version of a pentagonal diagram is shown in Fig. 4.3, in which the stepwise increase in oxidation states is shown in a clockwise direction. Water release, where appropriate, and the ligand in the sixth coordination position of the iron, where applicable, are shown in the diagram. Three species contain one more proton than the other two. Thus (acidic) ferrous enzyme, and compounds II and III have an additional proton compared to the native ferric enzyme and compound I. This additional proton is presumably located in the active site on a distal amino acid residue, and this active site location may play an important catalytic role.

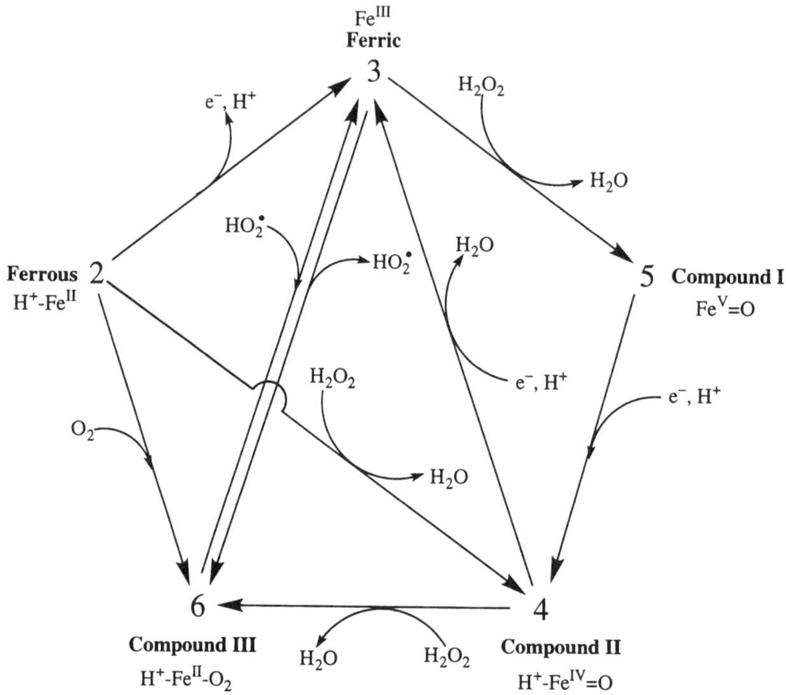

FIG. 4.3. Pentagonal diagram showing the five oxidation states of horseradish peroxidase. Correct stoichiometry is shown for all reactions, including proton balance. The Arabic numerals indicate formal oxidation states. [Based in part, with permission, on H. Yamada and I. Yamazaki, *Arch. Biochem. Biophys.* **165,** 728–738 (1974).]

One more reaction is shown in Fig. 4.3, the conversion of the ferrous enzyme to compound II by reaction with hydrogen peroxide,[13] which occurs with a rate constant of $10^5\ M^{-1}s^{-1}$:

$$\text{Ferrous HRP} + H_2O_2 \longrightarrow \text{Compound II} + H_2O \qquad (4.22)$$

The reaction appears to be analogous to the formation of compound I from native HRP and hydrogen peroxide.

The numbers 2–6 in Fig. 4.3 represent the total oxidation states of the five species. They are calculated relative to elemental iron and to an oxygen atom with its completed octet of electrons, as in water. Thus when native enzyme is converted to compound I, the increase in the formal oxidation state is assigned entirely to the iron, which increases from +3 to +5. In fact it is known that one of the electrons has been obtained from the porphyrin ring, and this is replaced when compound I is reduced to compound II. Since O_2 requires four electrons to reduce it to water, it contributes four oxidizing equivalents to compound III.

4.4. REDUCTION POTENTIALS OF HORSERADISH PEROXIDASE AND ITS COMPOUNDS

Figure 4.4 shows the reduction potential of ferric horseradish peroxidase as a function of pH.[14] This classic work by Henry Harbury was conducted on recrystallized HRP samples provided by Karl-Gustav Paul and Hugo Theorell.

Also shown in Fig. 4.4 are comparisons to other heme-containing species. Thus the iron(III) of cytochrome *c* is much easier to reduce to iron(II) than in the case of horseradish peroxidase, a reflection of nature's use of cytochrome *c* as an electron acceptor and of horseradish peroxidase compounds as oxidants. Similarly, methemoglobin and metmyoglobin are easier to reduce than horseradish peroxidase; upon reduction they contain iron(II) ions, which can

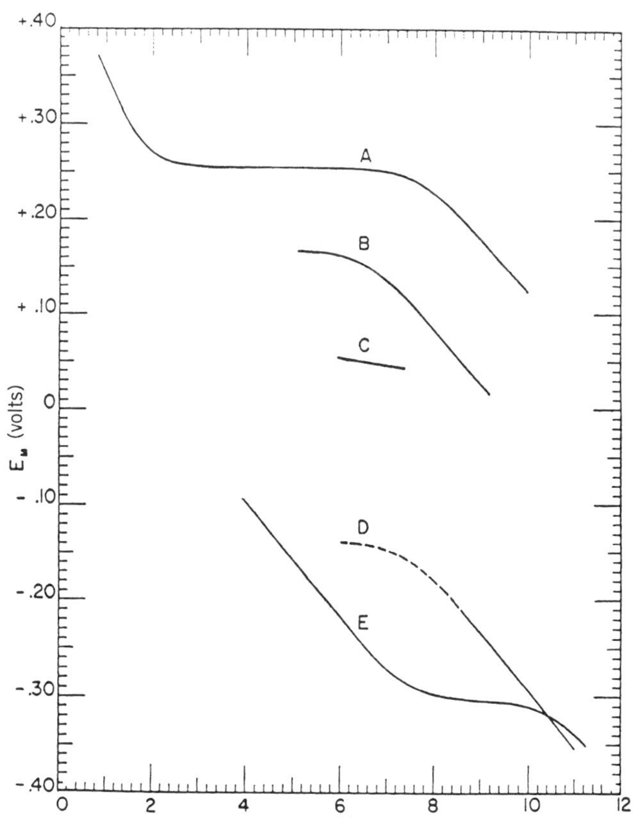

FIG. 4.4. One-electron reduction potentials as a function of pH: (A) Beef heart ferricytochrome *c*, (B) horse ferrimyoglobin, (C) horse heart ferrimyoglobin, (D) ferriprotoporphyrin IX, and (E) ferric horseradish peroxidase. [Reproduced with permission from H. A. Harbury, *J. Biol. Chem.* **225,** 1009–1024 (1957).]

accept dioxygen ligands. The greater difficulty in reduction of horseradish peroxidase compared to metmyoglobin has been attributed to different amino acid residues being hydrogen bonded to the proximal histidine side chain, found in the sixth coordination position of both species.[15] In the case of horseradish peroxidase an aspartate is hydrogen bonded, and the increased negative charge is believed to help stabilize higher oxidation states. In myoglobin, it is a carbonyl group in the peptide backbone.

Horseradish peroxidase is a sluggish electrochemical reactant so that mediators were used to accelerate response at a gold-plated platinum electrode. Methyl viologen and sodium dithionite were used as reducing agents and potassium ferricyanide and 2,6-dichlorophenol-indophenol as oxidizing agents. The equation that fits the experimental results, obtained at 30°C, is as follows:

$$\varepsilon = \varepsilon^0 + \frac{RT}{F} \ln \frac{[\mathrm{H}^+]\{[\mathrm{H}^+] + \mathrm{K}_r\}}{[\mathrm{H}^+] + \mathrm{K}_o} \tag{4.23}$$

The value of ε^0 is zero at zero pH. R is the universal gas constant, T the temperature in Kelvin, and F the faraday. The subscripts r and o stand for "reduced" and "oxidized." The value of pK_r is 7.38 and of pK_o 10.56. Thus at pH 7.38 the ferrous enzyme is 50% in a form where a heme-linked basic group is protonated, and 50% is deprotonated. The pK of 10.56 is that for the alkaline transition of the native ferric enzyme.

In later important work, the reduction potentials of the compound I–compound II and compound II–native HRP couples were determined.[16] Use was made of the fact that K_2IrCl_6 and $K_3Fe(CN)_6$ can act as reversible oxidants[17] and K_3IrCl_6 as a reversible reductant. The equations fit to the experimental data, obtained at 20°C, were the following.[16] For the compound I–compound II couple:

$$\varepsilon(\mathrm{I} - \mathrm{II}) = \varepsilon^0 + 0.058 \ln \{K_{\mathrm{II}} + [\mathrm{H}^+]\} \tag{4.24}$$

The value of ε^0 is 1.33 V and refers to pH 0. The subscript II refers to compound II. The value of pK_{II} is 8.6. The ionization of the acid group with the pK_a value of 8.6 makes compound II inactive. The same group with pK_a of 8.6 had been detected in several transient state kinetic studies of compound II reactions with reducing substrates.[18–21]

The compound II–ferric enzyme couple obeys the following equation:

$$\varepsilon(\mathrm{II} - \mathrm{ferric}) = \varepsilon^0 + 0.058 \ln \frac{K_{\mathrm{ferric}}[\mathrm{H}^+] + [\mathrm{H}^+]^2}{K_{\mathrm{II}} + [\mathrm{H}^+]} \tag{4.25}$$

Again, the value of ε^o is 1.33 V and refers to pH 0. Thus, despite different reactivities, compounds I and II have almost identical reduction potentials at lower pH. The value of pK_{II} is 8.6 as in the I–II couple, and of pK_{ferric} 11.0, which refers to the alkaline transition of the native enzyme. From both couples one can calculate the couple for the direct two-electron reduction of compound I to native enzyme:

$$\varepsilon(\text{I} - \text{ferric}) = 1.33 + 0.029 \log \{K_{\text{ferric}} [\text{H}^+] + [\text{H}^+]^2\} \qquad (4.26)$$

The important work in the group of Isao Yamazaki, described above,[9,12] laid to rest two myths about horseradish peroxidase: that native enzyme could not be oxidized directly to compound II and that compounds I and II have greatly different reduction potentials. It was shown that the native enzyme and compound II can be reversibly oxidized in one-electron reactions and that the reduction potentials of compounds I and II, despite the different reactivities of the two compounds, are very similar in acid and neutral solutions. At about the same time, oxidation of compound II to compound I, and native enzyme to compound II, was demonstrated using chlorite,[22,23] periodate, and also radical anions, such as $SO_4{\bullet}^-$, $CO_3{\bullet}^-$, $Cl_2{\bullet}^-$, and $Br_2{\bullet}^-$, the latter being produced in photolysis reactions.[24,25] In a more recent study, one-electron oxidations of native enzyme and compound II were accomplished using a ground-state ruthenium(III) bipyridine complex that was photochemically generated and then quenched.[26] The procedure facilitated fast time measurements. It was found that the native enzyme was oxidized first to a porphyrin cation radical, and the rate-controlling step was the addition of water in the sixth coordination position of the iron, which was followed by proton and

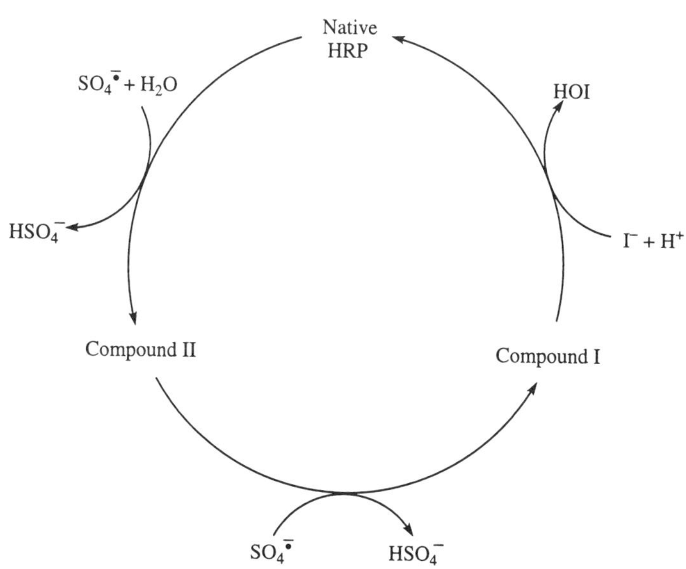

FIG. 4.5. Horseradish peroxidase cycle run in reverse. Each of the steps has been observed using transient-state kinetics.

electron transfers to convert it into a ferryl oxygen, completing formation of compound II. At high pH conversion of the hydroxylated form of the native enzyme to compound II was a faster simpler process. Compound II is converted to compound I with a rate constant of $1.1 \times 10^8\ M^{-1}s^{-1}$ at pH 10.3.

The direct reduction of compound I to native enzyme had been attained using iodide,[27] so that the reversal of every step in the peroxidase cycle had been demonstrated, as illustrated in Fig. 4.5. The reversal of the cycle has been accomplished one step at a time using transient-state kinetics. Running the total cycle in reverse, in the steady state, would appear to be an impossible task.

4.5. MECHANISM OF PEROXIDASE COMPOUND I FORMATION

4.5.1. Poulos–Kraut Mechanism

The original mechanism by Poulos and Kraut[28] was based on the 1980 crystal structure of yeast cytochrome *c* peroxidase. Two features of their mechanism for the transfer of an oxygen atom from hydrogen peroxide to iron(III) have stood the test of time. These are the following:

1. Distal histidine is a proton acceptor from hydrogen peroxide.
2. Distal arginine is a charge stabilizer.[29]

These two features have become accepted for other plant peroxidases because of the conservation of distal histidine and arginine residues.[30] The basic features of the mechanism of compound I formation for plant peroxidases are shown in Fig. 4.6.[31]

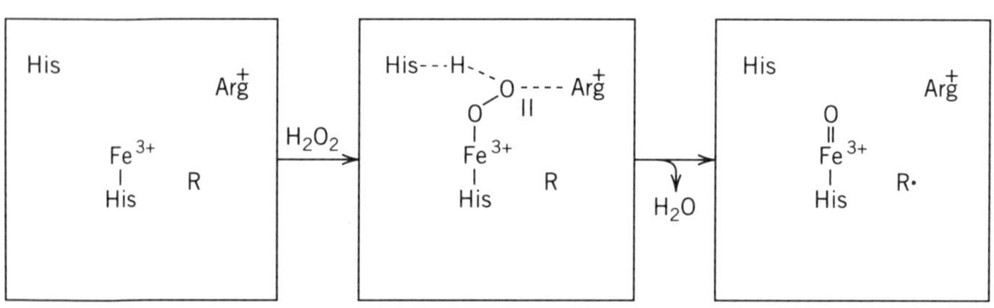

FIG. 4.6. Mechanism of peroxidase compound I formation. For horseradish peroxidase R is the porphyrin ring, for yeast cytochrome *c* peroxidase Trp191. (Reproduced with permission from T. L. Poulos and R. E. Fenna in *Metals in Biological Systems,* H. Siegel, ed., Marcel Dekker, New York, 1994, Vol. 30, pp. 25–75 by courtesy of Marcel Dekker Inc.).

4.5.2. pK_a Values for Imidazole Side Chains of Distal Histidines

The problem of the abnormally low pK_a values for the imidazole side chain of the distal histidine residue of yeast cytochrome *c* peroxidase and horseradish peroxidase has been discussed.[29] This discussion was based upon assignments made on the basis of the pH rate profiles for compound I formation, a pK_a value of 5.5 for cytochrome *c* peroxidase,[32] and the even lower assignment of 4.1 for horseradish peroxidase.[5] These assignments likely are too high; in other words the imidazole groups of the distal histidines are even more acidic, for reasons which are discussed below.

The kinetics of formation of compound I of horseradish peroxidase will be discussed in detail. The conclusions are also applicable to the yeast enzyme. The inflection in the plot of the log of the second-order rate constant versus pH indicated an acid group with a pK_a value of 4.1, which led to the conclusion that a carboxylate group might be involved.[33] The basic form of the native enzyme reacts with hydrogen peroxide, the acidic form does not. By use of a chemically modified heme prosthetic group, the propionate side chains of the heme were eliminated as the possible catalysts.[34]

It was belatedly discovered that nitrate anion, used to maintain constant ionic strength, binds to the enzyme at lower pH, as does acetate that was used as buffer.[6] Acetate binding causes little spectral change from that of the native horseradish peroxidase.[35]

Nitrate has a striking effect on the rate of compound I formation, which can be explained quantitatively in the following way. Binding of nitrate to the acidic form of the enzyme lowers the concentration of unbound acidic enzyme. This shifts the equilibrium between the acidic and basic forms of the enzyme, and less of the basic enzyme is available to react with hydrogen peroxide. The smaller the pH the more the concentration of basic enzyme is lowered, with a corresponding decrease in the apparent rate constant for compound I formation. Thus the apparent pK_a for the acid–base equilibrium, determined from the inflection in the pH rate profile, is too large. This effect, which has important consequences, has not been appreciated. In subsequent studies acetate buffer was replaced with citrate and nitrate with sulfate. The quantitative equations for the effect of nitrate follow:

$$
\begin{array}{ll}
E^- + H_2O_2 \xrightarrow{k_1} \text{Compound I} + H_2O & \\
K_A \updownarrow & \\
HE & (4.27) \\
K_{NO_3^-} \updownarrow & \\
HE{\bullet}NO_3^- &
\end{array}
$$

$$K_A = \frac{[E^-][H^+]}{[HE]} \qquad K_{NO_3^-} = \frac{[HE][NO_3^-]}{[HE \cdot NO_3^-]} \tag{4.28}$$

The total amount of enzyme present, $E_{tot,}$ is given by

$$[E]_{tot} = [E^-] + [EH] + [EH \cdot NO_3^-] \tag{4.29}$$

$$v = k_{app}[E]_{tot}[H_2O_2] = k_1[E^-][H_2O_2] \tag{4.30}$$

The middle portion of Eq. (4.30) is the operational rate equation defined in terms of an apparent rate constant, $k_{app,}$ and the total concentration of reactants. The right portion defines the mechanism in which hydrogen peroxide reacts only with the basic form of the enzyme. From Eq. (4.30)

$$k_{app} = \frac{k_1[E^-]}{[E]_{tot}} = \frac{k_1[E^-]}{[E^-] + [HE] + [HE \cdot NO_3^-]} \tag{4.31}$$

Introduction of the dissociation constants into Eq. (4.31) yields

$$k_{app} = \frac{k_1}{1 + \dfrac{[H^+]}{K_A}\left[1 + \dfrac{[NO_3^-]}{K_{NO_3^-}}\right]} = \frac{k_1}{1 + \dfrac{[H^+]}{K_{A,app}}} \tag{4.32}$$

An apparent equilibrium constant is introduced into Eq. (4.32), which is defined by

$$K_{A,app} = \frac{K_A}{1 + \dfrac{[NO_3^-]}{K_{NO_3^-}}} \tag{4.33}$$

Therefore, nitrate ion causes a decrease in $K_{A,app}$. The acid dissociation constant of the group in the active site of the native enzyme appears to be smaller than it actually is, and the pK_a value is overestimated.

Instead of a pK_a value of 4.1 for the key distal group in horseradish peroxidase, the correct value is approximately 2.5. The group is much more acidic than originally thought.

In summary, the binding of nitrate in the active site of horseradish peroxidase to an acidic group shifted the equilibrium so there was less of the basic form of the same group to participate in the reaction with hydrogen peroxide. The result was that the kinetic effect of dissociation of the distal group, not observable in the accessible pH range in the absence of nitrate, was pulled into the accessible range by the presence of nitrate. This result was tested on chloroperoxidase, not in its reaction with hydrogen peroxide, because of the instability of compound I, but for cyanide binding. Nitrate addition shifted an apparent pK_a value into the accessible pH range, as predicted (Section 14.1.4).

In the corresponding kinetic study of compound I formation in yeast cytochrome *c* peroxidase, the same experimental conditions were used: nitrate to maintain constant ionic strength and acetate buffer at lower pH. When the nitrate was replaced with phosphate, the inflection in the pH rate profile disappeared, and instead of a pK_a value of 5.5 for the active site distal group, a value of less than 4 is indicated.[36]

Therefore, the imidazole side chain of the distal histidine residue, conserved in all peroxidases, has an abnormally low pK_a value. How can its very low pK_a value be explained? It is now clear that the acid dissociation constants determined in aqueous solution for amino acids may have little relevance to their values in an enzyme structure. Hydrogen bond networks enable dispersal of charge, so a negatively charged group can exist at pH values that used to be regarded as abnormally low. Similarly, a positively charged group can exist at high pH values.

4.5.3. Kinetics of Compound I Formation

Accurate pH rate data for compound I formation from native horseradish peroxidase and hydrogen peroxide are shown in Figs. 4.7 and 4.8. Results over the pH range 2.5–8 are shown in Fig. 4.7. The spurious effects of acetate buffer and potassium nitrate at low pH are also shown in this figure. Results obtained in both H_2O and D_2O from pH 6–11.5 are shown in Fig. 4.8.

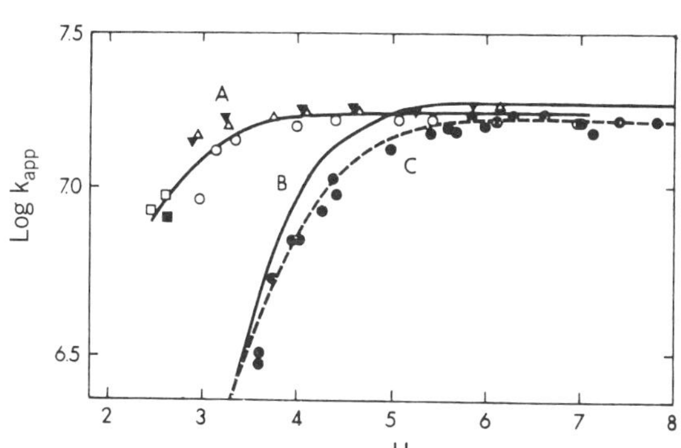

FIG. 4.7. pH rate profiles for the formation of compound I of horseradish peroxidase using hydrogen peroxide as oxidant. Curve A, obtained in the absence of nitrate and acetate, is accurate. Curve B shows the spurious effect of acetate and curve C the combined effect of nitrate and acetate at low pH. [Reproduced with permission from T. Araiso and H. B. Dunford, *Biochem. Biophys. Res. Commun.* **94,** 1177–1182 (1980).]

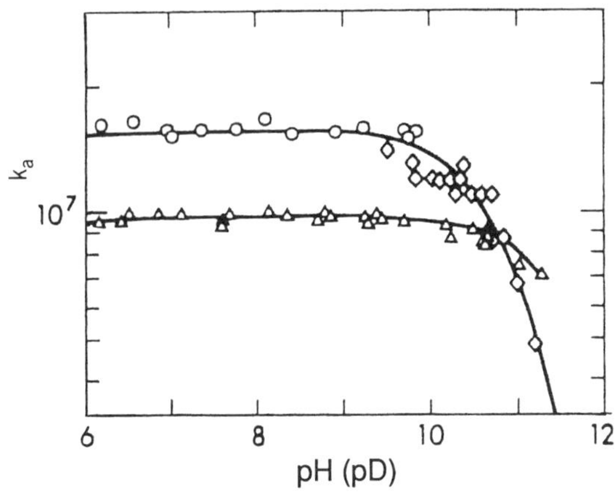

FIG. 4.8. Profiles for compound I formation from horseradish peroxidase and hydrogen peroxide over the pH/pD range 6–11.5. Upper curve obtained in H_2O, lower in D_2O. [Reproduced with permission from H. B. Dunford, W. D. Hewson, and H. Steiner, *Can. J. Chem.* **56,** 2844–2852 (1978).]

The mechanism that fits the experimental data over the pH range of 2.5–11 is the following.

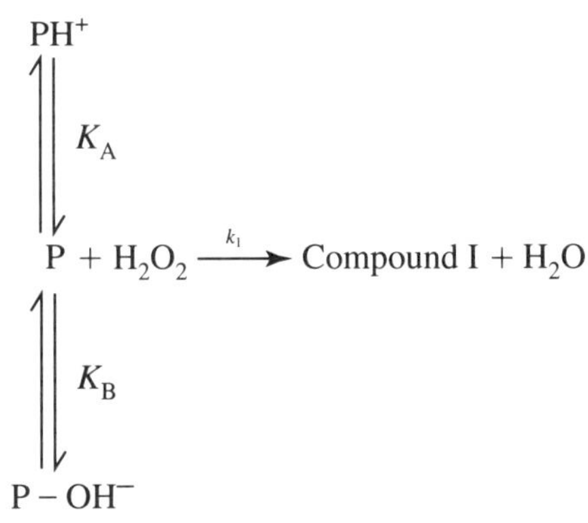

$$k_{app} = \frac{k_1}{1 + \dfrac{[H^+]}{K_A} + \dfrac{K_B}{[H^+]}}$$

The values for the three parameters that control the observed kinetics are: $k_1 = (1.7 \pm 0.1) \times 10^7\ M^{-1}s^{-1}$, $pK_A = 2.5$ and $pK_B = 10.9$. The value quoted for k_1 is the mean of three independent studies.[5,6] Thus, there is a very wide range of pH values where the rate of compound I formation is constant. Hydrogen peroxide reacts with a neutral form of peroxidase. In acid solutions, a key catalytic group becomes protonated and this acidic form of the enzyme is inactive. In basic solutions, the alkaline form of peroxidase is formed, which also is unreactive.

Compound I formation is not diffusion controlled. Its rate in solvents of high viscosity, either glycerol–water or sucrose–water remained unchanged.[37] When the reaction is studied in D_2O the value of k_1 decreases. The ratio of rate constants[5] in H_2O compared to D_2O, is $k_H/k_D = 1.6 \pm 0.1$. Therefore there is a proton transfer reaction in compound I formation that is partially rate-controlling.

The kinetics of compound I formation using *m*-chloroperbenzoic acid were thoroughly investigated.[38] It was shown that, in addition to being an oxidant, the peracid acts as a one-electron reductant of both compounds I and II. It also inactivates the enzyme. From homology modeling it was known that proximal Phe172 is near pyrolle ring C. In the Phe172Tyr mutant, it was demonstrated that the porphyrin π-cation radical form of compound I is initially formed, but it slowly converts into a protein cation radical, with the unpaired electron presumably located on replacement Tyr residue.[39] Replacement of either of the proximal,[40] or distal,[41] imidazoles by alanine greatly decreases or destroys the ability of the enzyme to form compound I. However, considerable enzyme activity is restored by adding exogenous imidazoles.

4.6. REACTIONS OF COMPOUND I

4.6.1. Transient-State Kinetics

Because of the huge molar absorptivities of native horseradish peroxidase and its compounds in the Soret region, transient-state kinetic measurements can be made on the enzyme species. Thus, direct information can be obtained on what is happening to the enzyme, rather than trying to infer what is happening to the enzyme by following rate of product formation, a limitation of steady-state kinetics.

Compounds I and II usually react with the same reducing substrate. Therefore, once the reaction between compound I and substrate is started, the formation of compound II and its reaction with substrate also begins. In order to eliminate influence from the compound II reaction on kinetic measurements, one makes spectrophotometric measurements at 411 nm, the isosbestic point between native enzyme and compound II. Then conversion of newly formed compound II to native enzyme has no influence on absorbance changes, the rate of which accurately reflects the rate of conversion of compound I to compound II.

Horseradish peroxidase forms a compound I that is very stable compared to most other peroxidase compounds I. With a good preparation of horseradish peroxidase, it will react with a stoichiometric amount of hydrogen peroxide to form compound I that is stable for 20 min or more, more than ample time to introduce it into one cuvette of a stopped flow apparatus, and study its reaction with a reducing substrate.

If the native enzyme preparation is impure, then an excess of hydrogen peroxide may be required to obtain a good yield of compound I. One does not want to have excess hydrogen peroxide left in the reaction mixture, because once compound II formation and reaction begins, the excess hydrogen peroxide would recycle the enzyme, and interfere with accurate measurement of the rate of the compound I reaction.

Typical kinetic measurements are made with approximately a 1-μM solution of horseradish peroxidase and a 10-fold excess of reducing substrate. If there is a trace of native enzyme left in the compound I preparation, it will not affect the accuracy of the measurement of rate of reduction of compound I: it is a property of pseudo-first-order kinetics that rate of the relative absorbance change yields as accurate a rate constant as that which would be obtained from absolute absorbance changes. The pseudo-first-order rate constant k_{obs} has units of s^{-1}. When it is studied as a function of substrate concentration, valuable information is obtained. If a plot of k_{obs} versus substrate concentration is linear, then the slope provides the most accurate determination of the second-order rate constant k_{app} in units of $M^{-1}s^{-1}$ (Fig. 4.9). A linear plot indicates that in practical terms the reaction is irreversible. A substrate–compound I complex must still be formed, but the rate of reaction of the complex to form products is considerably faster than the rate of dissociation of the complex to reform initial reactants.

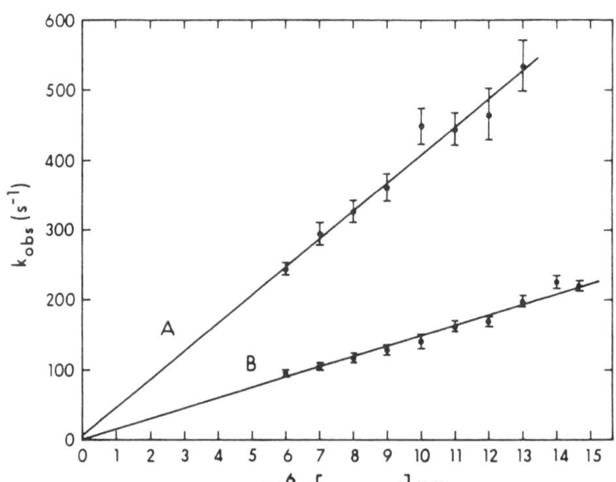

FIG. 4.9. Linear plots of the pseudo-first-order rate constant k_{obs} versus substrate concentration for the reaction of horseradish peroxidase compound I with *p*-cresol. (A) pH 6.0, (B) pH 10.5. [Reproduced with permission from W. D. Hewson and H. B. Dunford, *J. Biol. Chem.* **251,** 6036–6042 (1976).]

If the plot of k_{obs} versus substrate is curved upward, it indicates substrate activation of its own reaction with compound I. Downward curvature indicates substrate inhibition. In either case substrate binding, probably at a second site on compound I, is influencing the observed kinetics. The effects are examples of positive or negative cooperativity caused by the substrate on an enzyme consisting of a single subunit.[42]

4.6.2. Kinetics of the Reaction of *p*-Cresol with Compound I

pH-dependent behavior is observed in the reaction of *p*-cresol with compound I of horseradish peroxidase. At certain pH values plots of k_{obs} versus pH are linear (Fig. 4.9), and at others downward curvature (negative cooperativity) is observed (Fig. 4.10). The pH rate profile over the pH range 3.7–10.8 for the reaction of *p*-cresol with horseradish peroxidase[42] is shown in Fig. 4.11. The rate constant reaches a value of $5 \times 10^7\ M^{-1}s^{-1}$ at pH 9 and is in excess of $10^7\ M^{-1}s^{-1}$ over a wide range of pH. *p*-Cresol ionizes with a pK_a value of 10.1, and become unreactive at higher pH. This indicates that the unionized form of *p*-cresol is the reactant with compound I. Direct reduction of compound I to compound II by *p*-cresol shows that it is a one-electron reductant. Phenol reactivities correlate with Hammett σ values, not Brown–Okamoto σ^* values (see Chapter 6). Reduction of compound I to compound II requires uptake of a proton by the enzyme. The sum of all of these results can only be explained by a reaction in which a phenol is a hydrogen atom donor, not simply a one-electron donor, to compound I. The phenol donates both a single

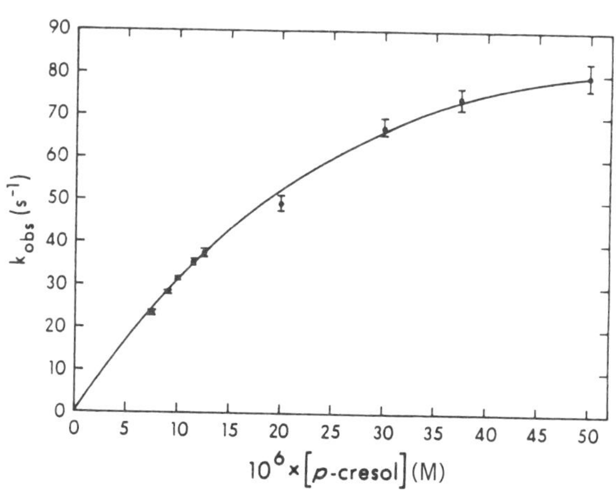

FIG. 4.10. Downward curvature at pH 3.7 in the plot of k_{obs} versus substrate concentration for the reaction of horseradish peroxidase compound I with *p*-cresol. [Reproduced with permission from W. D. Hewson and H. B. Dunford, *J. Biol. Chem.* **251,** 6036–6042 (1976).]

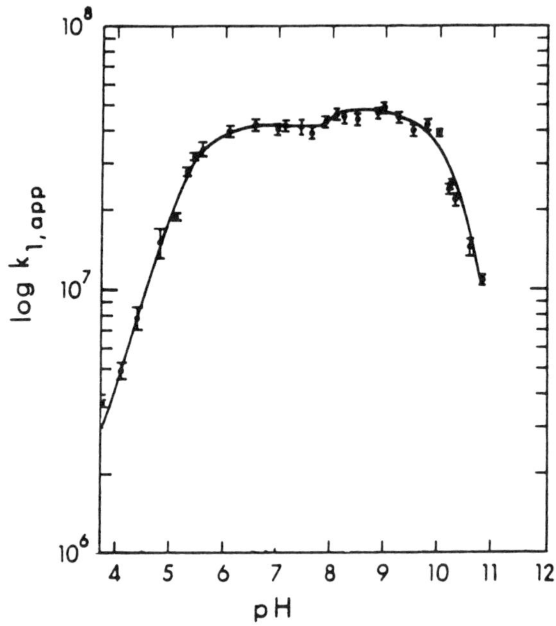

FIG. 4.11. pH rate profile for the reaction of *p*-cresol with horseradish peroxidase compound I. [Reproduced with permission from W. D. Hewson and H. B. Dunford, *J. Biol. Chem.* **251,** 6036–6042 (1976).]

electron and a single proton to compound I, converting it to compound II. It appears that all monophenols react by the same mechanism.

The pH rate profile for the *p*-cresol reaction with compound I can be fit by the following mechanism (simplified from the original article by neglecting data obtained at very low pH).

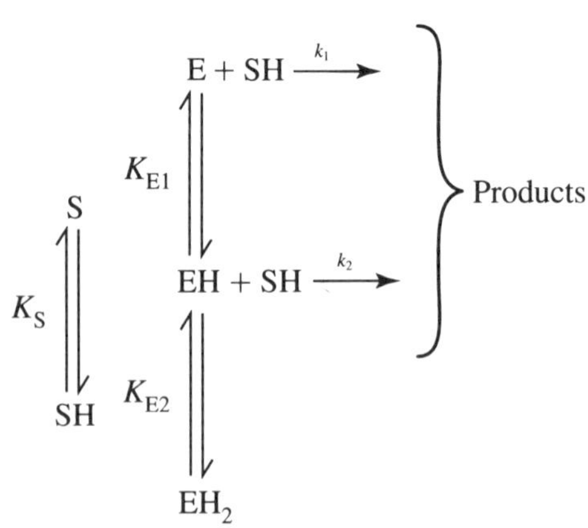

Following are the best-fit parameters for the above mechanism: $pK_S = 10.19 \pm 0.05$, $pK_{E1} = 8.0 \pm 0.3$, $pK_{E2} = 5.08 \pm 0.01$, $k_1 = (5.1 \pm 0.3) \times 10^7$ $M^{-1}s^{-1}$, and $k_2 = (4.0 \pm 0.2) \times 10^7$ $M^{-1}s^{-1}$. From comparison of the two rate constants, k_1 and k_2, it is apparent that the ionization of pK_a value 8.0 has only a small effect on the observed kinetics. The effect of the acid group with pK_a value of 5.08, in contrast, is dramatic. The species EH is highly reactive; EH_2 is unreactive with *p*-cresol.

p-Aminobenzoic acid appears to react with a basic form of compound I, the same as *p*-cresol, but much more slowly. A pK_a value of 4.9 for compound I was determined from the *p*-aminobenzoic acid kinetics.[21]

Hydrogen atom donation by *p*-cresol results in formation of a free radical with O-centered and C-centered resonance structures. The free radical may dimerize to form a biphenol (Fig. 4.12), or it may form Pummerer's ketone, shown below.

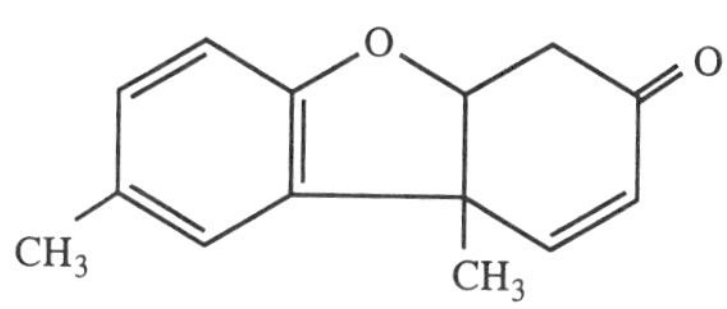

Formation of a biphenol would explain the unusual stoichiometry observed when compound I is titrated with a phenol.[43,44] One molar equivalent of phenol can convert compound I to native enzyme.

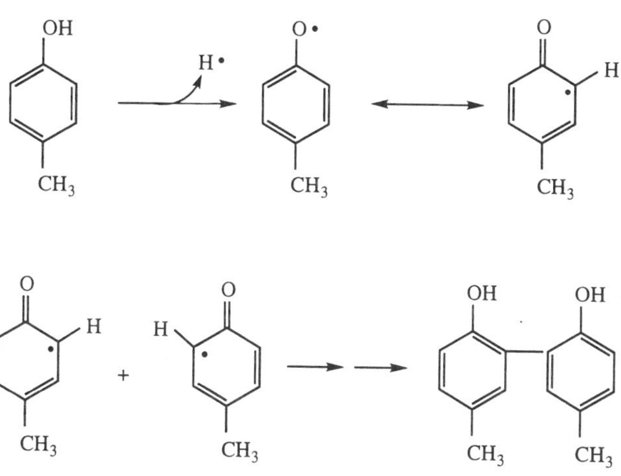

FIG. 4.12. Resonance structures for the free radical formed by oxidation of *p*-cresol. Dimerization of the carbon-centered free radicals leads to biphenol formation.

$$\text{Compound I} + \text{Phenol—OH} \longrightarrow \text{Compound II} + \text{Phenol—O}\bullet$$

$$\text{Phenol—O}\bullet \longrightarrow \tfrac{1}{2}\,\text{HO—Biphenol—OH}$$

$$\text{Compound II} + \tfrac{1}{2}\,\text{HO—Biphenol—OH} \longrightarrow \text{Native HRP} + \text{Products}$$

$$\text{Compound I} + \text{Phenol} = \text{Native HRP} + \text{Products}$$

Thus, the overall two-electron reduction of compound I by a stoichiometric amount of phenol is accounted for by the usual one-electron (hydrogen atom) donations to the enzyme. From a mechanistic standpoint, it is incorrect to describe the reaction as a two-electron reduction.[45]

4.6.3. The Reaction of Iodide with Compound I

The pH rate profile for the reaction of iodide with horseradish peroxidase compound I is dramatically different from that for *p*-cresol. The rate constant decreases linearly with decreasing pH until about pH 5, where it displays a brief tendency to level off and then increases further at still lower pH. pH jump experiments were not conducted at the lowest pH values, and there may be some change in compound I structure at low pH. The data over the pH range 4–10 can be fit by the equation

$$k_{\text{app}} = \frac{k}{1 + \dfrac{K_A}{[\text{H}^+]}}$$

where k, the pH-independent rate constant, is $1.3 \times 10^6\ \text{M}^{-1}\text{s}^{-1}$, and the value of a rather poorly defined K_A is 2.6×10^{-5} M ($\text{p}K_A = 4.6$). The acid group may be the same as the one affecting the *p*-cresol kinetics ($\text{p}K_A = 5.0$). As discussed below, iodide reacts with an acidic form of compound I, whereas *p*-cresol reacts with a basic form.

The really dramatic difference in behavior of iodide compared to *p*-cresol is that iodide is a two-electron reductant of compound I, the first known example. (When it is put in solution with compound II, it is a one-electron reductant, reacting slower than with compound I, as described in Section 4.7). The two-electron reduction of compound I was established in six different ways.

1. The difference spectrum was obtained between compound I and the enzyme species formed in the reaction with iodide, by transient-state measurements at five wavelengths. The enzyme product was not compound II but corresponded to native enzyme.

2. When a mixture of excess hydrogen peroxide, a smaller amount of iodide, and horseradish peroxidase is allowed to react, the compound I spectrum was obtained in experiments conducted using a conventional spectrophotometer. Normally, one obtains the compound II spectrum in a steady-state reaction. With the smaller amount of iodide the rate-controlling step in the steady-state cycle is conversion of compound I directly back to the native ferric enzyme. Thus, compound I is dominant in the steady state, with negligible amounts of native enzyme and compound II present.

3. In the steady-state reaction, with iodide in excess of hydrogen peroxide, a linear kinetic trace is obtained at 411 nm followed by an exponential decay. The duration of the linear portion is directly dependent upon the hydrogen peroxide concentration, but is always followed by an exponential decay with the same time constant. The rate constants obtained from the exponential decays are equal to the rate constant for the transient-state reaction of compound I with iodide. These results cannot be explained by a cycle that involves compound II but are readily explained by the cycle: native HRP → compound I → native HRP. The linear portion corresponds to the steady-state reaction, and when the hydrogen peroxide is exhausted, compound I is converted exponentially into native enzyme by its reaction with iodide. Then the reaction stops.

4. With an excess of iodide present, successive additions of an amount of hydrogen peroxide were made over a considerable period of time, equal to the concentration of native enzyme. These experiments were conducted at higher pH, where the iodide reaction with compound I is slow. The same exponential growth was obtained in successive experiments, equal to the rate constant for compound I formation. In the time between successive additions of hydrogen peroxide, iodide converted compound I back to the native enzyme. If compound II were part of the cycle, it would have continued to accumulate as the series of experiments progressed.

5. If compound II participated in the reaction cycle, with an excess of hydrogen peroxide over iodide, the rate-controlling step would be the conversion of compound II back to native enzyme. The rate constant for the steady-state rate-controlling step was determined over the pH range 5.4–7.0 and was 100 times greater than that for compound II reacting with iodide.

6. The stoichiometry between compound I reacted and triiodide formed was exactly 1 : 1, which corresponds to the reactions

$$\text{H}^+\text{–Compound I} + \text{I}^- \longrightarrow \text{Native HRP} + \text{HOI}$$
$$\text{HOI} + \text{I}^- + \text{H}^+ \longrightarrow \text{I}_2 + \text{H}_2\text{O}$$
$$\text{I}_2 + \text{I}^- \longrightarrow \text{I}_3^-$$

$$\text{H}^+\text{–Compound I} + 3\,\text{I}^- + \text{H}^+ = \text{Native HRP} + \text{I}_3^- + \text{H}_2\text{O}$$

In the presence of excess iodide the formation of molecular iodine and triiodide anion would be very fast. The exact 1 : 1 stoichiometry between compound I reacted and triiodide formed indicates that, on the time scale of the experiments, molecular iodine does not react with the enzyme.

The representation of compound I as H^+–compound I shows that iodide reacts with an acid form of compound I, in contrast to *p*-cresol, which reacts with a basic form. The proton required for HOI formation cannot be supplied by HI, since the latter is far too strong an acid, and it would appear not to be supplied by solvent or buffer, because of the inflection in the pH rate profile at low pH.

In acid solution the bisulfite anion, HSO_3^-, like iodide, reduces compound I directly to native HRP and reacts with an acidic form of compound I, with a pK_a value of 5.1.[46,47]

4.7. REACTIONS OF COMPOUND II

Unlike compound I, with its crucial active site pK_A value of 5.0, which may react in its alkaline form or its acidic form depending upon the substrate, compound II only reacts in its acidic form. The crucial pK_A value for compound II is 8.6, so for pH values increasing above 9, compound II becomes less and less reactive. This fact can be exploited to prepare compound II with little or no other forms of horseradish peroxidase present. The problem in attempting to prepare pure compound II is that it reacts with the same reducing substrates as does compound I. The normal route to compound II is via the sequence native enzyme → compound I → compound II, so that after addition of a stoichiometric amount of hydrogen peroxide to obtain compound I, an equivalent amount of reducing substrate is added to obtain compound II. As soon as some compound II forms, it is capable of reacting with some of the remaining reducing substrate. However, this capability is removed by having the solution at pH 9 or greater. In my group we have been able to prepare close to 100% compound II that is stable for up to 24 hours. Experiments on compound II at all pH values are then readily conducted using a pH jump method.

The reactivity of compound II toward various substrates as a function of pH has been examined in various ways. The reactivity in some cases appears to reflect qualitatively the compound II reduction potential and in others the total charge on the protein.

A titration curve for horseradish peroxidase is reproduced in Fig. 4.13.[48] At low pH where the positive charge on the enzyme is large, the reduction potential of compound II is highest. Compound II is easiest to reduce to native enzyme at low pH. It is easiest to add an electron to the highest total positive charge. Around neutral pH there are few ionizable groups so that addition of a small amount of acid or base causes a large pH change, with little change in protein charge. The considerable variation in pI values for HRP isoenzymes has been explained in terms of this broad region around pH 7. Addition or subtraction of a group that ionizes in this region could considerably affect the

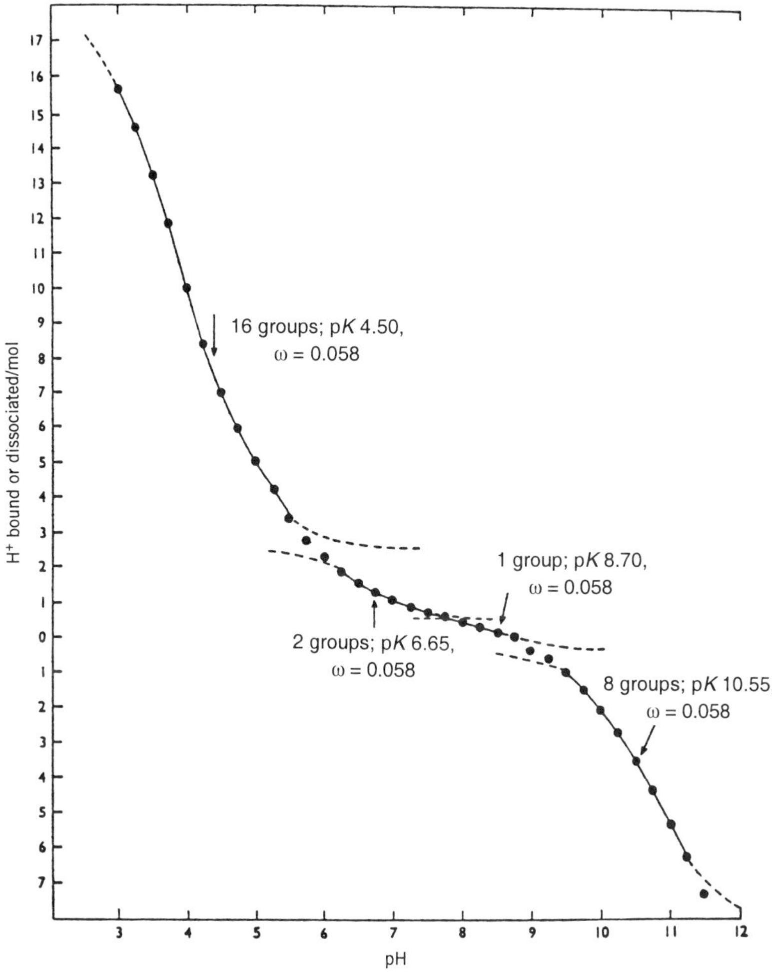

FIG. 4.13. Titration curve for horseradish peroxidase. The experimental data have been fit by assuming many different acid groups have the same pK_A values. [Reproduced with permission from C. Phelps, L. Forlani, and E. Antonini, *Biochem. J.* **124,** 605–614 (1971).]

pI value. At high pH where the protein has a large negative charge, the ease of reduction is lowest.

For a quantitative explanation of the pH rate profile for a particular substrate, neither reduction potential of HRP–II nor total protein charge appears to suffice. Each substrate may exhibit unique behavior. As in the case of reactions with HRP–I, *p*-cresol and iodide appear to typify extremes of behavior.

4.7.1. Reactions of Compound II as Typified by *p*-Cresol and Iodide

A pK_a value of 8.6 affects the *p*-cresol kinetics, as it does reduction potentials and the reactivities of other substrates with compound II, such as ferrocyanide, K_3IrCl_6, and *p*-aminobenzoic acid. It is the acidic form of compound II that is reactive. However, unlike these other substrates, the rate of reaction of *p*-cresol does not increase as the pH is lowered in the acid region. Rather, the rate constant drifts to somewhat lower values as the pH is decreased from 8 to 4. The reaction of *p*-cresol with compound II can be formulated in the following manner.

$$H^+{-}\text{HRP–II} + \text{HO}{-}\text{Phenol} \longrightarrow \text{Native HRP} + {\bullet}\text{O}{-}\text{Phenol} + H_2O$$

The oxygen atom in the product water is derived from the $Fe^{IV}{=}O$ group of compound II, and both compound II and substrate donate a proton.

Iodide is a one-electron reductant of HRP–II with one of the most unusual pH rate profiles in enzymology. From pH 2.7–9 the plot of the log of the second-order rate constant versus pH has a slope of -1 within very small experimental error. The value of the rate constant for the reaction of HRP–II with iodide is $10^5\ M^{-1}s^{-1}$ at pH 3 and $10^2\ M^{-1}s^{-1}$ at pH 6. In the original paper,[46] the reaction was attributed to an acid form of HRP-II with a pK_A value of approximately zero. A possible alternative explanation is the rate-controlling addition of a proton from bulk solution to the $Fe^{IV}{=}O$ group, which facilitates water release, a requirement for completion of conversion of compound II to native enzyme. Proton transfers are usually very fast, so that there would have to be a huge steric restriction to make proton transfer the rate-controlling step.

The reaction of iodide with compound II is about 100-fold slower than its reaction with compound I. In a steady-state mixture it will not occur at all unless there is a second reductant in addition to iodide present, so that some compound II is formed. Nature apparently sees merit in production of hypohalous acids but not in liberation of free halogen atoms.

Ferrocyanide and *p*-aminobenzoic acid typify behavior between the above two extremes. The acid group with pK_A of 8.6 directly affects both pH rate profiles, but the reactions also increase with $[H^+]$ at low pH. Both substrates have acid ionizations at low pH. For ferrocyanide, the fourth ionization has a pK_A value of 4.25.[49] Ferrocyanide reacts considerably faster than iodide, and its pH rate profile can be accounted for by a combination of two mechanisms. At lower pH, the acid form of ferrocyanide could be reactive, and at higher pH proton transfer from solvent or buffer to the $Fe^{IV}{=}O$ group might facilitate the reaction. pK_A values for *p*-aminobenzoic acid are 4.77 and 2.45.[21] *p*-Aminobenzoic acid is very unreactive compared to most aromatic amines. Its mechanism of reaction would appear to be the same as that of ferrocyanide. pH rate profiles for several substrates reacting with horseradish peroxidase compound II are displayed in Fig. 4.14.

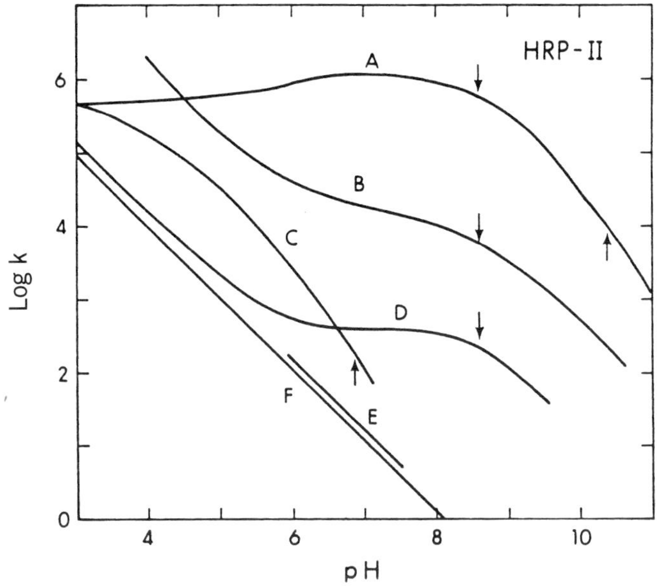

FIG. 4.14. Log of second-order rate constants ($M^{-1}s^{-1}$) as a function of pH for reactions with horseradish peroxidase compound II of (A) *p*-cresol, (B) ferrocyanide ion, (C) hydrogen sulfite ion, (D) *p*-aminobenzoic acid, (E) nitrous acid, and (F) iodide ion. Downward arrow indicates a pK_A of 8.6 on compound II, upward arrows pK_A's on substrates. [Reproduced with permission from H. B. Dunford and J. S. Stillman, *Coord. Chem. Revs.* **19,** 187–251 (1976).]

4.8. FERROUS PEROXIDASE

The first detailed report on the preparation and properties of ferrous horseradish peroxidase was published in 1937.[11] Upon reduction of peroxidase with sodium dithionite ($Na_2S_2O_4$) the solution turned from brown to bright red. Complexes with carbon monoxide, cyanide, methyl isocyanide, and nitric oxide were formed. Cyanide and nitric oxide are unique in that they bind to both the ferrous and ferric forms of the enzyme. In this early work only absorption bands in the visible region were detected. Formation of ferrous peroxidase was also accomplished by Harbury in his electrochemical experiments, reported in preliminary form in 1953.[50] Accurate Soret region and visible spectra were published in 1951.[51] It was shown that cyanide can be photodissociated from the ferrous enzyme in 1955.[52] Little further work was reported until 1965. There are problems in trying to work with the ferrous enzyme. A strong reducing agent such as sodium dithionite is required to obtain the ferrous enzyme and the reaction must be accomplished anaerobically. It was believed until 1966 that oxygen binding to the ferrous enzyme could not be observed because the complex was too unstable. The influence of the mutants His42Leu,

His42Arg, Arg38Leu, Arg38Lys, Arg38Ser, and Arg38Gly on the rates of dioxygen binding to the ferrous enzyme were measured.[53] The results showed that His42 and Arg38 impede the binding.

The equilibria and kinetics of carbon monoxide binding to ferrous horseradish peroxidase were studied in detail.[54] The binding is very tight with an equilibrium constant of $4.5 \times 10^6\ M^{-1}$. The binding rate constant is about $4 \times 10^3\ M^{-1}s^{-1}$. Upon binding of carbon monoxide to ferrous peroxidase the heme-linked pK_a value of 7.25 is shifted to 8.25.[55]

A unique preparation of ferrous peroxidase was obtained from the reaction of reduced nicotinamide adenine dinucleotide (NADH) with peroxidase.[56] This marked the discovery of the famous peroxidase–oxidase oscillatory reaction,[57] which is described in Chapter 5.

The reaction of the ferrous enzyme with molecular oxygen was reported in 1966.[58] A year later results on the kinetics of oxygen binding, studied using a stopped flow apparatus, were published.[59] At 20°C the rate constant for binding is $(5.8 \pm 0.4) \times 10^4\ M^{-1}\ s^{-1}$. The activation energy for the reaction is 10 kcal/mol. The complex of ferrous enzyme and O_2 is identical to that of horseradish peroxidase compound III and is spectrally similar to oxymyoglobin and oxyhemoglobin. Hence, compound III can also be called oxyperoxidase. However, unlike the oxyglobins, oxyperoxidase is unstable. In the painstaking study discussed here,[59] the excellent preparations of oxyperoxidase had half lives of 4 min at 20°C and 1 hour at 0°C.

When ferrous peroxidase, is titrated with oxygen, four moles of ferric enzyme are obtained per mole of oxygen,[60] which corresponds to the reactions

$$Fe^{II} + O_2 \longrightarrow Fe^{II}-O_2$$

$$Fe^{II}-O_2 + 3\ Fe^{II} + 4H^+ \longrightarrow 4\ Fe^{III} + 2\ H_2O$$

$$4\ Fe^{II} + O_2 + 4\ H^+ = 4\ Fe^{III} + 2\ H_2O$$

Thus, compound III is rapidly reduced by an excess of ferrous peroxidase.

The kinetics and equilibrium of the reaction of ethyl isocyanide with ferrous peroxidase have been studied.[61] Compound II is formed from the reaction of ferrous peroxidase with H_2O_2.[13]

4.9. COMPOUND III OR OXYPEROXIDASE

There are three routes to obtain compound III.

1. The earliest known method was by reaction of horseradish peroxidase with excess H_2O_2. Compound II is obtained as an intermediate.[11]

2. Reaction of the ferrous enzyme with O_2, described in Section 4.8.
3. Reaction of the ferric enzyme with superoxide $O_2\bullet^-$, or its protonated form, the hydroperoxyl radical $HO_2\bullet$.

Further detail is provided here on the formation of compound III, starting with compound II or the ferric enzyme.

4.9.1. Formation of Compound III from Compound II and H_2O_2

The replacement of one oxygen atom bound to iron by two is an intriguing reaction:

$$Fe^{IV}{=}O + H_2O_2 \longrightarrow Fe^{II}{-}O_2 + H_2O$$

and, it turns out, to be rather complex. In a detailed and comprehensive study, it was shown that there are two mechanisms operating simultaneously.[12]

There is a dissociative exchange of the ferryl oxygen with water.[62] In one mechanism this provides the opportunity for H_2O_2 to react with the Fe^{IV} ion:

$$H^+{-}Fe^{IV}{=}O + H^+ \rightleftharpoons Fe^{IV} + H_2O$$

$$Fe^{IV} + H_2O_2 \longrightarrow H^+{-}Fe^{III}{-}O_2^- + H^+$$

Compound III is shown as a ferric ion–superoxide complex that exists as a resonance structure:

$$H^+{-}Fe^{III}{-}O_2^- \longleftrightarrow H^+{-}Fe^{II}{-}O_2$$

There appears to be general agreement that the ferric ion–superoxide form is dominant.

The other mechanism involves formation of superoxide, which was demonstrated by use of tetranitromethane (TNM) an excellent superoxide scavenger:

$$H^+{-}Fe^{IV}{=}O + H_2O_2 \longrightarrow [Fe^{III}, H_2O, HO_2\bullet]$$

$$[Fe^{III}, H_2O, HO_2\bullet] \longrightarrow H^+{-}Fe^{III}{-}O_2^- + H_2O$$

An intermediate enzyme complex containing water and superoxide is formed, which rearranges to compound III and releases water. This pathway is quenched in the presence of TNM.

4.9.2. Formation of Compound III from Ferric Enzyme and Superoxide

The possibility of compound III formation from native peroxidase and superoxide was first proposed in 1963.[63] It occurs in several reactions where peroxidase appears to be acting as an oxidase. These are discussed in Chapter 5. Study of the direct reaction is difficult because in any superoxide generating system, whether radiolytic[64–67] or enzymatic,[68] superoxide dismutes to water and hydrogen peroxide.

In a study using xanthine–xanthine oxidase to produce superoxide, two steps were observed in compound III formation that were interpreted in terms of a loose complex that collapses to compound III.

$$\mathrm{Fe^{III}} + \mathrm{O_2}\bullet^- \longrightarrow \mathrm{Fe^{III}} \cdots \mathrm{O_2}\bullet^-$$

$$\mathrm{Fe^{III}} \cdots \mathrm{O_2}\bullet^- \longrightarrow \mathrm{Fe^{III}}{-}\mathrm{O_2}\bullet^-$$

The compound III, once formed, decays back to the ferric enzyme without detectable intermediates. The same direct decay was observed for compound III formed from the ferrous enzyme and molecular oxygen.[59]

4.10. COMPOUND X

In a search for a halogenating intermediate, horseradish peroxidase was reacted with chlorite ion. A compound stable at alkaline pH was formed,[69] which was very similar spectrally to compound II.[70,71] Further work led to the postulate that the chlorine is transferred to a heteroatom on an active site amino acid side chain.[72] The compound has different properties in acid and alkaline solution, leading to the designations compounds X_I and X_{II}.

4.11. ALKALINE FORM OF HORSERADISH PEROXIDASE

When the pH of horseradish peroxidase is changed from 7 to 12, the iron of the enzyme is converted from a high-spin to low-spin form.[73] The Soret region absorption band shifts from 403 to 420 nm. The conversion to the alkaline form has at various times been attributed to dissociation of water in the sixth coordination position of the heme iron(III) to form a hydroxide ligand, to displacement of water by an amino acid side chain, to occupancy of a vacant sixth position by hydroxide, and to various combinations of the above. It is now known that the sixth coordination position of the iron in the native enzyme is vacant, and that this position is occupied by hydroxide ion at high pH.[74–76] The pK value for the alkaline transition is in the range 10.8–11.1, the value being dependent upon ionic strength, temperature, and the particular enzyme preparation.

REFERENCES

1. S. Aibara, H. Yamashita, E. Mori, M. Kato, and Y. Morita, *J. Biochem.* **92,** 531–539 (1982).
2. H. B. Dunford, *J. Theor. Biol.* **46,** 467–479 (1974).
3. H. B. Dunford and R. A. Alberty, *Biochemistry* **6,** 447–451 (1967).
4. W. D. Ellis and H. B. Dunford, *Biochemistry* **7,** 2054–2062 (1968).
5. H. B. Dunford, W. D. Hewson, and H. Steiner, *Can. J. Chem.* **56,** 2844–2852 (1978).
6. T. Araiso and H. B. Dunford, *Biochem. Biophys. Res. Commun.* **94,** 1177–1182 (1980).
7. T. Araiso and H. B. Dunford, *J. Biol. Chem.* **256,** 10099–10104 (1981).
8. G. R. Schonbaum and S. Lo, *J. Biol. Chem.* **247,** 3353–3360 (1972).
9. H. Yamada and I. Yamazaki, *Arch. Biochem. Biophys.* **165,** 728–738 (1974).
10. B. C. Saunders, A. G. Holmes-Siedle, and B. P. Stark, *Peroxidase,* Butterworths, London, 1964, pp. 10–39.
11. D. Keilin and T. Mann, *Proc. Roy. Soc.* (*London*) **B122,** 119–133 (1937).
12. R. Nakajima and I. Yamazaki, *J. Biol. Chem.* **262,** 2576–2581 (1987).
13. R. W. Noble and Q. H. Gibson, *J. Biol. Chem.* **245,** 2409–2413 (1970).
14. H. A. Harbury, *J. Biol. Chem.* **225,** 1009–1024 (1957).
15. T. L. Poulos, S. T. Freer, R. A. Alden, S. L. Edwards, U. Skogland, K. Takio, B. Eriksson, N. Xuong, T. Yonetani, and J. Kraut, *J. Biol. Chem.* **255,** 575–580 (1980).
16. Y. Hayashi and I. Yamazaki, *J. Biol. Chem.* **254,** 9101–9106 (1979).
17. P. George, *Science* **117,** 220–221 (1953).
18. J. E. Critchlow and H. B. Dunford, *J. Biol. Chem.* **247,** 3703–3713 (1972).
19. B. B. Hasinoff and H. B. Dunford, *Biochemistry* **9,** 4930–4939 (1970).
20. M. L. Cotton and H. B. Dunford, *Can. J. Chem.* **51,** 582–587 (1973).
21. H. B. Dunford and M. L. Cotton, *J. Biol. Chem.* **250,** 2920–2932 (1975).
22. W. D. Hewson and L. P. Hager, *J. Biol. Chem.* **254,** 3175–3181 (1979).
23. W. D. Hewson and L. P. Hager, J. *Biol. Chem.* **254,** 3182–3186 (1979).
24. A. Nadezhdin and H. B. Dunford, *Can. J. Biochem.* **57,** 1080–1083 (1979).
25. A. Nadezhdin and H. B. Dunford, *Can. J. Chem.* **58,** 2504–2507 (1980).
26. J. Berglund, T. Pascher, J. R. Winkler, and H. B. Gray, *J. Am. Chem. Soc.* **119,** 2464–2469 (1997).
27. R. Roman and H. B. Dunford, *Biochemistry* **11,** 2076–2082 (1972).
28. T. L. Poulos and J. Kraut, *J. Biol. Chem.* **255,** 8199–8205 (1980).
29. T. L. Poulos. *Adv. Inorg. Biochem.* **7,** 1–36 (1987).
30. K. G. Welinder, *Curr. Opin. Struct. Biol.* **2,** 388–393 (1992).
31. T. L. Poulos and R. E. Fenna in *Metals in Biological Systems,* H. Siegel, ed., Marcel Dekker, New York, 1994, Vol. 30, pp. 25–75.

32. S. Loo and J. E. Erman, *Biochemistry* **14,** 3467–3470 (1975).
33. H. B. Dunford, *Adv. Inorg. Biochem.* **4,** 41–68 (1982).
34. T. Araiso, H. B. Dunford, and C. Chang, *Biochem. Biophys. Res. Commun.* **90,** 520–524 (1979).
35. D. Dolman, G. A. Newell, M. D. Thurlow, and H. B. Dunford, *Can. J. Biochem.* **53,** 495–501 (1975).
36. L. B. Vitello, M. Huang, and J. E. Erman, *Biochemistry* **29,** 4283–4288 (1990).
37. H. B. Dunford and W. D. Hewson, *Biochemistry* **16,** 2949–2957 (1977).
38. J. N. Rodriguez-Lopez, J. Hernández-Ruiz, F. Garcia-Cánovas, R. N. F. Thorneley, M. Acosta, and M. B. Arnao, *J. Biol. Chem.* **272,** 5469–5476 (1997).
39. V. P. Miller, D. B. Goodin, A. E. Friedman, C. Hartmann, and P. R. Ortiz de Montellano, *J. Biol. Chem.* **270,** 18413–18419 (1995).
40. S. L. Newmeyer, J. Sun, T. M. Loehr, and P. R. Ortiz de Montellano, *Biochemistry* **35,** 12788–12795 (1996).
41. S. L. Newmeyer and P. R. Ortiz de Montellano, *J. Biol. Chem.* **271,** 14891–14896 (1996).
42. W. D. Hewson and H. B. Dunford, *J. Biol. Chem.* **251,** 6036–6042 (1976).
43. M. Santimone, *Can. J. Biochem.* **53,** 649–657 (1975).
44. W. D. Hewson and H. B. Dunford, *J. Biol. Chem.* **251,** 6043–6052 (1976).
45. R. Z. Harris, S. L. Newmyer, and P. R. Ortiz de Montellano, *J. Biol. Chem.* **268,** 1637–1645 (1993).
46. R. Roman and H. B. Dunford, *Can. J. Chem.* **51,** 588–596 (1973).
47. T. Araiso, K. Miyoshi, and I. Yamazaki, *Biochemistry* **15,** 3059–3063 (1976).
48. C. Phelps, L. Forlani, and E. Antonini, E. *Biochem. J.* **124,** 605–614 (1971).
49. I. M. Kolthoff and W. J. Tomsicek, *J. Phys. Chem.* **39,** 955–958 (1935).
50. H. A. Harbury, *J. Am. Chem. Soc.* **75,** 4625 (1953).
51. D. Keilin and E. F. Hartree, *Biochem. J.* **49,** 88–104 (1951).
52. D. Keilin and E. F. Hartree, *Biochem. J.* **61,** 153–171 (1955).
53. J. N. Rodriguez-Lopez, A. T. Smith, and R. N. F. Thorneley, *J. Biol. Chem.* **272,** 389–395 (1997).
54. D. Kertesz, E. Antonini, M. Brunori, J. Wyman, and R. Zito, *Biochemistry* **4,** 2672–2676 (1965).
55. Y. Hayashi, H. Yamada, and I. Yamazaki, *Biochim. Biophys. Acta* **427,** 608–616 (1976).
56 I. Yamazaki and K. Yokota, *Biochem. Biophys. Res. Commun.* **19,** 249–254 (1965).
57. I. Yamazaki, K. Yokota, and R. Nakajima, *Biochem. Biophys. Res. Commun.* **21,** 582–586 (1965).
58. I. Yamazaki, K. Yokota, and M. Tamura in *Hemes and Hemoproteins,* B. Chance, R. W. Estabrook and T. Yonetani, eds., Academic, New York, 1966, pp. 319–325.
59. J. B. Wittenberg, R. W. Noble, B. A. Wittenberg, E. Antonini, M. Brunori, and J. Wyman, *J. Biol. Chem.* **242,** 626–634 (1967).
60. C. F. Phelps, E. Antonini, G. Giacometti, and M. Brunori, *Biochem. J.* **141,** 265–272 (1974).

61. C. F. Phelps, E. Antonini, and M. Brunori, *Biochem. J.* **128,** 377–382 (1972).
62. S. Hashimoto, Y. Tatsuno, and T. Kitagawa, *Proc. Natl. Acad. Sci. U.S.A.* **83,** 2417–2421 (1986).
63. I. Yamazaki and L. H. Piette, *Biochim. Biophys. Acta* **77,** 47–64 (1963).
64. B. H. Bielski, *Biochim. Biophys. Acta* **289,** 57–67 (1972).
65. B. H. Bielski, B. A. Comstock, A. Haber, and P. C. Chan, *Biochim. Biophys. Acta* **350,** 113–120 (1974).
66. B. H. Bielski and J. M. Gebicki, *Biochim. Biophys. Acta* **364,** 233–235 (1974).
67. K. Kobayashi and K. Hayashi in *Fast Methods in Physical Biochemistry and Cell Biology,* R. I. Shaafi and S. M. Fernandez, eds., Elsevier, New York, 1983, pp. 87–111.
68. D. Metodiewa and H. B. Dunford, *Arch. Biochem. Biophys.* **272,** 245–253 (1989).
69. S. Shahangian and L. P. Hager, *J. Biol. Chem.* **257,** 11529–115353 (1982).
70. W. D. Hewson and L. P. Hager, *J. Biol. Chem.* **254,** 3175–3181 (1979).
71. W. D. Hewson and L. P. Hager, *J. Biol. Chem.* **254,** 3182–3186 (1979).
72. Y. J. Suh and L. P. Hager, *J. Biol. Chem.* **266,** 22102–22109 (1991).
73. H. Theorell, *Arkiv Kemi Mineral. Geol.,* **16A,** No. 3, 1–11 (1942).
74. N. Foote, P. M. A. Gadsby, M. J. Berry, C. Greenwood, and A. J. Thomson, *Biochem. J.* **246,** 659–668 (1987).
75. A. J. Sitter, J. R. Shifflett, and J. Terner, *J. Biol. Chem.* **263,** 13032–13038 (1988).
76. A. Feis, M. P. Marzocchi, M. Paoli, and G. Smulevich, *Biochemistry* **33,** 4577–4583 (1994).

CHAPTER 5

Horseradish Peroxidase. II: Isoenzymes, Steady-State Kinetics, and Peroxidase–Oxidase Reaction

5.1. ISOENZYMES OF HORSERADISH PEROXIDASE

In the prechromatographic era, different forms of horseradish peroxidase were detected, beginning in 1942.[1,2] Beginning in 1952, by use of paper chromatography and filter paper electrophoresis, these forms were recognized as multiple components, not as modifications of a single component.[3,4] The modern letter designation for isoenzymes of horseradish peroxidase—A, B, C, and D—was first proposed by Paul in 1958.[5] He used carboxymethylcellulose chromatography to isolate the isoenzymes. In a later investigation, Paul and Stigbrand used a combination of several types of column chromatography, starch gel and polyacrylamide gel electrophoresis, and isoelectric focusing to separate the isoenzymes; and they switched to a different terminology.[6] However, the capital letter numbering from A to E is now universally accepted. The detection of multiple components in commercial preparations of horseradish peroxidase was the subject of several investigations.[7–12]

In a series of papers from 1967 to 1971 by Shannon and co-workers, properties of the A–E isoenzymes were characterized.[13–16] Seven isoenzymes were detected, labeled A1, A2, A3, B, C, D, and E. The acidic isoenzymes, A1–A3, had similar properties. The remaining isoenzymes were also thought to have similar properties, distinctly different from the acidic isoenzymes. From work in the group of Morita,[17] it is now known that isoenzymes B and C are neutral enzymes, and they contain five components: B1, B2, B3, C1, and C2. All components of the B and C families contain two calcium ions and 17–21% carbohydrate. They are essentially identical in physical and chemical properties except for their isoelectric points. D and E are basic enzymes. In a detailed study of the basic peroxidases, six isoenzymes were identified and labeled E1–E6.[18] The label isoenzyme D appears to have been dropped. E3–E6 are extremely basic, with pI values over 12. Thus, the current count on isoenzymes of horseradish peroxidase is three A, three B, two C, and six E for a total of 14.

Unless specified otherwise, commercial preparations of horseradish peroxidase are of the neutral enzyme, and are a mixture of B and C isoenzymes. The experience in my group has been that commercial preparations can be regarded as homogeneous for all practical applications. My recommendation is to buy an ammonium sulfate suspension of horseradish peroxidase, not a lyophilized powder.

By far the largest amount of work has been performed on the neutral form of horseradish peroxidase (HRP). In the discussion that follows, it can be assumed that the neutral forms of the enzyme are being described, unless otherwise specified.

5.2. STEADY-STATE KINETICS

The use of steady-state kinetics is far more widespread than the use of the transient state. One reason is obvious: A steady-state kinetic investigation can be conducted on a conventional spectrophotometer, whereas most transient-state experiments require a minimum of stopped flow instrumentation. For the transient-state investigation of unstable intermediates, a sequential mixing stopped flow is required. Assays for enzyme activity are a form of steady-state kinetics. In this section the basic equations for peroxidase steady-state kinetics are derived and described, and the important differences are discussed for the two-substrate reactions of peroxidases compared to those for most other enzymes.

The three basic steps of a conventional peroxidase cycle, and the sum of the three reactions, are as follows:

$$\text{HRP} + H_2O_2 \xrightarrow{k_1} \text{Compound I} + H_2O \tag{5.1}$$

$$\text{Compound I} + AH_2 \xrightarrow{k_2} \text{Compound II} + \bullet AH \tag{5.2}$$

$$\text{Compound II} + AH_2 \xrightarrow{k_3} \text{HRP} + \bullet AH + H_2O \tag{5.3}$$

$$H_2O_2 + 2AH_2 = 2\,\bullet AH + 2H_2O \tag{5.4}$$

Since one H_2O_2 reacts for two AH_2, then the expression for the velocity of the reaction, v, must reflect this stoichiometry:

$$v = -2\frac{d\,[H_2O_2]}{dt} = -\frac{d\,[AH_2]}{dt} \tag{5.5}$$

The steady-state equation that describes most of the observed peroxidase kinetics is as follows (see Section 5.4):

$$\frac{v}{2\,[\mathrm{E}]_{\mathrm{tot}}} = \frac{1}{\dfrac{1}{k_1\,[\mathrm{H_2O_2}]} + \dfrac{1}{k_3\,[\mathrm{AH_2}]}} \tag{5.6}$$

Steady-state peroxidase kinetics is usually governed by two rate constants, k_1 and k_3, the constants for the rate of formation of compound I from native enzyme and the oxidizing substrate hydrogen peroxide, and the rate of reduction of compound II back to native enzyme by the reducing substrate AH_2. The reduction of compound I is typically so fast that the steady-state concentration of compound I is negligibly small. In contrast the total enzyme concentration is partitioned mainly between native enzyme and compound II. If the concentration of hydrogen peroxide is very low, then compound I formation is rate-controlling, and the native enzyme is the dominant species in the steady state. If the concentration of reducing substrate is very low, then compound II is dominant in the steady state.

A convenient way to determine both k_1 and k_3 from a family of rectangular hyperbolae is to rearrange the steady-state equation into the following form:

$$\frac{v}{2\,[\mathrm{E}]_{\mathrm{tot}}} = \frac{k_3\,[\mathrm{AH_2}][\mathrm{H_2O_2}]}{\dfrac{k_3}{k_1}\,[\mathrm{AH_2}] + [\mathrm{H_2O_2}]} \tag{5.7}$$

$$\frac{v}{2\,[\mathrm{E}]_{\mathrm{tot}}} = \frac{A\,[\mathrm{H_2O_2}]}{B + [\mathrm{H_2O_2}]} \tag{5.8}$$

where $A = k_3[\mathrm{AH_2}]$ and $B = k_3[\mathrm{AH_2}]/k_1$. Each rectangular hyperbola can be fit by the two parameters A and B. A plot of A versus AH_2 yields a straight line with slope equal to k_3, and using this value of k_3 in a similar straight-line plot of B versus AH_2, the value of k_1 is obtained.

Peroxidase kinetics show similarities to classical ping-pong kinetics. If double reciprocal plots are employed, parallel line kinetics are observed in both cases. However, there are major differences, and the name peroxidase ping-pong should be employed to distinguish peroxidase kinetics from the classical case. Since peroxidase reactions are usually irreversible, there is no upper limit in rate. As long as substrate concentrations are increased, the observed rate will increase. The use of the usual steady state parameters, k_{cat} and K_M, is inappropriate for peroxidase kinetics. Only the two parameters k_1 and k_3 should be used, and they are adequate for normal ranges of substrate concentrations. Of course, for very large substrate concentrations the basic equations eventually will break down. Pure hydrogen peroxide is not an appropriate enzyme reaction medium.

Classical ping-pong kinetics refers to reversible steps in the overall enzyme-catalyzed reaction. When substrate concentrations are low, the rate-controlling step is in substrate binding; and when substrate concentrations are high, the rate controlling step is in product release. Thus, an upper limit in rate, defined by k_{cat}, is attained. Two additional parameters, a Michaelis constant for each substrate, for a total of three parameters, are required to describe the classical ping-pong mechanism. In the double reciprocal plot method of analysis, described in standard enzyme kinetics texts, the ordinate intercept of the secondary plot for classical ping-pong kinetics is equal to $1/k_{cat}$. If peroxidase ping-pong kinetics are described in terms of k_{cat} and K_M, then the secondary plot intercept is also equal to $1/k_{cat}$, but the value of this intercept is zero, so that the value of k_{cat} is infinity.

5.2.1. Some Steady-State Reactions with One-Electron Reductants

The steady-state oxidation of thiocyanate ion by horseradish peroxidase is 100-fold slower than by lactoperoxidase. The reaction occurs via a one-electron mechanism that generates the sulfur-centered thiocyanate radical. Dimerization of the radical followed by hydrolysis generates cyanide that inactivates the enzyme.[19]

α-Tocopherol reacts with compounds I and II of horseradish peroxidase as a one-electron reductant. By combining high-pressure liquid chromatography (HPLC) with electron paramagnetic resonance (EPR) spectroscopy, it was shown that the α-tocopherol radical can oxidize isolated low-density lipoprotein.[20]

A free radical mechanism occurs in the coupling of diiodotyrosine catalyzed by horseradish peroxidase.[21] The oxidation of tyrosine-containing peptides and their subsequent polymerization was investigated by steady-state kinetics.[22]

Sulfides are converted in part to sulfoxides in reactions with horseradish peroxidase. There is agreement that the reaction proceeds via sulfide cation radicals in one-electron reactions.[23–25] Details of the subsequent steps were in question.[26–28] Direct transfer of the ferryl oxygen atom of horseradish peroxidase compound II to the sulfide cation radical was favored in a study using water-soluble reactive sulfides capable of undergoing C—H and C—S bond cleavage, but perhaps a direct transfer from chloroperoxidase compound I occurs.[29]

5.2.2. Steady-State Kinetics for Two-Electron Reduction of Compound I

The first example of a two-electron reduction, converting compound I directly back to native enzyme, was the reaction of hydrogen peroxide with iodide catalyzed by horseradish peroxidase. The mechanism of the reaction is:

$$\text{HRP} + H_2O_2 \xrightarrow{k_1} \text{Compound I} + H_2O \tag{5.9}$$

$$\text{Compound I} + I^- \xrightarrow[H^+]{k_2} \text{HRP} + \text{HOI} \tag{5.10}$$

The source of the proton is discussed in Section 4.6.3. The sum of the above two reactions gives the stoichiometry:

$$H_2O_2 + I^- + H^+ = HOI + H_2O \tag{5.11}$$

By the standard derivation using the steady-state approximation and the conservation relation for total enzyme concentration, one obtains the following rate expression:

$$\frac{v}{[E]_{tot}} = \frac{1}{\dfrac{1}{k_1[H_2O_2]} + \dfrac{1}{k_2[I^-]}} \tag{5.12}$$

The latter equation can be rearranged into the form

$$\frac{v}{[E]_{tot}} = \frac{A\,[H_2O_2]}{B + [H_2O_2]} \tag{5.13}$$

where A is now equal to $k_2[I^-]$ and B to $k_2[I^-]/k_1$. Thus k_2 and k_1 can be obtained from the fits to a series of recangular hyperbolas obtained using steady-state kinetics. A striking difference between the oxidation of iodide and the oxidation of one-electron substrates such as phenols is that compound II is absent from the steady state. By using a very low concentration of iodide, the reduction of compound I becomes rate controlling, and compound I is the dominant species in the steady state. A very accurate spectrum of compound I was obtained for this steady-state condition.[30] If the steady-state reaction is allowed to proceed to longer times, there are interesting complications, as discussed below.

5.2.3. Catalatic Activity of Horseradish Peroxidase and the Iodine Clock

The hypoiodous acid HOI generated in the hydrogen peroxide–iodide–horseradish peroxidase system

$$H_2O_2 + I^- + H^+ \rightarrow HOI + H_2O \tag{5.14}$$

can undergo further reactions. It can oxidize H_2O_2 to oxygen:

$$HOI + H_2O_2 \rightarrow O_2 + H_2O + H^+ + I^- \qquad (5.15)$$

The sum of the latter two reactions is as follows:

$$2H_2O_2 \rightarrow O_2 + 2H_2O \qquad (5.16)$$

which is the reaction catalyzed by catalase, the *catalatic* reaction. Thus, in the presence of iodide, horseradish peroxidase displays catalatic activity.

The hypoiodous acid will also react with excess iodide to form I_2, which further reacts with iodide to form I_3^-.

$$HOI + I^- + H^+ \rightleftarrows I_2 + H_2O \qquad (5.17)$$

$$I_2 + I^- \rightleftarrows I_3^- \qquad (5.18)$$

Thus, there is a competition for the HOI. Iodide reacts with it to form triiodide, and hydrogen peroxide reacts with it to generate O_2 and regenerate I^-.

At lower pH triiodide formation occurs first, but the reversible equilibria maintains a supply of HOI that is later siphoned off by reaction with hydrogen peroxide. Therefore, a maximum in triiodide concentration is reached, which then gradually decreases.

At pH 6.5 the catalatic reaction dominates initially, but as soon as H_2O_2 is consumed the remaining HOI reacts with excess I^- to form I_3^-. Therefore a clock reaction occurs, with the length of time to the appearance of the color of I_3^- proportional to the concentration of H_2O_2.[31] This appears to be the only known example of an enzyme-catalyzed clock reaction.

5.3. PEROXIDASE–OXIDASE REACTION

An oxidase is an enzyme that transfers electrons to molecular oxygen. The best known example is cytochrome *c* oxidase, which accepts four electrons from cytochrome *c* and transfers them, along with four protons, to O_2, reducing the O_2 to two molecules of H_2O. Iron(II) is required to bind O_2 in iron-containing enzymes. In 1965, in a search for the route via ferrous peroxidase to compound III, or oxyperoxidase, Yamazaki and Yokota reacted horseradish peroxidase with molecular oxygen in the presence of reduced nicotamide adenine dinucleotide (NADH).[32] They were able to demonstrate compound III formation. The compound III disappeared after complete consumption of oxygen, and a mixture of ferric and ferrous peroxidase was obtained. By addition of O_2 to this mixture, they were able to regenerate compound III. Later the same year, the group of Yamazaki showed that in an open system (continuous supply of O_2) damped oscillations occurred in the concentrations of compound III and of O_2 in the solution.[33] This was the first known example of oscillations in a single-enzyme system, and to this day it remains as the only known example. It provides the simplest model to help understand various forms of oscillatory and chaotic behavior in more complex biological systems.

The possible role of compound III in the reduction of O_2 remained obscure. It is now known that an important electron transfer route does not occur directly from peroxidase to O_2. Both compounds I and II oxidize NADH to the free radical NAD• and this liberated free radical reduces O_2 to superoxide.[34] Since there is no direct participation of the enzyme in the reduction of O_2, the terminology of an oxidase for the action of peroxidase on NADH in the presence of O_2 is incorrect. However, the peroxidase–oxidase reaction is firmly embedded in the literature, and it is usually understood to describe the reactions in the system peroxidase–O_2–NADH. Similar oscillations have been reported in other horseradish peroxidase–O_2 systems, where the third component is indole-3-acetic acid, dihydroxfumarate or triose reductone.[35] The discussion here is restricted to the aerobic reaction of peroxidase with NADH and it shall be referred to as the peroxidase–oxidase reaction. The overall reaction is as follows:

$$2\text{NADH} + \text{O}_2 + 2\text{H}^+ \rightarrow 2\text{NAD}^+ + 2\text{H}_2\text{O} \qquad (5.19)$$

After the initial discovery of damped oscillations, bistability,[36] and sustained oscillations,[37] and chaos[38] were reported.

5.3.1. Oscillations

The discoverers of oscillations in the peroxidase–oxidase reactions, Yamazaki and Yokota drew an analogy to the siphon action of an old-fashioned pipette washer.[39] Oxygen is reacted faster than it diffuses from the gas phase into solution so that it is totally consumed. This stops the reaction until the supply of oxygen is replenished. The analogy is accurate in describing the oxygen oscillations, which are in phase with the oscillations in the concentration of compound III. In an open system, the total concentration of horseradish peroxidase is divided mainly between native enzyme and compound III, so that the concentration of native enzyme also oscillates, but 180° out of phase with the oxygen and compound III oscillations. However, the siphon analogy tells us nothing about the mechanism of the reaction.

The first detailed experimental investigation of the peroxidase–oxidase reaction was conducted by Yokota and Yamazaki in a closed system at pH 5.6.[34] Because their system was closed, only one oscillation was observed, with the reaction being stopped when oxygen consumption was complete. Some of their experimental and simulated results are shown in Fig. 5.1. Several features of their mechanism for consumption of oxygen and conversion of native enzyme to compound III have stood the test of time. They include the following reactions:

$$\text{NADH} + \text{O}_2 + \text{H}^+ \rightarrow \text{NAD}^+ + \text{H}_2\text{O}_2 \qquad (5.20)$$

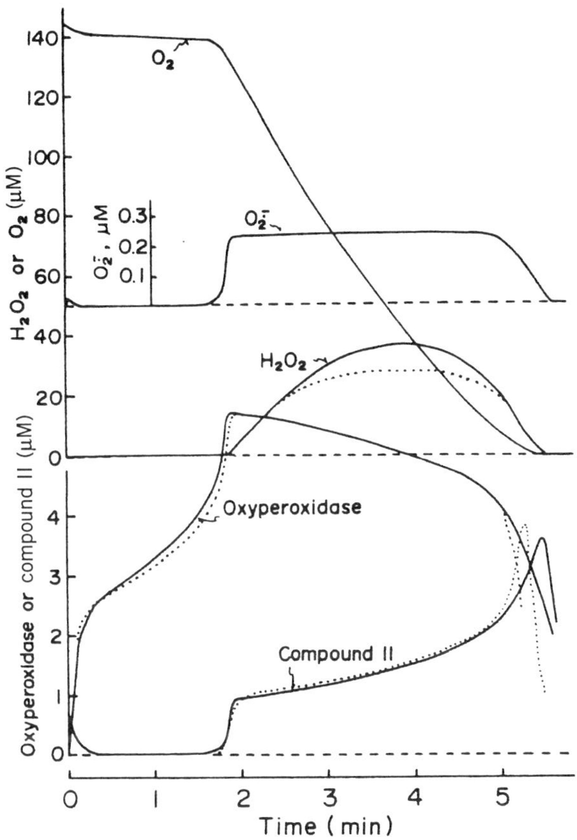

FIG. 5.1. Experimental (dotted lines) and simulated (solid lines) results on the peroxidase–oxidase reaction in a closed system at pH 5.6. Initial concentrations are the following: enzyme 6.8 μM, NADH 0.43 mM, and oxygen 0.15 mM. [Reproduced with permission from K. Yokota and I. Yamazaki, *Biochemistry* **16,** 1913–1920 (1977).]

NADH is unstable in acid solution and is capable of generating a trace of hydrogen peroxide to start the reaction. There is a long induction period during which the enzyme is slowly turning over, the rate limited by the supply of H_2O_2.

$$\text{Peroxidase} + H_2O_2 \rightarrow \text{Compound I} + H_2O \tag{5.21}$$

$$\text{Compound I} + \text{NADH} \rightarrow \text{Compound II} + \bullet\text{NAD} \tag{5.22}$$

$$\text{Compound II} + \text{NADH} \rightarrow \text{Peroxidase} + \bullet\text{NAD} + H_2O \tag{5.23}$$

The pH dependence was investigated for reactions (5.21) and (5.22), compounds I and II oxidizing NADH to •NAD.[40] At pHs above 7 there is a large difference in rate, compound I reacting faster. At pH 5.7 the difference is small: For compound I reacting with NADH the rate constant is $(4.0 \pm 0.1) \times 10^4\ M^{-1}s^{-1}$, and for compound II $(2.6 \pm 0.1) \times 10^4\ M^{-1}s^{-1}$. The pH rate profile for the reaction of compound I with NADH is shown in Fig. 5.2.

The sum of one cycle is:

$$H_2O_2 + 2NADH \rightarrow 2\ \bullet NAD + 2H_2O \tag{5.24}$$

The liberated •NAD radicals reduce O_2:

$$2\ \bullet NAD + 2O_2 \rightarrow 2NAD^+ + 2O_2\bullet^- \tag{5.25}$$

Dismutation of the superoxide generates hydrogen peroxide:

$$2O_2\bullet^- + 2H^+ \rightarrow H_2O_2 \tag{5.26}$$

Therefore, one H_2O_2 is regenerated so the peroxidase cycle can be repeated. This is a tenuous chain reaction since alternative nonproductive reactions of •NAD or superoxide radicals decrease the yield of H_2O_2. A more efficient production of H_2O_2 would be via the reaction

$$O_2\bullet^- + H^+ + NADH \rightarrow H_2O_2 + \bullet NAD \tag{5.27}$$

if it occurs. Only one superoxide radical, instead of two, is required to produce one H_2O_2. This reaction was postulated by Yokota and Yamazaki and later

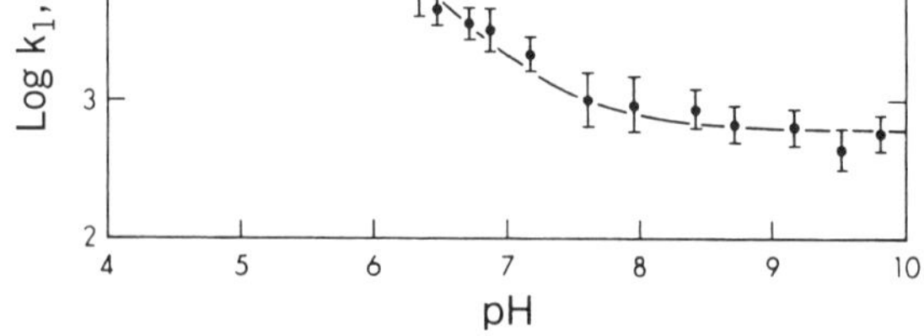

FIG. 5.2. pH dependence of the rate constant for the one-electron reduction of compound I of horseradish peroxidase by NADH to form compound II and the NAD• radical. [Reproduced with permission from M. A. Kashem and H. B. Dunford, *Biochem. Cell Biol.* **64,** 323–327 (1986).]

was investigated by flash photolysis.[41] It turns out that superoxide is unreactive, but its protonated form, the hydroperoxyl radical $HO_2\bullet$, is highly reactive with NADH.

$$HO_2\bullet + NADH \rightarrow H_2O_2 + \bullet NAD \tag{5.27}$$

The rate constant for the reaction is $(1.8 \pm 0.2) \times 10^5\ M^{-1}s^{-1}$ at 20°C. Since the pK_a value for the dissociation of $HO_2\bullet$ is 4.9, of the total amount of $O_2\bullet^-$ and $HO_2\bullet$ in acid solution, an appreciable fraction is in the form of $HO_2\bullet$. Reaction (5.27) has been ignored in later models of the peroxidase–oxidase reaction.

Ironically, Yokota and Yamazaki[34] ignored the formation of ferrous horseradish peroxidase. Its formation and reaction with O_2 were the original reasons for investigating the reaction of horseradish peroxidase with NADH in an aerated solution. They also did not include a chain termination step, which now appears to be accepted as

$$2\ \bullet NAD \rightarrow (NAD)_2 \tag{5.28}$$

Compound III formation is a nonproductive reaction:

$$\text{Peroxidase} + O_2\bullet^- + H^+ \rightarrow \text{Compound III} \tag{5.29}$$

The rate constant[42] for reaction (5.29) is $1.5 \times 10^5\ M^{-1}s^{-1}$. Compound III does not participate in the normal peroxidase cycle. The return of the enzyme into a reactive form for the peroxidase–oxidase reaction was postulated to occur via the following reaction[34]:

$$\text{Compound III} + \bullet NAD \xrightarrow{H^+} \text{Compound I} + NAD^+ + H_2O \tag{5.30}$$

The most detailed experimental investigation of the peroxidase–oxidase reaction up to 1995 was that conducted in the group of Alexander Scheeline.[43,44] The experiments were conducted in a system, referred to as a semibatch reactor, in which O_2 diffused from the gas phase into a well-stirred solution, and NADH was added at a constant rate. Concentrations of native enzyme, compound III, O_2, and NADH were continuously monitored. A meticulous examination was made of the correct way to treat the oxygen diffusion kinetics from the gas phase into solution.[45] Under these conditions the total enzyme concentration of 7.9 μM was initially accounted for by native enzyme and compound III. As the reaction progressed Pigment 670 (P670), an inactivated form of the enzyme, accumulated. The results were modeled using, wherever possible, measured rate constants of elementary reactions. Reaction (5.27), the reaction of $HO_2\bullet$ with NADH to generate H_2O_2, was neglected. It was found that without the chain termination step of dimerization of $\bullet$NAD radicals [reaction (5.28)], that the experimentally observed oscillations could not be simulated. The oscillations in the two dominant forms of the enzyme, native and compound III, obtained by deconvolution of the observed spectra, are shown in Fig. 5.3.

Oscillations were observed over a 6-hour period, which would meet the most stringent definitions of sustained oscillations. The slow formation of P670 occurs in the following reaction[46]:

$$\text{Compound I} + H_2O_2 \rightarrow \text{P670} \qquad (5.31)$$

Modeling was able to fit the observed oscillations reasonably well, and the model is called the Urbanalator, named after the location of the investigations.

5.3.2. More Complicated Behavior

By the addition of two reactions to the Scheeline mechanism, the formation of ferrous peroxidase and its reaction with O_2 to form compound III is accounted for:

$$\text{Peroxidase} + \bullet\text{NAD} \xrightarrow{H^+} \text{Ferrous peroxidase} + \text{NAD}^+ \qquad (5.32)$$

$$\text{Ferrous peroxidase} + O_2 \rightarrow \text{Compound III} \qquad (5.33)$$

Bronnikova et al.[47] were able to model both simple and complex periodic oscillations, period doubling bifurcations, and chaos. Some of their results are shown in Fig. 5.4. With all other factors held constant, very small variations in the rate constant for reaction (5.29), the reduction of compound III to compound I by •NAD, caused the observed multiplicity of effects.

5.3.3. Return of Compound III into the Peroxidase Cycle

The route to complete O_2 consumption and compound III formation is now well documented. However, the only route by which compound III is converted into a form that participates in the normal peroxidase cycle is postulated to be reaction (5.30), conversion of compound III into compound I by reaction with •NAD. The rate constant for this reaction has never been measured. Detailed experimental investigations of compound III reactions would be desirable.

Compound III converts slowly and spontaneously directly back to the native ferric enzyme.[48] The process is accelerated by addition of superoxide dismutase. It would appear that the reverse of compound III formation from superoxide occurs:

$$\text{Compound III} \rightarrow \text{Peroxidase} + O_2\bullet^- + H^+ \qquad (5.34)$$

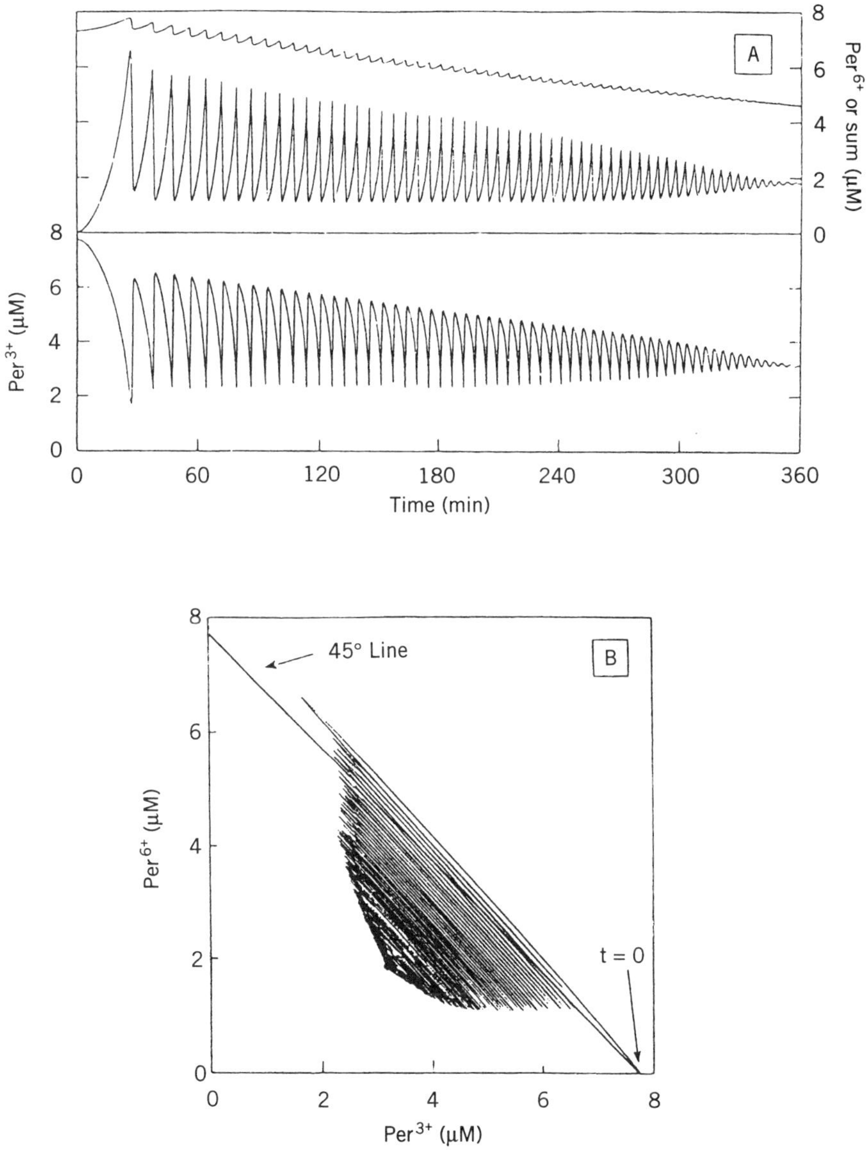

FIG. 5.3. Oscillations in the peroxidase–oxidase reaction in an open stirred system. (A) Upper curve sum of compound III (Per^{6+}) and native enzyme (Per^{3+}), middle curve compound III, and lower curve native enzyme. (B) Plot of compound III versus native enzyme concentrations. The plot oscillates back and forth until the reaction stops. The 45° line indicates idealized interconversion involving only the two forms of the enzyme. [Reproduced with permission from D. L. Olson, E. P. Williksen, and A. Scheeline, *J. Am. Chem. Soc.* **117,** 2–15 (1995).]

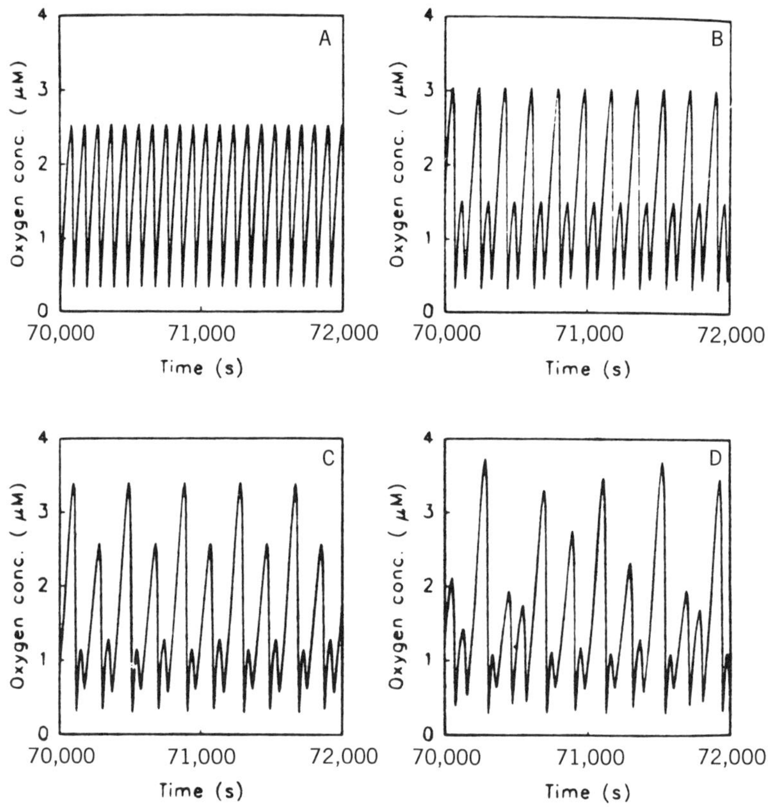

FIG. 5.4. Modeled oscillations in oxygen concentration obtained by varying the rate constant ($M^{-1}s^{-1}$) for the reduction of compound III to compound I by the NAD• radical: (A) 1.1×10^8, (B) 1.2×10^8, (C) 1.317×10^8, and (D) 1.35×10^8. [Reproduced with permission from T. V. Bronnikova, V. R. Fed'kina, W. M. Schaffer, and L. F. Olsen, *J. Phys. Chem.* **99,** 9309–9312 (1995).]

If one adds the following known reactions to reaction (5.34)

$$O_2{\bullet}^- + H^+ \rightleftarrows HO_2{\bullet}$$

$$HO_2{\bullet} + \text{NADH} \rightarrow {\bullet}\text{NAD} + H_2O_2$$

$$\text{Peroxidase}^{3+} + H_2O_2 \rightarrow \text{Compound I} + H_2O$$

one obtains the sum

$$\text{Compound III} + \text{NADH} \rightarrow \text{Compound I} + {\bullet}\text{NAD} + H_2O \qquad (5.35)$$

Thus, there is more than one possibility for the mechanism by which compound III reenters the peroxidase cycle.

5.3.4. Effect of Methylene Blue and Dichlorophenol

In the initial investigations of the peroxidase–oxidase reaction, markedly damped oscillations were observed with variable induction periods. It was found that addition of methylene blue to the system would activate the oscillations, and dichlorophenol would cause the system to switch from damped to sustained oscillations.[49] It has been shown that methylene blue reacts directly with NADH.[50,51] The simplest explanation of the results is in terms of two-electron transfers involving oxidized methylene blue, MB^+, and completely reduced methylene blue, MBH:

$$\mathrm{NADH} + \mathrm{MB}^+ \xrightarrow{\text{slow}} \mathrm{NAD}^+ + \mathrm{MBH} \tag{5.36}$$

$$\mathrm{MBH} + \mathrm{O}_2 + \mathrm{H}^+ \xrightarrow{\text{fast}} \mathrm{MB}^+ + \mathrm{H}_2\mathrm{O}_2 \tag{5.37}$$

If the concentration of NADH exceeds 5×10^{-4} M, saturation kinetics is observed in reaction (5.35). The H_2O_2 generated in reaction (5.37) would initiate the peroxidase–oxidase reaction, cause larger oscillations, and eliminate irreproducible induction periods observed in the absence of added methylene blue.

When dichlorophenol is added, the amplitude of the oscillations is reduced, which can be explained because phenols are excellent peroxidase substrates. They compete for compounds I and II and so curtail the peroxidase–oxidase reaction. The sustained oscillations observed when dichlorophenol is added can be explained by elimination or curtailment of P670 formation [reaction (5.30)] because the steady-state concentration of compound I is reduced.

5.3.5. Effect of Aromatic Amines and Other Phenols

Aromatic amines and phenols other than 2,4-dichlorophenol can also induce sustained oscillations and complex dynamics in the peroxidase–oxidase reaction.[52] Experimental results obtained using 4-hydroxybenzoic acid are shown in Fig. 5.5. As the concentration of NADH is increased, the amount of compound III changes from repetitive oscillations, called 1^0, to a pattern consisting of a large oscillation followed by a small one, 1^1. This is followed by one large and two small oscillations, 1^2, then by chaotic behavior. The final frame shows continuous small oscillations, 0^1, which is followed by a true steady state. The total behavior of increased NADH concentration is called a period-adding sequence.

5.3.6. Reaction in a Continuous-Flow Stirred Tank Reactor

The peroxidase–oxidase reaction has also been studied in a continuous-flow stirred tank reactor (CSTR).[53,54] In these experiments glucose-6-phosphate and glucose-6-phosphate dehydrogenase were added to the system to regenerate NADH from NAD^+.

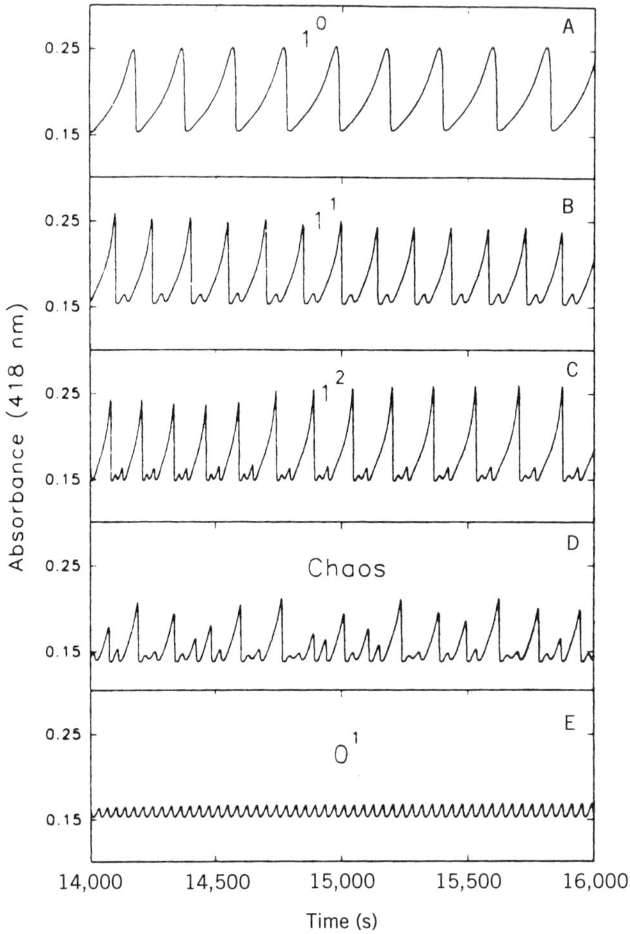

FIG. 5.5. Effect of increased concentrations of NADH on the concentration of compound III monitored at 418 nm. Experimental conditions: A stirred semibatch reactor is used at pH 6.3, 1.3 μM horseradish peroxidase, 750 μM 4-hydroxybenzoic acid, 0.1 μM methylene blue. Average NADH concentrations in μM: (A) 73.0, (B) 91.6, (C) 94.0, (D) 95.6, and (E) 97.8. For other conditions consult the original reference. [Reproduced with permission from U. Kummer, M. J. B. Hauser, K. Wegmann, L. F. Olsen, and G. Baier, *J. Am. Chem. Soc.* **119,** 2084–2087 (1997).]

5.3.7. Theoretical Models

Abstract mathematical models have been used to explain the peroxidase–oxidase reaction. Part of the reason was lack of sufficient experimental data on the mechanism of the reaction and rate constants for the relevant elementary reactions. All of these models use at least one of the following to explain the experimentally observed oscillations or more complex behavior: autocatalysis, self-inhibition, or delayed feedback inhibition. It is difficult to compare an abstract model with one based on known reactions and rate constants. Part of the problem is that calculations are often performed using dimensionless parameters that cannot be directly compared to first- or second-order rate constants. A review of mathematical models has been published.[55] Gradually the role of the enzyme and actual mechanistic steps have been introduced into the mathematical models[56] so that the gap between theory and experiment is being closed.[57–60] The state of knowledge on the peroxidase–oxidase reaction has been reviewed.[61,62]

5.4. APPENDIX: DERIVATION OF STEADY-STATE RATE EQUATIONS

There are five basic types of equations required for derivation of steady-state equations. Following is the derivation for conventional peroxidase kinetics:

1. Mechanism

$$\text{HRP} + \text{H}_2\text{O}_2 \xrightarrow{k_1} \text{Compound I} + \text{H}_2\text{O} \tag{5.1}$$

$$\text{Compound I} + \text{AH}_2 \xrightarrow{k_2} \text{Compound II} + \bullet\text{AH} \tag{5.2}$$

$$\text{Compound II} + \text{AH}_2 \xrightarrow{k_3} \text{HRP} + \bullet\text{AH} + \text{H}_2\text{O} \tag{5.3}$$

2. Stoichiometry

$$\text{H}_2\text{O}_2 + 2\text{AH}_2 = 2\bullet\text{AH} + 2\text{H}_2\text{O} \tag{5.4}$$

3. Steady-State Approximation

$$-\frac{d[\text{HRP}]}{dt} = k_1\,[\text{HRP}][\text{H}_2\text{O}_2] - k_3\,[\text{Compound II}][\text{AH}_2] = 0 \tag{5.38}$$

$$+\frac{d\,[\text{Compound I}]}{dt} = k_1\,[\text{HRP}][\text{H}_2\text{O}_2] - k_2\,[\text{Compound I}][\text{AH}_2] = 0 \tag{5.39}$$

$$+\frac{d\,[\text{Compound II}]}{dt} = k_2\,[\text{Compound I}][\text{AH}_2] - k_3\,[\text{Compound II}][\text{AH}_2] = 0 \tag{5.40}$$

4. Conservation

The total protein concentration $[E]_{tot}$ is given by

$$[E]_{tot} = [HRP] + [\text{Compound I}] + [\text{Compound II}] \tag{5.41}$$

From Eq. 5.4

$$[HRP] = \frac{k_3[AH_2]}{k_1[H_2O_2]}[\text{Compound II}] \tag{5.42}$$

$$[\text{Compound I}] = \frac{k_3}{k_2}[\text{Compound II}] \tag{5.43}$$

Therefore, the conservation relation (5.41) can be reformulated as

$$[\text{Compound II}] = \frac{[E]_{tot}}{1 + \dfrac{k_3}{k_2} + \dfrac{k_3[AH_2]}{k_1[H_2O_2]}} \tag{5.44}$$

5. Velocity

Defined in terms of the rate of disappearance of AH_2, the velocity is given by

$$v = -\frac{d[AH_2]}{dt} \tag{5.45}$$

$$= k_2\,[\text{Compound I}][AH_2] + k_3\,[\text{Compound II}][AH_2]$$

Substitution of (5.43) into (5.45) yields

$$v = -\frac{d\,[AH_2]}{dt} = 2k_3\,[\text{Compound II}][AH_2] \tag{5.46}$$

Insertion of (5.44) into (5.46) converts it to

$$v = \frac{2k_3\,[AH_2][E]_{tot}}{1 + \dfrac{k_3}{k_2} + \dfrac{k_3\,[AH_2]}{k_1\,[H_2O_2]}} \tag{5.47}$$

which can be rearranged into the form

$$\frac{v}{2\,[E]_{tot}} = \frac{1}{\dfrac{1}{k_1\,[H_2O_2]} + \dfrac{k_2 + k_3}{k_2k_3\,[AH_2]}} \tag{5.48}$$

Since k_2 is typically greater than k_3 by a factor of 10, Eq. (5.48) can be reduced to

$$\frac{v}{2\,[\mathrm{E}]_{\mathrm{tot}}} = \frac{1}{\dfrac{1}{k_1\,[\mathrm{H_2O_2}]} + \dfrac{1}{k_3\,[\mathrm{AH_2}]}} \tag{5.49}$$

REFERENCES

1. H. Theorell, *Ark. Kemi Mineral. Geol.* **16A,** No. 2, 1–11 (1942).
2. D. Keilin and E. F. Hartree, *Biochem. J.* **49,** 88–104 (1951).
3. M. A. Jermyn, *Nature* **169,** 487–489 (1952).
4. M. A. Jermyn and R. Thomas, *Biochem. J.* **56,** 631–639 (1954).
5. K. G. Paul, *Acta Chem. Scand.* **12,** 1312–1318 (1958).
6. K. G. Paul and T. Stigbrand, *Acta Chem. Scand.* **24,** 3607–3617 (1970).
7. C. J. Wilder, *J. Food Sci.* **27,** 567–573 (1962).
8. M. H. Klapper and D. P. Hackett, *Biochim. Biophys. Acta* **96,** 272–282 (1965).
9. H. E. Kasinsky and D. P. Hackett, *Phytochem.* **7,** 1147–1150 (1968).
10. H. Delincée and B. J. Radola, *Biochim. Biophys. Acta* **200,** 404–407 (1970).
11. H. Delincée and B. J. Radola, *Eur. J. Biochem.* **52,** 321–330 (1975).
12. K. G. Paul, P. I. Ohlsson, B. Norden, and M. L. Smith, *Acta Chem. Scand.* **B39,** 507–508 (1985).
13. L. M. Shannon, E. Kay, and J. Y. Lew, *J. Biol. Chem.* **241,** 2166–2172 (1966).
14. E. Kay, L. M. Shannon, and J. Y. Lew, *J. Biol. Chem.* **242,** 2470–2473 (1967).
15. E. H. Strickland, E. Kay, L. M. Shannon, and J. Horwitz, *J. Biol. Chem.* **243,** 3560–3565 (1968).
16. J. H. C. Shih, L. M. Shannon, E. Kay, and E. J. Y. Lew, *J. Biol. Chem.* **246,** 4546–4551 (1971).
17. S. Aibara, H. Yamashita, E. Mori, M. Kato, and Y. Morita, *J. Biochem.* **92,** 531–539 (1982).
18. S. Aibara, T. Kobayashi, and Y. Morita, *J. Biochem.* **90,** 489–496 (1981).
19. S. Adak, A. Mazumdar, and R. K. Banerjee, *J. Biol. Chem.* **272,** 11049–11056 (1997).
20. P. K. Witting, J. M. Upston, and R. Stocker, *Biochemistry* **36,** 1251–1258 (1997).
21. D. R. Doerge, A. Taurog, and M. L. Dorris, *Arch. Biochem. Biophys.* **315,** 90–99 (1994).
22. T. Michon, M. Chenu, N. Kellershon, M. Desmadril, and J. Guéguen, *Biochemistry* **36,** 8504–8513 (1997).
23. S. Kobayashi, M. Nakano, T. Kimura, and A. P. Schaap, *Biochemistry* **26,** 5019–5022 (1987).
24. U. Pérez and H. B. Dunford, *Biochemistry* **29,** 2757–2763 (1990).
25. D. R. Doerge, N. M. Cooray, and M. E. Brewster, *Biochemistry* **30,** 8960–8964 (1991).

26. U. Pérez and H. B. Dunford, *Biochim. Biophys. Acta* **1038,** 98–104 (1990).
27. S.-I. Ozaki and P. R. Ortiz de Montellano, *J. Am. Chem. Soc.* **116,** 4487–4488 (1994).
28. S.-I. Ozaki and P. R. Ortiz de Montellano, *J. Am. Chem. Soc.* **117,** 7056–7064 (1995).
29. E. Baciocchi, O. Lanzalunga, S. Malandrucco, M. Ioele, and S. Steeken, *J. Am. Chem. Soc.* **118,** 8973–8974 (1996).
30. R. Roman and H. B. Dunford, *Biochemistry* **11,** 2076–2082 (1972).
31. U. Kummer, K. Wegmann, and G. Baier in *Plant Peroxidases. Biochemistry and Physiology,* C. Obinger, U. Burner, R. Ebermann, C. Penel and H. Greppin, eds., University of Geneva Press, Geneva, 1996, pp. 101–105.
32. I. Yamazaki and K. Yokota, *Biochem. Biophys. Res. Commun.* **19,** 249–254 (1965).
33. I. Yamazaki, K. Yokota, and R. Nakajima, *Biochem. Biophys. Res. Commun.* **21,** 582–586 (1965).
34. K. Yokota and I. Yamazaki, *Biochemistry* **16,** 1913–1920 (1977).
35. L. F. Olsen and H. Degn, *Biochim. Biophys. Acta* **523,** 321–334 (1978).
36. H. Degn, *Nature* **217,** 1047–1050 (1968).
37. S. Nakamura, K. Yokota, and I. Yamazaki, *Nature* **222,** 794 (1969).
38. L. Olsen and H. Degn, *Nature* **267,** 177–178 (1977).
39. I. Yamazaki and K. Yokota in *Biological and Biochemical Oscillators,* B. Chance, E. K. Pye, A. K. Ghosh, and B. Hess, eds., Academic, New York, 1973, pp. 109–114.
40. M. A. Kashem and H. B. Dunford, *Biochem. Cell Biol.* **64,** 323–327 (1986).
41. A. Nadezhdin and H. B. Dunford, *J. Phys. Chem.* **83,** 1957–1961 (1979).
42. L. F. Olsen, *Biochim. Biophys. Acta* **527,** 212–220 (1978).
43. D. L. Olson and A. Scheeline, *J. Phys. Chem.* **99,** 1212–1217 (1995).
44. D. L. Olson, E. P. Williksen, and A. Scheeline, *J. Am. Chem. Soc.* **117,** 2–15 (1995).
45. D. L. Olson and A. Scheeline, *J. Phys. Chem.* **99,** 1204–1211 (1995).
46. M. B. Arnao, M. Acosta, J. A. del Rio, R. Varon, and F. Garcia-Canovas, *Biochim. Biophys. Acta* **1041,** 43–47 (1990).
47. T. V. Bronnikova, V. R. Fed'kina, W. M. Schaffer, and L. F. Olsen, *J. Phys. Chem.* **99,** 9309–9312 (1995).
48. D. Metodiewa and H. B. Dunford, *Arch. Biochem. Biophys.* **272,** 245–253 (1989).
49. I. Yamazaki and K. Yokota, *Mol. Cell Biochem.* **2,** 39–52 (1973).
50. P. Sevcik and H. B. Dunford, *J. Phys. Chem.* **95,** 2411–2415 (1991).
51. P. Sevcik and H. B. Dunford, *Int. J. Chem. Kinetics* **27,** 925–928 (1995).
52. U. Kummer, M. J. B. Hauser, K. Wegmann, L. F. Olsen, and G. Baier, *J. Am. Chem. Soc.* **119,** 2084–2087 (1997).
53. J. G. Lazar and J. Ross, *Science* **247,** 189–192 (1990).
54. J. G. Lazar and J. Ross, *J. Chem. Phys.* **92,** 3579–3589 (1990).
55. R. Larter, L. F. Olsen, C. G. Steinmetz, and T. Geest in *Chaos in Chemistry and Biochemistry,* R. J. Field and L. Györgyi, eds., World Scientific, London, 1993, Chapter 7.
56. V. R. Fed'kina, F. I. Ataullakhanov, and T. V. Bronnikova, *Biophys. Chem.* **19,** 259–264 (1984).
57. B. D. Aguda and B. L. Clarke, *J. Chem. Phys.* **87,** 3461–3470 (1987).

58. B. D. Aguda and R. Larter, *J. Am. Chem. Soc.* **112,** 2167–2174 (1990).
59. B. D. Aguda and R. Larter, *J. Am. Chem. Soc.* **113,** 7913–7916 (1991).
60. S. Alexandre and H. B. Dunford, *Biophys. Chem.* **40,** 189–195 (1991).
61. A. Scheeline, D. L. Olson, E. P. Williksen, G. A. Horras, M. L. Klein, and R. Larter, *Chem. Rev.* **97,** 739–756 (1997).
62. A. C. Møller, M. J. B. Hauser, and L. F. Olsen, *Biophys. Chem.* **72,** 63–72 (1998).

CHAPTER 6

Horseradish Peroxidase. III: Reaction with Indole-3-Acetic Acid, Light Emission, and Quantitative Structure–Activity Relationships

6.1. REACTION OF HORSERADISH PEROXIDASE WITH INDOLE-3-ACETIC ACID

Substances promoting plant growth, called auxins, have been recognized since the 1920s. Their inactivation by water extracts of plant material was demonstrated in 1934, and some information on the nature of the inactivator was obtained in 1936.[1,2] The principal auxin is indole-3-acetic acid (Fig. 6.1), and the enzyme that metabolizes it was known as indoleacetic acid oxidase. Work on auxins has been extensively reviewed over the years.[3,4] In 1955 the ability of horseradish peroxidase to degrade indole-3-acetic acid in the absence of added hydrogen peroxide was reported.[5] It is now accepted that peroxidases are the indoleacetic acid oxidases, but use of the older terminology has persisted.

Several papers have concentrated on the products of indole-3-acetate degradation by peroxidase,[6,7] including a study on the corn enzyme[8] and a study using labeled oxygen.[9] Major products are indole-3-methanol, indole-3-aldehyde, and indole-3-methylhydroperoxide, the latter being an oxidizing substrate for the enzyme (Fig. 6.1). Formation of these products indicate a mechanism in which the indole-3-acetate is decarboxylated, a result confirmed using ^{14}C-labeled substrate.[10]

Other studies have concentrated upon the role of the enzyme[11] and of inhibitors.[12] The most comprehensive pioneering enzyme mechanistic studies are those by the groups of Ricard[13–14] and Yamazaki.[16–18] Different conclusions were reached. At pH 4 only ferrous peroxidase and compound III formation was observed; at pH 6 stable compound II was observed after the transient appearance of ferrous enzyme and compound III.[14] In contrast, at pH 4.4 both compounds I and II were detected.[16] The latter workers concluded that the ferrous enzyme was formed upon reduction of the ferric enzyme by a free radical of indoleacetic acid, not by indoleacetic acid itself. In a later study

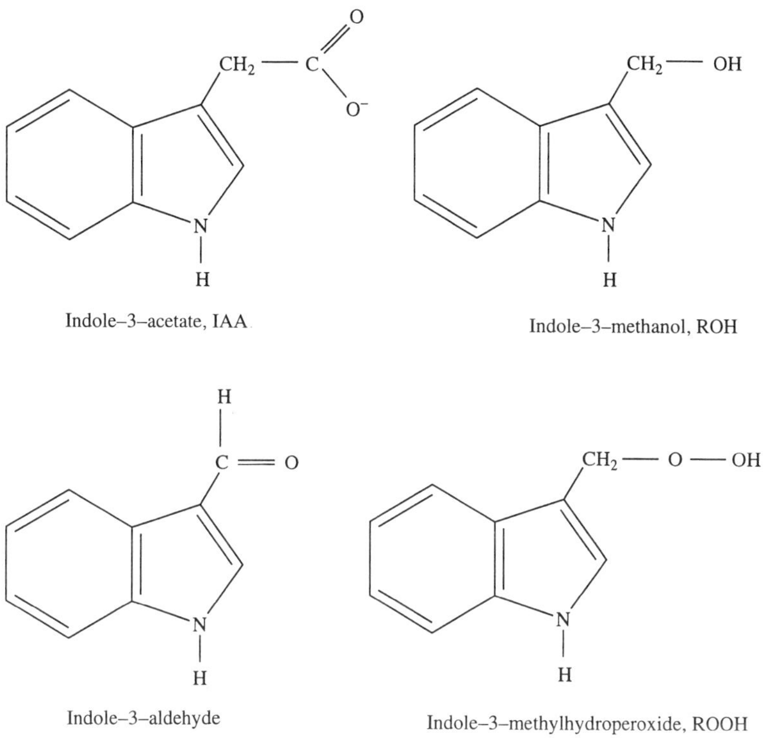

FIG. 6.1. Structures of indole-3-acetate and three major products of its reaction with horseradish peroxidase: indole-3-methanol, indole-3-aldehyde, and indole-3-methylhydroperoxide. The hydroperoxide and methanol derivatives are formed early in the reaction, whereas the aldehyde is formed later from a slow oxidation of the methanol derivative. These products show that the indole-3-acetate is decarboxylated.

conducted mainly at pH 5 it was concluded that either the normal peroxidase cycle involving compounds I and II or the ferrous-compound III shuttle occurred, depending mainly on enzyme–substrate ratios.[19] This rather unsatisfactory state of affairs has not been completely resolved at lower pHs. However, at pH 7.4 there is clearly no ferrous enzyme or compound III formation, so the possible competition of two different mechanisms has been removed.

6.1.1. Mechanism of Indoleacetate Degradation at pH 7.4

One of the difficult facts to establish was the mechanism of reaction initiation. Hydrogen peroxide is not required. Therefore, at pH 7.4 the obvious choice would appear to be trace autoxidation of indoleacetate to form a peroxide that is sustained in a chain reaction. I was suspicious of this choice because

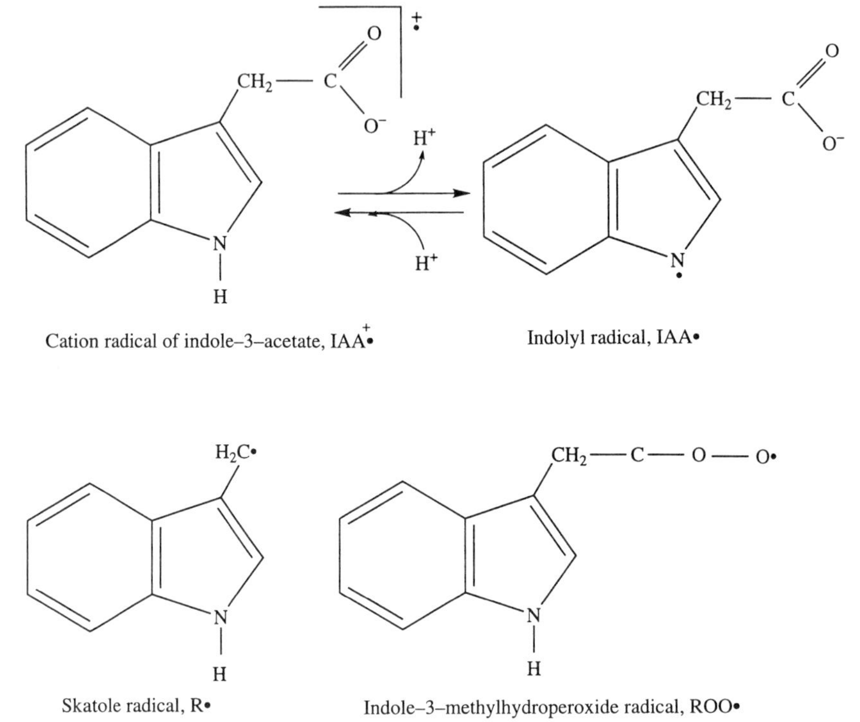

FIG. 6.2. Structures of radicals formed during the reaction of horseradish peroxidase with indole-3-acetate.

careful purification of indole-3-acetic acid, either by recrystallization or sublimation, had no apparent effect on reaction initiation.[20] Finally, both kinetic and modeling studies have established that the autoxidation occurs by a highly efficient branched-chain mechanism, not a simple chain reaction.[21,22]

Important contributions have come from electron paramagnetic resonance (EPR) and radiation chemistry studies. The skatole radical was trapped and detected.[23] The one-electron oxidation of indoleacetic acid was shown to yield its radical cation with a pK_a value of 5.09, so that at neutral pH it exists largely in its deprotonated form.[24,25] The small amount of cation radical decarboxylates, providing the route to the skatole radical. Detection of these free radical intermediates, shown in Fig. 6.2, helped greatly in formulating an overall reaction mechanism. Also shown in Fig. 6.2 are the species used and produced in compound I formation. A simplified version of the reaction mechanism is shown in Fig. 6.3, which differs slightly from the versions of Krylov and Dunford.[21,22] Slow production of a trace of the cation radical IAA$\bullet^+$ (rate constant 3×10^{-7} s^{-1})

$$\text{IAA} \longrightarrow \text{IAA}\bullet^{+}$$

sets off the highly efficient branched-chain reaction.[26] Relevant reactions and rate constants, either measured directly or inferred from modeling studies for the detailed model are numbered and listed in Table 6.1.[27]

The branched-chain reaction is obvious in Fig. 6.3. Each enzymatic cycle produces two cation radicals of indoleacetate, IAA•$^+$ (reactions 2 and 3), and formation of indole-3-methyl hydroperoxide ROOH produces IAA• (reaction 7). Each cation radical is capable of producing another hydroperoxide, via the route of decarboxylation to form the skatole radical R• (reaction 5), scavenging of the skatole radical by molecular oxygen to form the hydroperoxyl radical ROO• (reaction 6), and hydrogen atom abstraction form indoleacetate to form the hydroperoxide molecule (reaction 7). The latter reaction would produce the deprotonated form of the indoleacetate cation radical, the indolyl radical IAA•, which is in rapid equilibrium with the cation radical.

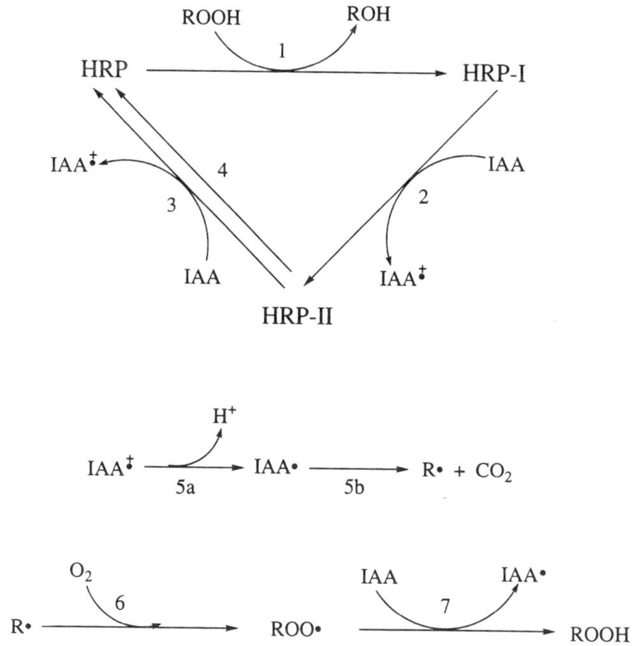

FIG. 6.3. Branching chain reaction of indole-3-acetate (IAA) with horseradish peroxidase. HRP native enzyme, HRP-I compound I, HRP-II compound II, IAA•$^+$ indoleacetic acid cation radical, IAA• indolylyl radical, R• skatolyl radical, ROO• indolemethyl hydroperoxyl radical, ROOH indolemethylhydroperoxide. Other abbreviations defined in Figs. 6.1 and 6.2. Not shown are the inactivation steps or the very slow spontaneous conversion IAA to IAA•$^+$.

TABLE 6.1. Reactions Involved in the HRP-catalyzed Oxidation of IAA at pH 7.4.[a]

No.	Reaction	Rate Constants	Ref.
1	HRP + ROOH → HRP-I + ROH	$k_1 = 2.0 \times 10^6\ M^{-1}s^{-1}$	17,21
2	HRP-I + IAA → HRP-II + IAA•⁺	$k_2 = 2.3 \times 10^3\ M^{-1}s^{-1}$	21,22
3	HRP-II + IAA → HRP + IAA•⁺	$k_3 = 2.05 \times 10^2\ M^{-1}s^{-1}$	21,22
4	HRP-II → HRP	$k_4 = 1.75 \times 10^{-3}\ s^{-1}$	21
5	IAA•⁺ → R• + CO_2	$k_5 = 90\ s^{-1}$	24
6	R• + O_2 → ROO•	$k_6 = 2.0 \times 10^8\ M^{-1}s^{-1}$	24
7	ROO• + IAA → ROOH + IAA•	$k_7 = 1.0 \times 10^6\ M^{-1}s^{-1}$	21
8	IAA → IAA•⁺	$k_8 = 3 \times 10^{-7}\ s^{-1}$	26
9	HRP-I + ROOH → P-670	$k_9 = 6.2 \times 10^2\ M^{-1}s^{-1}$	21
10	IAA•⁺ + IAA•⁺ → P_1	$k_{10} = 5.3 \times 10^8\ M^{-1}s^{-1}$	24
11	R• → P_2	$k_{11} = 1.8 \times 10^4\ s^{-1}$	21
12	ROO• → P_3	$k_{12} = 5.62\ s^{-1}$	21
13	ROOH → P_4	$k_{13} = 1.45 \times 10^{-3}\ s^{-1}$	17,21

[a] Abbreviations are defined in the captions to Figs. 6.1 and 6.2.

Each hydroperoxide molecule can initiate another enzyme cycle (reaction 1). Formation of the indoleacetate cation radical by reduction of compound II (reaction 3) is so slow that the spontaneous decay of compound II becomes important (reaction 4).

According to the mechanism shown in Fig. 6.3, the only stable product is indole-3-methanol. The latter molecule can react very slowly with compounds I and II to form indole-3-aldehyde, shown in Fig. 6.1.[28] Thus, indole-3-methanol is a major product, and the later formation of indole-3-aldehyde is accounted for. The other major final product, 3-methylene oxindole, appears to be formed in a nonenzymatic pathway from indole-3-methyl hydroperoxide[8] (Fig. 6.4). This may be reaction 13 in Table 6.1.

A major route to enzyme inactivation is the reaction of indole-3-methyl hydroperoxide with compound I to form P670.[10,18] Each free radical may be converted into a stable molecule, thus removing it from the chain process, and estimates for the rates of these chain termination steps are also listed in Table 6.1.

Phenols are also substrates for peroxidase, and their presence complicates the competitive reaction of indole-3-acetic acid in unpredictable ways.[29–32] The inhibiting effect of caffeic acid[26] and accelerating effect of umbelliferone[27] on the reaction of indole-3-acetic acid with peroxidase have been investigated in detail.

6.1.2. Mechanism of Indoleacetate Degradation at pH 5.0

The dominant mechanism in acid solution, according to some of the literature,[15,19] is the oxidase pathway: native ferric enzyme → ferrous enzyme →

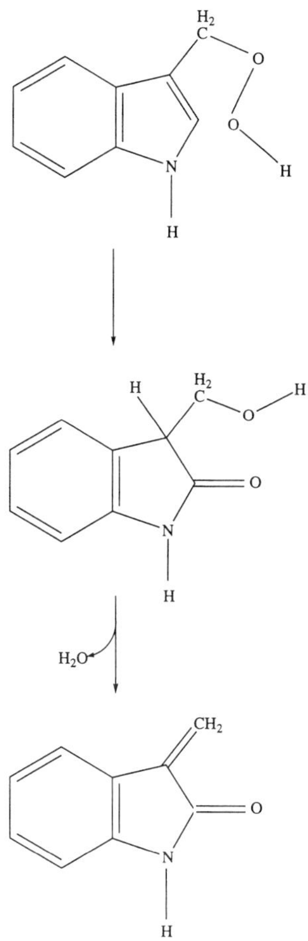

FIG. 6.4. Nonenzymatic pathway for the conversion of indole-3-methylhydroperoxide to 3-methylene oxindole.

compound III → ferrous enzyme. The reduction of ferric enzyme to ferrous has been assumed to start the cycle. In very careful experiments using ascorbate–ascorbate oxidase or glucose–glucose oxidase in a Thunberg tube to remove the last trace of oxygen, it was shown that there was no reduction of ferric enzyme by indoleacetic acid at pH 4.4.[17] As soon as either oxygen or hydrogen peroxide was introduced, formation of the ferrous enzyme was observed. A similar result was obtained 17 years later, where an artifactual reduction initiated by irradiated light was also noted.[33] Claims that indoleacetic acid would reduce the ferric enzyme appear to have been based upon experiments where oxygen was not totally removed.[15,19] The correct explanation for formation of the ferrous enzyme was postulated to be reduction by a free radical of indoleacetic acid,[17] most likely the skatolyl radical.[24]

$$\text{R}\bullet + \text{Peroxidase} \longrightarrow \text{Ferrous peroxidase} + \text{R}^+$$

The skatole cation R^+ would be expected to hydrolyze rapidly to form indole-3-methanol, ROH.

$$\text{R}^+ + \text{H}_2\text{O} \longrightarrow \text{ROH} + \text{H}^+$$

Once the ferrous enzyme is formed, it would be rapidly scavenged by molecular oxygen to form compound III:

$$\text{Ferrous peroxidase} + \text{O}_2 \longrightarrow \text{Compound III}$$

and since the ferric enzyme–superoxide resonance form of compound III is dominant, dissociation of superoxide, or more correctly at acid pH, of the hydroperoxyl radical, could occur.

$$\text{Compound III} \rightleftharpoons \text{Peroxidase} + \text{HO}_2\bullet$$

Indoleacetic acid can accelerate release of the hydroperoxyl radical.[34,35] Thus, all of the attributes of the oxidase pathway observed at low pH can be initiated by skatolyl free radical reduction of ferric to ferrous peroxidase. Formation of the indolyl radical in reactions 5a and 7, plus decarboxylation of the latter in reaction 5b (Fig. 6.3), leads to formation of the skatolyl radical. The latter radical reduces the ferric enzyme to ferrous, and thus starts the oxidase cycle. In brief the peroxidase cycle starts the oxidase cycle.

The opposite assumption, that the oxidase cycle initiates the peroxidase cycle, was made by Gazarian et al.,[33] despite confirming the reductive capacity of the skatolyl radical.[17] They postulated the existence of a ternary complex of native ferric peroxidase, indoleacetic acid, and molecular oxygen, which leads directly to compound III:

$$\text{Peroxidase} + \text{O}_2 + \text{IAA} \longrightarrow$$
$$[\text{Peroxidase}{-}\text{O}_2\bullet^-{-}\text{IAA}\bullet^+] \longrightarrow \text{Compound III} + \text{IAA}\bullet^+$$

There is no evidence for such a complex. A major reason for their postulate of a ternary complex involving indoleacetic acid is that the 2-methyl derivative of the latter substrate is unreactive at pH 4.4, despite being a better reactant with compound I. A number of alternative explanations are possible, such as an inefficient free radical chain reaction for oxidation of the 2-methyl derivative to the corresponding hydroperoxide.

In conclusion, and with benefit of considerable hindsight, I feel that the weight of evidence supports a peroxidase mechanism, native enzyme → compound I → compound II → native enzyme, with an extremely efficient branched-chain reaction tied into the peroxidase cycle. This appears to explain the observed results at all pHs. In acid solution, there is no doubt about formation of ferrous enzyme, compound III, and superoxide or hydroperoxyl radical, but their formation can be explained as a spin-off from the peroxidase mechanism, as originally stated by Nakajima and Yamazaki.[17] There are still many elementary reactions that are not understood and for which there are no rate constants. Stay tuned for future developments!

6.1.3. Light Emission from Oxidation of Indoleacetate

In the initial observation of the spontaneous oxidation of indoleacetate by oxygen with horseradish peroxidase as catalyst, light emission was also reported.[5] Emission was detected at 420, 465, and 535 nm.[36] Singlet oxygen was detected as a product,[37] and its formation could be explained by disproportionation of two hydroperoxyl radicals, either by the Russell mechanism or an alternative.[24] The overall reaction, which produces one indole-3-aldehyde, one indole-3-methanol, and one singlet oxygen molecule from two hydroperoxyl radicals, is shown in Fig. 6.5.

An alternative route to electronic excitation is the formation of a free radical formed on a tertiary carbon atom. This radical was trapped in the

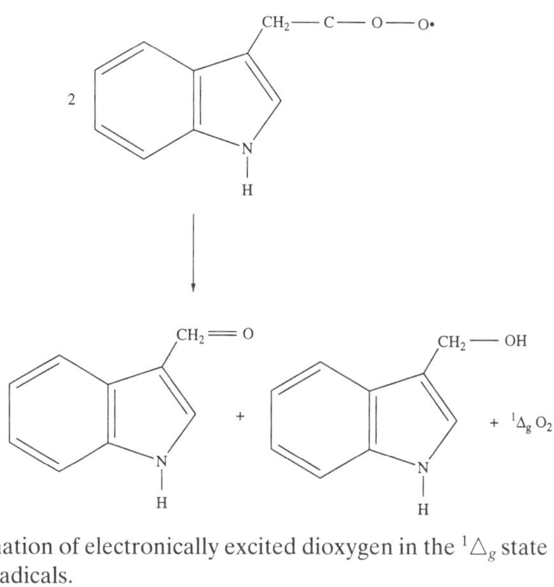

FIG. 6.5. Formation of electronically excited dioxygen in the $^1\Delta_g$ state from two indole hydroperoxyl radicals.

reaction of an indoleacetate ester with peroxidase, but was postulated to be formed also when the unesterified indoleactetate was reacted.[38] The subsequent reaction led to formation of a dioxetane intermediate and indole ring opening.

The route or routes to light emission in the reaction of indoleacetate with oxygen, catalyzed by peroxidase, are clearly not part of the main reaction pathways. Considerable effort has been expended to study reactions involving a second substrate that quenches the light emission. Two alternative explanations have been proposed. One is that electronic energy is transferred to the second substrate, leading to subsequent chemical reactions that resemble those that are photo-induced but in which no light emission occurs. This mechanism is called photobiochemistry without light.[39–41] An alternative explanation is that co-oxidation of the second substrate by the primary radicals is occurring in reactions so efficient that the light-emitting reactions do not occur.[42,43]

6.1.4. Nondecarboxylation Pathways

Nondecarboxylated products, such as oxindole and dioxindole (Fig. 6.6), have been detected,[44] and these are formed independently of the peroxidase-catalyzed reaction.[4] The enzyme or enzymes that catalyze these reactions have not been isolated. Nondecarboxylative transformation of indoleacetate in dormant apple seeds by a nonheme enzyme has been studied and may have a role in controlling dormancy.[45]

6.1.5. Some Comments on the Physiology of Indole-3-Acetic Acid

For a comprehensive overview of the role of peroxidases and indoleacetate in plant physiology, consult the 1982 volume[46] and the series of conference proceedings from 1986 to 1996 published by the University of Geneva.[47,48]

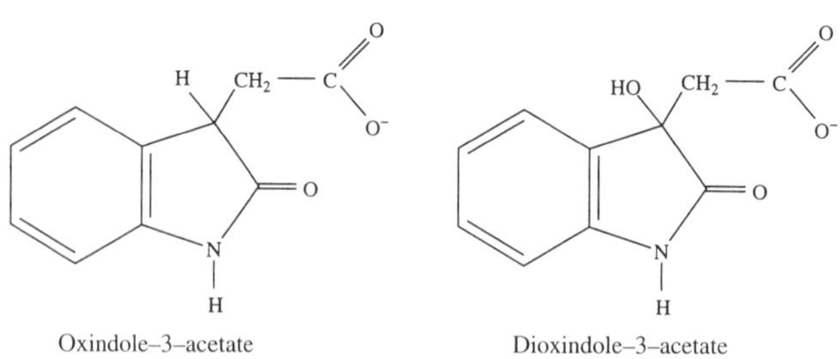

FIG. 6.6. Nondecarboxylated products of the reaction of indoleacetate with horseradish peroxidase. Since decarboxylated products also are found, there must be more than one reaction mechanism.

Indoleacetic acid is synthesized either from tryptophan or a tryptophan precursor.[4] It is stored either in an ester or amide form. The esters are sugar conjugates and the amides either an amino acid or peptide conjugate. Little is known about the enzymes that catalyze the release of indoleacetate.

Studies on the anionic peroxidase in transgenic tobacco plants shows that growth is inhibited in plants that overproduce the enzyme and is greater than normal in underproducers.[49] The conclusion is that the primary function of the anionic peroxidase is in defence mechanisms against disease or insect predators and in wound healing. The destruction of indoleacetate is a secondary consequence of the enzyme activity.[49] Whether this conclusion is valid for other peroxidases remains to be determined.

The mechanism of indoleacetate degradation by peroxidases is very complicated, and the primary degradative regulation of indoleacetate levels by nonheme enzymes is a virtually unexplored possibility. Since lignin biosynthesis, required for growth, is catalyzed by peroxidases, and degradation of indoleacetate, which inhibits growth, is also catalyzed by peroxidases, either a delicate balance is required or the functions are separated by compartmentalization within the cell. Calcium ion plays an important role in cell wall binding of peroxidases.[50–52]

6.2. OTHER LIGHT-EMITTING PEROXIDASE REACTIONS

Light emission from the peroxidase-catalyzed aerobic oxidation of indole-3-acetic acid was described in the previous section. The complicated emission patterns from the indoleacetic acid are not well understood, partly because they are not part of the main reaction pathways. In this section, two different types of light-emitting reactions catalyzed by horseradish peroxidase are described, which are part of the main reaction pathway, and which are much better understood. First, some background information.

The oxidation of luciferin [*D*-(−)-2-(6′-hydroxy-2′-benzothiazolyl)-Δ^2-thiazoline-4-acid][53] and related compounds[54,55] causes strong light emission, reactions utilized in fascinating ways by some organisms. The light emission can be turned on and off, and selected wavelengths for emission maxima can be chosen. Thus, both time and color can be used to signal other members of the same species. Furthermore, predators have learned to mimic these signals. Thus the female *Photuris* species of firefly can generate the appropriate response light signal to that used by *Photinus* males as a sexual attractant, lure the innocent male to come hither for the biggest and last surprise of his life. He is devoured on the spot.[56]

The enzyme luciferase[57] is utilized by the North American firefly to oxidize luciferin, the structure of which is shown in Fig. 6.7. The enzymatic reaction requires adenosine 5′-triphosphate (ATP), O_2 and Mg^{2+}, so the enzyme is not a peroxidase.[58] An abbreviated light emission scheme for luciferin is shown in Fig. 6.7, including a postulated dioxetanone intermediate.[59] Extracts of

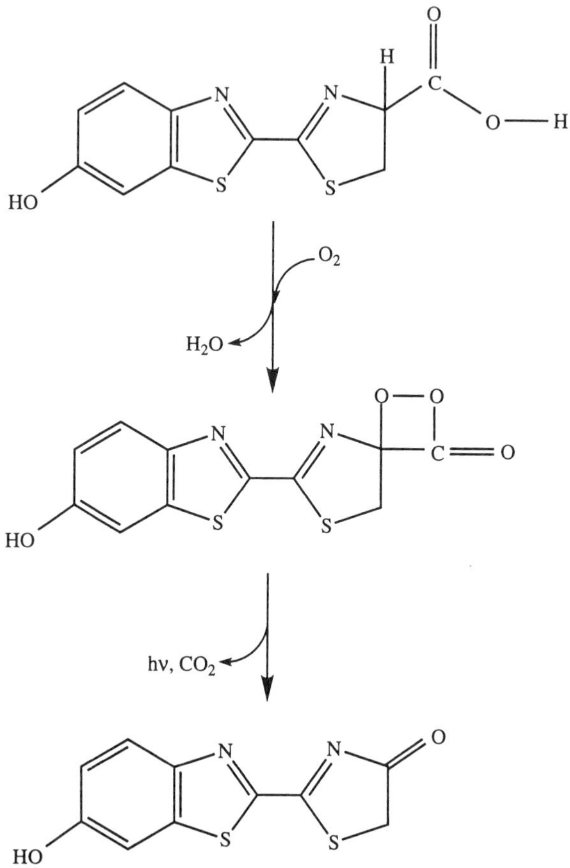

FIG. 6.7. Structure of luciferin, top, and a schematic pathway of its light-emitting reaction.

the marine acorn worm, *Balanaglossus biminiensis,* were found to oxidize a luciferin, and hydrogen peroxide was required.[60] The luciferase was found to function as a peroxidase, and horseradish peroxidase could be substituted effectively. 5-Amino-2,3-dihydro-phthalazine-1,4-dione could be substituted for the luciferin, and there was no oxygen requirement.[61]

6.2.1. Light Emission from Luminol Catalyzed by Horseradish Peroxidase

It is now known that it is the monoanion of 5-amino-2,3-dihydro-phthalazine-1,4-dione that reacts to emit light, and this monoanion is now the species that is called luminol. The structures of luminol and its parent acid form are shown in Fig. 6.8. The sodium salt of luminol, called sodium luminol, can be prepared in highly purified form that provides the best source of high-intensity, long-lasting emission.[62] The peroxidase kinetics of the oxidation of luminol have been well studied,[61] as have the kinetics of the subsequent reactions leading to luminescence.[63–65]

The species involved in the main pathways from luminol to luminescence are shown in Fig. 6.9. Luminol, LH^-, reacts with both compounds I and II of horseradish peroxidase yielding a hydrogen atom and being converted to the anion radical $L\bullet^-$. I have shown the hydrogen atom abstraction to occur from the 5-amino group, by analogy with the reactions of other aromatic amines. A tautomeric equilibrium would then yield $L\bullet^-$. This anion radical can then react by three pathways. First it can yield an electron to compound I to be converted to the species L. (If the phenol conjugate acid of the anion radical,$LH\bullet$, is the reactive species, then the substrate would be a hydrogen atom donor, typical of both phenols and aromatic amines.) Although the reaction of $L\bullet^-$ with compound II has not been demonstrated, there appears to be no reason why it cannot occur. Second, $L\bullet^-$ can react with molecular oxygen to form L and superoxide anion. Third, it can react with hydroperoxyl radical (the conjugate acid of superoxide) to yield 3-aminophthalic acid, the light emitter, and N_2. Thus, although oxygen is not required for luminescence, it provides an alternative pathway via superoxide formation. The species L can react directly with hydrogen peroxide to yield the same products as $L\bullet^-$ and hydroperoxyl radical, including the light emission. The peroxide adduct, shown as the final structure in Fig. 6.9, LO_2H^-, is described as an intermediate species that decomposes to the final products: 3-aminophthalic acid (AP), nitrogen, and light.[64] It would appear more likely to be a transition state species, rather than an intermediate.

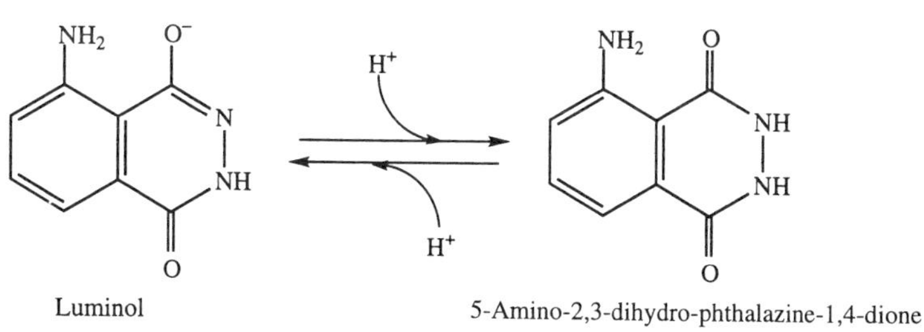

FIG. 6.8. Structure of luminol and its conjugate acid.

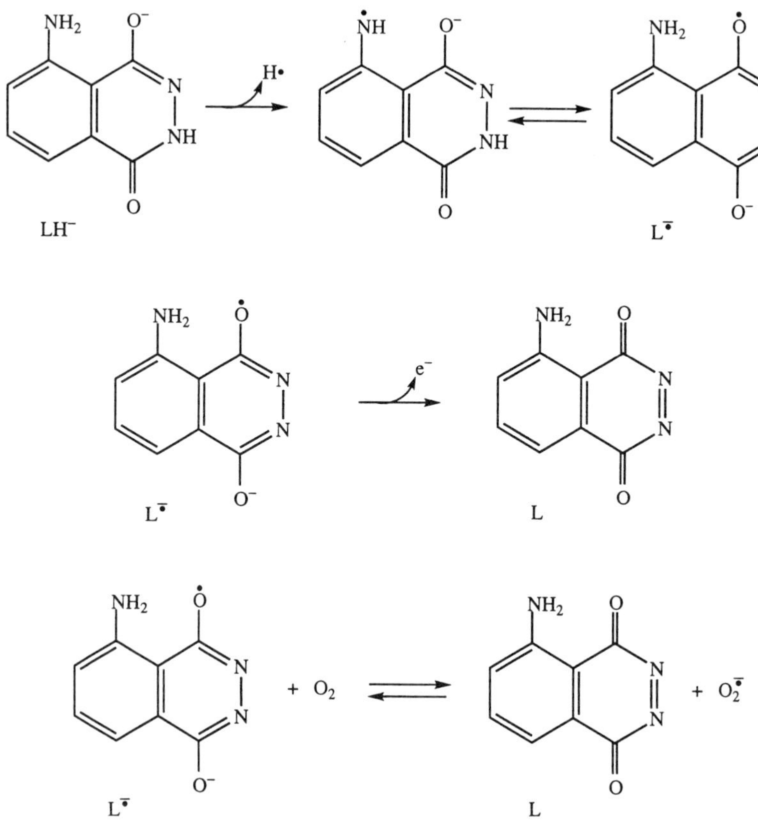

(A)

FIG. 6.9. Reaction pathway for the emission of light catalyzed by the reaction of horseradish peroxidase with luminol, LH^-. AP is 3-aminophthalic acid. The last structure is a postulated intermediate or transition-state species that decomposes to emit light. Figure continued on next page.

A summary of all of the elementary reactions, their rate constants, and references is given in Table 6.2.[66] Hydroxide ion converts L into a non-emitting species, amino-2-formyl benzoate (AFB). The conjugate acid LH• of $L\bullet^-$ is converted into LH^- by $O_2\bullet^-$.

6.2.2. Enhanced Light Emission from Luminol

Some phenols can enhance the luminescence of the peroxidase-catalyzed oxidation of luminol, and others can inhibit, effects that have been intensively investigated.[67,68] Earlier explanations were that enhancer phenols react faster with the horseradish peroxidase compounds than is predicted by Hammett

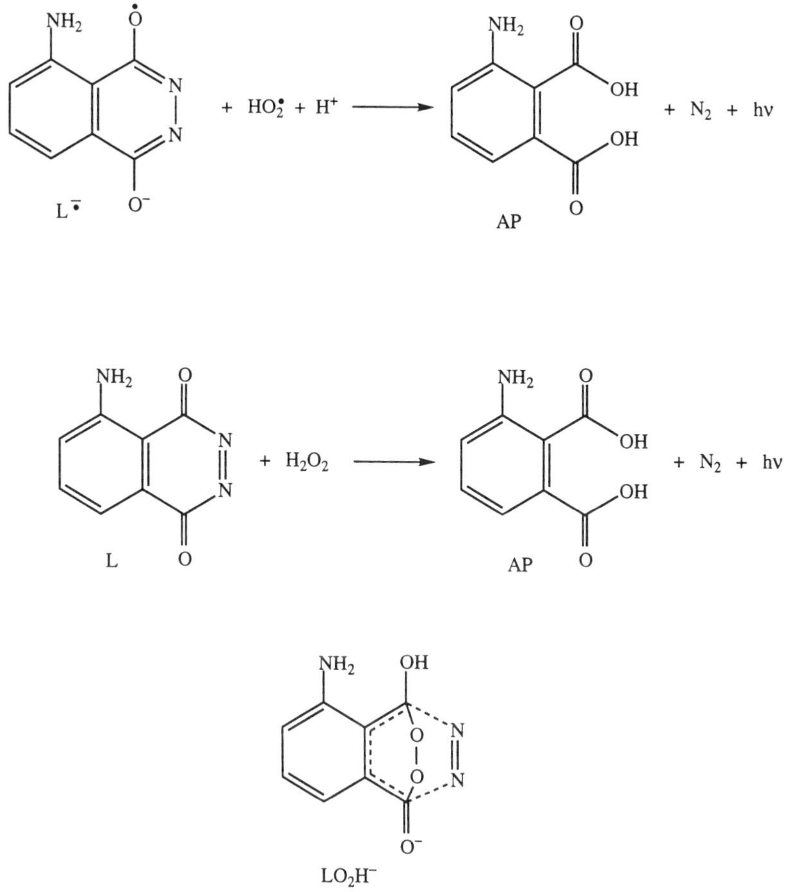

(B)

FIG. 6.9. (*Continued*)

correlations, and that they reduce compound II faster than does luminol so that the enzyme recycles faster[69] (see Section 7.1). Inhibitor molecules were believed to approach close to the active site, blocking the reaction of luminol.[70] The latest and so far, at least, the best explanation is based upon redox potentials of the pairs luminol and its anion radical, and the phenol and its phenoxyl radical.[71] A reversible equilibrium occurs:

$$LH^- + Ph{-}O\bullet \rightleftharpoons L\bullet^- + Ph{-}OH$$

The efficiency of luminescent enhancement versus reduction potential of the phenoxyl radical was a bell-shaped curve with a maximum at approximately 0.8 V. Enhanced conversion of luminol LH^- into its anion radical $L\bullet^-$ enhances

TABLE 6.2. Reactions Involved in the HRP-catalyzed Luminescence of Luminol.[a]

No.	Reaction	Rate Constants	Ref.
1	HRP + ROOH → HRP-I + ROH	$k_1 = 1.8 \times 10^7\ M^{-1}s^{-1}$	95
2	HRP-I + LH^- → HRP-II + $L\bullet^-$	$k_2 = 2.3 \times 10^6\ M^{-1}s^{-1}$	61
3	HRP-II + LH^- → HRP + $L\bullet^-$	$k_3 = 7.2 \times 10^4\ M^{-1}s^{-1}$	61
4	HRP-I + $L\bullet^-$ → HRP-II + L	$k_4 = 1.3 \times 10^6\ M^{-1}s^{-1}$	61
5	HRP-II + $L\bullet^-$ → HRP + L	$k_5 = ?$	
6	$L\bullet^- + O_2 \rightarrow L + O_2\bullet^-$	$k_6 = 5.5 \times 10^2\ M^{-1}s^{-1}$	65
7	$L + O_2\bullet^- \rightarrow L\bullet^- + O_2$	$k_7 = 2.5 \times 10^9\ M^{-1}s^{-1}$	65
8	$L\bullet^- + HO_2\bullet \rightarrow LO_2H^-$	$k_8 = 2.3 \times 10^8\ M^{-1}s^{-1}$	65
9	$L + H_2O_2 \rightarrow LO_2H^- + H^+$	$k_9 = 5.0 \times 10^7\ M^{-1}s^{-1}$	65
10	$LO_2H^- \rightarrow AP + N_2 + h\nu$	$k_{10} = 2.0 \times 10^5\ M^{-1}s^{-1}$	64
11	$2L\bullet^- + H^+ \rightarrow L + LH^-$	$k_{11} = 5.0 \times 10^8\ M^{-1}s^{-1}$	65
12	$LH\bullet + O_2\bullet^- \rightarrow LH^- + O_2$	$k_{12} = 7.1 \times 10^9\ M^{-1}s^{-1}$	64
13	$L + OH^- \rightarrow AFB$	$k_{13} = 4.0 \times 10^6\ M^{-1}s^{-1}$	65

[a] Abbreviations are defined in Section 6.2.1.

luminescence since the anion radical is the precursor of light emission. Quencher phenols are efficient scavengers of L^-. Although enhancers react faster with peroxidase compounds I and II than does luminol, the correlation with rate is not as good as with reduction potential. Substances other than phenols are also enhancers of luminol bioluminescence.[72]

6.2.3. Peroxidase-Catalyzed Light Emission from Isobutyraldehyde

Intense light emission can be observed from the reaction of isobutyraldehyde with horseradish peroxidase. All that is to required is to mix the two reactants together in an aerated solution and light emission can be observed for up to 10 min. The light emitter is triplet state acetone,[73,74] a fact not easily accepted in some quarters because it was felt that solvent would quench the emission. Perhaps the answer is that the enzyme provides some type of protection against quenching. The basic steps in the light emission process are shown in Fig. 6.10.

There is a keto-enol tautomerism for isobutyraldehyde, and it is the enol form from which hydrogen atom abstraction occurs with compounds I and II of horseradish peroxidase. (In addition there is an unreactive hydrate form of the aldehyde, for which a concentration correction is readily applied.) The keto-enol tautomerism is catalyzed by phosphate buffer.[75] There is an analogy between the reaction mechanisms of enols with that of phenols.

A key question is what initiates the peroxidase cycle? The answer is the autoxidation of isobutyraldehyde to produce a peracid, the mechanism of which is well understood.[76] The peracid oxidizes native enzyme to compound I and a normal peroxidase cycle follows. If the isobutyraldehyde is very carefully

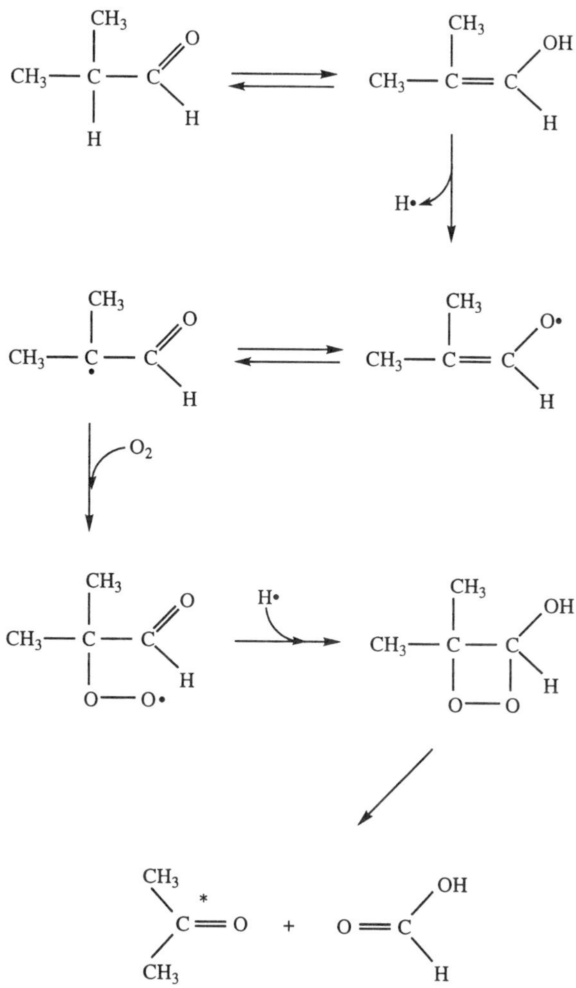

FIG. 6.10. Reaction pathway for light emission in the reaction of isobutyraldehyde with horseradish peroxidase. The enol form of the aldehyde reacts with both compounds I and II of the enzyme to form the free radical partially centered on each of an O and C atom. Triplet state acetone and formic acid are the final products.

vacuum distilled and kept out of contact with air, then no reaction occurs. Hydrogen peroxide can be used to initiate the peroxidase cycle.

The product of the enol reactions with compounds I and II is an enol (oxygen centered) radical, which exists in tautomeric equilibrium with an α-carbon free radical. The latter is rapidly scavenged by molecular oxygen to form a peroxyl radical, which abstracts a hydrogen atom from an unknown source to form a peroxide. The peroxide may cyclize to a dioxetane intermediate that decomposes to formic acid and the light emitter, triplet state acetone.

TABLE 6.3. Equilibrium Constant and Rate Constants for Keto-enol Tautomerism of Isobutyraldehyde.[a]

K_{enol} = [enol]/[keto]	$(1.7 \pm 0.3) \times 10^{-4}$
k_{enol}	$(8.7 \pm 0.6) \times 10^{-5}\ M^{-1}s^{-1}$
k_{keto}	$(5 \pm 1) \times 10^{-1}\ M^{-1}s^{-1}$
k_2 (compound I + enol)	$(8 \pm 1) \times 10^{6}\ M^{-1}s^{-1}$
k_3 (compound II + enol)	$(1.3 \pm 0.3) \times 10^{6}\ M^{-1}s^{-1}$

[a] Rate constants for enol reactions with compounds I and II. All data at 35°C, pH 7.4, phosphate buffer 0.1 M, and total ionic strength 0.67 M.
Reproduced with permission from C. Bohne, I. D. MacDonald, and H. B. Dunford, *J. Am. Chem. Soc.* **108,** 7867–7868 (1986).

If the peroxyl radical abstracts a hydrogen atom from the aldehyde, it could contribute to further autoxidation. If abstraction is from the enzyme, the enzyme must have a good repair mechanism to be able to catalyze the reaction for a long period of time. Some key parameters for the isobutyraldehyde reaction are listed in Table 6.3.[77] Compound I of horseradish peroxidase can be used to measure both rates and equilibria of keto-enol tautomerism of aldehydes.[78]

6.2.4. Other Peroxidase-Catalyzed Light-Emitting Reactions

Pyrogallol,[79] linear aldehydes,[80] phenylacetic aldehyde,[81] tenoxicam,[82] and *N*-methylcarbazole[83] are all examples of compounds that produce light-emitting species when reacted with horseradish peroxidase. However, none of these reactions has been as intensively studied as the luminol and isobutyraldehyde reactions described above, and will not be discussed here.

6.3. QUANTITATIVE STRUCTURE–ACTIVITY RELATIONSHIPS

6.3.1. Introduction

The original quantitative structure–activity relationship (QSAR) was the $\rho\sigma$ correlation established by Louis Hammett for derivatives of benzoic acid.[84] He based his correlation entirely on experimentally measured constants. The original name was a linear free energy relationship (LFER), a name that did not include the activity, or kinetic, half of the correlation. The constant σ was obtained from the relation:

$$\log (K/K_0) = \sigma$$

where K_0 is the ionization constant of benzoic acid and K of a benzoic acid derivative with a substituent in the para or meta position of the benzene ring. Ortho substituents usually have a dominant steric effect whereas the quantitative correlations for para and meta groups are electronic in nature. Thus, a derivative with an electron-withdrawing substituent that further polarized the O—H bond of the carboxyl group, and hence increased the acidity of the compound, would have a positive σ value. An electron-donating group would have a negative σ value. In this way values were tabulated for meta and para substituents. Hammett then went on to measure the rate constants for hydrolysis of esters of the corresponding benzoic acid derivatives. He established the correlation:

$$\log (k/k_0) = \rho\sigma$$

where k_0 is the rate constant for hydrolysis of the ester of benzoic acid, and k the rate constant for the same ester of the substituted benzoic acid. Upon plotting $\log (k/k_0)$ versus σ for a family of esters, a straight line was obtained, the slope of which gave the value of ρ. A positive value of ρ indicates that electron-withdrawing substituents accelerate the rate of the reaction, and a negative value that electron-donating groups accelerate the rate.

Hammett then went on to establish linear correlations for many other reactions.[85] His accomplishment was truly remarkable since often the correlations were valid over millionfold changes in rate. By 1953 there were over 400 applications of the Hammett equation,[86] and by 1979 between 5000 and 10,000.[87] The original Hammett equation was applied to reactions centered on the side chain of an aromatic ring, where resonance interactions between the ring substituent and the reaction center do not occur.

The aromatic ring is electron-rich and so an attacking reagent is electrophilic. The Hammett relation was found to be valid for rates of direct attack on aromatic rings with meta substituents, but not in the para position. The reason is that resonance interactions with the ring reaction center are negligible for meta substituents but can be very large for groups in the para position. A clever way to correct for the para effect was devised by H.C. Brown.[88] By determining an accurate value of ρ based entirely on reactions for compounds with meta substituents and rates of reaction for para-substituted compounds, he used the equation

$$\sigma^+ = 1/\rho \log (k/k_0)$$

to calculate a new type of sigma value, σ^+, which corrected for the resonance effect of para substituents. Then by comparing known σ^+ and σ values, he devised an empirical correlation form which unknown σ^+ values could be calculated from known σ constants.[88] The results led to a different linear free energy correlation now known as the Brown–Okamoto relation:

$$\log (k/k_0) = \rho\sigma^+$$

It can be seen that the Brown–Okamoto equation is a close relative of the Hammett equation, devised to obtain QSARs for reactions occurring on aromatic rings.

Because many aromatic compounds have limited solubility in water, often mixed solvents have been necessary to study QSARs. A debate occurred about whether different sets of the various types of σ values should be obtained for different solvent systems. The practical solution, used by the person most responsible for the latest tabulations of σ values, Dr. Corwin Hansch,[88,89] was to take average values, with occasional judicious choices of best values, which can be applied universally. The large gain in convenience more than offsets the small loss in accuracy.

A different type of structural parameter measures hydrophobicity. A hydrophobic molecule by definition is not very water-soluble but has high solubility in a nonpolar solvent. From a large number of trial-and-error experiments, it was concluded that *n*-octanol/water provides a good universal model system to determine hydrophobicity.[89,90] A distribution coefficient P can be measured that defines the ratio of solubility in octanol compared to the solubility in water. From the distribution coefficients, hydrophobicity parameters π for substituents are defined as follows:

$$\pi = \log (P/P_0)$$

where P_0 is the distribution of the unsubstituted parent compound, and P is the distribution coefficient of the compound with a substituent group. A positive value of π indicates a preference of a substituent for the octanol phase compared to water, and a negative π indicates a preference for the water phase compared to water. Hydrophobicity parameters provide a valuable tool in the investigation of membrane reactions and reactions in enzyme hydrophobic active sites. Interestingly, some substituents exhibit a hydrophobic effect in one position and none in another.

The above sojourn into physical organic chemistry provides background to appreciate QSARs for reactions of peroxidases with aromatic compounds.

6.3.2. Reactions of Peroxidases with Phenols and Aromatic Amines

The first study of a QSAR for reactions of a peroxidase with oxidizing substrates was published in 1976.[91] Rate constants of horseradish peroxidase compound I with 15 phenols and nine aromatic amines were obtained over the pH range of 6.4–7.8. The results were correlated using the Hammett $\rho\sigma$ relation, and ρ values of -6.9 for phenols and -7.0 for anilines were obtained at pH 7.0. *p*-Chlorophenol and 3-methyl-4-chlorophenol reacted faster than predicted by the Hammett relation, and *p*-aminophenol reacted slower, probably because the physical process of diffusion control was becoming dominant. The rate constant for the *p*-amino compound was $2.3 \times 10^8\ M^{-1}s^{-1}$. In 1979 better σ values were published,[87] and a reanalysis of the same date yielded a smaller value of ρ, −5.5, for compound I reacting with phenols.[92] The Hammett correlation for compound I reactions has stood the test of time. One would expect electron-donating substituents on the phenols to accelerate the rate of the reaction, which is electron transfer from the phenol to compound I, and therefore the value of ρ would be negative. The very large negative value of ρ indicates high sensitivity of rate to electron-donating substituents. Hammett correlations have also been obtained for horseradish peroxidase compound II reactions,[93,94] and a ρ value of −4.5 was obtained.[93]

REFERENCES

1. K. V. Thimann, *J. Gen. Physiol.* **18,** 23–34 (1934).
2. P. Larsen, *Planta* **25,** 311–314 (1936).
3. S. A. Gordon, *Ann. Rev. Plant Physiol.* **5,** 341–378 (1954).
4. R. S. Bandurski, J. D. Cohen, J. P. Slovin, and D. M. Reinecke in *Plant Hormones,* 2nd ed., P. J. Davies, ed., Kluwer Academic, Dordrecht, 1995, pp. 39–65.
5. R. H. Kenten, *Biochem. J.* **59,** 110–121 (1955).
6. R. L. Hinman and J. Lang, *Biochemistry* **4,** 144–158 (1965).
7. H. J. Grambow and B. Langenbeck-Schwich, *Planta* **157,** 131–137 (1983).
8. J. N. BeMiller and W. Colilla, *Phytochem.* **11,** 3393–3402 (1972).
9. S. Kobayashi, K. Sugioka, H. Nakano, M. Nakano, and S. Tero-Kubota, *Biochemistry* **23,** 4589–4597 (1984).
10. M. Acosta, M. B. Arnao, J. A. del Río, and F. García-Cánovas, *Biochim. Biophys. Acta* **996,** 7–12 (1989).
11. L. R. Fox, W. K. Purves, and H. I. Nakada, *Biochemistry* **4,** 2754–2763 (1965).
12. D. A. Gelinas, *Plant Physiol.* **51,** 967–972 (1973).
13. J. Ricard and J. Nari, *Biochim. Biophys. Acta* **132,** 321–329 (1967).
14. J. Ricard, D. Job, M. Santimone, and H. Cambe, *C.R. Acad. Sci. Paris* **269,** 1445–1448 (1969).
15. J. Ricard and D. Job, *Eur. J. Biochem.* **44,** 359–374 (1974).
16. H. Yamazaki and I. Yamazaki, *Arch. Biochem. Biophys.* **154,** 147–159 (1973).
17. R. Nakajima and I. Yamazaki, *J. Biol. Chem.* **254,** 872–878 (1979).
18. R. Nakajima and I. Yamazaki, *J. Biol. Chem.* **255,** 2067–2071 (1980).
19. A. M. Smith, W. L. Morrison, and P. J. Milham, *Biochemistry* **21,** 4414–4419 (1982).
20. H. B. Dunford, *Quimíca Nova* **16,** 350–353 (1993).
21. S. N. Krylov and H. B. Dunford, *J. Phys. Chem.* **100,** 913–920 (1996).
22. S. N. Krylov and H. B. Dunford, *Biophys. Chem.* **58,** 325–334 (1996).
23. C. Mottley and R. P. Mason, *J. Biol. Chem.* **261,** 16860–16864 (1986).
24. L. P. Candeias, L. K. Folkes, M. F. Denis, K. B. Patel, S. A. Everett, M. R. L. Stratford, and P. Wardman, *J. Phys. Chem.* **98,** 10131–10137 (1994).

25. L. P. Candeias, L. K. Folkes, M. Porssa, J. Parrick, and P. Wardman, *Biochemistry* **35,** 102–108 (1996).
26. S. N. Krylov and H. B. Dunford, *Photochem. Photobiol.* **63,** 735–741 (1996).
27. S. N. Krylov and H. B. Dunford, *J. Phys. Chem.* **100,** 19719–19727 (1996).
28. M. B. Arnao, J. Sanchez-Bravo, and M. Acosta in *Plant Peroxidases: Biochemistry and Physiology,* K. G. Welinder, S. K. Rasmussen, C. Penel, and H. Greppin, eds., University of Geneva Press, Geneva 1993, pp. 181–184.
29. D. A. Gelinas, *Plant Physiol.* **51,** 967–972 (1973).
30. H. J. Grambow in *Molecular and Physiological Aspects of Plant Peroxidases,* H. Greppin, C. Penel, and Th. Gaspar, eds., University of Geneva Press, Geneva, 1986, pp. 31–41.
31. S. N. Krylov, S. M. Krylova, I. G. Chebotarev, and A. B. Chebotareva, *Phytochem.* **36,** 263–267 (1994).
32. S. N. Krylov, B. D. Aguda, and M. L. Ljubimova, *Biophys. Chem.* **53,** 213–218 (1995).
33. I. G. Gazarian, L. M. Lagrimini, G. A. Ashby, and R. N. F. Thorneley, *Biochem. J.* **313,** 841–847 (1996).
34. D. Metodiewa, M. Pires de Melo, J. A. Escobar, G. Cilento, and H. B. Dunford, *Arch. Biochem. Biophys.* **296,** 27–33 (1992).
35. M. Pires de Melo, J. A. Escobar, D. Metodiewa, H. B. Dunford, and G. Cilento, *Arch. Biochem. Biophys.* **296,** 34–39 (1992).
36. S. N. Krylov, V. V. Lazarev, and L. B. Rubin, *Dokl. Biophys.* **310,** 28–31 (1990).
37. J. R. Kanofsky, *J. Biol. Chem.* **263,** 14171–14175 (1988).
38. J. A. Escobar, J. Vásquez-Vivar, and G. Cilento, *Photochem. Photobiol.* **55,** 895–902 (1992).
39. G. Cilento, *Photochem. Photobiol. Rev.* **5,** 109–228 (1980).
40. G. Cilento, *Pure Appl. Chem.* **56,** 1179–1190 (1984).
41. G. Cilento and W. Adam, *Photochem. Photobiol.* **48,** 361–368 (1988).
42. S. N. Krylov and A. B. Chebotareva, *FEBS Lett.* **324,** 6–8 (1993).
43. S. N. Krylov and S. M. Krylova, *Photochem. Photobiol.* **63,** 621–627 (1996).
44. E. C. Schuytema, M. P. Hargie, I. Merits, J. R. Schenk, D. J. Siehr, M. S. Smith, and E. L. Varner, *Biotech. Bioeng.* **8,** 275–286 (1966).
45. K. Dziewanowska and S. Lewak, *Biologia Plantarum* **29,** 110–117 (1987).
46. Th. Gaspar, C. L. Penel, T. Thorpe, and H. Greppin, eds., *Peroxidases,* 1970–1980, University of Geneva Press, Geneva 1982.
47. H. Greppin, C. Penel, and Th. Gaspar, eds. *Molecular and Physiological Aspects of Plant Peroxidases,* University of Geneva Press, Geneva 1986.
48. C. Obinger, U. Burner, R. Ebermann, C. Penel, and H. Greppin, eds. *Plant Peroxidases. Biochemistry and Physiology,* University of Geneva Press, Geneva 1996.
49. L. M. Lagrimini in Ref. 48, pp. 235–242.
50. C. Penel, H. Greppin, *Phytochem.* **18,** 29–33 (1979).
51. S. Kiefer, C. Penel, and H. Greppin, *Plant Sci.* **39,** 37–43 (1985).
52 C. Penel, M. Crevecoeur and H. Greppin in Ref. 48, pp. 259–263.
53. E. H. White, F. McCapra, and G. F. Field, *J. Am. Chem. Soc.* **85,** 337–343 (1963).

54. J. W. Hastings, *Ann. Rev. Biochem.* **37,** 597–630 (1968).
55. M. A. DeLuca and W. D. McElroy, eds., *Bioluminescence and Chemiluminescence* in *Methods in Enzymology,* Academic, New York, Vol. 133B, 1986, pp. 3–603.
56. W. Agosta, *Bombardier Beetles and Fever Trees.* Addison-Wesley, Reading, MA, 1996, pp. 43–44.
57. M. DeLuca and W. D. McElroy in *Methods in Enzymology,* M. A. DeLuca, ed., Vol. LVII, Academic, New York, 1978, pp. 3–15.
58. M. J. Cormier and J. R. Totter, *Ann. Rev. Biochem.* **33,** 431–458 (1964).
59. F. McCapra, *Chem. Commun.* 155–156, (1968).
60. L. S. Dure and M. J. Cormier, *J. Biol. Chem.* **238,** 790–793 (1963).
61. M. J. Cormier and P. M. Prichard, *J. Biol. Chem.* **243,** 4706–4714 (1968).
62. R. A. W. Stott and L. J. Kricka in *Bioluminescence and Chemiluminescence. New Perspectives,* J. Schölmerich, R. Andreesen, A. Kapp, M. Ernst, and W. G. Woods, eds., Wiley, New York, 1987, pp. 237–240.
63. G. Merényi, and J. Lind, *J. Am. Chem. Soc.* **102,** 5830–5835 (1980).
64. J. Lind, G. Merényi, and T. E. Eriksen, *J. Am. Chem. Soc.* **105,** 7655–7661 (1983).
65. G. Merényi, J. Lind, and T. E. Eriksen, *J. Phys. Chem.* **88,** 2320–2323 (1984).
66. A. Lundin and L. O. B. Hollander, in Ref. 62, pp. 555–558.
67. G. H. G. Thorpe, L. J. Kricka, S. B. Moseley, and T. P. Whitehead, *Clin. Chem.* **31,** 1335–1341 (1985).
68. G. H. G. Thorpe and L. J. Kricka in Ref. 62, pp. 199–208.
69. M. Hodgson and P. Jones, *J. Biolumin. Chemilumin.* **3,** 21–25 (1989).
70. S. B. Vlasenko, A. A. Arefyev, A. D. Klimov, B. B. Kim, E. L. Gorovits, A. P. Osipov, E. M. Gavrilova, and A. M. Yegorov, *J. Biolumin. Chemilumin.* **4,** 164–176 (1989).
71. P. Easton, A. Simmonds, A. Rakishev, A. M. Egorov, and L. P. Candeias, *J. Am. Chem. Soc.* **118,** 6619–6624 (1996).
72. W. Sun, X. Ji, L.J. Kricka, and H. B. Dunford, *Can. J. Chem.* **72,** 2159–2162 (1994).
73. N. Durán, O. M. M. Faria Oliveira, M. Haun, and G. Cilento, *J. Chem. Soc. Chem. Commun.* 442–443 (1977).
74. E. J. H. Bechara, O. M. M. Faria Oliveira, N. Durán, R. Casadei de Baptista, and G. Cilento, *J. Biol. Chem.* **253,** 4707–4712 (1978).
75. W. J. Baader, C. Bohne, G. Cilento, and H. B. Dunford, *J. Biol. Chem.* **260,** 10217-10225 (1985). Correction: *J. Biol. Chem.* **261,** 2994 (1986).
76. G. E. Zaikov, J. A. Howard, and K. U. Ingold, *Can. J. Chem.* **51,** 582–587 (1973).
77. C. Bohne, I. D. MacDonald, and H. B. Dunford, *J. Biol. Chem.* **262,** 3572–3578 (1987).
78. C. Bohne, I. D. MacDonald, and H. B. Dunford, *J. Am. Chem. Soc.* **108,** 7867–7868 (1986).
79. G. Ahnström and R. Nilsson, *Acta Chem. Scand.* **19,** 313–316 (1965).
80. A. Campa, L. Nassi, and G. Cilento, *Photochem. Photobiol.* **40,** 127–131 (1984).
81. A. L. T. O. Nascimento and G. Cilento, *Biochim. Biophys. Acta* **843,** 254–260 (1985).
82. A. L. T. O. Nascimento, J. A. Escobar, and G. Cilento, *Photochem. Photobiol.* **57,** 362–366 (1993).

83. M. Pires de Melo and G. Cilento, *Photochem. Photobiol.* **59,** 677–682 (1994).
84. L. P. Hammett, *Chem. Rev.* **17,** 125–136 (1935).
85. L. P. Hammett, *Physical Organic Chemistry,* McGraw-Hill, New York, 1940, pp. 184–199.
86. H. H. Jaffé, *Chem. Rev.* **53,** 191–261 (1953).
87. C. Hansch and A. Leo, *Substituent Constants for Correlation Analysis in Chemistry and Biology,* Wiley, New York, 1979, p. 2.
88. H. C. Brown and Y. Okamoto, *J. Am. Chem. Soc.* **80,** 4979–4987 (1958).
89. T. Fujita, J. Iwasa, and C. Hansch, *J. Am. Chem. Soc.* **86,** 5175–5180 (1964).
90. A. Leo, C. Hansch, and D. Elkins, *Chem. Rev.* **71,** 525–626 (1971).
91. D. Job and H. B. Dunford, *Eur. J. Biochem.* **66,** 607–614 (1976).
92. H. Zhang and H. B. Dunford, *Can. J. Chem.* **71,** 1990–1994 (1993).
93. H. B. Dunford and A. J. Adeniran, *Arch. Biochem. Biophys.* **251,** 536–542 (1986).
94. J. Sakurada, R. Sekiguchi, K. Sato, and T. Hosoya, *Biochemistry* **29,** 4093–4098 (1990).
95. H. B. Dunford, W. D. Hewson, and H. Steiner, *Can. J. Chem.* **56,** 2844–2852 (1978).

CHAPTER 7

Spectroscopy of Horseradish Peroxidase. I: Optical, Resonance Raman, Magnetic Circular Dichroism, X-ray Absorption, and Diffraction

Peroxidases cannot be understood without the application of spectroscopic techniques. The common theme in this chapter is the interaction with peroxidases of photons from all but the microwave and radio-frequency regions of the electromagnetic spectrum. Brief introductions are given for each technique, followed by applications to horseradish peroxidase. Magnetic circular dichroism, which combines use of light and a magnetic field, is described in this chapter. In the following chapter various other magnetic spectroscopic methods are described, as well as Mössbauer spectroscopy, which complements electron paramagnetic resonance studies.

7.1. OPTICAL ABSORPTION SPECTRA

7.1.1. Introduction

Peroxidase absorption bands in the visible and near ultraviolet (Soret) region are of two types: They are due to either $\pi \rightarrow \pi^*$ or charge-transfer transitions, the nature of which is described below. First is a description of energy levels of the *d* electrons of iron, then of the π electrons of the porphyrin, and finally of the interaction between the two. The iron centered in the heme of peroxidases plays a central role in determining spectral properties.

The properties of the five *d* electrons of iron(III) with six ligands provide a good starting point for discussion. In the absence of external influence, the five *d* electrons are equivalent in energy, and according to Hund's rule they will each occupy a separate *d* orbital (Fig. 7.1*a*). When the six ligands are identical, then all bond strengths and lengths are equal, and the geometry is described as octahedral, since the central atom and its ligands can be enclosed in an eight-sided figure. The effect on the energy levels of the *d* electrons is

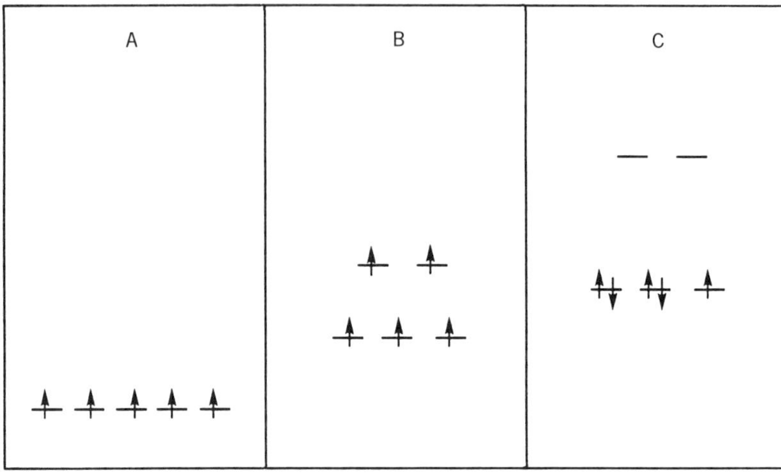

FIG. 7.1. Effect of weak and strong ligand fields on the five *d* electrons of iron(III) in an octahedral ligand field. (A) No field, (B) weak ligand field, (C) strong ligand field.

most simply explained in terms of crystal field theory. Each ligand has a lone pair of electrons available for bonding to the iron, and the lone pairs repel the *d* electrons, with different strengths of repulsion, depending upon the location of the orbital occupied by each *d* electron (Fig. 7.1*b*). The ligands are located on the *x, y,* and *z* axes. The electrons in *d* orbitals ($d_{x^2-y^2}$ and d_{z^2}), located along the axes of approach of the ligands, will be the more strongly repelled than those located off the axes (d_{xy}, d_{xz}, and d_{yz}). If the repulsion energy is not too strong, then Hund's rule still applies, since the energy saved by an electron dropping into a lower energy orbital is more than offset by the repulsion energy of two *d* electrons occupying the same orbital. This case, illustrated in Fig. 7.1*b*, is called a weak ligand field. If the ligand field is strong, then the *d*-orbital energies are more widely separated, and electrons locate themselves in the orbitals of lowest possible energy (Fig. 7.1*c*). Thus, in a weak ligand field, ferric ion has five unpaired electrons (high spin), which corresponds to a spin-only magnetic moment of 5.92 Bohr magnetons. In a strong ligand field there is only one unpaired electron (low spin), which corresponds to a spin–orbital magnetic moment in the range 1.7–2.8.

If the iron has four identical ligands in a plane, with two other ligands at right angles, then a tetragonal rather than an octahedral field exists. The tetragonal field provides an appropriate description for the iron of peroxidases, whether five- or six-coordinate. The four pyrrole nitrogen atoms of the heme are equivalent, but the influence of the fifth and sixth coordination sites are different. For determining whether the iron of a peroxidase complex is low-

or high-spin, the simple crystal field arguments used to construct Fig. 7.1 are satisfactory. However, for a discussion of spectral properties, one must switch to a more realistic *d*-orbital energy level diagram (Fig. 7.2).

The four pyrrole nitrogen atoms are located in the *xy* plane, which also contains the $d_{x^2-y^2}$ and d_{xy} orbitals of the iron. Electrons in the $d_{x^2-y^2}$ orbital are pointed toward the bonding pair of electrons in each pyrrole nitrogen atom, and hence are strongly repelled; those in the d_{xy} orbital less so. The d_{xz} and d_{yz} orbitals are also called d_π orbitals since they are oriented along the *z* axis at right angles to the plane of the four pyrrole nitrogen atoms. Electrons in the d_π orbitals are repelled a bit more strongly than those in the d_{xy} orbitals. The d_{z^2} orbital is sensitive to the fifth and sixth ligands, both of which are located along the *z* axis. In the case of the native ferric enzyme, the sixth ligand is largely absent and so repulsion occurs only from the proximal histidine bonding pair of electrons. Thus, the energy of the d_{z^2} orbital is lower than that of the $d_{x^2-y^2}$ orbital (Fig. 7.2). If one considers only the immediate environment of the ferric ion in native peroxidase, then the system is described as having pseudo-tetragonal geometry. This has been simplified to the notation for square planar geometry where the symmetry is D_{4h}. The labeling of the energy levels on the left of Fig. 7.2 is for D_{4h} symmetry. It turns out that D_{4h} symmetry is also a fairly adequate description for the entire heme, even though the periphery of the heme does not contain symmetrical substituents.

Two of the resonance structures of the conjugated π system of an iron heme are shown in Fig. 7.3. There are three electronic bands observed, which are π-π^* transitions. In order of increasing energy and intensity, these bands are called Q_{00} or α, Q_{0v} or β, and B or Soret (or γ in the

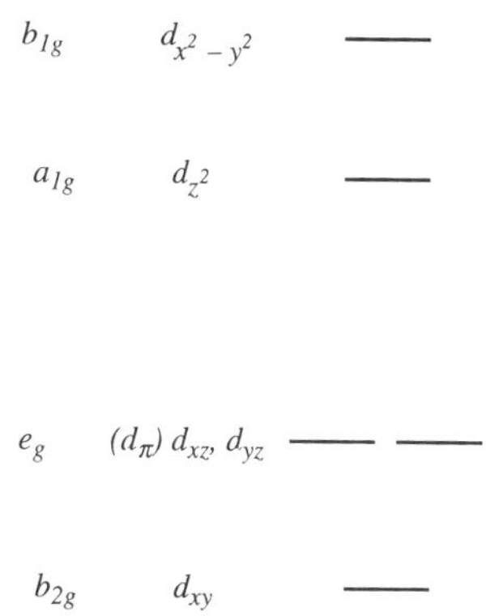

FIG. 7.2. Further splitting of the five *d*-orbital energy levels, shown in Fig. 7.1, in a tetragonal ligand field, as occurs in peroxidases.

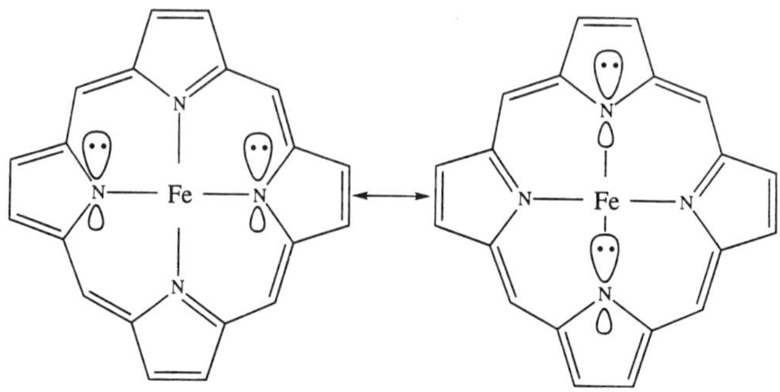

FIG. 7.3. The 18 π-electron system of an iron porphyrin. [Reproduced with permission from F. Adar in *The Porphyrins,* D. Dolphin, ed., Academic, New York, 1978, Vol. IIIA, pp. 167–209].

early literature). The Q_{00} (α) transition originates from the a_{2u} state to the upper π^* state, which is labeled e_g (Fig. 7.4). The Q_{0v} (β) band has the same origin but goes to vibrationally excited levels of the e_g state, and so are of slightly higher energy. (The Q_{0v} band was originally thought to go to the first vibrational level only of the electronically excited state, rather than to an envelope of vibrationally excited states, and so it was formerly called the Q_{01} transition.) The B (Soret) band originates from the a_{1u} ground state, goes to the same e_g excited state, and so has the highest observed energy.

If the iron is absent from the porphyrin, then two protons are attached to pyrrole nitrogen atoms, the symmetry is reduced from D_{4h} to D_{2h}, and band splitting occurs.[1] If a metal ion that has filled shells is in the center of the porphyrin, the same three α, β, and Soret bands are observed. These are called regular metalloporphyrins. When the metal ion has only partially filled d orbitals as in the case of iron, then one has an irregular metalloporphyrin and extra transitions can occur (Fig. 7.4). These are charge-transfer transitions in which electrons are transferred from the highest energy-filled orbitals of the porphyrin to the lowest energy-unfilled orbitals of the iron. In addition the porphyrin orbitals and iron orbitals interact, and stronger transitions may occur than for regular porphyrins. Thus, the α and β bands may be largely obscured by charge-transfer transitions. The optical spectra of peroxidases can also be explained in terms of the four-orbital model illustrated in Fig. 7.4. The transitions among the four orbitals, a_{1u}, a_{2u}, and e_g of the porphyrin and $d\pi$ (e_g) for the iron were first described by Gouterman,[2] who used iterative extended Hückel calculations. His model was successful, where other supposedly more sophisticated theoretical treatments could not reproduce the observed spectra.

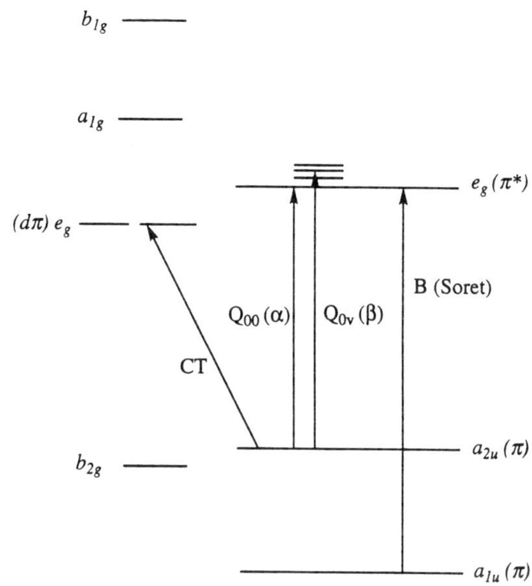

FIG. 7.4. Absorption bands typically observed for a peroxidase. CT, charge transfer; Q_{00} transitions from the zeroth vibrational level of the a_{2u} state to the zeroth vibrational level of the e_g state; Q_{0v} to excited vibrational levels. Energy levels on the left for the iron, on the right for the porphyrin. (Based upon, with permission, G. Smulevich in *Plant Peroxidases. Biochemistry and Physiology,* C. Obinger, U. Burner, R. Ebermann, C. Penel, and H. Greppin, eds., University of Geneva Press, Geneva 1996, pp. 13–19.)

7.1.2. Optical Spectra for Horseradish Peroxidase

The fifth site of the iron in native horseradish peroxidase is occupied by the proximal imidazole side chain of His170 and the sixth position is essentially vacant. (A more detailed discussion is given in the section on resonance Raman spectra.) Thus, native horseradish peroxidase has mainly a five-coordinate ferric ion. It contains a small amount of six-coordinate high-spin heme, and its band at 643 nm is sensitive to the amount of this species. Examples of pure six-coordinate iron are the fluoride complex, where the ligand field is weak, and the cyanide complex, where the ligand field is strong. Thus, the iron of the fluoride complex has five unpaired *d* electrons and the cyanide complex only one unpaired electron.

The optical spectra of native ferric horseradish peroxidase and its compounds I and II are shown in Fig. 7.5. In order of increasing energy, the readily discernible bands of the native ferric horseradish peroxidase are located at 643, 498, and 403 nm. The two of longer wavelength are charge-transfer bands.

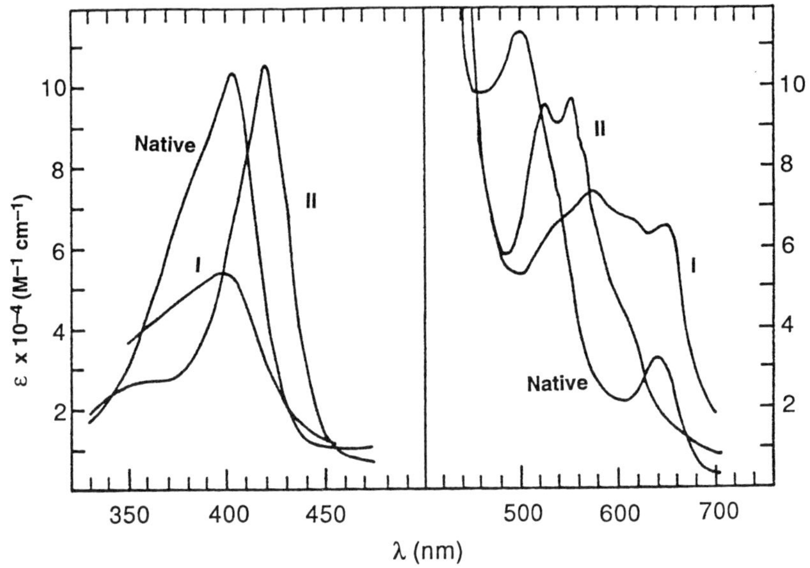

FIG. 7.5. Molar absorptivities of native horseradish peroxidase and its compounds I and II as a function of wavelength. [Reproduced with permission from H. B. Dunford, *Adv. Inorg. Biochem.* **4,** 41–68 (1982), with permission from Elsevier Science.]

A splitting of the d (e_g) energy levels may explain the existence of two charge-transfer bands. The source of a charge-transfer band may be different from an a_{2u}-filled porphyrin orbital, such as a'_{2u} (π) or b_{2u} (π).[3] The Q_{00} and Q_{0v} bands of native horseradish peroxidase are largely obscured by the charge-transfer bands. There is a shoulder to the Soret band of native horseradish peroxidase at 380 nm. To the best of my knowledge the source of this shoulder has not been explained. Charge-transfer bands disappear from the hydroxide and cyanide low-spin complexes of horseradish peroxidase. If a porphyrin π^* electron drops into a d orbital of the iron, then the d orbital is entirely filled and the charge-transfer transition is blocked (Fig. 7.4).

Wavelengths and molar absorptivities of the five bands of native enzyme are shown in Table 7.1, along with spectral data on the compounds and complexes of the enzyme in its various oxidation states. The data in Table 7.1 are from many sources. Since isoenzymes B and C appear to be identical in all respects except for electrophoretic mobility, data obtained from commercial sources was not distinguished from, for example, preparations of isoenzyme C2. However, data obtained from impure preparations [reinheitzahl (RZ) number < 3.0] were ignored. Value judgments were sometimes made where slightly different data exist in the literature. For example, since compound I is formed from native enzyme in a reaction that decreases the absorbance, the lowest molar absorptivity was chosen for the Soret maximum of compound

TABLE 7.1. Optical Absorption Bands of Horseradish Peroxidase Isoenzymes B and C, Its Complexes and Compounds[a]

Species	CT1	$Q_{00}(\alpha)$	$Q_{0v}(\beta)$	CT2	B (Soret)	Mag. Mom.	Ref.
Native HRP[b]	643 (3.23)	~580 (~2.8)	~530 (~8.4)	498 (11.25)	402.5 (102)	5.0	4–11
Fluoride	612 (7.7)	560 (5.4)	~530 sh.[c]	488 (8.5)	404 (146)	5.92	7,8,10
Azide	635 (1.9)	568 (6.3)	535 (9.2)	~495 sh.	416 (128)	2.17	7,8,12,13
Hydroxide[d]	None	571 (8.6)	542 (10.7)	None	416 (110)	2.66	7,8,14
Cyanide	None	~568 (~8.4)	539 (11.4)	None	423 (105)	2.67	6,7,8,10
Nitric oxide		572 (19.0)	534 (15.6)	~490 sh.	419 (171)		12,15
Ferrous		~585 sh.	557 (13.0)	~510 sh.	439 (85)		6,8,11
Ferrous—CO		572 (12.5)	542		423		6,8,11

TABLE 7.1. (*Continued*).

Species	CT1	$Q_{00}(\alpha)$	$Q_{0v}(\beta)$	CT2	B (Soret)	Mag. Mom.	Ref.
Ferrous–CN		566 (19)	536 (14.5)		432?		6
Ferrous–ethyl isocyanide	None	563 (12.2)	535 (14.1)	None	433 (168)		16
Ferrous–O_2 (Compound III)	None	580 (10.0)	544 (11.5)	None	417 (108)		11,17[e]
Compound II		555 (9.65)	527 (9.5)		420 (105)	3.53	5,11,12,18
Compound I	651,622 (6.6), sh.	577 (7.4)	~525 sh.		400 (53.8)	3.99	4,5,11,16,19

[a] Wavelengths of band maxima in nm, and corresponding molar absorptivities (in brackets) in units of mM^{-1} cm^{-1}. Magnetic moments are in Bohr magnetons.
[b] HRP is horseradish peroxidase. Where necessary data have been corrected from the old reference point of 91 mM^{-1} cm^{-1} for the Soret band maximum of native horseradish peroxidase to the correct value of 102 mM^{-1} cm^{-1} (Refs. 4,9).
[c] Shoulder.
[d] A double correction has been applied to data from Ref. 8, first for incomplete formation of the hydroxide complex (Ref. 12) and second for a better reference point as described above in footnote[b].
[e] There is also a band labeled δ at 352 nm with a molar absorptivity of 25 mM^{-1} cm^{-1}. The band for HRP-III at 673 nm shown in Ref. 8 is most likely due to P670 formed because of the presence of excess hydrogen peroxide.

I; higher values usually indicate incomplete compound I formation. In other cases, higher values are probably more accurate. Where there was no basis for a value judgment, average values are quoted.

7.2. RESONANCE RAMAN SPECTROSCOPY

7.2.1. Introduction

To a first approximation a chemical bond may be regarded as a simple harmonic oscillator. The two atoms at each end of the chemical bond oscillate back and forth. This oscillation continues even at 0 K, an illustration of the Heisenberg uncertainty principle which states in one of its forms that both position and energy cannot be defined precisely. This lowest energy oscillation is defined by a vibrational quantum number v of 0. Even at room temperature most bonds occupy the vibrational level defined by $v = 0$. Irradiation by the resonance infrared frequency will cause transitions to the next highest vibrational level, defined by $v = 1$, provided the selection rule for the transition is obeyed. The energy of the vibration is related to the strength of the chemical bond; the stronger the bond the higher the energy of the vibration. Thus, infrared spectra provide a means of identifying the structure of a molecule. For example, a carbon–carbon single bond will have a lower stretching frequency than the stronger carbon–carbon double bond.

For a diatomic molecule only a stretching mode is observed. In a polyatomic molecule there are also other types of vibrational motion. The hydrogen atom and oxygen atom of the OH group of phenol will undergo stretching and contraction, just as in a diatomic molecule. In addition the OH group can wag back and forth, characterized by a bending frequency. A symmetric polyatomic molecule can undergo symmetric stretches akin to a breathing motion. The same molecule can vibrate asymmetrically, in which one bond is contracting while its symmetric counterpart is stretching. Thus, interpretation of vibrational spectra can become complicated. Nevertheless, in the hands of experts great advancements have been made in the field, and extremely detailed structural information has been obtained, unique to the discipline.

Isotopic substitution changes a vibrational frequency because of the change in the reduced mass. For example, substitution of ^{2}H (deuterium) for ^{1}H causes a substantial lowering of vibrational frequency. Substitution of ^{54}Fe for ^{56}Fe produces a small increase, of the order of 1 or 2 cm^{-1}. The effect is enhanced if compared to ^{57}Fe substitution. ^{18}O shifts are of the order of 20–30 cm^{-1}. Thus, isotopic substitution is a powerful tool to identify vibrational bands.

One way to obtain a vibrational spectrum is to scan the sample by continuously changing the frequency of the incident infrared radiation. Absorption bands will be observed at each of the resonance frequencies. Modern infrared spectrometers use continuous radiation, and the bands are detected using

Fourier transforms, as in modern nuclear magnetic resonance (NMR) spectroscopy.

Raman spectroscopy provides a variation on infrared spectroscopy. Some of the differences between infrared and Raman spectroscopies are the following. In Raman spectroscopy,[20] a high-intensity monochromatic light source is used that is of higher energy than that of the vibrational resonance frequencies. A partial transfer of energy occurs from the light source photons to the compound being studied, causing vibrational excitation from the $v = 0$ to the $v = 1$ level. Stokes lines, which are at a lower frequency than that of the source radiation are observed in the scattered light. The difference in frequency between the Stokes lines and the source radiation corresponds to the resonance frequency. In addition anti-Stokes lines occur, in which vibrationally excited molecules add their vibrational energy to the incident photons, and return to the $v = 0$ level. Thus the first Stokes line corresponds to the $v = 1 \leftarrow v = 0$ transition and the first anti-Stokes line to the $v = 0 \leftarrow v = 1$ transition, and the energy difference between them to twice the energy difference between the lowest two vibrational levels. Because of a Boltzmann distribution of energies, most molecules are in their vibrational ground state. The development of lasers, powerful monochromatic light sources, has greatly increased the use of Raman spectroscopy.

A further large increase in sensitivity has occurred because of the development of resonance Raman spectroscopy. In the latter technique a laser, with a frequency corresponding to the frequency of an electronic transition, is used to excite the transition in the molecules of interest. This greatly enhances the intensity of vibrational spectra for vibrations that are coupled to the electronic transition. Thus, in a resonance Raman study of a heme protein, the laser may be tuned to the region of the Soret (~400 nm) electronic transition to obtain a large enhancement of some vibrational bands. By use of a different wavelength of excitation light, which causes a different electronic transition, different vibrational bands will be enhanced.

Studies of compounds of biological interest, dissolved in water, have been greatly facilitated by Raman spectroscopy. The infrared absorption by the solvent water is so intense that there are limited windows for observing absorption bands of the solute. In contrast, the interference of water is largely eliminated in Raman spectroscopy. Furthermore, the selection rules for Raman-induced transitions are different from those for infrared transitions, so that the available information is greatly extended.

7.2.2. Some Details of Resonance Raman Spectroscopy of Heme Proteins

Laser light is not only monochromatic, it is plane-polarized. There are three extremes in the different types of scattering observed for this plane-polarized light, measured by a depolarization ratio $\rho = I_{\perp}/I_{\parallel}$, where I is the intensity of the scattered light, polarized either at right angles or parallel to the plane of the incident light.[21–23] If $\rho < 0.75$, then the scattered light remains polarized in the same plane as the incident light. If ρ is 0.75, then the scattered light is depolarized. The third case is called anomalous or inverse depolarization, where the value of $\rho > 0.75$. Anomalous depolarization is observed in certain asymmetric vibrations associated with the tetragonal geometry of heme proteins. The depolarization ratio gives information about the symmetry of the enhanced vibrations; the ratio is dependent upon the frequency of the excitation light. A very large advantage of resonance Raman over infrared spectroscopy is that band polarizations can be determined in solution, which greatly facilitates unambiguous band assignments. Microscopic resonance Raman spectroscopy on single crystals can provide direct correlations with X-ray crystal structures.[24]

A major breakthrough in the analysis of vibrational spectra of porphyrins occurred when as many as possible of the fundamental vibrations of nickel(II) octaethylporphyrin were measured and a satisfactory model was produced.[25,26] The molecule has D_{4h} planar symmetry in the crystalline state, which was assumed to be valid in solution. The eight ethyl side chains were approximated as point masses. Normal coordinate analysis led to a model in reasonable agreement with experimental data and also to predictions of the frequencies of previously unobserved bands.

There are 35 Raman-active bands in the nickel porphyrin, numbered according to their symmetry. Those from ν_1 to ν_9 have A_{1g} symmetry, from ν_{10} to ν_{18} B_{1g} symmetry, ν_{19} to ν_{26} A_{2g} symmetry, and ν_{27} to ν_{35} B_{2g} symmetry. In addition there are 18 vibrational frequencies of E_u symmetry that are infrared active. Because of the lower symmetry in protoheme, the bands of E_u symmetry split into two components with A and B symmetry and become Raman active.

Studies were extended to nickel(II) protoporphyrin IX,[27] and several iron protoporphyrin IX complexes. Thus, the influence of changing from nickel to iron and of the vinyl side chains could be assessed. The assumption of D_{4h} symmetry accounts for the majority of the observed bands. However, lowering of the D_{4h} symmetry by the vinyl side chains leads to additional A bands (polarized) and B bands (depolarized). The numbering scheme developed for the fundamental frequencies of the nickel porphine skeleton was adopted for vibrational bands of hemes.

The labeling of carbon atoms of the porphine skeleton and of the vinyl side chains of a metal protoporphyrin IX is shown in Fig. 7.6. The resonance Raman spectrum of a heme protein is characterized by fairly strong bands in the 1300–1650 cm^{-1} region. These bands are highly sensitive to the size of the metal-binding site, the core size. (A core size change of only 0.01 Å produces a shift of 5–6 cm^{-1}.) The bands are called core size marker bands and are indicative of the oxidation state, coordination number, and spin state of the metal ion.[23] One particular vibrational frequency of the porphyrin, ν_4, shown in Fig. 7.7, is the strongest band under Soret excitation. It was used as a marker for oxidation state of the metal ion and other effects.[29] The correlations using ν_4 were extended to other skeletal frequencies.[30]

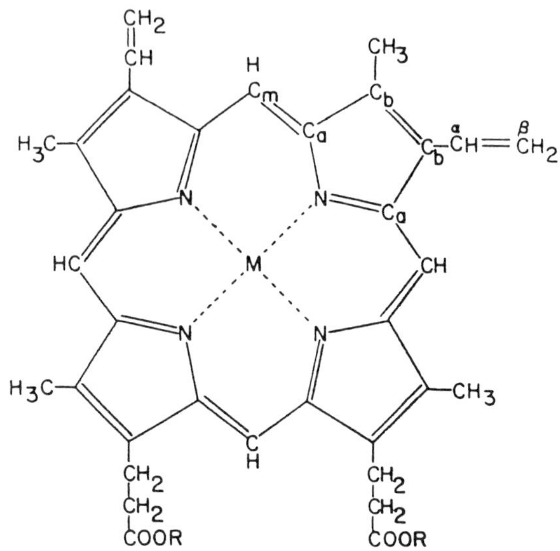

FIG. 7.6. Structure of protoporphyrin IX showing atom labeling. The pyrrole carbon atoms are labeled C_a and C_b, the vinyl carbons C_α and C_β, and the methene carbons C_m. [Reproduced with permission from S. Choi, T. G. Spiro, K. C. Langry, and K. M. Smith, *J. Am. Chem. Soc.* **104,** 4337–4344 (1982).]

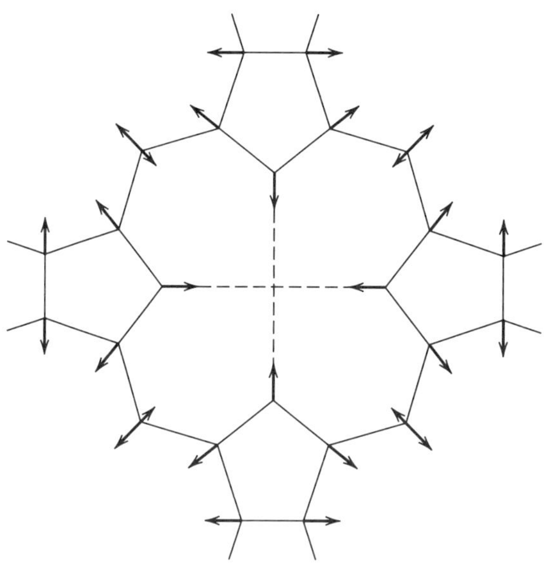

FIG. 7.7. Breathing vibration, ν_4, of a porphyrin. The motions of the pyrrole nitrogen atoms in this symmetric vibration, which occurs at 1383 cm^{-1} for octaethylporphinato–Ni^{II}, make the vibrational frequency sensitive to the oxidation state of the central metal ion. [Reproduced with permission from M. Abe, T. Kitagawa, and Y. Kyogoku, *J. Chem. Phys.* **69,** 4526–4534 (1978).]

The vinyl substituents, in positions 2 and 4 of the protoporphyrin IX, give rise to $\nu_{C=C}$ stretching modes in the 1615–1635 cm^{-1} region, identified by vinyl deuteration.[27] Their frequencies are indicative of the degree of conjugation of the vinyl double bond with the porphyrin macrocycle, and they also affect the wavelength of electronic absorption bands.[31]

7.2.3. Shape of Peroxidase Hemes

Metalloporphyrins and other tetrapyrolle macrocycles can have many different conformations. The nonplanar distortions can be described as saddling, ruffling, doming, waving (in x and y directions), and propellering (Fig. 7.8). In the remarkable research program of John Shelnutt and co-workers, the extents of these distortions have been analyzed for families of metalloproteins. The principal tools being used are resonance Raman spectroscopy and normal coordinate analysis. X-ray structures, molecular modeling, and optical spectra are also utilized. The analysis utilizes a new normal-coordinate structural decomposition (NSD) for characterizing and quantifying heme distortions in terms of these fundamental symmetric deformations.[32] By use of the NSD analysis, an accurate static X-ray structure can be deconvoluted into the relative contributions of fundamental vibrational frequencies. The NSD analysis leads to a tabulation, expressed in terms of plus/minus displacements in angstroms (Å) from planarity, caused by each of the six types of distortion. Thus, a static X-ray structure can be converted into a simple picture related to the heme vibrational motions.

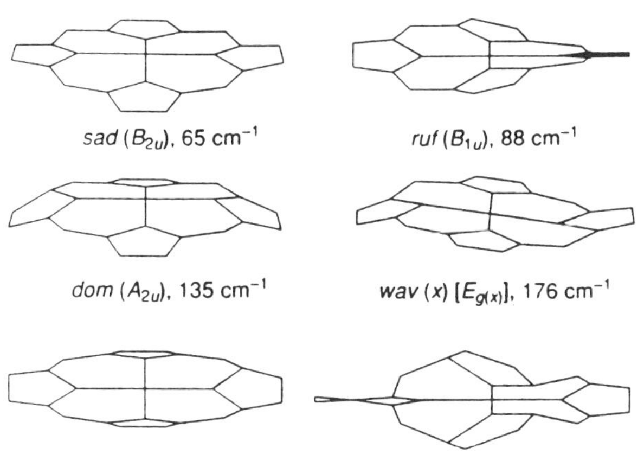

FIG. 7.8. Symmetric normal-coordinate deformations of hemes for saddling (sad), ruffling (ruf), doming (dom), waving (wav), and propellering (pro). Displacements of 1 Å are shown. [Reproduced with permission from J. A. Shelnutt, X.-Z. Song, J.-G. Ma, S.-L. Jia, W. Jentzen, and C. J. Medforth, *Chem. Soc. Rev.* **27,** 31–41 (1998).]

The heme of cytochrome *c* peroxidase was shown to be markedly different from that of a wide variety of other heme proteins.[33] Whereas the nonperoxidase hemes had mainly positive deviations, those of the peroxidase heme were mainly negative. More importantly, the major contribution was from the saddling mode. Other peroxidase hemes have similar saddle shapes.[34] The saddle shape of peroxidase hemes appears to be an intrinsic property. The five-coordination is a result, not a cause, of the heme shape since the saddle shape remains even upon the addition of exogenous sixth ligands or even removal of the proximal histidine by genetic engineering. Almost all of the over 80 peroxidases and their mutants for which X-ray structures are available exhibit strong saddling as their predominant deformation. Not a single one of the hemes of the 84 plant peroxidases in classes I, II, and III lacks saddling as the predominant heme deformation. (Exceptions are some of the prostaglandin synthase structures.) Even the structurally unrelated heme of myeloperoxidase, which is covalently bound to the protein, is highly saddled.

7.2.4. Resonance Raman Spectra of Horseradish Peroxidase

Resonance Raman spectral studies on horseradish peroxidase began in 1973,[35] when ferrous forms of the enzyme were compared to cytochrome *c* and microperoxidase. In a more comprehensive investigation, spectra for the native, fluoro, and cyano derivates of the ferric enzyme and carbonmonoxy and cyano derivates of the ferrous enzyme were obtained.[36] Oxidation state, coordination, and spin state changes have been detected as a consequence of the changes in frequency of the core size marker bands.

Laser Raman studies were extended to the oxidized forms of peroxidases and a catalase.[37,38] For horseradish peroxidase a band at 1576 cm^{-1} in the native enzyme shifts to 1587 cm^{-1} in compound II and is consistent with the iron moving from about 0.4–0.5 Å out of the porphyrin plane into the plane (Fig. 7.9). This movement is now known to correspond to the change from a five- to six-coordinate species. Native horseradish peroxidase bands at 1500, 1575, and 1630 cm^{-1} are close to those for five-coordinate model compounds.[39] By means of isotopic iron substitution the iron–proximal histidine stretching frequency of horseradish peroxidase was detected.[40] Binding of benzhydroxamic acid, but not hydroquinone, changed the iron(III) from five- to six-coordinate, with likely binding of water in the sixth position.

The controversy over the nature of the alkaline form of native horseradish peroxidase was finally resolved by resonance Raman spectroscopy.[41,42] The transition to the alkaline form occurs at pH 11.0. It consists of a change from five-coordinate high-spin iron(III) to six-coordinate low-spin iron(III) with a hydroxide ion occupying the sixth coordination position of the iron. The assignment was made iron-clad by the use of ^{18}O, ^{2}H, ^{54}Fe, and ^{57}Fe substitutions. ^{18}O results are shown in Fig. 7.10. Hydrogen bonding from the hydroxide to the distal histidine causes the lowering of the ferryl/ferric reduction potential

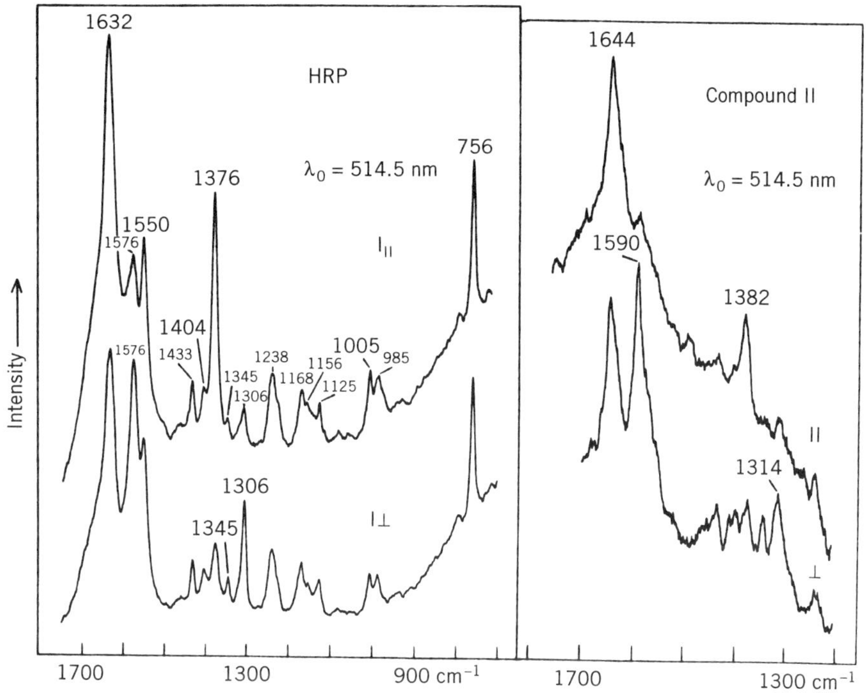

FIG. 7.9. Parallel and perpendicular resonance Raman spectra of native horseradish peroxidase and compound II. The shift of the 1576–1587 cm^{-1} band indicates movement of the iron into the porphyrin plane. [Reproduced with permission from R. H. Felton, A. Y. Romans, N.-T. Yu, and G. R. Schonbaum, *Biochim. Biophys. Acta* **434,** 82–89 (1976) with permission from Elsevier Science.]

at high pH.[43] In a recent paper, the effects of various mutants were studied at alkaline pH.[44] It appeared that Arg38 is essential for stabilization of the OH ligand, while His42 acts as a hydrogen bond acceptor. Similar results were found for fluoride binding.[31]

Horseradish peroxidase compound III, or oxyperoxidase, has been investigated by resonance Raman spectroscopy[45] and compared to oxymyoglobin.[46] The conclusions were that the porphyrin core of the oxyperoxidase is slightly expanded and the iron closer to the porphyrin plane than in oxymyoglobin. There is a stronger iron–His bond that leads to greater iron-to-oxygen π^* back-bonding in oxyperoxidase. This in turn is believed to weaken the O—O bond in oxyperoxidase.

Photoreduction of compounds I and II, observed in visible–ultraviolet spectra,[47,48] were confirmed in resonance Raman spectra.[49] The photolability of compound I complicates optical spectral studies.

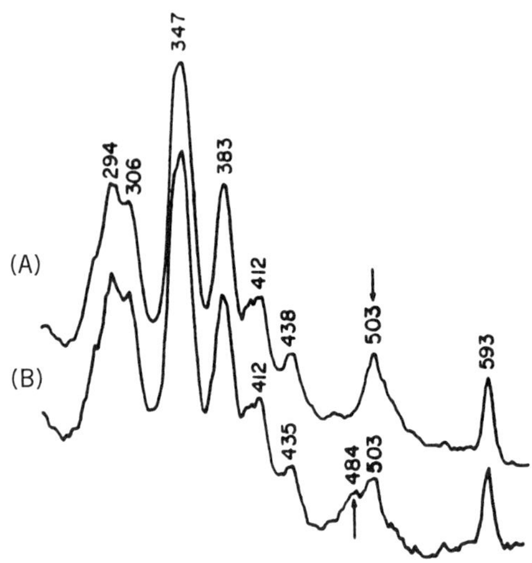

FIG. 7.10. Resonance Raman spectra of the alkaline form of horseradish peroxidase (pH 12.0). The shift of the 503 cm^{-1} band in $H_2{}^{16}O$ (upper spectrum) to 484 cm^{-1} in $H_2{}^{18}O$ (lower) shows that the $\nu_{Fe(III)-OH}$ frequency is being observed. [Reproduced with permission from A. J. Sitter, J. R. Shifflett and J. Terner, *J. Biol. Chem.* **263,** 13032–13038 (1988).]

Properties of horseradish peroxidase compound II were examined by resonance Raman spectroscopy in a series of papers.[50–52] The $Fe^{IV}{=}O$ stretching frequency for isoenzyme A1 is at 779 cm^{-1}, shifted to 745 cm^{-1} by ^{18}O substitution.[51] The same bands for isoenzymes B and C occur at 776 and 743 cm^{-1} (all measurements at pH 7.0). Alkaline transitions are observed that shift the $Fe^{IV}{=}O$ stretching frequency of the A1 and A2 isoenzymes to 790 cm^{-1} with the ^{18}O substitution changing it to 755 cm^{-1}. For isoenzymes B and C the frequency for the alkaline form is 788 cm^{-1}, shifted to 756 cm^{-1} by ^{18}O. The pK_a values for these alkaline transitions have been accurately determined: 6.9 for isoenzyme A2[53] and 8.5 ± 0.1 for isoenzymes B and C[54,55] (Fig. 7.11). The changes are attributed to the ionization of an amino acid side chain at higher pH, the imidazole group of the distal His residue, breaking the hydrogen bond to the ferryl oxygen atom. The alkaline forms of the compounds II are unreactive. The ferryl oxygen atom of the acidic form of the compound II derived from isoenzyme C rapidly exchanges with water.[55]

Because of the inherent instability and photosensitivity of horseradish peroxidase compound I, either low-temperature[56] or flow systems were used.[57–59] One system was designed so that both resonance Raman and optical spectra could be monitored on the same flowing sample.[58] In definitive studies,[60] it was shown that unless near-ultraviolet excitation was used, porphyrin π-cation

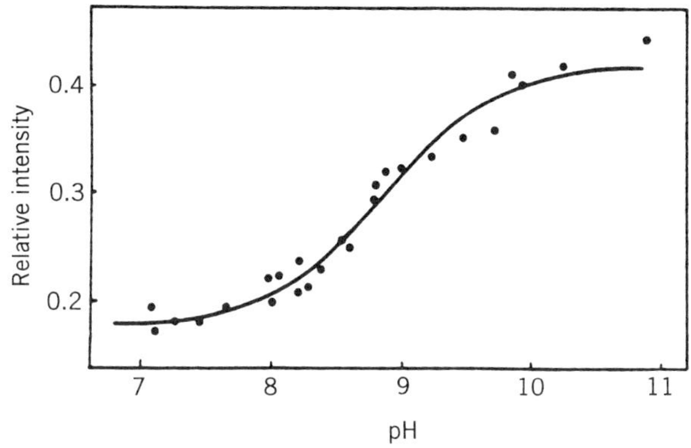

FIG. 7.11. pH dependence of the $Fe^{IV}{=}O$ stretching frequency. The ratios of intensities at 787 cm^{-1} compared to 680 cm^{-1} are plotted as a function of pH. The results indicate that the $Fe^{IV}{=}O$ group is not hydrogen bonded at high pH. The hydrogen bond is required for compound II reactivity and for exchange of the ferryl oxygen atom with water. [Reproduced with permission from S. Hashimoto, Y. Tatsuno, and T. Kitagawa, *Proc. Natl. Acad. Sci. U.S.A.* **83,** 2417–2421 (1986).]

radical resonance Raman spectra were obscured by enhancement of other bands. With low-power laser excitation in the vicinity of 360 nm, bands characteristic of porphyrin π-cation radical model compounds were detected (Fig. 7.12). Compared to compound II, the v_4 frequency, primarily C_a–N, is downshifted by 20 cm^{-1}. The v_2, v_{11}, and v_{37} frequencies, primarily C_b–C_b, are upshifted 10–20 cm^{-1}.

An Fe=O stretching frequency of 791 cm^{-1} at pH 7 has been reported for compound I.[59] This has been shown to consist of two components:[61] near pH 7 it has a frequency of 795 cm^{-1}; at pH 5 it shifts to 786 cm^{-1}. These correspond to the H-bonded and non-H-bonded values seen for compound II.[62] A value of 790 cm^{-1} was obtained for the Fe=O stretching frequency of compound I, derived both from native enzyme and enzyme reconstituted with mesoheme. It was concluded that the ground state of compound I was $^2A_{1u}$.The importance of near-ultraviolet excitation was demonstrated.[63]

Further details on ferric horseradish peroxidase as well as information on its aromatic donor complexes have been obtained by resonance Raman spectroscopy.[64] At room temperature two sets of skeletal porphyrin vibrations are observed that correspond to a mixture of five- and six-coordinate high-spin forms, with the five-coordinate species being dominant. As the temperature is lowered to 20 K intermediate- and low-spin states are obtained. Two different C=C stretching modes are observed for both the enzyme and its complexes, indicating that the vinyl groups are not equivalent. Hydroquinone has little

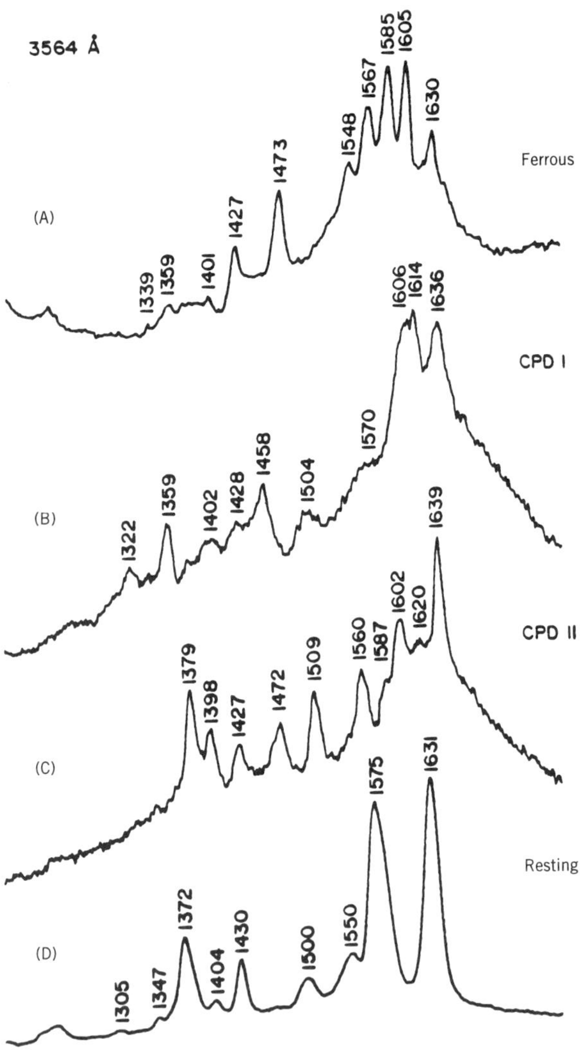

FIG. 7.12. Resonance Raman spectra obtained with ultraviolet excitation of (A) ferrous enzyme, (B) compound I, (C) compound II, and (D) resting horseradish peroxidase. [Reproduced with permission from V. Palaniappan and J. Terner, *J. Biol. Chem.* **264,** 16046–16053 (1989).]

effect on the observed spectra, but benzhydroxamic acid shifts the predominantly five-coordinate spectra of the native enzyme to six-coordinate high-spin, and one of the vinyl modes is enhanced.

Dramatic changes occur to the native enzyme when the pH is lowered to 3.1 as detected by resonance Raman and electronic absorption spectra of ferric, ferrous, and ferrous–CO forms.[65] The native ferric enzyme is first converted to a six-coordinate, then a final five-coordinate high-spin species. Water is first bound in the sixth coordination position, and then the proximal Fe–imidazole bond is cleaved.

Recombinant horseradish peroxidase C, and three mutants Phe41Trp, Phe-41Val, and Arg38Lys were examined by resonance Raman spectroscopy (Fig. 7.13).[66] The active enzymes were obtained by *in vitro* refolding after expression in *Escherichia coli.* Spectra of the iron(II) and iron(III) forms and complexes with benzhydroxamic acid were recorded at neutral pH. The mutations, all of which are on the distal side, dramatically affect the ability to bind benzhydroxamic acid. The substitution at position 38 and of Trp at position 41

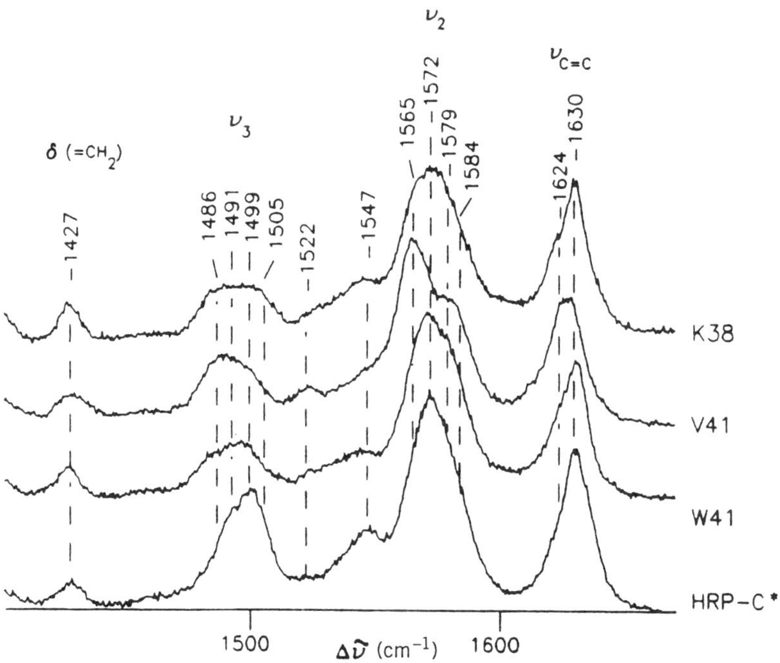

FIG. 7.13. Resonance Raman spectra of recombinant horseradish peroxidase and the mutants F41W, F41V, and R38K; pH 7.1, 406.7 nm excitation. [Reproduced with permission from G. Smulevich, M. Paoli, J. F. Burke, S. A. Sanders R. N. F. Thorneley, and A. T. Smith, *Biochemistry* **33,** 7398–7407 (1994).]

eliminate binding altogether, whereas that of Val at position 41 leads to a twofold increase in affinity. These studies were recently extended to His42Leu, His42Arg, Arg38Gly, and Arg38Leu mutants.[44] The results support the analysis of the iron of the wild-type enzyme as being predominantly five-coordinate high-spin, as well as containing two forms of six-coordinate high-spin, one with water coordinated to the distal Arg and the other to the distal His. Resonance Raman spectra of horseradish peroxidase have been compared to those of cytochrome *c* and *Coprinus cinereus* peroxidases.[67] The influence of distal mutants His42Leu, His42Arg, Arg38Gly, and Arg38Leu on resonance Raman and electronic absorption spectra were examined.[44] The relative populations of different spin states and five- and six-coordination iron species were found to vary markedly.

Infrared spectra of carbonyl horseradish peroxidase and the denatured enzyme have been obtained.[68,69] The resonance Raman, electronic absorption and circular dichroism spectra of the ferric, ferrous, and ferrous–CO forms of horseradish peroxidase have been obtained at pH 3.1.[70] The final form of horseradish peroxidase C has a five-coordinate iron in which the proximal histidine has been replaced by a water molecule.

7.3. MAGNETIC CIRCULAR DICHROISM

7.3.1. Introduction

A description of magnetic circular dichroism (MCD) hinges upon an understanding of circular dichroism (CD) and optical rotatory dispersion (ORD). Plane polarized light, as obtained in a polarimeter, consists of left (counterclockwise when viewed toward the light source) and right (clockwise) circularly polarized components that are equal in intensity. When the plane polarized light is passed through a solution containing a optically active chromophore, the refractive index of the light changes in the region of the absorption band. Furthermore, different changes occur in the refractive indices of the left and right components. As a result the left and right components are transmitted through the solution with different velocities. Upon recombination after passing through the solution, the result is an elliptically polarized beam that is rotated either to the left or right by an amount characteristic of the solute. The difference in refractive index between left and right as a function of wavelength leads to an ORD spectrum,[71] which can have both peaks and troughs, as illustrated in Fig. 7.14. Although discovered in 1817 by J. B. Biot, the practical use of ORD can be dated from 1953 with the introduction of reliable commercial instruments capable of measurements in the ultraviolet region. Great success was attained in the characterization of steroids.[72] While valuable information was obtained on peptides and proteins using ORD,[73] attempts to predict the amount of α-helix and β-pleated sheet structure in a protein were notoriously unreliable.

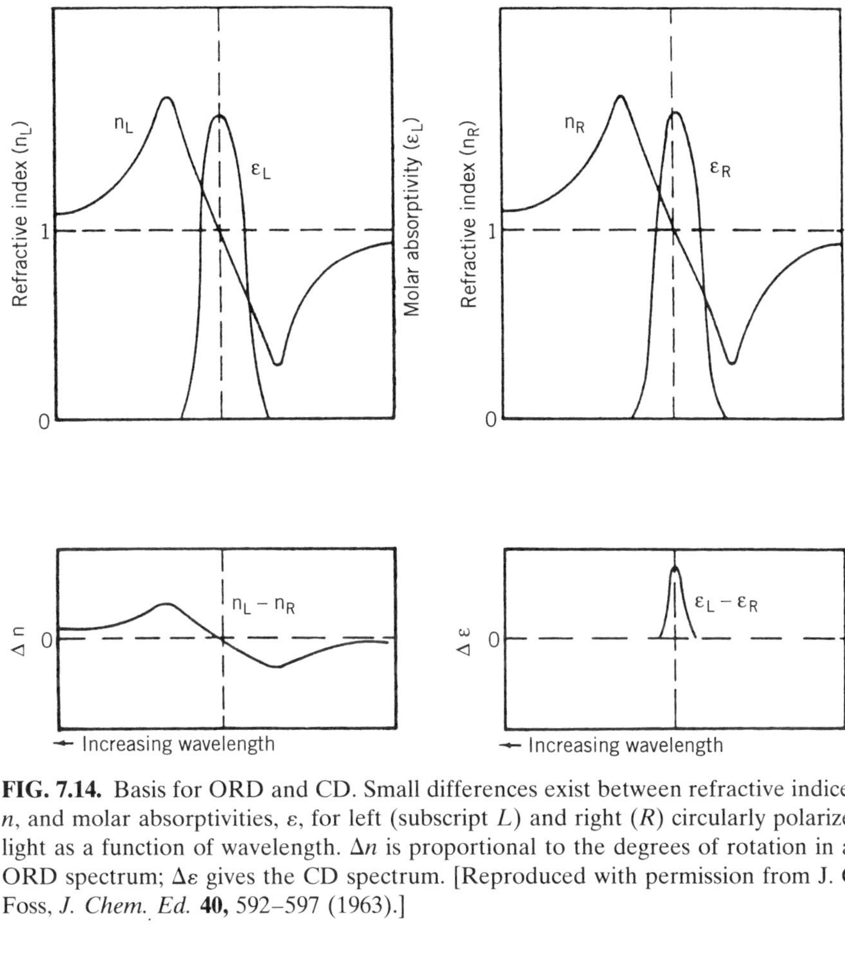

FIG. 7.14. Basis for ORD and CD. Small differences exist between refractive indices, n, and molar absorptivities, ε, for left (subscript L) and right (R) circularly polarized light as a function of wavelength. Δn is proportional to the degrees of rotation in an ORD spectrum; $\Delta\varepsilon$ gives the CD spectrum. [Reproduced with permission from J. G. Foss, *J. Chem. Ed.* **40,** 592–597 (1963).]

Because the left and right circularly polarized components travel with different velocities, they have different extinction coefficients. The two components can be separated and measured independently. This is the basis of CD and its magnetic counterpart MCD. The device used for this separation is called a quarter-wave plate; an example is a thin sheet of mica.[71] With a 45° orientation to the light beam one component is separated; with the opposite orientation the other component is obtained. The difference in molar absorptivity between the left and right components, $\Delta\varepsilon = \varepsilon_L - \varepsilon_R$, can be either positive or negative; the measurement of $\Delta\varepsilon$ as a function of wavelength is CD spectroscopy, which can yield important information about the optically active chromophore responsible for the absorption band (Fig. 7.14). The units of $\Delta\varepsilon$ are M^{-1} cm^{-1}. The results may be expressed in terms of the molar ellipticity, $[\theta]$, which is $\Delta\varepsilon$ multiplied by a constant. The definition is $[\theta] = 2.303 \times (\Delta\varepsilon) \times 100 \times 180°/4\pi$.

In 1845 Michael Faraday discovered that, in a magnetic field, plane polarized light was rotated when passed through a solution that had no optical activity in the absence of the field. In other words, optical activity can be induced by a magnetic field, now known as the Faraday effect. The magnetic field orientation is in the direction of the light path. Thus, there are magnetic counterparts to the spectroscopic techniques of ORD and CD, given the acronyms MORD and MCD, where M means magnetic. MORD is not a practical technique since the magnetic field induces contributions to optical rotation in everything in the path of the light such as cell windows and solvent. MCD is a valuable technique since its basis is measurement of changes in molar absorptivity and thus is similar to conventional optical spectroscopy.

Circular dichroism results are expressed in terms of $\Delta\varepsilon$; MCD results in terms of $\Delta\varepsilon/B$, where B is the magnetic flux density. B has SI units of the Tesla, T, or cgs units of the Gauss, G ($1\ T = 10^4\ G$). Thus, $\Delta\varepsilon/B$ is commonly expressed in units of $M^{-1}\ cm^{-1}\ T^{-1}$. The symbol H for magnetic field is sometimes substituted for B, which is an error since the units of H are ampere-meter^{-1}. Results may also be expressed in terms of the magnetic molar ellipticity $[\theta]_M$ which is $2.303 \times (\Delta\varepsilon) \times 100 \times 180°/4\pi B$.

Both theory[74,75] and experimental technique[76] for MCD became highly developed in the 1960s. Experiments were facilitated by the use of superconducting solenoids to generate high-strength magnetic fields. Detailed reviews are available.[77,78]

Different types of MCD spectra are observed, described in terms of Faraday constants A, B, and C. An A term line shape is shown in Fig. 7.15. In the example transitions occur from a nondegenerate ground state to a doubly degenerate excited state. A photon has one unit of angular momentum, which in a magnetic field is either negative (left polarized) or positive (right polarized). Thus, the allowed transitions are those in which the angular momentum of the absorber changes by ± 1. The two transitions occur with equal probability since they originate from the same level.

The origin of a Faraday C term is illustrated in Fig. 7.16. The selection rules remain the same. However, if the ground state is doubly degenerate, then there is a population difference between the two levels described by a Boltzmann factor. This leads to differences in intensity between the left and right components, and the result is a single slightly distorted peak. Thus, the Faraday C term is temperature-dependent. A band described by a Faraday B term is caused by magnetic mixing of states. It has the same line shape as that described by a Faraday C term but is temperature-independent.

There are two approaches that can be used in the interpretation of MCD spectra. One is to analyze the spectra from fundamental principles (see, e.g., a description of the MCD spectrum of $IrCl_6^{2-}$ and other inorganic transition metal complexes).[79] The analysis becomes increasingly difficult with increased

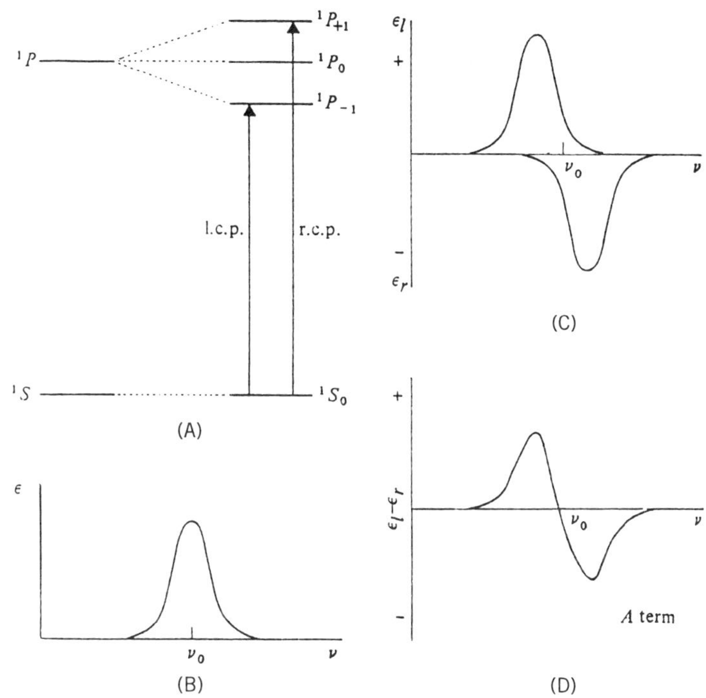

FIG. 7.15. Basis for an MCD spectrum described by a Faraday *A* term. (A) Allowed transitions between a singlet ground state and triplet excited state for photons either left circularly polarized (lcp) or right circularly polarized (rcp). (B) Optical spectrum in the absence of the magnetic field. (C) Separation of left and right components. (D) Resultant MCD spectrum. [Reproduced with permission from P. N. Schatz and A. J. McCaffery, *Quart. Rev.* **23,** 552–584 (1969).]

complexity of the system, for example, from overlapping bands. The alternative is to use the MCD spectra of appropriate model compounds of known structure to interpret the MCD spectra of metalloproteins. The latter approach has been used successfully by the group of Dawson[80] and is the one emphasized here. The absorption and MCD spectra of a five-coordinate iron(III) porphyrin are shown in Fig. 7.17.

7.3.2. Magnetic Circular Dichroism Spectra of Horseradish Peroxidase

The first report on MCD spectra for horseradish peroxidase and some of its complexes at pH 7.0 was published in 1976,[81] with another the following year

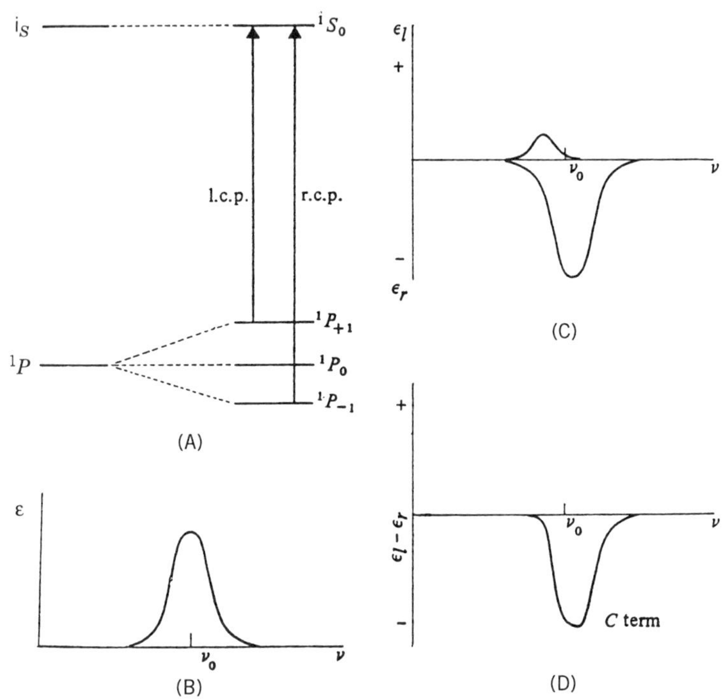

FIG. 7.16. Basis for an MCD spectrum described by a Faraday *C* term. Same notation as in Fig. 7.14, except the transitions are between a triplet ground state and singlet excited state. [Reproduced with permission from P. N. Schatz and A. J. McCaffery, *Quart. Rev.* **23,** 552–584 (1969).]

on the alkaline transition.[82] The MCD spectra for the alkaline transition is shown in Fig. 7.18. The spectra were obtained on impure samples of horseradish peroxidase (RZ of 1.48). Comparison can be made for the pH 7.0 results with a more recent study using highly purified enzyme.[83]

The MCD spectra of native enzyme, compounds II and III, their iron mesoporphyrin derivatives, and the corresponding myoglobin species are shown in Figs. 7.19–7.21. Substitution of the mesoporphyrin for protoporphyrin consistently causes a blue shift of up to 12 nm. Large differences are apparent between five-coordinate ferric horseradish peroxidase and six-coordinate ferric myoglobin. In contrast, the azide derivatives, both of which have six-coordinate iron(III), are very similar (data not shown). The spectrum for horseradish peroxidase compound I, for which there is no myoglobin analog, is shown in Fig. 7.22. A conclusion from this work is that mesoporphyrin derivates, which are more robust, can be substituted for ferriprotoporphyrin IX.

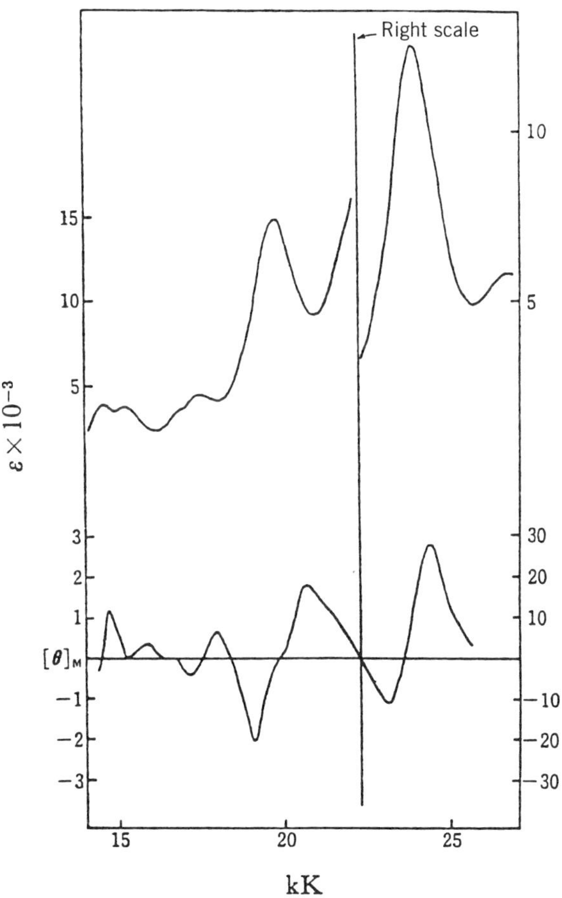

FIG. 7.17. Absorption (top) and MCD (bottom) spectra for the iron(III) chloride of tetraphenylporphin in benzene. [Reproduced with permission from H. Kobayashi, M. Shimizu, and I. Fujita, *Bull. Chem. Soc. Japan* **43,** 2335–2341 (1970).]

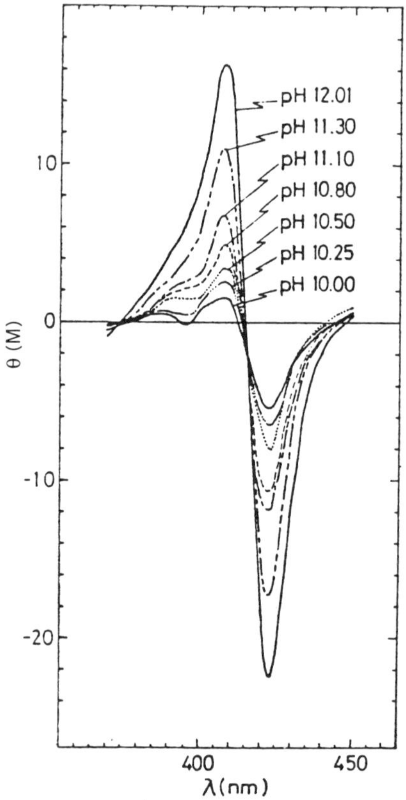

FIG. 7.18. MCD spectrum for the alkaline transition of horseradish peroxidase obtained at 0°C. At pH 11 there is a 50–50 mixture of the acid and alkaline forms. [Reproduced with permission from N. Kobayashi, T. Nozawa, and M. Hatano, *Biochim. Biophys. Acta* **493,** 340–351 (1977).]

The MCD spectra of horseradish peroxidase compound I in frozen glycerol glasses has been studied from 273 K down to liquid helium temperature.[84,85] A strong temperature dependence occurs from 4.2 to 30 K, dominated by Faraday C terms, which disappears from 30 to 273 K. The MCD spectra of compound I from 1.6 to 75 K are shown in Fig. 7.23. The results indicate strong coupling between the $S = 1$ iron and the $S = \frac{1}{2}$ porphyrin cation radical electron at low temperatures, which becomes weak at higher temperatures. Care must be taken to prevent photolysis of compound I, which produces striking changes in spectra (Fig. 7.24).

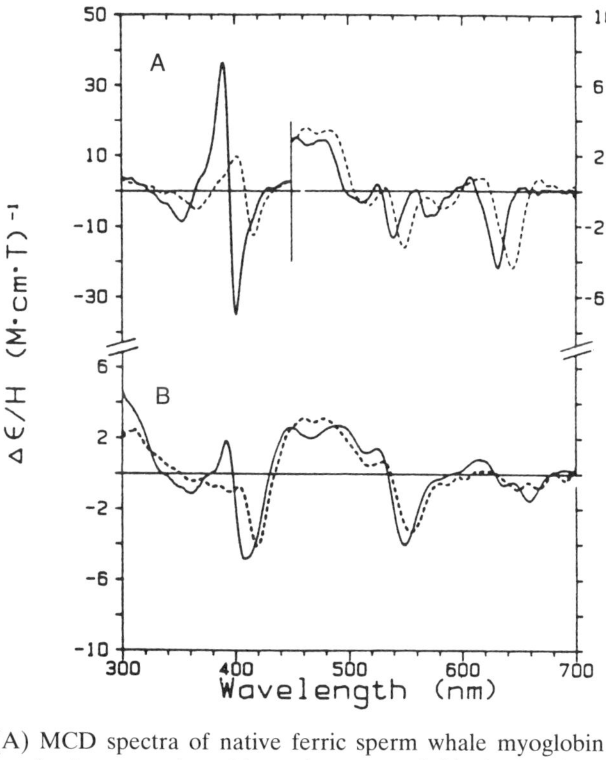

FIG. 7.19. (A) MCD spectra of native ferric sperm whale myoglobin (– – –) and ferric mesoporphyrin–reconstituted horse heart myoglobin (—). (B) MCD spectra of native ferric horseradish peroxidase (– – –) and the corresponding mesoporphyrin reconstituted derivative (—) . [Reproduced with permission from J. H. Dawson, S. Kadkhodayan, C. Zhuang, and M. Sono, *J. Inorg. Biochem.* **45,** 179–192 (1992) with permission from Elsevier Science.]

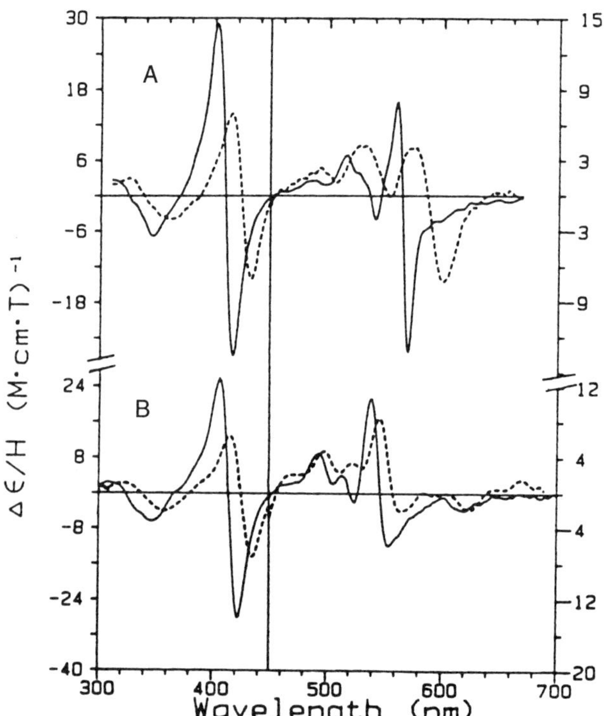

FIG. 7.20. (A) MCD spectra for the compound II of myoglobin (– – –) and its mesoporphyrin derivative (—). Same conditions as in Fig. 7.6 (B) MCD spectra for compound II of horseradish peroxidase (– – –) and its mesoporphyrin derivative (—). [Reproduced with permission from J. H. Dawson, S. Kadkhodayan, C. Zhuang, and M. Sono, *J. Inorg. Biochem.* **45,** 179–192 (1992) with permission from Elsevier Science.]

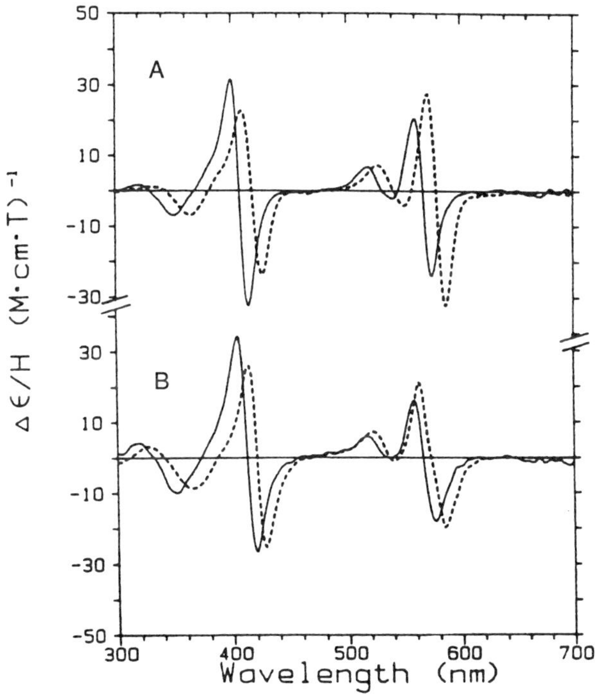

FIG. 7.21. (A) MCD spectra for compound III of horseradish peroxidase (– – –) and its mesoporphyrin derivative (—). (B) MCD spectra for the compound III of myoglobin (– – –) and its mesoporphyrin derivative (—). [Reproduced with permission from J. H. Dawson, S. Kadkhodayan, C. Zhuang, and M. Sono, *J. Inorg. Biochem.* **45,** 179–192 (1992) with permission from Elsevier Science.]

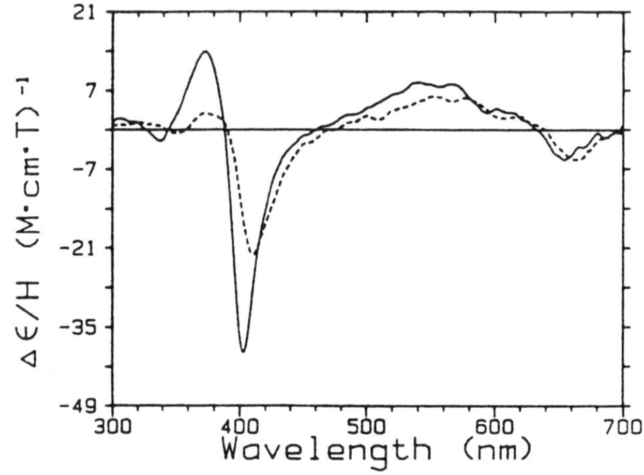

FIG. 7.22. MCD spectra for compound I of horseradish peroxidase (– – –) and its mesoporphyrin derivative (—). [Reproduced with permission from J. H. Dawson, S. Kadkhodayan, C. Zhuang, and M. Sono, *J. Inorg. Biochem.* **45,** 179–192 (1992) with permission from Elsevier Science.]

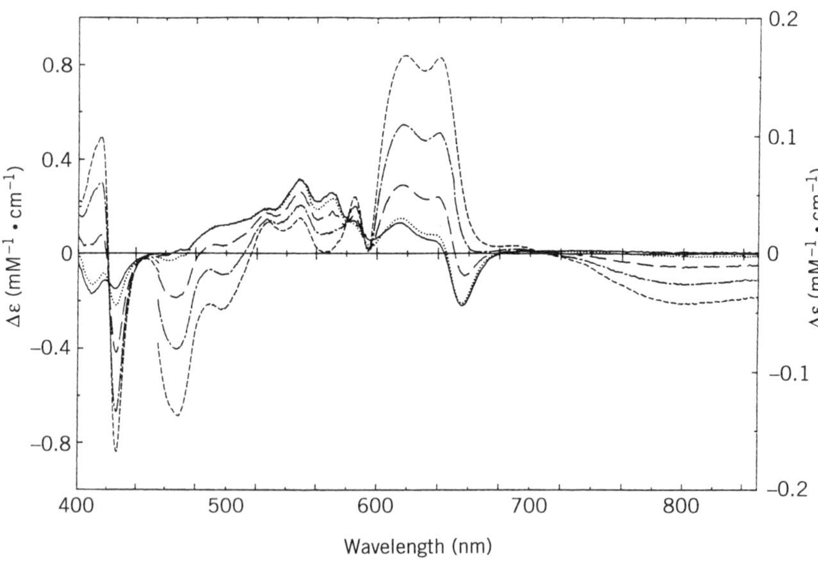

FIG. 7.23. Temperature dependence of the near-UV/visible region MCD spectrum of horseradish peroxidase compound I. Magnetic flux density 5 T; (– – – –) 1.6 K, (— – —) 4.2 K, (— —) 10 K, (· · · · ·) 25 K, (—) 75 K. The change in ordinate scales occurs at ~ 465 nm. [Reproduced with permission from N. Foote, P. M. A. Gadsby, M. J. Berry, C. Greenwood, and A. J. Thomson, *Biochem. J.* **246,** 659–668 (1987).]

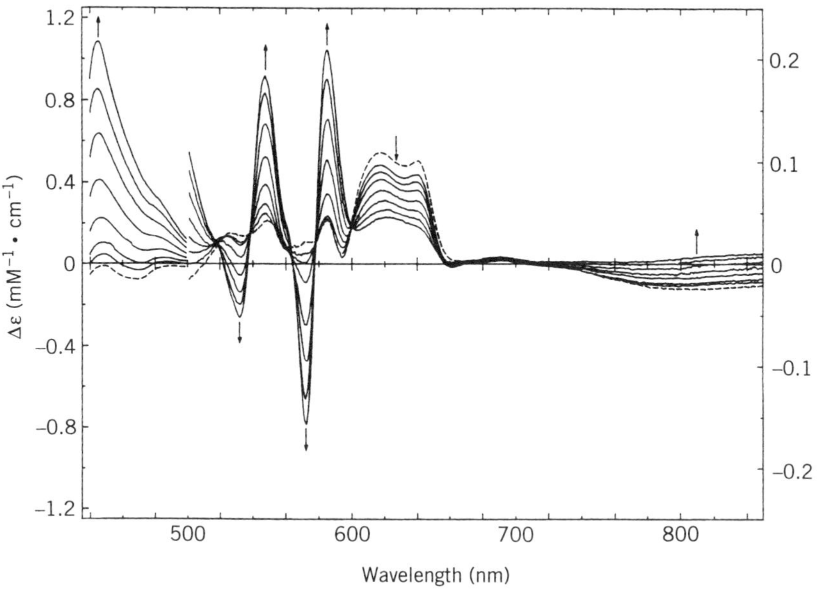

FIG. 7.24. Effect of photolysis at 4.2 K on the visible region MCD spectrum of horseradish peroxidase compound I. Magnetic flux density 5 T; (– – – –) compound I before photolysis, (—) the same sample after irradiation with white light for 10, 20, 40, 80, 160, 300, and 540 min. The change in ordinate scale occurs at 500 nm. [Reproduced with permission from N. Foote, P. M. A. Gadsby, M. J. Berry, C. Greenwood, and A. J. Thomson, *Biochem. J.* **246,** 659–668 (1987).]

7.4. X-RAY ABSORPTION SPECTROSCOPY

7.4.1. Introduction

A neophyte needs a definition of the relevant acronyms and their meaning. These include XAS, EXAFS, XANES, and finally XAFS. X-ray absorption spectroscopy (XAS) refers to the experimental technique of obtaining X-ray absorption data for a compound of interest, which can be in solution, frozen, amorphous, crystalline, or even in the gas phase. The discussion here is concentrated upon data obtained from a central metal ion. The XAS data can be divided into two parts, the edge component and the absorption that occurs at higher energies (Fig. 7.25). Instead of X-ray energies for plots of XAS data, the common convention is to use a function of energy, *k,* which is the photoelectron wave vector and is defined as follows.

$$k = [(2m/\hbar^2)(E - E_0)]^{1/2}$$

where $\hbar$ is $h/2\pi$, h is Planck's constant, m the mass of the electron, E the energy of the X-ray photon, and E_0 the threshold energy for absorption. The units of k are (distance)$^{-1}$; it is commonly expressed in Å^{-1}.

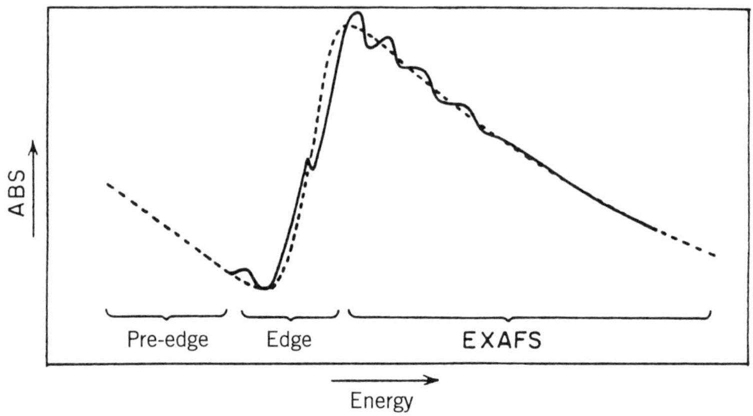

FIG. 7.25. Schematic representation of data from X-ray absorption spectroscopy. [Reproduced with permission from S. P. Cramer and K. O. Hodgson, *Prog. Inorg. Chem.* **25,** 1-39 (1979).]

K edge absorption is determined by the transition energy of the innermost 1*s* electrons to unoccupied higher energy orbitals of the same ion, such as 3*d* orbitals. *L* edge absorption originates from 2*s* and 2*p* electrons. The edge absorption extends over a range of ~20 electron volts (eV). At somewhat higher energies these inner electrons are liberated from the metal ion. The liberated electrons are called photoelectrons. The electrons exhibit wave as well as particle properties; the waves interact with nearest-neighbor atoms in such a way that backscattering occurs. If the backscattering is in phase, then wave reinforcement occurs; out-of-phase scattering causes troughs. Thus, oscillations are superimposed on the high-energy absorption (Fig. 7.25). The backscattering can be measured up to 1000 eV energy in excess of the energy of the absorption edge. The backscattering extended over this 1 keV range yields information about fine structure, such as bond lengths to ±0.02 Å and coordination numbers to ±20%, hence the name for the technique: EXAFS, for extended X-ray absorption fine structure. Quantitative interpretation of the EXAFS region proceeded more rapidly than of the edge region. Later analysis called XANES (X-ray absorption near edge structure) was developed. The location of the near edge gives information about the electron density on the metal ion and hence about its oxidation state. Thus, higher energies are required to obtain the edge spectrum of a metal ion with a higher positive charge. Today, comprehensive analysis of both the edge and higher energy regions is called XAFS (X-ray absorption fine structure).[86]

The phenomenon of EXAFS has been known since 1931, but little progress was made until the availability of synchrotron radiation, capable of delivering X-ray beams up to one million times more intense than obtained for a conventional X-ray tube. The first report on the use of synchrotron radiation for

EXAFS was published in 1975.[87] EXAFS is a unique technique in that theory and experiment were developed simultaneously by the same groups of researchers, and to a large extent in the study of problems of biological interest.[88] There are several general reviews.[89–91] and more recent ones that cover problems in data analysis.[86,92]

The quantity known in EXAFS as χ is the difference between the total oscillating signal and the smoothed absorption curve (Fig. 7.25). This difference is a damped sine wave. The sensitivity in determining this wave is improved by using fluorescent detection.[93] The wave is defined in terms of its amplitude and its phase shift. The amplitude varies with the inverse of the square of the distance, R_{as}, between absorber a and the scatterer s. The phase shift is a function of both the absorber and scattering atoms. The problem is simplest if all of the backscattering atoms are identical and have identical environments, an occurrence not found in peroxidases. Each number of equivalent scattering atoms, N_s, makes its own unique contribution. Errors are introduced if the analysis includes an insufficient number of the damped oscillations.

Another complication is multiple backscattering, especially important for colinear or nearly colinear atoms. Thus, a wave contribution may be obtained from the sequence: absorber $\rightarrow$ scatterer$_1$ $\rightarrow$ scatterer$_2$ $\rightarrow$ scatterer$_1$ $\rightarrow$ absorber.[86] Multiple backscattering from imidazole has been discussed.[93] Multiple backscattering is especially important at lower energies, so it can affect XANES spectra and the most intense part of an EXAFS spectrum.

EXAFS analysis may use *ab initio* calculations of wave and phase functions[94] or data from model compounds. Use of the former is unreliable, as is the latter if the model compounds are not chosen with care.[86] It is unlikely that porphyrin geometry can be determined by EXAFS.[95] Thus, it can be seen that obtaining accurate results from EXAFS is an art form. Other data, such as X-ray crystallographic results, if available, can provide an excellent starting point. Results on rubredoxin[96] and Cu,Zn-superoxide dismutase[97] are good examples. Knowing in advance the total number of ligands and their identity provides a huge advantage. Whereas crystallographic data provides comparatively poorly defined coordinates for the entire structure, EXAFS is restricted to the central metal ion and its immediate environment, where much better precision and accuracy is possible.

7.4.2. XAFS of Horseradish Peroxidase

The first report on XAS of horseradish peroxidase analyzed edge data and contained a preliminary analysis of the EXAFS region.[98] The edge data were consistent with iron(IV) being present in both compounds I and II. Oxygen and nitrogen ligands could not be distinguished in the EXAFS data, but in accord with other data, a short bond length of ~1.6 Å was assigned to Fe=O and ~2.0 Å to the Fe—N bonds connecting the metal ion to the porphyrin. Comparisons were made to several model compounds.

A year later EXAFS results on compound I indicated a Fe=O distance of 1.64 Å in compound I but a much longer bond length of 1.93 Å for the Fe—O distance in compound II.[99] The latter result appears inconsistent with other spectroscopic data relevant to bond lengths in compound II; also with subsequent EXAFS data in which a bond length of 1.62 ± 0.02 Å was obtained for Fe=O in both compounds I and II by two different methods of analysis.[100] K_2FeO_4 was used as a model for Fe=O EXAFS. Results were consistent with both EXAFS and crystallographic data on a high valent oxo-chromium porphyrin and two Fe^{IV}=O derivatives of porphyrins, models for compound II.[101,102] The EXAFS data on compounds I and II and their Fourier transforms are shown in Fig. 7.26. The data could not be fit by a single shell of nitrogen atoms, but inclusion of the oxygen from the short-distance Fe=O bond as a second shell of ligands led to an excellent fit. The Fe—N bond length involving the imidazole fifth ligand cannot be distinguished from the Fe—N bond lengths with the pyrrole nitrogens, but must be included in an average for all Fe—N bonds, which is 2.00 ± 0.01 Å in both compounds I and II.[100]

The XANES data are consistent with iron(IV) in both compounds I and II and model compounds.[99,100] The EXAFS studies on the oxidation product of metmyoglobin and hydrogen peroxide indicate an Fe=O bond length of

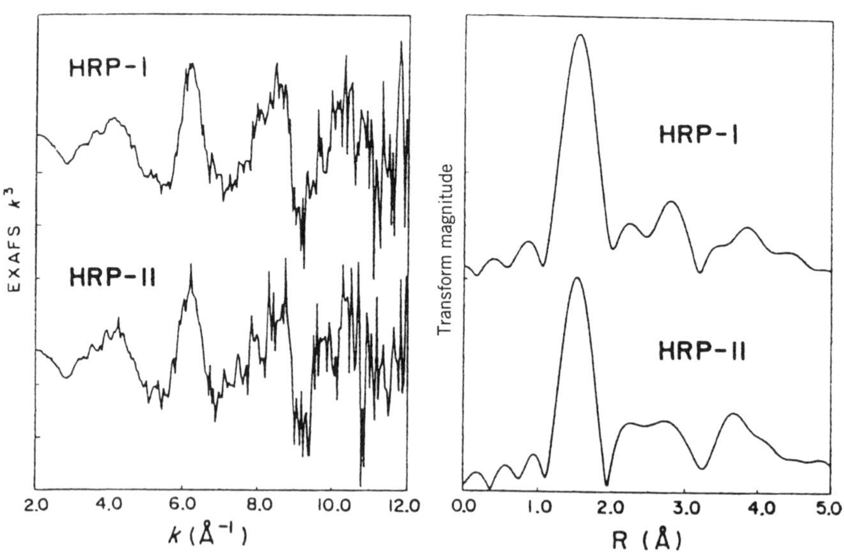

FIG. 7.26. EXAFS spectra (left) and Fourier transforms (right) for compounds I and II of horseradish peroxidase. Data were obtained on 5 mM peroxidase samples prepared at pH 7 at 0°C and frozen to –80°C. EXAFS data are weighted by k^3 to enhance the amplitude of high k oscillations. [Reproduced with permission from J. E. Penner-Hahn, K. M. Eble, T. J. McMurry, M. Renner, A. L. Balch, J. T. Groves, J. H. Dawson, and K. O. Hodgson, *J. Am. Chem. Soc.* **108,** 7819–7825 (1986).]

1.69 ± 0.03 Å.[103] The contradiction in bond lengths reported for Fe=O in horseradish peroxidase compound II remains unexplained. Whether a hydrogen bond to the Fe^{IV}=O group, unique to compound II, could account for the long Fe=O bond length is an important point that needs to be clarified.

7.5. X-RAY DIFFRACTION: CRYSTAL STRUCTURE OF HORSERADISH PEROXIDASE

An introduction to X-ray structural determinations was given in Section 3.3. The long-awaited crystal structure of horseradish peroxidase was published in 1997.[104] Direct contributions to the structure, determined in Denmark, involved a total of four countries. The techniques for cloning horseradish peroxidase in *E. coli* were established in the United Kingdom. The only other available class III plant peroxidase structural determination, that of peanut peroxidase obtained in the United States (Section 13.1.2), was used to help solve the horseradish peroxidase structure. Finally the large-scale production of peanut peroxidase occurred in Canada. Publication of the barley peroxidase structure (Section 13.2) followed shortly.

Horseradish peroxidase isoenzyme C was expressed in *E. coli* from the synthetic gene comprising 308 amino acids plus the Met required to initiate translation. Obtained at 2.15 Å resolution, 306 residues were visible in the crystal structure; the N-terminal Met and the two C-terminal residues were

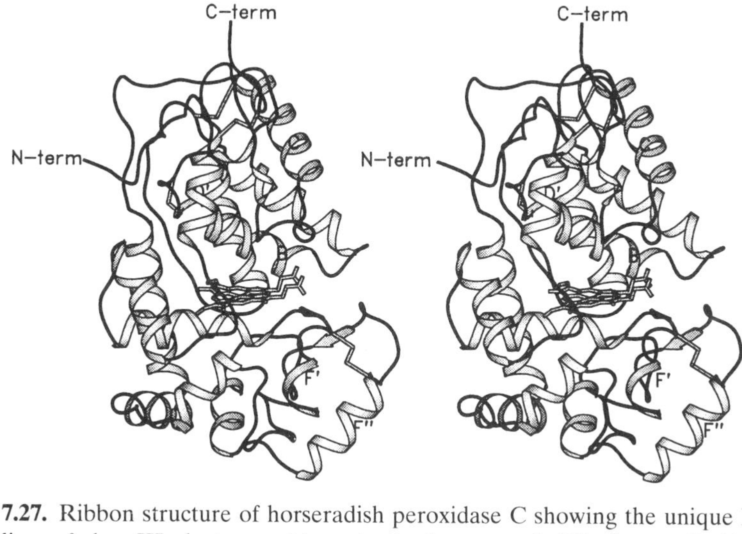

FIG. 7.27. Ribbon structure of horseradish peroxidase C showing the unique F′ and F″ helices of class III plant peroxidases in the foreground. (Kindly supplied by Drs. M. Gajhede and A. Henriksen.)

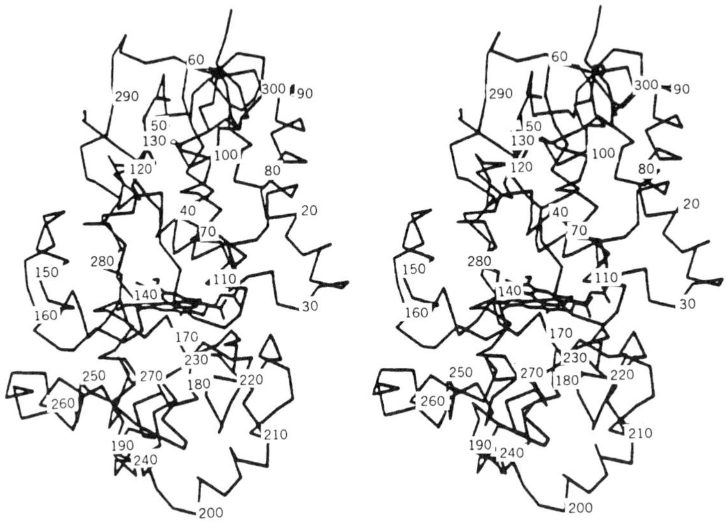

FIG. 7.28. Stereo carbon skeleton of horseradish peroxidase. [Reproduced with permission from M. Gajhede, D. J. Schuller, A. Henriksen, A. T. Smith, and T. L. Poulos, *Nature Struct. Biol.* **4,** 1032–1038 (1997).]

distorted. The major difference between peanut and horseradish peroxidases (class III) and the class I and II plant peroxidases is the addition of three helices, D′, F′ and F″, to the class III enzymes. A ribbon representation of the horseradish peroxidase structure, shown in Fig. 7.27, shows the F′ and F″ helices in the foreground. The carbon skeleton of horseradish peroxidase is shown in Fig. 7.28.

The outer substrate access channel is hydrophobic and contains three phenylalanine residues (68, 142, and 179) near the C18 methyl at the heme edge. Two of the phenylalanines are absent or have different orientations in the peanut enzyme. These aromatic residues contribute to an aromatic substrate binding site in horseradish peroxidase.

REFERENCES

1. M. Gouterman in D. Dolphin, ed., *The Porphyrins,* Academic, New York, Vol. IIIA, 1978, pp. 12–16.
2. M. Gouterman in Ref. 1, pp. 87–165.
3. F. Adar in Ref. 1, pp. 167–209.
4. G. R. Schonbaum and S. Lo, *J. Biol. Chem.* **247,** 3353–3360 (1972).
5. H. B. Dunford and J. S. Stillman, *Coord. Chem. Rev.* **19,** 187–251 (1976).
6. W. E. Blumberg, J. Peisach, B. A. Wittenberg, and J. B. Wittenberg, *J. Biol. Chem.* **243,** 1854–1862 (1968).

7. A. S. Brill and R. J. P. Williams, *Biochem. J.* **78,** 246–253 (1961).
8. D. Keilin and E. F. Hartree, *Biochem. J.* **49,** 88–104 (1951).
9. P.-I. Ohlsson and K.-G. Paul, *Acta Chem. Scand.* **B30,** 373–375 (1976).
10. H. Theorell, *Ark. Kemi Miner. Geol.* **16A,** No. 3, 1–11 (1942).
11. R. Makino and I. Yamazaki, *J. Biochem.* **72,** 655–664 (1972).
12. J. H. Dawson, S. Kadkhodayan, C. Zhuang, and M. Sono, *J. Inorg. Biochem.* **45,** 179–172 (1992).
13. M. Tamura, *Biochim. Biophys. Acta* **243,** 239–248 (1971).
14. W. D. Ellis and H. B. Dunford, *Arch. Biochem. Biophys.* **133,** 313–317 (1969).
15. K. Kobayashi, M. Tamura, K. Hayashi, H. Hori, and H. Morimoto, *J. Biol. Chem.* **255,** 2239–2242 (1980).
16. C. Phelps, E. Antonini, and M. Brunori, *Biochem. J.* **128,** 377–382 (1972).
17. J. B. Wittenberg, R. W. Noble, B. A. Wittenberg, E. Antonini, M. Brunori, and J. Wyman, *J. Biol. Chem.* **242,** 626–634 (1967).
18. H. Theorell and A. Ehrenberg, *Arch. Biochem. Biophys.* **41,** 442–461 (1952).
19. J. S. Stillman, M. J. Stillman, and H. B. Dunford, *Biochemistry* **14,** 3183–3188 (1975).
20. A. T. Tu, *Raman Spectroscopy in Biology. Principles and Applications,* Wiley, New York, 1982, pp. 3–61.
21. Ref. 1, pp. 346–349.
22. T. G. Spiro, *Biochim. Biophys. Acta* **416,** 169–189 (1975).
23. T. G. Spiro and X.-Y. Li in *Biological Applications of Raman Spectroscopy,* T. G. Spiro, ed., Wiley, New York, 1988, Vol. 3, pp. 1–37.
24. G. Smulevich and T. G. Spiro, *Methods Enzymol.* **226,** 397–408 (1993).
25. T. Kitagawa, M. Abe, and H. Ogoshi, *J. Chem. Phys.* **69,** 4516–4525 (1978).
26. M. Abe, T. Kitagawa, and Y. Kyogoku, *J. Chem. Phys.* **69,** 4526–4534 (1978).
27. S. Choi, T. G. Spiro, K. C. Langry, and K. M. Smith, *J. Am. Chem. Soc.* **104,** 4337–4344 (1982).
28. S. Choi, T. G. Spiro, K. C. Langry, K. M. Smith, D. L. Budd, and G. N. La Mar, *J. Am. Chem. Soc.* **104,** 4345–4351 (1982).
29. L. D. Spaulding, C. C. Chang, N.-T. Yu, and R. H. Felton, *J. Am. Chem. Soc.* **97,** 2517–2525 (1975).
30. N. Parthasarathi, C. Hansen, S. Yamaguchi, and T. G. Spiro, *J. Am. Chem. Soc.* **109,** 3865–3871 (1987).
31. F. Neri, D. Kok, M. A. Miller, and G. Smulevich, *Biochemistry* **36,** 8947–8953 (1997).
32. J. A. Shelnutt, X.-Z. Song, J.-G. Ma, S.-L. Jia, W. Jentzen, and C. J. Medforth, *Chem. Soc. Rev.* **27,** 31–41 (1998).
33. W. Jentzen, X.-Z. Song, and J. A. Shelnutt, *J. Phys. Chem.* **101,** 1684–1699 (1997).
34. J. A. Shelnutt, results presented at the Pre-Inorganic Chemistry-Bioionorganic Chemistry Meeting, Okazaki, Japan, July 25–26, 1997.
35. T. M. Loehr and J. S. Loehr, *Biochem. Biophys. Res. Commun.* **55,** 218–223 (1973).
36. G. Rakshit and T. G. Spiro, *Biochemistry* **13,** 5317–5323 (1974).

37. W. H. Woodruff and T. G. Spiro, *Appl. Spectrosc.* **28,** 576–578 (1974).
38. R. H. Felton, A. Y. Romans, N.-T. Yu, and G. R. Schonbaum, *Biochim. Biophys. Acta* **434,** 82–89 (1976).
39. T. G. Spiro, J. D. Stong, and P. Stein, *J. Am. Chem. Soc.* **101,** 2648–2655 (1979).
40. J. Teraoka and T. Kitagawa, *J. Biol. Chem.* **256,** 3969–3977 (1981).
41. A. J. Sitter, J. R. Shifflett, and J. Terner, *J. Biol. Chem.* **263,** 13032–13038 (1988).
42. A. Feis, M. P. Marzocchi, M. Paoli, and G. Smulevich, *Biochemistry* **33,** 4577–4583 (1994).
43. Y. Hayashi and I. Yamazaki, *J. Biol. Chem.* **254,** 9101–9106 (1979).
44. B. D. Howes, J. N. Rodriguez-Lopez, A. T. Smith, and G. Smulevich, *Biochemistry* **36,** 1532–1543 (1997).
45. H. E. Van Wart and J. Zimmer, *J. Biol. Chem.* **260,** 8372–8377 (1985).
46. T. G. Spiro and T. C. Strekas, *J. Am. Chem. Soc.* **96,** 338–345 (1974).
47. J. S. Stillman, M. J. Stillman, and H. B. Dunford, *Biochemistry* **14,** 3183–3188 (1975).
48. A. D. Nadezhdin and H. B. Dunford, *Photochem. Photobiol.* **29,** 899–903 (1979).
49. H. E. Van Wart and J. Zimmer, *J. Am. Chem. Soc.* **107,** 3379–3381 (1985).
50. J. Terner, A. J. Sitter, and C. M. Reczek, *Biochim. Biophys. Acta* **828,** 73–80 (1985).
51. A. J. Sitter, C. M. Reczek, and J. Terner, *J. Biol. Chem.* **260,** 7515–7522 (1985).
52. R. Makino, T. Uno, Y. Nishimura, T. Iizuka, M. Tsuboi, and Y. Ishimura, *J. Biol. Chem.* **261,** 8376–8382 (1985).
53. I. Yamazaki, T. Araiso, Y. Hayashi, H. Yamada, and R. Makino, *Adv. Biophys.* **11,** 249–281 (1978).
54. J. E. Critchlow and H. B. Dunford, *J. Biol. Chem.* **247,** 3703–3713 (1975).
55. S. Hashimoto, Y. Tatsuno, and T. Kitagawa, *Proc. Natl. Acad. Sci. U.S.A.* **83,** 2417–2421 (1986).
56. S. Hashimoto, Y. Tatsuno, and T. Kitagawa, *J. Am. Chem. Soc.* **109,** 8096–8097 (1987).
57. K.-J. Paeng and J. R. Kincaid, *J. Am. Chem. Soc.* **110,** 7913–7915 (1988).
58. T. Ogura and T. Kitagawa, *J. Am. Chem. Soc.* **109,** 2177–2179 (1987).
59. W. A. Oertling and G. T. Babcock, *Biochemistry* **27,** 3331–3338 (1988).
60. V. Palaniappan and J. Terner, *J. Biol. Chem.* **264,** 16046–16053 (1989).
61. V. Palaniappan, M. M. Fitzgerald, A. M. Sullivan, J. R. Shifflett, and J. Terner in *Time Resolved Vibrational Spectroscopy V,* H. Takahashi, ed., Springer, Berlin, 1992, pp. 32–35.
62. J. Terner, Department of Chemistry, Virginia Commonwealth University, Richmond, VA, personal communication, March 11, 1997.
63. J. R. Kincaid, Y. Zheng, J. Al-Mustafa, and K. Czarnecki, *J. Biol. Chem.* **271,** 28805–28811 (1996).
64. G. Smulevich, A. M. English, A. R. Mantini, and M. P. Marzocchi, *Biochemistry* **30,** 772–779 (1991).
65. G. Smulevich, M. Paoli, G. De Sanctis, A. R. Mantini, F. Ascoli, and M. Coletta, *Biochemistry* **36,** 640–649 (1997).

66. G. Smulevich, M. Paoli, J. F. Burke, S. A. Sanders, R. N. F. Thorneley, and A. T. Smith, *Biochemistry* **33,** 7398–7407 (1994).
67. G. Smulevich, *Biochem. Soc. Trans.* **23,** 240–243 (1995).
68. I. E. Holzbaur, A. M. English, and A. A. Ismail, *J. Am. Chem. Soc.* **118,** 3354–3359 (1996).
69. I. E. Holzbaur, A. M. English, and A. A. Ismail, *Biochemistry* **35,** 5488–5494 (1996).
70. G. Smulevich, M. Paoli, G. De Sanctis, A. R. Mantini, F. Ascoli, and M. Coletta, *Biochemistry* **36,** 640–649 (1997).
71. J. G. Foss, *J. Chem. Ed.* **40,** 592–597 (1963).
72. C. Djerassi, *Optical Rotatory Dispersion,* McGraw-Hill, New York, 1960.
73. E. R. Blout in Ref. 72, pp. 238–273.
74. Y. R. Shen, *Phys. Rev.* **133,** A511–A515 (1964).
75. A. D. Buckingham and P. J. Stephens, *Ann. Rev. Phys. Chem.* **17,** 399–432 (1966).
76. P. N. Schatz, A. J. McCaffrey, W. Suëtaka, G. N. Henning, A. B. Ritchie, and P. J. Stephens, *J. Chem. Phys.* **45,** 722–734 (1966).
77. P. N. Schatz and A. J. McCaffery, *Quart. Rev.* **23,** 552–584 (1969).
78. S. B. Piepho and P. N. Schatz, *Group Theory in Spectroscopy with Applications to Magnetic Circular Dichroism,* Wiley, New York, 1983.
79. Ref. 8, pp. 480–532.
80. J. H. Dawson and D. M. Dooley in *Iron Porphyrins, Part 3,* A. B. P. Lever and H. B. Gray, eds., VCH, New York, 1989, pp. 1–135.
81. T. Nozawa, N. Kobayashi, and M. Hatano, *Biochim. Biophys. Acta* **427,** 652–662 (1976).
82. N. Kobayashi, T. Nozawa, and M. Hatano, *Biochim. Biophys. Acta* **493,** 340–351 (1977).
83. J. H. Dawson, S. Kadkhodayan, C. Zhuang, and M. Sono, *J. Inorg. Biochem.* **45,** 179–192 (1992).
84. W. R. Browett, Z. Gasyna, and M. J. Stillman, *Inorg. Chim. Acta* **79,** 113–114 (1983).
85. N. Foote, P. M. A. Gadsby, M. J. Berry, C. Greenwood, and A. J. Thomson, *Biochem. J.* **246,** 659–668 (1987).
86. P. J. Riggs-Gelasco, T. L. Stemmler, and J. E. Penner-Hahn, *Coord. Chem. Rev.* **144,** 245–286 (1995).
87. B. M. Kincaid and P. Eisenberger, *Phys. Rev. Lett.* **34,** 1361–1364 (1975).
88. R. G. Shulman, P. Eisenberger, and B. M. Kincaid, *Ann. Rev. Biophys. Bioeng.* **7,** 559–578 (1978).
89. S. P. Cramer and K. O. Hodgson, *Prog. Inorg. Chem.* **25,** 1–39 (1979).
90. P. A. Lee, P. H. Citrin, P. Eisenberger, and B. M. Kincaid, *Rev. Mod. Phys.* **53,** 769–806 (1981).
91. R. A. Scott, *Methods Enzymol.* **117,** 414–459 (1985).
92. C. D. Garner, *Adv. Inorg. Chem.* **36,** 303–339 (1991).
93. S. P. Cramer and R. A. Scott, *Rev. Sci. Instrum.* **52,** 395–399 (1981).
94. B.-K. Teo and P. A. Lee, *J. Am. Chem. Soc.* **101,** 2815–2832 (1979).

95. J. E. Penner-Hahn, Department of Chemistry, University of Michigan, personal communication, January 28, 1997.
96. R. G. Shulman, P. Eisenberger, B.-K. Teo, B. M. Kincaid, and G. S. Brown, *J. Mol. Biol.* **124,** 305–321 (1978).
97. N. J. Blackburn, S. S. Hasnain, N. Binsted, G. P. Diakun, C. D. Garner, and P. F. Knowles, *Biochem. J.* **219,** 985–990 (1984).
98. J. E. Penner-Hahn, T. J. McMurry, M. Renner, L. Latos-Grazynsky, K. M. Eble, I. M. Davis, A. L. Balch, J. T. Groves, J. H. Dawson, and K. O. Hodgson, *J. Biol. Chem.* **258,** 12761–12764 (1983).
99. B. Chance, L. Powers, Y. Ching, T. Poulos, G. R. Schonbaum, I. Yamazaki, and K. G. Paul, *Arch. Biochem. Biophys.* **235,** 596–611 (1984).
100. J. E. Penner-Hahn, K. M. Eble, T. J. McMurry, M. Renner, A. L. Balch, J. T. Groves, J. H. Dawson, and K. O. Hodgson, *J. Am. Chem. Soc.* **108,** 7819–7825 (1986).
101. D. H. Chin, G. N. La Mar, and A. L. Balch, *J. Am. Chem. Soc.* **102,** 4344–4350 (1980).
102. M. Schappacher, R. Weiss, R. Montiel-Montoya, and A. Trautwein, *J. Am. Chem. Soc.* **107,** 3736–3738 (1985).
103. M. Chance, L. Powers, C. Kumar, and B. Chance, *Biochemistry* **25,** 1259–1265 (1986).
104. M. Gajhede, D. J. Schuller, A. Henriksen, A. T. Smith, and T. L. Poulos, *Nature Struct. Biol.* **4,** 1032–1038 (1997).

CHAPTER 8

Spectroscopy of Horseradish Peroxidase. II: Nuclear Magnetic Resonance, Electron Paramagnetic Resonance, Electron Nuclear Double Resonance, Mössbauer Spectroscopy, and Theoretical Studies

8.1. NUCLEAR MAGNETIC RESONANCE SPECTROSCOPY

8.1.1. Introduction

Since the discovery of nuclear magnetic resonance (NMR) in 1946, (which resulted in the Nobel prize in physics being awarded to Konrad E. Bloch and Edward M. Purcell in 1952), a multibillion-dollar industry has developed, with NMR spectrometers becoming standard equipment in chemistry, biochemistry, and medical laboratories. There are excellent reviews showing the development of NMR spectroscopy, of which I cite a few examples.[1–6] The field has grown and changed so rapidly that a nonexpert has trouble keeping up with new techniques. Since the development of NOE (Nuclear Overhauser Enhancement), there appears to be competition for the cutest acronym: COSY (COrrelated SpectroscopY), ROESY (Rotating frame Overhauser Enhancement SpectroscopY), TOCSY (TOtal Correlation SpectroscopY), HOHAHA (HOmonuclear HArtmann HAhn spectroscopy), WEFT (Water-Eliminated Fourier Transform), and DEFT (Driven Equilibrium Fourier Transform).[7] Definite losers in this competition are BURP and INEPT-INADEQUATE. A rough rule-of-thumb on the cost of instruments was a kilodollar per megahertz; thus a 300-MHz NMR machine would cost about $300,000. However, the cost is now outpacing the strength of the magnet by a large margin. Few laboratories can afford a 600-MHz instrument, let alone one of 800-MHz. Although few of us will become specialists in NMR, it is important to understand the information that NMR can provide.

By far the most common type of NMR experiment is concerned with proton spectra. The ^1H nuclei can be thought of as tiny magnets with spin quantum number I of $\frac{1}{2}$. Since these tiny magnets may align themselves with or opposed to a strong external magnetic field, indicated by magnetic quantum numbers m_I of $\pm\frac{1}{2}$, their energy levels are split in two, α and β, in the presence of an external magnetic field. The spins that align themselves with the magnetic field, β, are of lower energy and those that oppose the field, α, are of higher energy. The populations of the two energy levels follow a Boltzmann distribution; so there is a slight excess population in the lower level. Thus, the sample possesses a net magnetization. The magnetic component of electromagnetic radiation of appropriate frequency will cause protons to flip from the lower to the higher level. This is called the *resonance frequency* (ν), and it occurs in the radio-frequency region of the electromagnetic spectrum. Thus, the basic requirements for an NMR experiment are a strong magnet, a radio-frequency transmitter and receiver, and a data aquisition system.

The stationary field (magnetic flux density) provided by the strong magnet is designated B_0, and its direction is defined as along the z axis. The net magnetization of the sample, M_0, aligns itself with B_0. In a basic NMR experiment a 90° pulse is applied from the radio-frequency transmitter, creating an additional magnetic field, B_1, at right angles to B_0. The result is a complete transfer of the M_0 magnetization into the xy plane, now called M_1, which oscillates (precesses) at the resonance frequency ν. For example, ν is 500 MHz for protons in a field of 11.75 T. The oscillating magnetization vector M_1 induces a signal that is detected in a receiver coil, amplified, and recorded as a time-dependent signal known as free induction decay (FID). By means of a mathematical process, Fourier transformation (FT), the time-dependent signal is transformed into frequencies (NMR peaks). Thus, the resonance energies of protons in different environments are recorded.

Protons are shielded in different ways from the magnetic field by the electron clouds in their immediate environment. Hence, resonance frequencies are unique for each unique environment. The shifts in resonance frequencies are called *chemical shifts* (δ), and they may provide unambiguous identification of each proton. Thus, in a poorly resolved NMR spectrum of ethanol the three equivalent methyl protons have a unique NMR peak, as do the two methylene protons and the single hydroxyl proton. The positions of these peaks identify the chemical shifts. The peak areas are proportional to the number of protons: 3, 2, and 1. In a high-resolution NMR spectrum, interactions called *spin–spin coupling* can be observed of protons attached to one atom with protons attached to an adjacent atom. The spin–spin coupling splits the signals into several components, doublets, triplets, quartets, and so forth. The strength of the coupling is defined by a *coupling constant* J. In the high-resolution NMR spectrum of ethanol, the spin–spin coupling between the methyl and methylene protons splits the energy levels of the

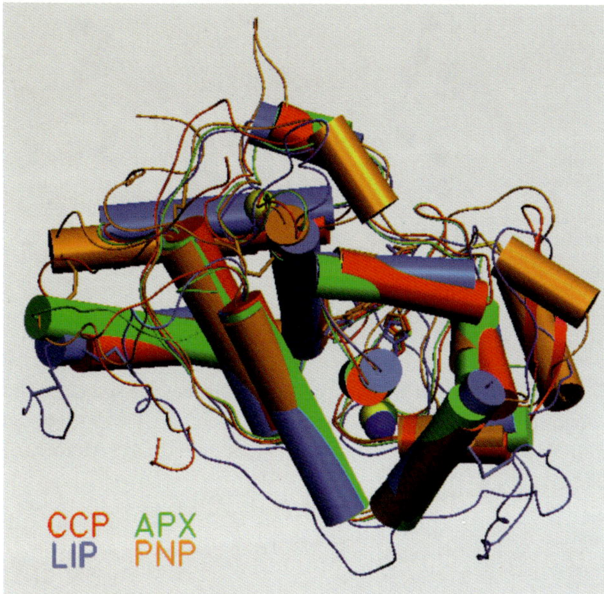

Fig. 3.3. Ribbon and helical structures for representative peroxidases (see p. 38). Class I, yeast cytochrome *c* peroxidase (CCP) and ascorbate peroxidase (APX). Class II, lignin peroxidase. (LTP) Class III, peanut peroxidase (PNP). [Reproduced with permission from D.J. Schuller, N. Ban, R.B. van Huystee, A. McPherson, and T. L. Poulos, *Structure* **4,** 311-321 (1996).]

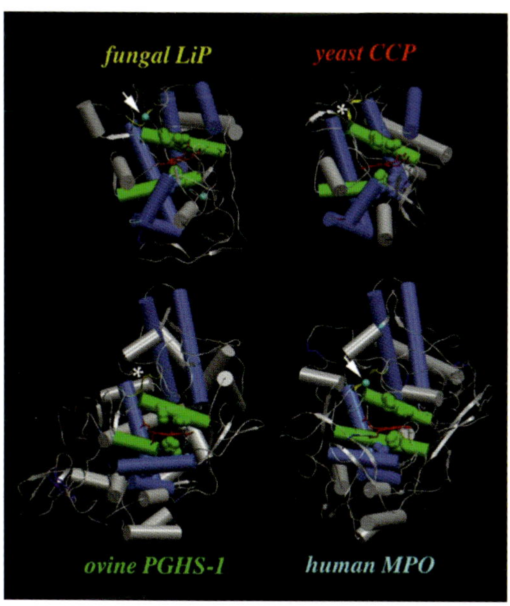

Fig. 3.5. Representations of four different peroxidase structures obtained in four different laboratories: those of lignin, yeast, and human myelo-peroxidases, and bottom left, that of ram prostaglandin H synthase isoenzyme 1 (see p. 41). The arrows point to calcium ion binding sites and the asterisks locate corresponding vestigal, but nonfunctional sites. (Figure kindly supplied by Dr. Michael Garavito, Department of Biochemistry, Michigan State University.)

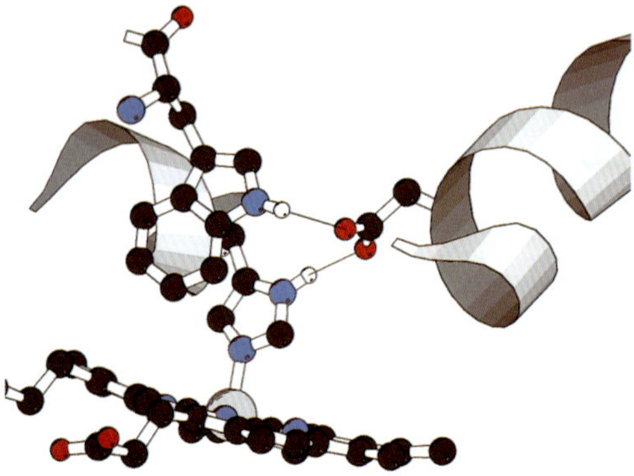

Fig. 3.10. View of the Asp-His-Fe triad of yeast cytochrome *c* peroxidase. The thin lines indicate hydrogen bonds between Asp235 and both His175 and the Trp191 free radical center (see p. 46). [Reproduced with permission from D.B. Goodin and D.E. McRee, *Biochemistry* **32,** 3313-3324 (1993).]

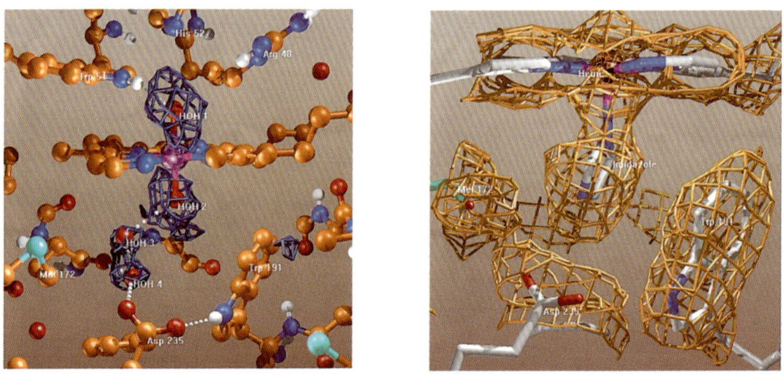

Fig. 3.11. Crystal structures of the His175Gly mutant and its imidazole complex in the vicinity of the heme. Left, the bisaquo complex of the mutant. Right, the imidazole complex (see p. 47). [Reproduced with permission from D. E. McRee, G. M. Jensen, M. M. Fitzgerald, H. A. Siegel, and D. B. Goodin, *Proc. Natl. Acad. Sci. U.S.A.* **91,** 12847-12851 (1994).]

Fig. 3.13. Cavity created in the Trp191Gly mutant. The surface of the cavity, in grey, is cut away to show the interior, colored yellow. Locations of water molecules bound in the cavity are shown in blue (see p. 50). [Reproduced with permission from M. M. Fitzgerald, M. J. Churchill, D. E. McRee and D. B. Goodin, *Biochemistry* **33,** 3807-3818 (1994).]

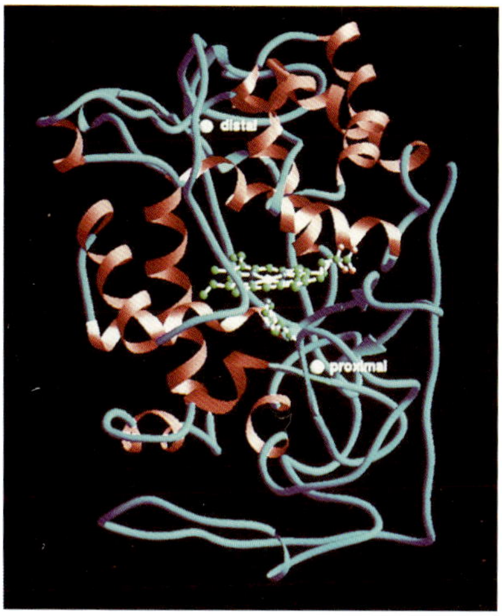

Fig. 12.14. Diagram of the lignin peroxidase structure showing the four disulfide bonds, the heme, proximal histidine ligand, and N and C terminal ends (see p. 294). [Reproduced with permission from T. L. Poulos, S. L. Edwards, H. Wariishi, and M. H. Gold, *J. Biol. Chem* **268,** 4429-4440 (1993).]

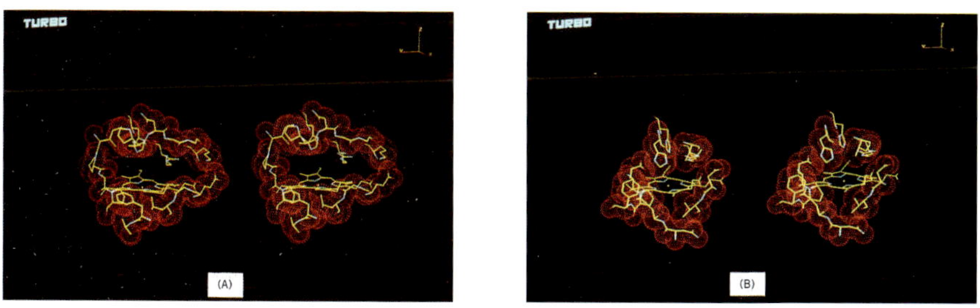

Fig. 12.19. Channels leading to the distal sides of the hemes of *(a) Arthromyces ramosus* and *(b)* lignin peroxidases (see p. 301). For *(a),* residues Lys48, Pro90, Gly94, Ile152, Pro155, Glu189, Leu191, Phe229 and Ala230 (corrected numbering) are involved. For *(b),* His82, Ile85, Glu146, Phe148, Asp183, Val184, Asn221 and Gln222. [Reproduced from N. Kunishima, K. Fukuyama, H.Matsubara, H. Hatanaka, Y. Shibano, and T. Amachi, *J. Mol. Biol.* **235,** 331-344 (1994) by permission of the publisher Academic Press Limited, London.]

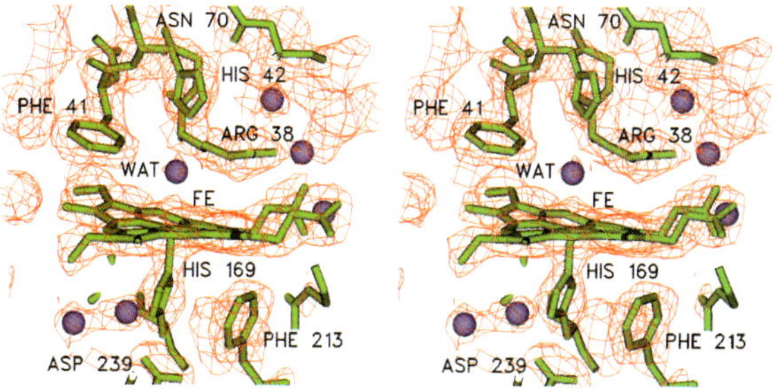

Fig. 13.3. Heme environment of peanut peroxidase (see p. 312). [Reproduced with permission from D. J. Schuller, N. Ban, R. B. van Huystee, A. McPherson, and T. L. Poulos, *Structure* **4,** 311-321 (1996).]

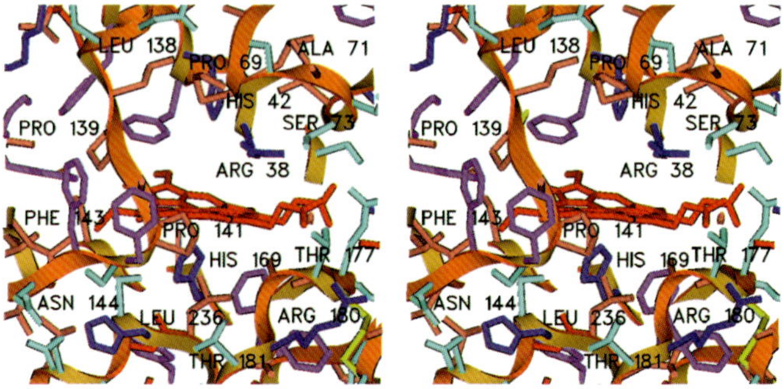

Fig. 13.4. The substrate access channel to the heme of peanut peroxidase (see p. 312). [Reproduced with permission from D. J. Schuller, N. Ban, R. B. van Huystee, A. McPherson, and T. L. Poulos, *Structure* **4,** 311-321 (1996).]

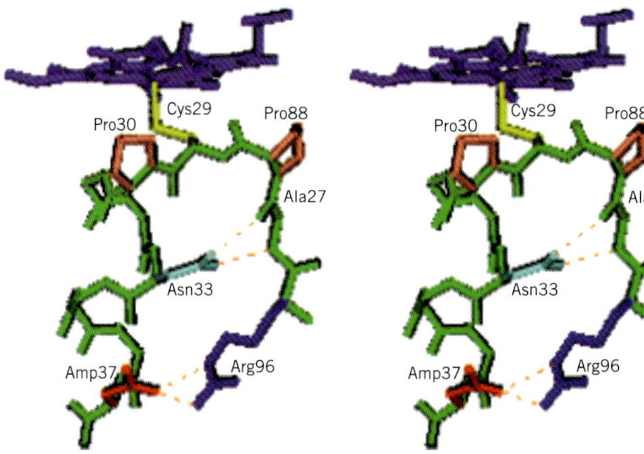

Fig. 14.10. Stereo diagram showing the ligation of Cys29 to the heme iron of chloroperoxidase and hydrogen bonds (dashed lines) which help to stabilize cysteine-ligand loop (see p. 332). [Reproduced with permission from M. Sundaramoorthy, J. Terner, and T. L. Poulos, *Structure* **3,** 1367-1377 (1995).]

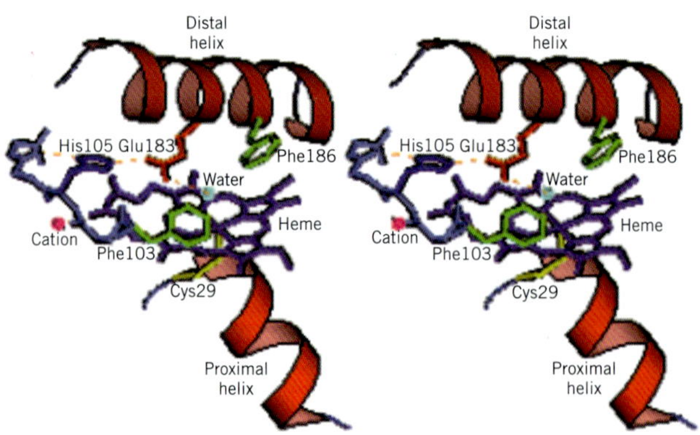

Fig. 14.11. Relation of the proximal (A) and distal (F) helices to the heme of chloroperoxidase. The water in the distal cavity is too far removed to be a ligand (see p. 333). [Reproduced with permission from M. Sundaramoorthy, J. Terner and T. L. Poulos, *Structure* **3,** 1367-1377 (1995).]

methylene protons into a quartet and of the methyl protons into a triplet (Fig. 8.1), a result obtained using the simple statistics of Pascal's triangle.[8] These splittings give rise to a *multiplet structure.* The OH proton of ethanol exchanges rapidly with other solvent molecules and so exhibits a broad single resonance. The total peak areas for the methyl triplet, the methylene quartet and the hydroxyl proton singlet maintain the ratio of 3:2:1. Nuclear magnetic resonance experiments are more difficult to carry out in H_2O since the intense signal from the water protons obscures other peaks. This problem can be circumvented by using deuterium oxide, D_2O, as solvent; alternatively there are pulse sequences that can be used to eliminate the water proton signal.

An NMR spectrum is displayed relative to a standard, the resonance frequency of a compound with a single sharp peak such as tetramethyl silane (TMS) or the water-soluble *2,2*-dimethyl-*2*-silapentane-*5*-sulfonate (DSS). Chemical shifts are recorded in parts per million according to the equation

$$\delta = 10^6(\nu_{obs} - \nu_{ref})/\nu_{ref}$$

where ν_{obs} and ν_{ref} are the resonance frequencies of the protons of the compound of interest and of the reference compound. Electron-donating groups such as aliphatic groups provide strong shielding, and small downfield chemical shifts are observed relative to the reference compound. Electronegative groups, on the other hand, withdraw electrons so the protons in their immediate environment are less shielded. These protons will have large downfield shifts. The NMR spectrum of the diamagnetic compound Zn^{2+}–porphin is shown in the top part of Fig. 8.2, in which it can be seen that all of the proton resonances are shifted downfield compared to the standard.

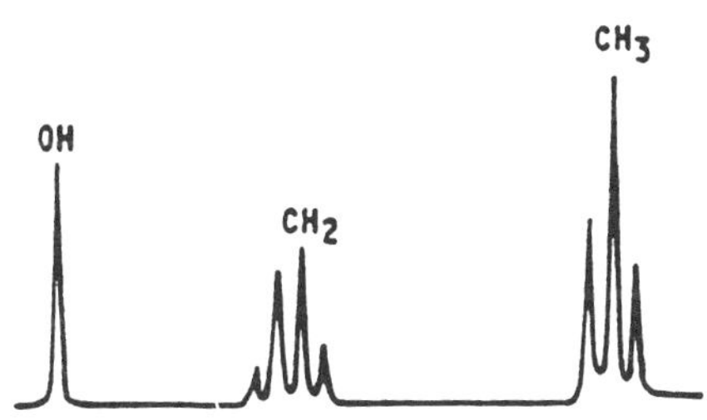

FIG. 8.1. NMR spectrum of ethanol. The electronegative oxygen atom tends to remove the electron cloud from its attached hydrogen atom, so the proton has least screening from the influence of the magnetic field, and its resonance frequency is shifted the farthest downfield. The methyl group is an electron donor, so its proton resonances are most highly shielded and occur upfield compared to those for the methylene and hydroxyl protons.

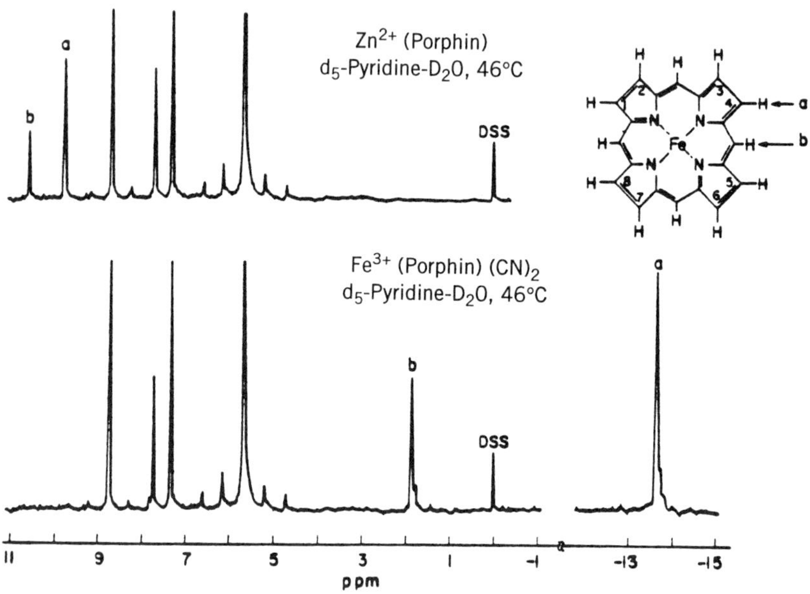

FIG. 8.2. ^1H NMR spectra of Zn^{2+}–porphin and Fe^{3+}–porphin–$(CN)_2$ at 220 MHz. The resonances *a* and *b* were identified from their relative intensities. Other resonances correspond to solvent resonances and their spinning side bands. The paramagnetic low-spin iron(III) causes a dramatic upfield shift of the *a* and *b* resonances. (Reproduced with permission from K. Wüthrich, *NMR in Biological Research: Peptides and Proteins,* North-Holland, Amsterdam, 1976, p. 228, with permission from Elsevier Science.)

The appearance of NMR spectra, indeed the entire NMR theory, is dominated by two relaxation mechanisms: longitudinal, characterized by a relaxation time T_1, and transverse, characterized by T_2. First a discussion of tranverse relaxation and T_2.

When the M_1 magnetization is created by the 90° pulse, magnetization vectors of the same type of protons, but originating from different molecules, are coherently aligned in the *xy* plane. As they precess, the coherence is gradually lost and the magnetization spreads over the entire 360°, resulting in a loss of signal. The speed of this relaxation process is characterized by T_2; and the value of T_2 directly affects the resonance signals. Signals with short T_2 are broad, and in extreme cases are so broad they are undetectable. On the other hand signals with long T_2 values are sharp. Assuming a homogeneous (i.e., well shimmed) magnetic field, T_2 can be estimated from the inverse of the line width at half-height.

On the other hand, T_1 is the time taken for M_1 magnetization to return to its original alignment with the *z* axis, where it is labeled M_0. Relaxation of T_1 follows an exponential rate and so the time of at least five T_1's must have elapsed for the complete return of M_1 to the z axis. In order to increase the intensity of an NMR spectrum, sequential 90° pulses may be applied. If the delay time between these pulses is too short, then inaccurate (partially saturated) spectra are obtained since the M_1 magnetization has not completely decayed back to M_0 before the next pulse.

8.1.2. Nuclear Overhauser Enhancement

^{13}C also has a nuclear spin of $\frac{1}{2}$. Because of its small natural abundance and lower NMR sensitivity, it has an NMR signal roughly 10^{-4} that of a proton. However, it exhibits chemical shifts in the range of 0–200 ppm compared to 0–10 ppm for protons. Modern spectrometers can detect natural abundance ^{13}C, and cells can be fed an enriched ^{13}C diet, so that ^{13}C-enriched proteins can be isolated. Thus, ^{13}C NMR has become much easier and very important.

The example of a C—H bond will be used to illustrate an NOE experiment. The ^1H and ^{13}C of a C—H bond will interact strongly because they are in close proximity. Multiplet splitting results from the application of the radio-frequency field B_1, which decreases the intensity of the ^{13}C resonance. The multiplet splitting can be eliminated by saturation of the proton resonance with a second fixed radio-frequency field, B_2, effectively destroying it. This also shifts the Boltzmann distribution of the two ^{13}C energy levels. Thus, a single ^{13}C peak is observed with up to three times its original intensity. Since two fields, B_1 and B_2, have been applied, a double resonance experiment has been performed.

Despite the example of atoms attached by a bond, an NOE experiment gives "through-space," not "through-bond," information. Its magnitude is dependent on $1/r^6$, where r is the internuclear distance. An NOE experiment can also be performed on nuclei of the same type. One can determine which of a pair of protons is nearest neighbor to a third proton, which is irradiated with its resonance frequency. The one with the greatest peak intensity enhancement is the nearest neighbor. Thus, important structural and conformational information can be obtained.

8.1.3. Two-Dimensional Nuclear Magnetic Resonance

Two-dimensional NMR has existed for over 25 years. Its development is largely a result of the work of Richard R. Ernst for which he received the Nobel prize in chemistry in 1991. Its increasing use has been dependent upon improved instrumentation and especially upon the availability of computers and software packages capable of handling large amounts of data. The basic technique is to use a series of pulse sequences, and to collect data as a function of an incremented delay time between pulses. By proper choice of pulses and delay times one eliminates extraneous data and collects only the results of interest. The use of these incremented pulse sequences is very sophisticated compared to a basic NMR experiment; the techniques are very different from using 90° pulses at well-spaced constant time intervals to enhance peak intensities.

One type of two-dimensional experiment is based on through-bond or scalar interactions. These experiments are usually one of the COSY, TOCSY, or HOHAHA type. A COSY proton NMR experiment is comparable to a one-dimensional double resonance or spin–spin decoupling experiment. Without going into the experimental details, the results can be expressed in a two-dimensional contour diagram in which there are two frequency axes and the peaks are viewed end-on. The conventional one-dimensional NMR spectrum is displayed along the diagonal. Thus, the location of the peak for one proton H_1 can be described by coordinates ν_1, ν_1, and that of the peak for a second proton H_2 by ν_2, ν_2. The existence of cross-peaks or off-diagonal peaks at ν_1, ν_2 and ν_2, ν_1 provides valuable correlation data from which structural information is obtained. An example of a two-dimensional NMR spectrum is shown in Fig. 8.3.

In another type of two-dimensional experiment through-space or dipolar interactions are measured by the NOESY technique, which is the two-dimensional equivalent of NOE. Data can be plotted in the same type of contour map used for COSY results. The appearance of cross-peaks indicate close proximity in space of the nuclei involved, generally ≤ 5 Å.

8.1.4. Influence of a Paramagnetic Center

The magnetic moment of an unpaired electron is about 660 times as great as that of a proton. Therefore, if a molecule has an unpaired electron (paramagnetic center), there will be profound effects on its NMR spectrum.[9] Whereas a paramagnetic contaminant is the bane of NMR spectroscopists, valuable

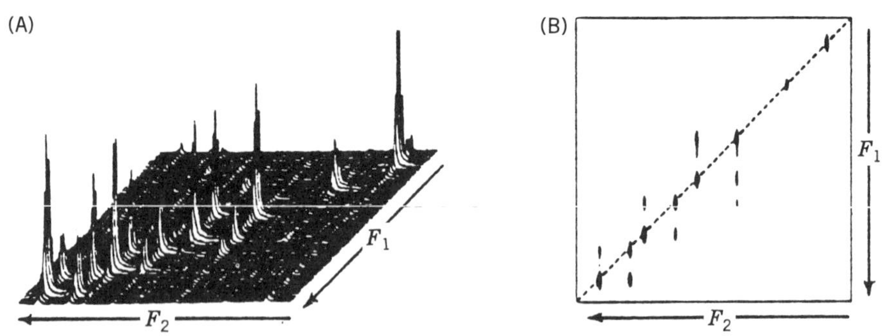

FIG. 8.3. Different representations of a 2D NMR spectrum: (*a*) as a stacked plot, and (*b*) the same spectrum as a contour plot. F_1 and F_2 are the two frequency axes. The cross (off-diagonal) peaks are readily apparent in the contour plot. (Reproduced with permission from H. Günther, *NMR Spectroscopy,* 2nd ed., Wiley, New York, 1995, p. 277).

information can be gained from the effect of an unpaired electron, which either is in a constituent atom of the molecule of interest or is present in a deliberately added externally added species that binds to the molecule.

The shifts caused by a paramagnetic center, called *hyperfine shifts,* can be extremely valuable in obtaining structural information about proteins. Both the ^{13}C and ^{1}H NMR spectra of a protein contain so many overlapping peaks that the identification of each peak is impossible. However, by using a shift reagent, peaks from nearby nuclei are shifted out from under the massive envelope of diamagnetic overlapping peaks. When combined with other information, such as amino acid sequence data, the shifted peaks may be identified and their distances from the shift reagent quantified.

The NMR spectrum of paramagnetic Fe^{3+}(porphin) $(CN)_2$ in Fig. 8.2 shows hyperfine shifts. The pyrrole and methene bridge proton resonances are shifted upfield by the paramagnetic center, as shown in the lower part of Fig. 8.2. In a heme protein there are two kinds of shift due to the paramagnetic center: contact (scalar) and dipolar or pseudo-contact (through space).

8.1.5. Nuclear Magnetic Resonance Spectra of Horseradish Peroxidase

In 1975 several NMR studies on horseradish peroxidase were published. In one study, ahead of its time, it was concluded from proton NMR spectra that there is no water in the sixth coordination position of the ferric ion.[10] The values of the relaxivity, $1/T_1$ expressed on a per mole basis, for the water protons in a native horseradish peroxidase solution were about the same as for the cyanide complex of the enzyme. No enhancement of proton relaxivity was observed in contrast to cytochrome *c* and metmyoglobin, where it is known that iron-coordinated water is present (Fig. 8.4). The indolepropionic acid complex with horseradish peroxidase was studied by ^{1}H NMR.[11,12] Proton NMR spectra were recorded for native high spin, as well as the low-spin cyanide complex of horseradish peroxidase and several turnip peroxidase isoenzymes. It was concluded that histidine is the proximal ligand,[13] which was later confirmed in a study that showed that the imidazole proton does not exchange with solvent.[14]

In a series of studies from the laboratory of Isao Morishima, comparisons of horseradish peroxidase and myoglobin structures,[15] aromatic donor binding,[16] effect of pressure,[17] and influence of calcium binding[18,19] were investigated by NMR. The downfield portion of the NMR spectrum of calcium-free and native horseradish peroxidase is shown in Fig. 8.5. The four heme methyl groups have resonances at 77.6, 70.1, 67.0, and 50.1 ppm for the native enzyme, which are shifted to 70.9, 63.3, 61.1, and 46.1 ppm upon removal of the calcium. The results indicate that one of the two calcium ions in horseradish peroxidase is bound near the heme and is necessary to maintain the structural integrity of the enzyme. Furthermore, titration of the calcium-free enzyme with one equivalent of calcium ion restores the NMR spectrum of the native enzyme.

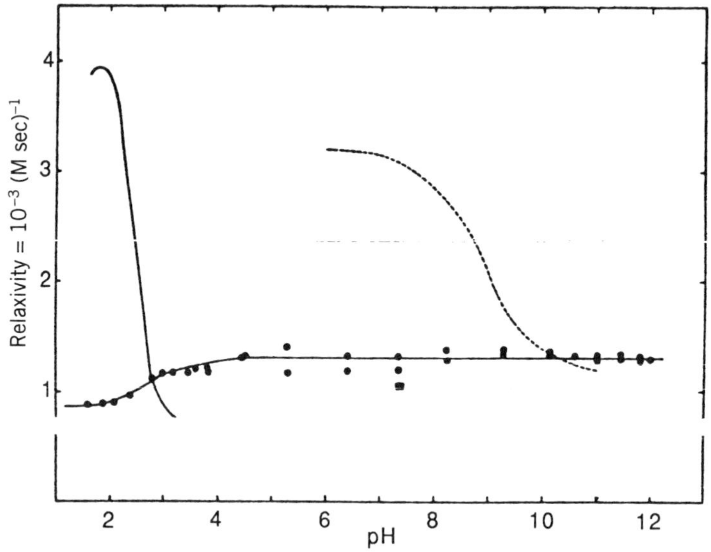

FIG. 8.4. Proton molar relaxivity as a function of pH for horseradish peroxidase (—•—•—), cytochrome *c* (solid line), and metmyoglobin (dashed line), peroxidase cyanide complex (■), reduced peroxidase(▲). The enhanced relaxivity caused by cytochrome *c* and metmyoglobin, where iron-coordinated water is known to be present, compared to the results for horseradish peroxidase, indicate that the iron(III) of the peroxidase does not have water in the sixth coordination position. [Reproduced with permission from A. Lanir and A. Schejter, *Biochem. Biophys. Res. Commun.* **62,** 199–203 (1975).]

The NMR studies on compounds I and II of horseradish peroxidase led to the erroneous conclusion that the free radical of compound I is located on an amino acid residue.[20,21] A later NMR study was consistent with the porphyrin π-cation radical structure.[22]

The binding of iodide,[23] azide,[24] and thiocyanate[25] to horseradish peroxidase has been investigated by NMR.

Comprehensive NMR investigations of horseradish peroxidase have come from the laboratory of Gerd La Mar.[26] By means of deuterium labels in key positions of ferriprotoporphyrin IX and deuteroheme, many of the hyperfine shifted resonances of reconstituted holoenzyme were assigned using a 360-MHz instrument.[26] The study with labeled hemes was extended to compound I, and further evidence for a porphyrin π-cation radical was obtained.[27]

It was shown that the nuclear Overhauser effect could be utilized, which when combined with the isotopic labeling and saturation transfer spectroscopy, led to the most comprehensive assignment of resonances in horseradish peroxidase up to 1987.[28] All heme resonances were assigned. The numbering scheme for these results is shown in Fig. 8.6 and a portion of these results in Fig. 8.7.

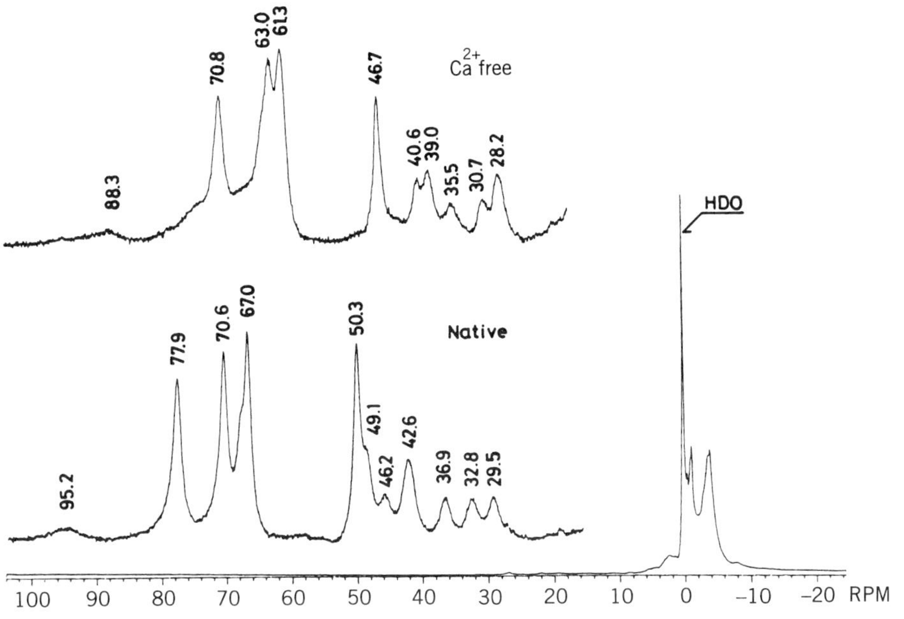

FIG. 8.5. Downfield shifts of the calcium-free horseradish peroxidase (upper NMR spectrum) and native enzyme (lower spectrum). [Reproduced with permission from Y. Shiro, M. Kurono, and I. Morishima, *J. Biol. Chem.* **261,** 9382–9390 (1986).]

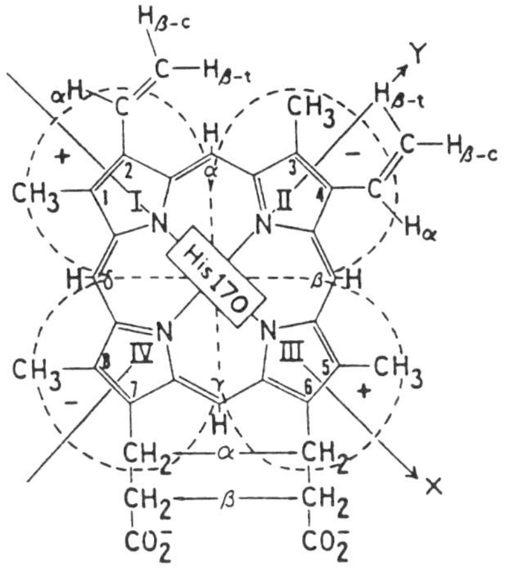

FIG. 8.6. Structure and numbering system for hemin in the cyanide complex of horseradish peroxidase with 2,4-vinyl groups in trans and cis orientations. The orientation of the proximal histidine residue is also shown along the x axis. The rhombic magnetic anisotropy yields a cloverleaf pattern with lobes alternating in sign and passing through the meso hydrogen atoms. [Reproduced with permission from V. Thanabal, J. S. de Ropp and G. N. La Mar, *J. Am. Chem. Soc.* **109,** 265–272 (1987).]

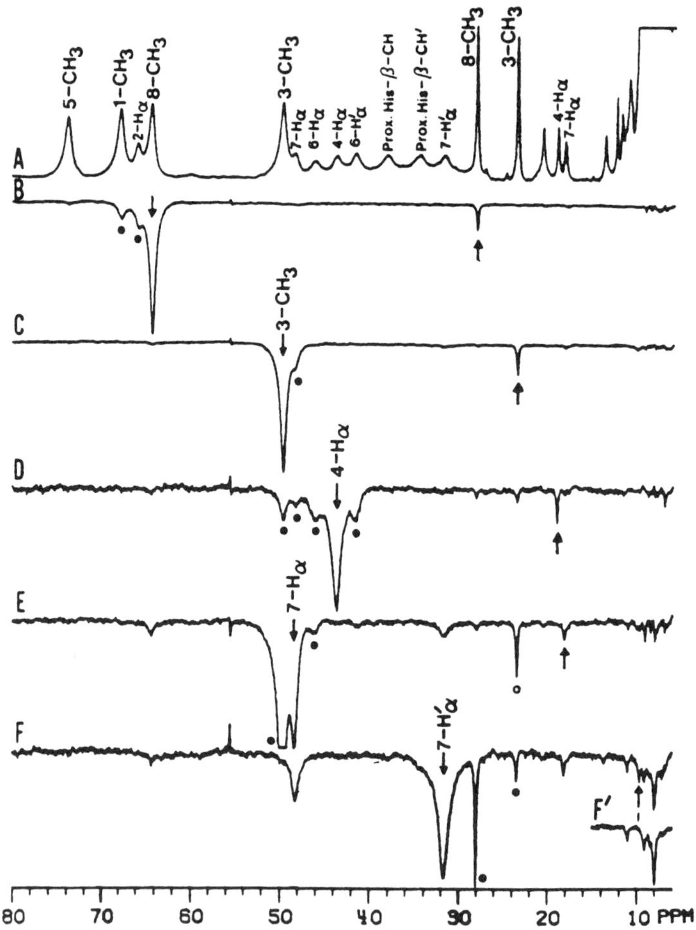

FIG. 8.7. The hyperfine shifted portions of the 360 MHz ^{1}H NMR spectrum of (A) a mixture of native horseradish peroxidase and its cyanide complex in the ratio 65 : 35 in D_2O at 55°C and pH 7.0. (B)–(F) The saturation transfer difference spectra. Downward arrows indicate the peak being saturated and upward arrows saturation transfer. (F′) is a portion of the spectrum in the absence of cyanide. [Reproduced with permission from V. Thanabal, J. S. de Ropp, and G. N. La Mar, *J. Am. Chem. Soc.* **109,** 265–272 (1987).]

Resonances of amino acid residues were also assigned[29]: on the proximal side, His170, Leu237, and Tyr185, and on the distal side, Arg38, His42, and Phe41. The results were consistent with the imidazole proton of His170 being transferred to a nearby residue less than 1 Å away. The assignments of Phe41, Leu237, and Tyr185 were later shown to be incorrect. Leu237 was reassigned to an Ile side chain,[30] later suggested to be Ile244.[31] The Phe41 resonances were reassigned to Phe152, and the Tyr185 resonances to a Phe residue.[32]

In a key paper in 1991 it was shown that all of the peak assignments obtained using deuterated hemes combined with apo-horseradish peroxidase could also be obtained using two-dimensional (2D) NMR.[33] It was a major accomplishment to show that 2D NMR could be applied to a protein containing a strongly paramagnetic metal ion. Some of the results of this investigation are shown in Fig. 8.8.

By comparison with other peroxidases, conservation of distal histidine and arginine residues was noted. Upon binding of hydrocyanic acid to horseradish peroxidase, the acid proton is transferred to the imidazole group of the distal

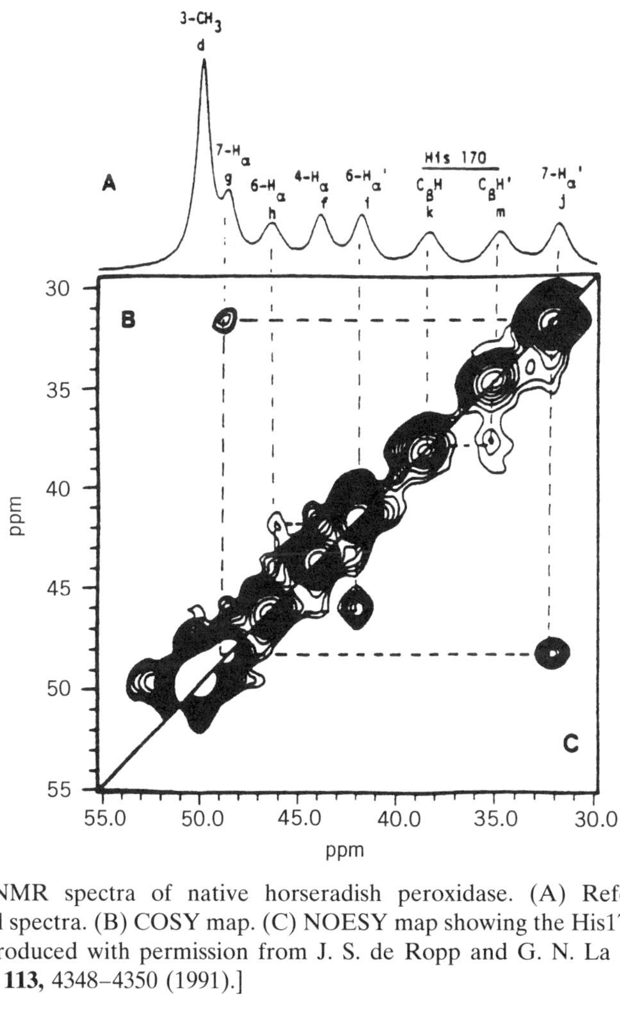

FIG. 8.8. NMR spectra of native horseradish peroxidase. (A) Reference one-dimensional spectra. (B) COSY map. (C) NOESY map showing the His170 $C_\beta H$ cross peak. [Reproduced with permission from J. S. de Ropp and G. N. La Mar, *J. Am. Chem. Soc.* **113,** 4348–4350 (1991).]

histidine, which then hydrogen bonds to the iron-bound cyanide.[34] Further data on interproton distances has been obtained, both for the cyanide complex[35] and substrate binding to the latter.[36] Although there are limitations to scalar correlations using COSY,[37] a comprehensive 2D investigation was made of the cyanide complex,[38] using TOCSY for through-bond correlations and NOESY combined with WEFT and DEFT for through-space correlations. Amino acid residue peak identifications were extended on both the distal and proximal sides of the heme. A detailed review and summary of various types of NMR spectra of the cyanide complex of horseradish peroxidase obtained using a 500-MHz instrument has been published.[39] All proton resonances within 6.5 Å of the iron were assigned.

Homology modeling of the horseradish peroxidase active site was conducted based upon the crystal structures of several peroxidases and 2D NMR spectra of horseradish peroxidase.[40]

Studies of the alkaline transition were interpreted in terms of the distal imidazole side chain of His42 occupying the sixth coordination position of the iron.[41] Proof that the alkaline transition consists of hydroxide ion occupying the sixth coordination position of the iron was obtained from resonance Raman spectroscopy[42,43] and an indirect method, a combined electron paramagnetic resonance–magnetic circular dichroism (EPR–MCD) study on photolyzed compound I.[44]

In a series of papers from different laboratories, complexes of the cyanide-bound and native enzyme with aromatic donor molecules were investigated. In their optical spectral investigation of the binding of aromatic molecules, Paul and Ohlsson classified the molecules according to whether they red- or blue-shifted the native peroxidase spectrum.[45] Sakurada et al. chose one molecule of each type: resorcinol and 2-methoxy-4-methylphenol.[46] They also studied the binding of the competitive inhibitor benzhydroxamic acid. They observed the transferred NOE only when the 8-methyl proton signal from the heme was irradiated. Their conclusion was that all the molecules were bound near the 8-methyl group, with possible hydrophobic interactions with what was then thought to be Tyr185, and hydrogen bonding to nearby residues such as Arg183. The benzhydroxamic acid does not coordinate directly to the iron. They estimated the distances from the iron to the phenol ring protons to be 8.4–11.0 Å. A two-dimensional NMR investigation was made of the indole-3-propionic acid complex with horseradish peroxidase using both COSY and NOESY. Two phenylalanine side chains located close to the heme were found to have considerable mobility. *p*-Cresol binding was studied both by NMR[47] and computer modeling.[48] Aromatic donor molecule binding was later reviewed.[49] In the latter reference the binding of benzhydroxamic acid by horseradish peroxidase isoenzymes A2 and C was studied and the resultant NMR spectra are shown in Fig. 8.9.

The interactions of benzhydroxamic acid and derivatives with horseradish peroxidase are unusual. The iron(III) of the native enzyme is five-coordinate

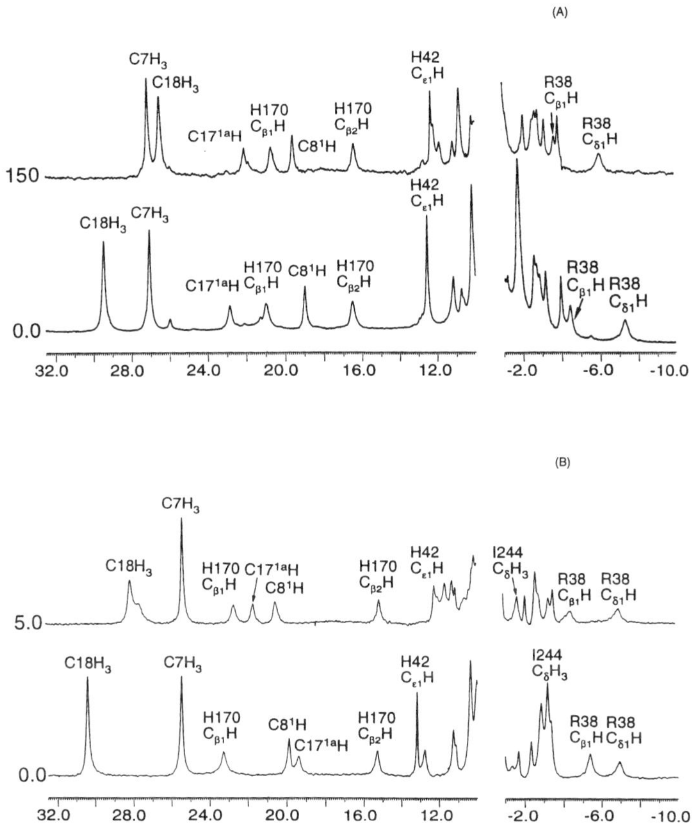

FIG. 8.9. Comparison of the NMR spectra of cyanide-ligated horseradish peroxidase isoenzymes A2 (A) and C (B) in the presence and absence of benzhydroxamic acid. Upper curves of each pair are obtained in the presence of benzhydroxamic acid. [Reproduced with permission from N. C. Veitch, *Biochem. Soc. Trans.* **23,** 232–240 (1995).]

and has a magnetic moment of 5.27 Bohr magnetons. Upon binding of benzhydroxamic acid, this is shifted to that for pure high-spin iron(III), 5.97 Bohr magnetons. Both NMR[50] and resonance Raman[51] measurements indicate that the binding of benzhydroxamic acid converts the iron to a six-coordinate species with water occupying the sixth position. Both kinetics and equilibria in the ternary system cyanide–benzhydroxamic acid–peroxidase were investigated.[52] Benzhydroxamic acid slows both the cyanide on rate and the exchange rate of the His42 imidazole proton with water. In a comprehensive investigation of peroxidase interactions with benzhydroxamic acid and a series of methyl derivatives,[53] two-dimensional ^1H spectra were obtained. The binding of benzhydroxamic acid disrupts or alters the hydrogen bond network present in the protein. The 2D NOESY spectrum of the binding of 3-methyl benzhydroxamic acid to horseradish peroxidase is shown in Fig. 8.10. In this spectrum the substrate methyl group is unambiguously assigned at 2.40 ppm, which leads to an NOE connectivity between it and His42 $C\varepsilon_1H$. The binding of benzhydroxamic acid to a series of Phe mutants of horseradish peroxidase is being investigated and shows that Phe residues peripheral to the binding site affect the dynamics of the binding process.[54,55] Phe179 is located in an insertion between helices F and G. The NMR studies of the mutants Phe179Ala, Phe179-His, and Phe179Ser showed that the Phe179 residue plays a key role in the binding of aromatic substrates, as tested using benzhydroxamic acid. Variations were noted in the resonances of heme methyl $C18H_3$ and heme propionate $C17^1H_2$.[56]

A series of NMR investigations have been made of horseradish peroxidase isoenzymes and mutants. In one study[57] the NMR spectra of the native acidic and neutral isoenzymes, A_1, A_2, A_3, B, and C and their cyanide complexes were investigated. The isoenzymes possess a common axial histidine ligand.

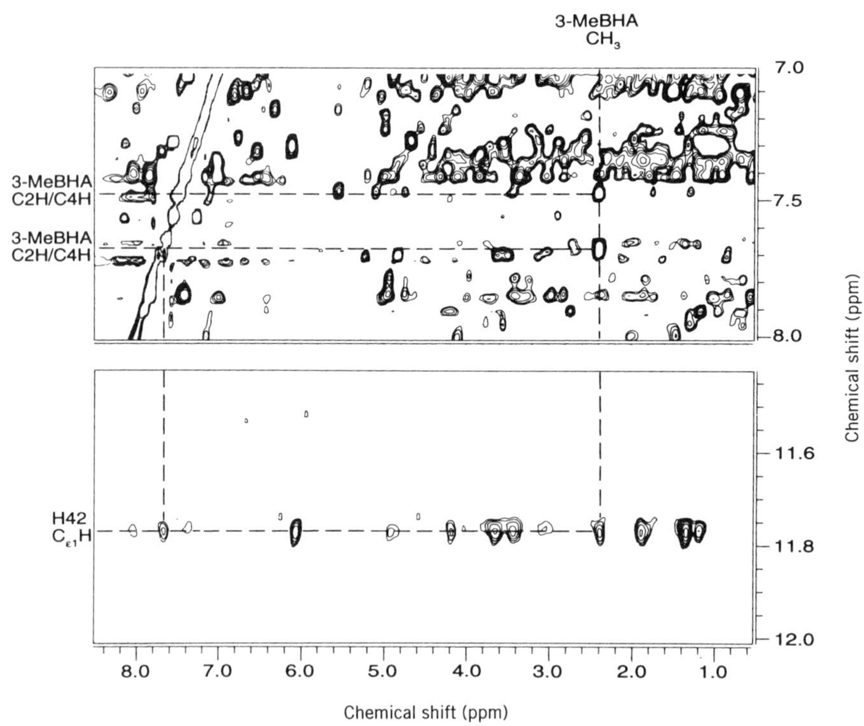

FIG. 8.10. NOESY NMR spectrum of cyanide-ligated 3-methylbenzhydroxamic acid-bound horseradish peroxidase. Dashed lines indicate key NOE connectivities. [Reproduced with permission from N. C. Veitch and R. J. P. Williams, *Eur. J. Biochem.* **229,** 629–640 (1995).]

Resonances outside the diamagnetic envelope are identical for the A_1/A_2 and B/C isoenzymes. The alkaline transitions occur at pH 9.3, 9.4, 9.8, and 10.9, the latter being the same for the B and C isoenzymes. Investigations of the A_1 and A_2 species were extended in a two-dimensional study.[58] The NMR shifts caused by the substitutions Phe41Val and Arg38Lys are shown in Fig. 8.11.[59] The overall fold of the mutant appears to be the same as that of the native enzyme. Unusual large upfield shifts of the 8-CH_3 and 4-H_α resonances occur, indicating that electronic properties of the enzyme are altered. The

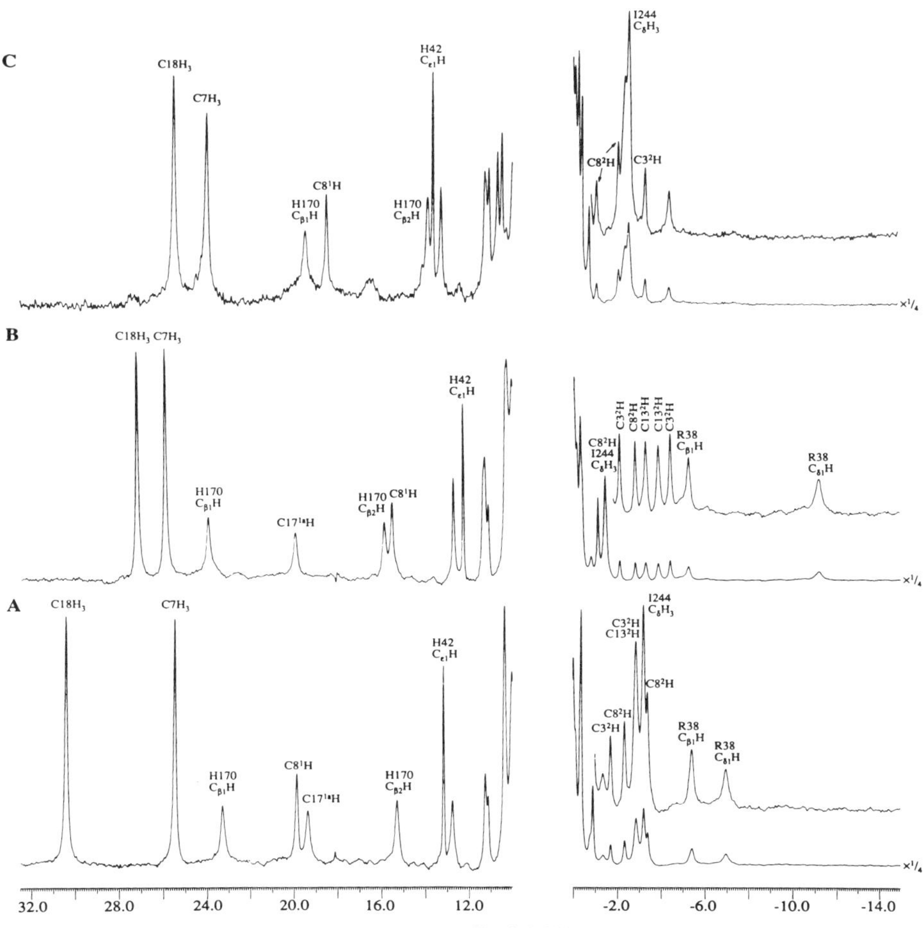

FIG. 8.11. NMR spectra of cyanide-ligated native horseradish peroxidase (C), the Phe-41Val (B), and the Arg38Lys (A) mutants. [Reproduced with permission from N. C. Veitch, R. J. P. Williams, R. C. Bray, J. F. Burke, S. A. Sanders, R. N. F. Thorneley, and A. T. Smith, *Eur. J. Biochem.* **207,** 521–531 (1992).]

Arg38Lys substitution was also investigated.[31] Recent NMR papers have investigated Asn70Val, Asn70Asp,[60] and His42Glu mutants.[61]

8.2. ELECTRON PARAMAGNETIC RESONANCE SPECTROSCOPY

8.2.1. Introduction

Electron paramagnetic resonance (EPR) spectroscopy, also called electron spin resonance (ESR) spectroscopy, was discovered by E. Zavoisky in 1945.[62] It provides a means of detecting and studying the properties of molecules and ions containing unpaired electrons. An unpaired electron defines the unique properties of a free radical; and one or more unpaired electrons are often found in transition-metal ions, such as iron. Thus, EPR spectroscopy is a very important tool for studying the properties of peroxidases.

EPR spectroscopy is based upon the interaction of the magnetic properties of the unpaired electrons with an external magnetic field. The technique is similar to NMR spectroscopy, but the magnitudes of the resonance energies of observed transitions are very different. In both EPR and NMR, the sample is placed in a static magnetic field of flux density B_0, and an oscillating magnetic field is applied to excite transitions. In NMR the oscillating magnetic field is radio-frequency radiation, while microwave radiation is required in EPR. Thus, an EPR absorption of energy occurs when the frequency of the microwave radiation and the magnetic flux satisfy the resonance condition

$$h\nu_0 = g\beta B$$

where h is Planck's constant, ν_0 is the resonance frequency, β is the Bohr magneton, and B is the magnetic flux density. The factor g is determined by the location and environment of the unpaired electron. For a free electron $g = 2.0023$, and this is well approximated by a free radical electron in a highly aromatic molecule. The experiment is conducted by slowly increasing the magnetic field to find the resonance condition.

Standard cavities are designed to resonate at a frequency of about 9 GHz (X band) or about 35 GHz (Q band). First-derivative spectra are recorded so that instead of an absorption peak above a baseline, a curve with positive and negative components is obtained (Fig. 8.12). The reason a derivative curve is obtained is the following. The slow sweep to detect the resonance frequency is subject to large noise levels. In order to minimize the noise, the slow sweep of the magnetic field is modulated by a high-frequency signal that samples the absorption curve in very small increments, thereby providing tangents to, and hence the derivative of, the absorption curve.

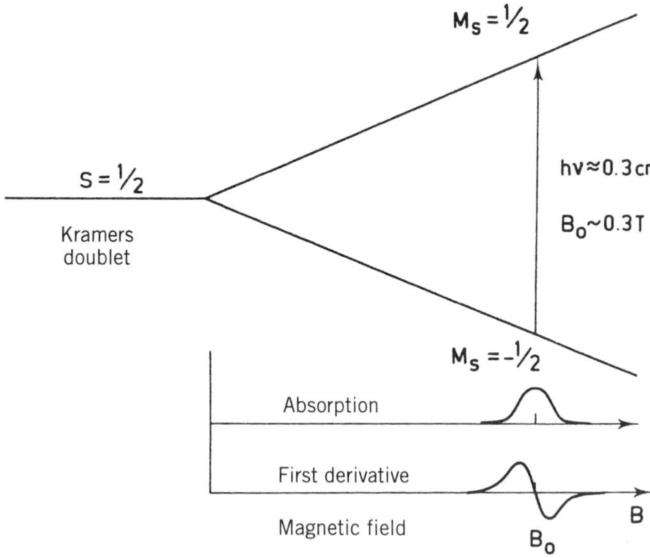

FIG. 8.12. Magnetic splitting of the energy level for one unpaired electron, $S = \frac{1}{2}$, into two. Thus, a Kramers doublet is obtained, one in which degeneracy is removed by a magnetic field. The absorption of energy occurs for the resonance condition. Spectra are recorded as the first derivative of absorption.

There are four factors that determine the nature of an EPR signal: the intensity, the linewidth, the *g* value, and multiplet splitting. Ideally, the intensity (area under the peak) is proportional to the number of detected electrons. Since the EPR signal is a derivative curve, a double integration is required to obtain the area. The linewidth is influenced by many factors, but simplistically it can be described in terms of the reciprocals of relaxation times T_1 and T_2, as in NMR spectroscopy. In transition metal ions T_1 may be so short that the linewidth is too large for the signal to be detected. The linewidth may be dramatically decreased at temperatures approaching 0 K.

The *g* values may range from about 1.4 to 10. The method of reporting the positions of resonance frequencies differs in EPR compared to NMR spectroscopy. For NMR the *g* values of the nuclei are assumed to be fixed, and different resonance frequencies are attributed to different shielding effects caused by variations in the electron densities in the vicinity of the nuclei. For EPR the different locations of the resonance frequencies are attributed entirely to *g* values, which therefore are empirical parameters.

Multiplet splitting can arise from two sources. Zero field splitting is the name given to splitting caused by interactions of two or more unpaired electrons within the same molecule or ion; they do not require the external magnetic field. A transition-metal ion with more than one unpaired electron will exhibit zero field splitting because of the interaction of the magnetic moments of the unpaired electrons. Hyperfine interactions, which occur between electron and nuclear spins, also lead to multiplet structure. They are characterized by a hyperfine splitting constant a. A hydrogen atom, which has a single proton and a single electron, provides the simplest example of hyperfine splitting. Because the proton of the hydrogen atom can take two different orientations, its interaction with the electron causes the resonance frequency to split in two, so a doublet spectrum is observed. Hyperfine splitting is restricted to the interaction of an electron with the nuclei within its orbital. The general rule for the splitting of the EPR spectrum of an electron that interacts with a single nucleus is:

$$\text{Number of lines} = 2I + 1$$

where I is the spin quantum number of the nucleus. For the proton $I = \frac{1}{2}$ so a doublet is observed for an unpaired electron attached to a proton. For a nitrogen atom $I = 1$. Therefore, an unpaired electron interacting only with a nitrogen atom would have a triplet EPR spectrum.

In practice, the free radical electron feels the influence of several nuclei because it occupies a molecular orbital that is not localized on a single nucleus. An illustration of hyperfine splitting is shown in Fig. 8.13, the EPR spectrum of the $\cdot CH_2OH$ radical. The ^{12}C, with six protons and six neutrons, has no net nuclear spin and hence causes no hyperfine splitting. However, the unpaired electron interacts with the two equivalent protons, which split the EPR signal into a triplet with the expected 1:2:1 intensities.

A π electron may feel the influence of many nuclei, in which case its resonance signal will be split into many components. If all the nuclei are identical, then the extent of the hyperfine splitting can be predicted using Pascal's triangle. A different splitting constant exists for each different nucleus. A more complex example is the EPR spectrum of the oxidation product of 3,3′,5,5′-tetramethylbenzidine reacting with horseradish peroxidase and hydrogen peroxide, the tetramethylbenzidine cation radical. The molecule possesses two amine N nuclei, four amine protons, four protons in the 2 and 4 positions, and 12 methyl protons. Of a theoretical number of 1625 lines, about 31 were resolved; the spectrum is shown in Fig. 8.14.

8.2.2. Electron Paramagnetic Resonance Spectral Anisotropy

In a single crystal all of the free radicals might be oriented in the same direction. If the external magnetic field is applied parallel to the molecular symmetry axis, then the g factor would be equal to $g_{\parallel}$, at right angles it would be equal to $g_{\perp}$. The other extreme to a single crystal is a fluid or frozen sample in which all orientations occur. In the latter case the orientation of each dipole can be resolved into a parallel and perpendicular component, and an envelope of all absorptions with a corresponding derivative curve would be obtained. The principal axis z is defined along the direction of the

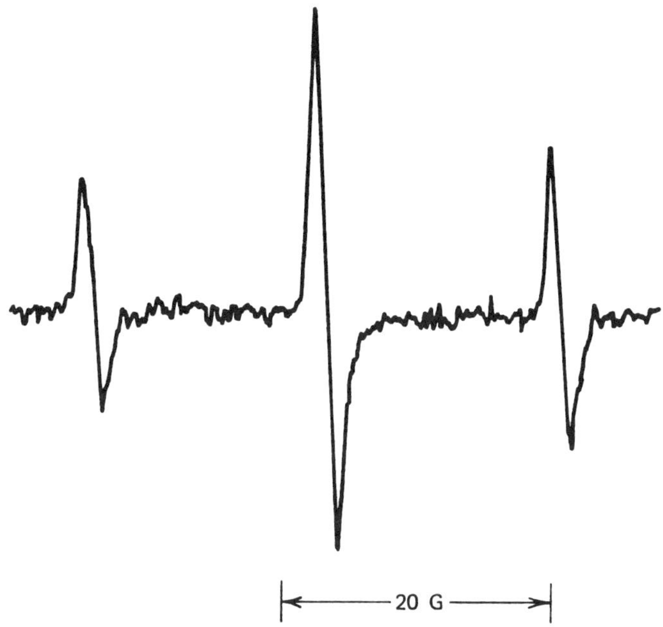

FIG. 8.13. First-derivative EPR spectrum of the •CH_2OH radical at pH 1.03. Only the protons attached to carbon influence the spectrum. (Reproduced with permission from J. R. Bolton in *Biological Applications of Electron Spin Resonance,* H. M. Swartz, J. R. Bolton and D. C. Borg, eds., Wiley-Interscience, New York, 1972, pp. 11–60.)

molecular symmetry axis and the corresponding g value to g_{zz}. For axial symmetry $g_{zz} \neq g_{yy} = g_{xx}$, in which case the resultant spectrum is well defined in terms of $g_{\|}$ and $g_{\perp}$. For lower symmetry $g_{zz} \neq g_{yy} \neq g_{xx}$, and all three g values are required to define the spectrum. Idealized spectra are shown in Fig. 8.15.

If an electron is in an s orbital, then its electron cloud has a finite density at the nucleus; and the interaction of the electron spin with the nuclear spin is called a Fermi contact interaction. The interaction does not vary with the orientation of the molecule; it is called isotropic. For an electron in a p or d orbital, the cloud density at the nucleus is zero, and the interaction between electron and nuclear magnetic moments is dipolar. The strength of the interaction depends on the orientation of the molecule; and so the electron–nucleus interaction is anisotropic.

8.2.3. Spin Trapping

A free radical localized on a single atom may be so short-lived that it cannot be detected by EPR. The possibility of detection can be greatly increased by

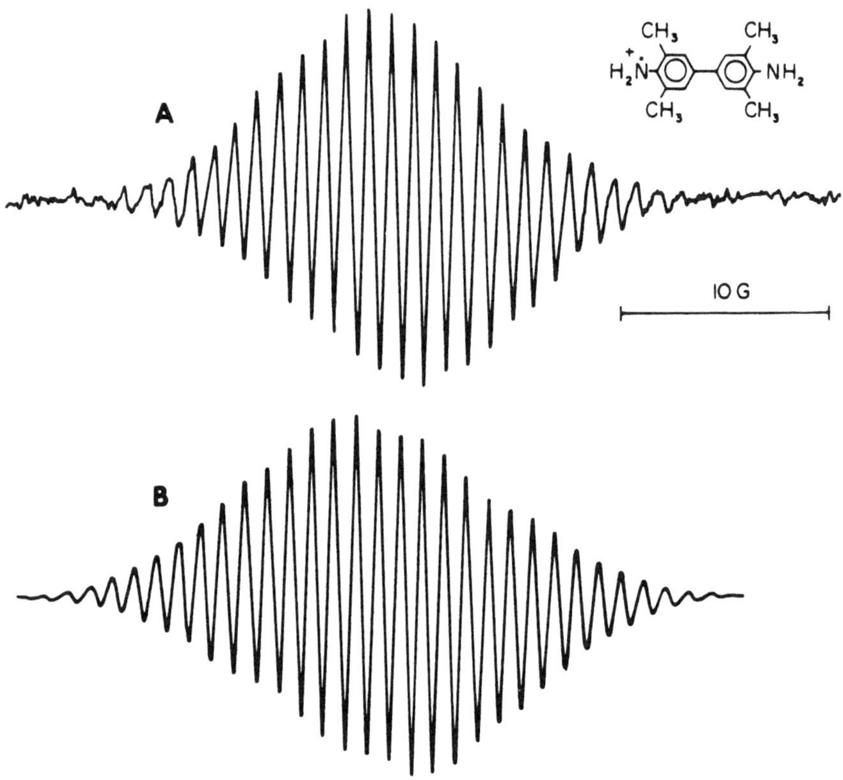

FIG. 8.14. (A) Experimental spectrum and (B) computer simulation of the EPR spectrum of the tetramethylbenzidine cation radical. Hyperfine splitting constants are $a^{N}_{NH2} = 3.09$, $a^{H}_{NH2} = 3.33$, $a^{H}_{NHC} = 1.08$, and $a^{H}_{ring} = 1.08$; linewidth 0.45, all values in G. [Reproduced with permission from P. D. Josephy, T. Eling, and R. P. Mason, *J. Biol. Chem.* **257,** 3669–3675 (1982).]

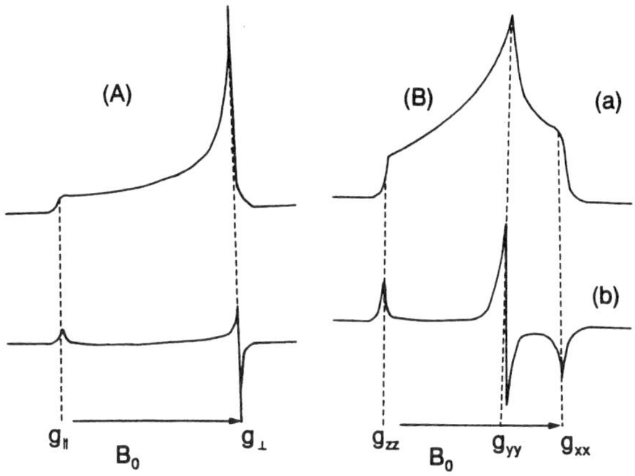

FIG. 8.15. (A) Idealized absorption and (B) derivative EPR spectra for randomly oriented molecules exhibiting g anisotropy. Axial symmetry, $g_{zz} = g_{\parallel}$, $g_{yy} = g_{xx} = g\perp$ (A), and lower symmetry, $g_{zz} \neq g_{yy} \neq g_{xx}$ (B). (Reproduced with permission from I. D. Campbell and R. A. Dwek, *Biological Spectroscopy,* Benjamin/Cummings, Menlo Park, 1984, pp. 179–215.)

using a spin trap S, a molecule that reacts with the free radical of interest R• to form a stable free radical RS•, which can be studied at leisure (Fig. 8.16).[63,64] The technique is highly relevant to peroxidase chemistry since free radicals are commonly produced by oxidation of substrates by both compounds I and II. If these radicals are too unstable or reactive then their detection by EPR must rely upon a spin trap. The spin-trapping technique was elevated to an art form to a large extent by the group of Edward Janzen.[65] Spin trapping is not to be confused with spin labeling. A spin label already possesses a stable free radical. It is attached chemically to a specific point on the molecule to be studied.

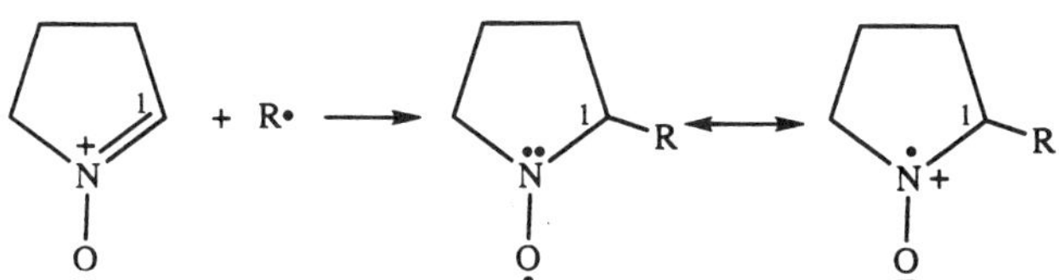

FIG. 8.16. Use of a pyrrolidine nitroxide as a spin trap. The free radical adds at C_1. Bulky substituents around the ring may help stabilize the trapped free radical, but perhaps at the expense of slowing the trapping reaction.

8.2.4. Electron Paramagnetic Resonance Studies on Horseradish Peroxidase

A landmark study was that published in 1960, the first EPR detection of free radicals in a biological system.[66] Semiquinone, ascorbate, and dihydroxyfumarate radicals were detected in reactions catalyzed by Japanese radish peroxidase. In the following year studies were extended to horseradish peroxidase.[67] The EPR spectrum of the free radical, obtained from the steady-state reaction of horseradish peroxidase with hydrogen peroxide and reductic acid in a slowly flowing system, is shown in Fig. 8.17. Subsequently, free radicals were detected in EPR studies on horseradish peroxidase reactions with dihydroxyfumarate, triose reductone, ascorbic acid, hydroqinone, pyrogallol, *p*-phenylenediamine, and catechol. Results were presented in tabular form.[68]

The sustained EPR studies of Ronald Mason and co-workers are particularly noteworthy. Free radical products from substrate oxidation have been detected in many other horseradish-peroxidase-catalyzed reactions. These include free radicals from *N,N*-dimethylaniline, *N,N*-dimethyl-*p*-toluidine,[69] aminopyrine,[70] tetramethylhydrazine,[71] benzidine,[72] tetramethylbenzidine,[73] *p*-aminophenol,[74] L-cysteine,[75] nitrosonaphthols,[76] the phenacitin metabolites *p*-phenetidine and acetaminophen,[77] Desferal,[78] bisulfite,[79,80] the antioxidant food additives 2-*t*-butyl-4-methoxyphenol (BHA) and 2,6-di-*t*-butyl-4-methylphenol (BHT),[81] porphyrins oxidized to porphyrin π-cation radicals,[82] crystal violet,[83] from the metabolite of the antitumor drug mitoxantrone and from

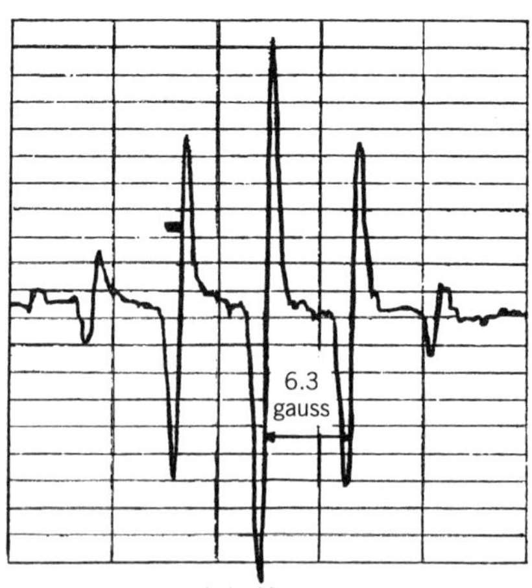

FIG. 8.17. EPR spectrum of the free radical obtained from the oxidation of reductic acid (2,3-dihydroxy-2-cyclopenten-2,3-diol-1-one) by horseradish peroxidase. [Reproduced with permission from I. Yamazaki and L. H. Piette, *Biochim. Biophys. Acta* **50,** 62–69 (1961) with permission from Elsevier Science.]

N,N′-diacetylmitoxantrone,[84] and the acetaminophen analog 3′,5′-dimethyl-4′-hydroxyacetanilide.[85] The spectrum of the free radical from 3,3′,5,5′-tetramethylbenzidine is shown in (Fig. 8.14). In all of these studies the free radicals were detected directly; spin traps were not used.

Examples of the use of spin traps are detection of radicals formed by horseradish-peroxidase-catalyzed oxidations of azide,[86] glutathione,[87] cyanide,[88,89] and N-substituted aromatic amines.[90] Transfer of electrons between free radicals was demonstrated in the reduction of the acetaminophen radical by glutathione and ascorbate.[91]

Attention is now turned to direct EPR examinations of horseradish peroxidase, its compounds, and the effect of complex formation on its spectrum.[92] No EPR signal was obtained from compounds II or III.[93] Many studies were made on the native enzyme.[94,95] Comparisons of the A2 and C2 isoenzymes were made.[96,97] In the most comprehensive study,[97] it was concluded that at a temperature of 5 K the native enzyme is a mixture of high-spin and intermediate-spin species (Fig. 8.18). In one study the authors found it impossible to obtain a reproducible EPR spectrum of the native enzyme.[98] The problem was almost certainly redox contaminants present in the water, buffer, or some other component of the solution. The EPR spectrum of the low-spin alkaline form is shown in Fig. 8.19.

In two EPR studies the coordination number of the iron(III) in horseradish peroxidase was addressed. Substitution of $H_2{}^{17}O$ for $H_2{}^{16}O$ caused line broadening for metmyoglobin, known to have water in the sixth coordination position of its iron, but no broadening for the peroxidase.[99] The conclusion: The iron is five-coordinate in the peroxidase. In a more complex, but elegant study, the low-temperature properties of the nitric oxide derivatives of metmyoglobin and ferric horseradish peroxidase, their photolyzed products, and the original ferric proteins were compared.[100] When the metmyoglobin complex with NO was photolyzed, the simplest explanation was that the NO could not return to its iron site, and a new five-coordinate iron(III) species was obtained, different from the original globin with water in the sixth coordination position. With horseradish peroxidase, the photolyzed product and the original enzyme were identical, again indicating that the native enzyme has five-coordinate iron(III).

The fluoride complex of horseradish peroxidase is more purely high-spin than is the native enzyme and the cyanide complex more low-spin.[92] Nitrite reacts with the native enzyme in a pH-dependent reaction.[101] The benzhydroxamic acid complex changes the iron(III) to six-coordinate,[98] unlike other aromatic substrates.[97]

Turning now to the elusive EPR signal for compound I of horseradish peroxidase, in one early study the $g = 2$ signal for compound I was detected but attributed to an artifact.[92] In another no signal was detected.[93] The first signal that was correlated with the amount of compound I formed was published in 1975.[102] The disappearance of both the intermediate-spin and high-spin forms of the native enzyme was correlated linearly with the appearance

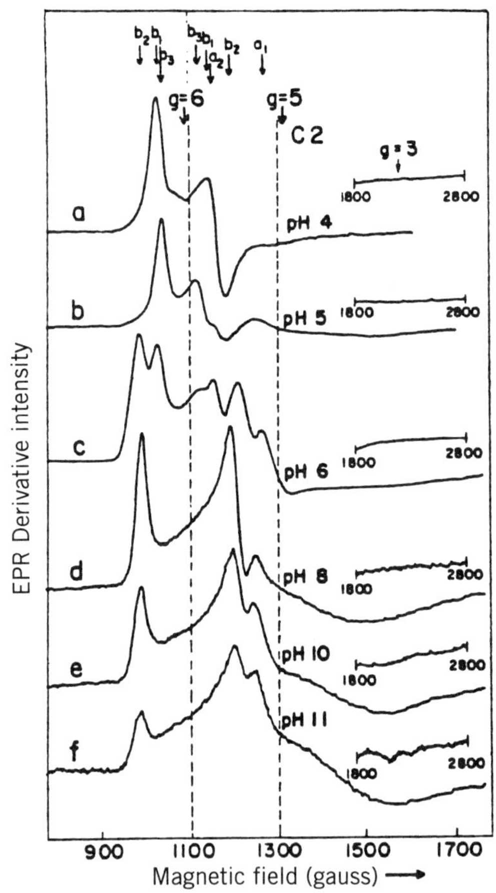

FIG. 8.18. EPR spectrum of native horseradish peroxidase isoenzyme C2 as a function of pH taken at 5 K. A total of five peaks are observed over the pH range 4–11: three split peaks in the vicinity of $g = 6.4$ and 5.5 corresponding to the high-spin species and two in the vicinity of $g = 5.0$ corresponding to the intermediate spin species. [Reproduced with permission from M. M. Maltempo, P.-I. Ohlsson, K.-G. Paul, and A. Ehrenberg, *Biochemistry* **18,** 2935–2941 (1979).]

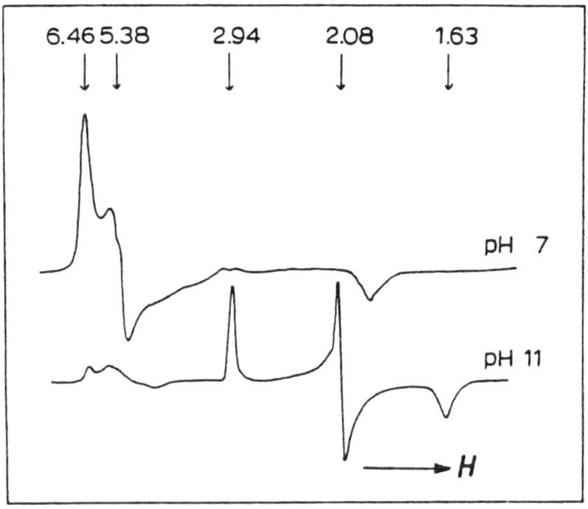

FIG. 8.19. Low-spin spectrum of the alkaline form of horseradish peroxidase isoenzyme C2 (pH 11) compared to the high-spin spectrum of the native enzyme (pH 7). The g value is listed at the top of the figure. [Reproduced with permission from M. Tamura and H. Hori, *Biochim. Biophys. Acta* **284,** 20–29 (1972) with permission from Elsevier Science.]

of the weak low-spin $g = 2$ signal from compound I formation as hydrogen peroxide was added to the native enzyme.[97] Compound I was subjected to a comprehensive investigation by both Mössbauer and EPR spectroscopy.[98,103] Both compounds I and II have a small isomer shift, $\delta\text{Fe} \approx 0.06$ mm s^{-1}, positive quadrupole splitting $\Delta E = 1.4$ mm s^{-1}, and large zero field splitting, $D/k \approx 35$ K, characteristic of $Fe^{IV}{=}O$ species. Whereas compound II is EPR silent, the iron of compound I weakly couples to the porphyrin radical with spin S' of $\frac{1}{2}$. The coupling leads to high- and low-field wings superimposed upon the $g = 2$ signal. When the signal of these wings is taken into account, an integrated signal intensity of 0.7 ± 0.4 spins/heme is obtained.[103] The EPR signal is strongly temperature-dependent: It is lost at 30 K but can be saturated at 5 K. The EPR spectra are shown in Fig. 8.20.

When a frozen solution of compound I is photolyzed, a doublet EPR signal centered on $g = 2$ is generated.[104,105] Apparently an electron is liberated causing reduction to iron(III), and leaving an unpaired electron on the protein. The photolyzed product thus has the oxidizing equivalents of compound II, but in a different configuration.

Electron paramagnetic resonance has been used to follow the kinetics of peroxidase reactions.[106] The method relies upon a subsequent reaction of the radicals generated in the normal peroxidase cycle. In the presence of the hydroxylamine derivative of Tempol, the radicals, for example, from a phenolic substrate, reduce the derivative to Tempol, a paramagnetic nitroxide. The long-lived Tempol radicals can be readily detected by EPR. The method is

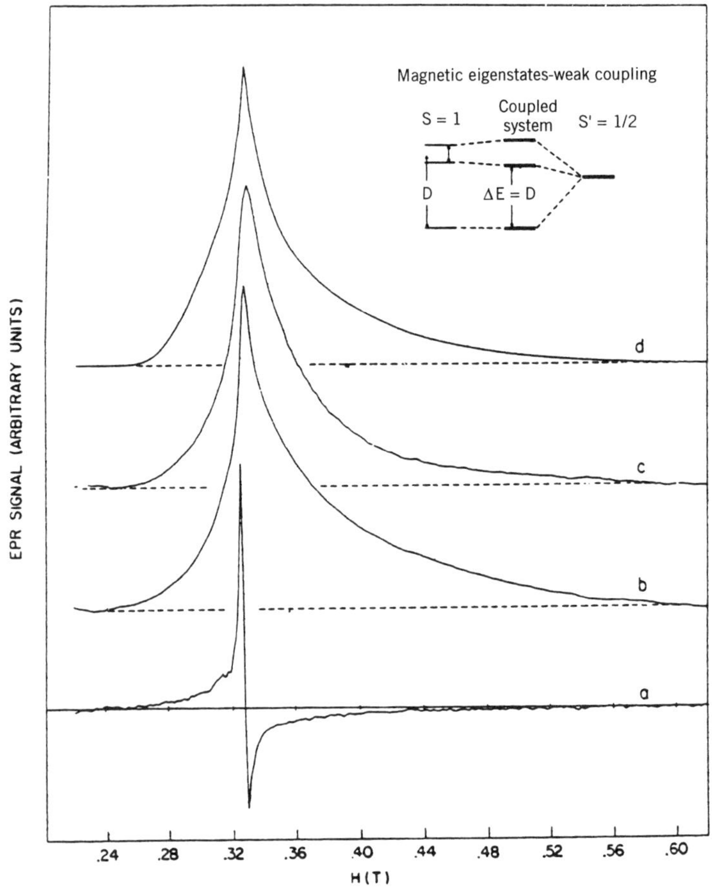

FIG. 8.20. X-band EPR spectrum of horseradish peroxidase compound I at 3.4 K. (A) Derivative spectrum showing the broad wings to the central signal. (B) Integration of (A). (C) Rapid passage yields an absorption-type spectrum. (D) Simulated absorption spectrum. [Reproduced with permission from C. E. Schulz, P. W. Devaney, H. Winkler, P. G. Debrunner, N. Doan, R. Chiang, R. Rutter, and L. P. Hager, *FEBS Lett.* **103,** 102–105 (1979) with permission from Elsevier Science.]

sensitive and can be applied to systems not amenable to optical spectroscopy, such as turbid solutions and possibly to reactions inside membranes.

8.3. ELECTRON NUCLEAR DOUBLE RESONANCE SPECTROSCOPY

8.3.1. Introduction

Electron Nuclear DOuble Resonance (ENDOR) spectroscopy is NMR spectroscopy performed on an EPR machine. If one understands the principles of NMR and EPR, then it is easy to extrapolate to an understanding of ENDOR, but not to performing ENDOR experiments.

The technique of ENDOR was first proposed by G. Feher in 1956[107] and was shown to work the same year.[108] It was extended to free radicals in solution[109,110] and to powdered samples containing paramagnetic ions.[111] Further instrumental improvements were made.[112] ENDOR provides a method of extracting fundamental information about nuclei involved in hyperfine interactions where the EPR signal is inhomogeneously broadened without structure. Such an apparently single EPR signal may be a spin packet, consisting of many overlapping signals. These overlapping EPR signals may be resolved in ENDOR. Furthermore, even though no improvement in resolution may be apparent, ENDOR still offers the advantage of greatly simplifying an EPR spectrum. Thus, the triphenyl phenoxyl radical has 4050 EPR transitions. The corresponding ENDOR spectrum has 14 lines.[113]

In order to perform an ENDOR experiment, one needs a radio-frequency field as well as a microwave field, both at right angles to the external magnetic field. The basis for an ENDOR experiment is illustrated in Fig. 8.21 for a system consisting of a single nucleus with spin I = one half and a single electron ($S = \frac{1}{2}$). The nucleus could be, for example, a proton or ^{57}Fe. The reversible arrows indicate both transitions to higher energy states and relaxation processes to lower energy states. The selection rules for an EPR transition are $\Delta m_s = \pm 1$ and $\Delta m_I = 0$. In words, the electron flips but the proton does not. By tuning the spectrometer to the electron-flip transition from level 1 to level 4 and increasing the microwave power, partial saturation is obtained, and hence nearly equal populations of levels 1 and 4. The microwave power

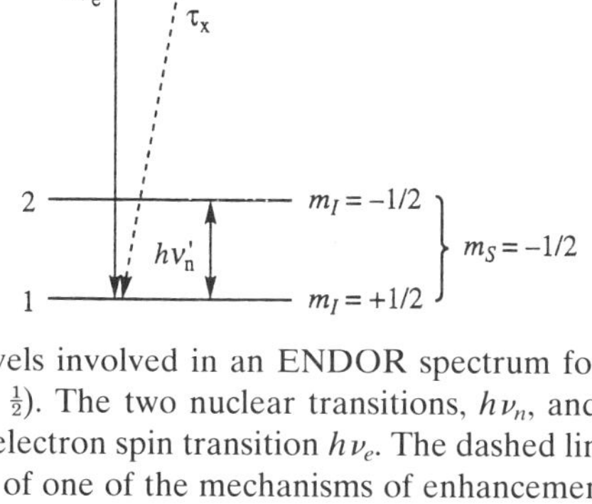

FIG. 8.21. Energy levels involved in an ENDOR spectrum for a nucleus with $I = \frac{1}{2}$ and an electron ($S = \frac{1}{2}$). The two nuclear transitions, $h\nu_n$, and $h\nu_n'$, are detected as enhancements of the electron spin transition $h\nu_e$. The dashed line with relaxation time τ_x shows the pathway of one of the mechanisms of enhancement of the population of level 1.

is kept constant so a constant EPR signal is obtained. Then, by saturating the upper nuclear transition with radio-frequency power, the populations of levels 3 and 4 are equalized. An ENDOR signal now results by two different mechanisms.

Both mechanisms involve changing in a favorable way the relative populations in levels 1 and 4. Either by decreasing the population in level 4 or increasing the population in level 1 means that there will be an increased absorption of microwave power and hence an enhanced EPR signal. A decrease in the population in level 4 occurs when the nuclear transition from $m_I = +$ one half to $m_I = -$one half is induced by the saturating radio frequency.

Because of the increased population in level 3, a new relaxation process becomes important, indicated by the dashed arrow in Fig. 8.21. It is a spin–lattice relaxation, with time τ_x, which helps to repopulate level 1, and hence further offset the partial microwave saturation. The net result is an increase in the EPR signal obtained at the radio frequency corresponding to the upper nuclear transition involving levels 3 and 4, and this is the ENDOR signal.

The above considerations are applicable to an isotropic (Fermi contact) hyperfine interaction. If the hyperfine interaction is anisotropic, then further relaxation processes become important. Direct transitions from level 4 to level 2 occur (not shown in Fig. 8.21). When the saturating radio frequency corresponding to $h\nu_n$ is changed, then the first ENDOR signal disappears. When the radio frequency corresponds to $h\nu'_n$, populations of levels 1 and 2 are equalized, and a new mechanism for repopulating level 1 has occurred. Thus a second ENDOR signal is obtained.

The energy of the upper nuclear transition $h\nu_n$, *is given by* $A/2 - g_N\beta_N B_0$, and the lower $h\nu'_n$, by $A/2 + g_N\beta_N B_0$, where A is the electron-nuclear hyperfine coupling constant and $g_N\beta_N B_0$ is the nuclear resonance frequency. The difference in transition energies is given by $2g_N\beta_N B_0$. Since $g_N\beta_N$ is known for all stable nuclei, the spacing between the two NMR signals, one on each side of the frequency of the hyperfine coupling A/h, provides a fingerprint identifying the nuclear species responsible for the hyperfine interaction. Because different relaxation mechanisms are involved, it would be a coincidence if the two peaks obtained in the ENDOR spectrum would have equal intensities. For the example shown in Fig. 8.21, $|A| > g_N\beta_N B_0$.

Many reviews on ENDOR may be consulted for further information,[114–116] including discussions of applications to iron- and heme-containing proteins.[117,118]

8.3.2. Electron Nuclear Double Resonance Spectroscopy of Horseradish Peroxidase Compound I

The ENDOR spectra of compound I of horseradish peroxidase are shown in Figs. 8.22 and 8.23.[119] The spacing and shift with field strength of the main two peaks in Fig. 8.22 identifies them as proton peaks. They are centered on the Larmor precession frequency of a free proton. In this case $|A| < g_N\beta_N B_0$,

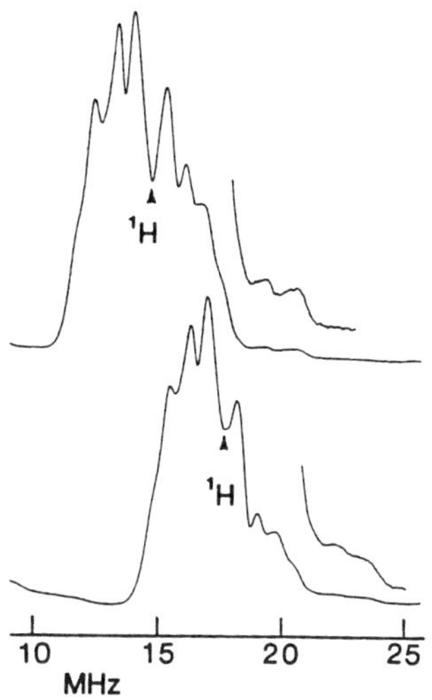

FIG. 8.22. ^{1}H ENDOR spectra of horseradish peroxidase compound I taken at $g \approx 2$. Microwave frequencies (B) 11.66 GHz and (A) 9.60 GHz. Arrows on the spectra indicate ν_H, the free proton Larmor frequency. [Reproduced with permission from J. E. Roberts, B. M. Hoffman, R. Rutter, and L. P. Hager, *J. Biol. Chem.* **256,** 2118–2121 (1981).]

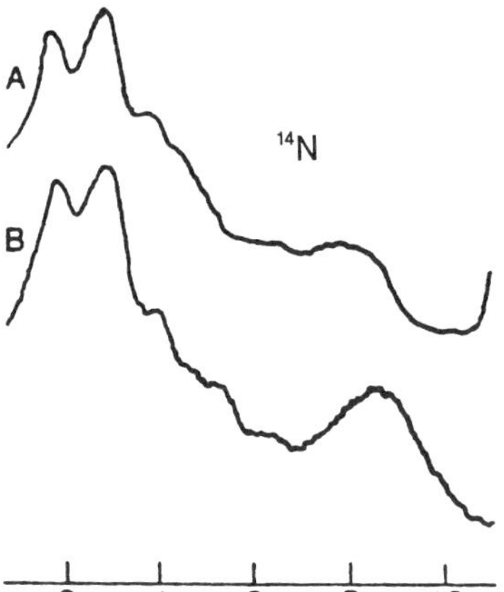

FIG. 8.23. ^{14}N ENDOR spectra of horseradish peroxidase compound I. Except for radio-frequency field strength, same conditions as for Fig. 8.22. [Reproduced with permission from J. E. Roberts, B. M. Hoffman, R. Rutter, and L. P. Hager, *J. Biol. Chem.* **256,** 2118–2121 (1981).]

and the spacing is given by $g_N \beta_N B_0 \pm A/2$. Furthermore, from their strength and shape they are further identified as β-protons, those attached to carbon atoms one removed from spin-bearing carbon. It appears that at least some other resonances in Fig. 8.22 are superpositions from more than one set of protons with similar coupling constants. At lower field ^{14}N ENDOR spectra are obtained (Fig. 8.23). Unlike proton ENDOR signals the center of the ^{14}N pattern is independent of microwave frequency. The ENDOR spectra confirm that compound I contains a single spin-coupled center and prove that the free radical portion of the center is a porphyrin radical.

8.4. MÖSSBAUER SPECTROSCOPY

8.4.1. Introduction

Mössbauer spectroscopy, also called recoil-free nuclear resonance absorption spectroscopy, was discovered in 1957 by Rudi L. Mössbauer, for which he received the Nobel prize in physics in 1961. Mössbauer discovered that in a crystalline solid a large amount of the recoil from γ emission is absorbed by the entire crystal, in which case the direct recoil of the emitting nucleus is zero. Similarly an absorbing atom remains motionless. The technique is applicable to any radioactive element that is a γ emitter with energy of the order of 100 keV or less. By far the most research has been performed on iron, so the technique is very suitable for heme protein research on frozen samples.[120,121] For reasons outlined below the Mössbauer signal is obtained from the ^{57}Fe isotope, which has a 2% natural abundance. For the simplest spectra on a ^{57}Fe-enriched sample a good signal-to-noise ratio can be obtained in a scanning time of ~12 hours. For more complex spectra, for example, high-spin iron(III) in an applied magnetic field, longer data aquisition time is required. A high content of ^{57}Fe can be obtained either by chemical substitution or by growing organisms using a ^{57}Fe source. Even with ^{57}Fe enrichment, millimolar concentrations of enzyme are required to obtain a satisfactory signal-to-noise ratio. The first excited state of the ^{57}Fe nucleus is 14.4 keV above the ground state, and the Mössbauer spectrum of an iron-containing frozen sample is obtained by monitoring the transition from the ground state to this excited state.

The application of Mössbauer spectroscopy consists of using a source, a γ emitter, and a counter to detect the amount of absorption of the γ radiation by the sample. For biological applications, the γ source is usually ^{57}Co, which decays into an excited state of ^{57}Fe by the process of electron capture. Subsequent decays of the excited ^{57}Fe nucleus to its ground state provides the required γ photons. If the sample contains a ^{57}Fe nucleus, it has an exceptionally large cross section to capture the emission from the source.

Since the ^{57}Fe nuclei in the source and in the sample have different chemical environments, the energies of emission and absorption are different. By deliberately moving the source at the appropriate speed, the source emission can be tuned to the resonance absorption frequency of the sample, a use of the Doppler effect. The spectral results are commonly expressed in terms of the velocity of the source in mms^{-1}, one of the most unique set of units in spectroscopy.

8.4.2. Hyperfine Shifts

Isomer or Chemical Shift

The term chemical shift is more appropriate for the biochemical applications of Mössbauer spectroscopy. The shift, δE, arises from the Coulombic interaction of the nucleus and the electronic charge at the nucleus, which arises from the *s* electrons. Because the nucleus changes size upon absorption of γ radiation, the Coulombic interaction changes, causing the shift, which is usually measured with respect to metallic iron (Fig. 8.24*a* and *b*). The chemical shift defines the position of the center of absorption and for a given isotope provides clues to its oxidation state and spin state.

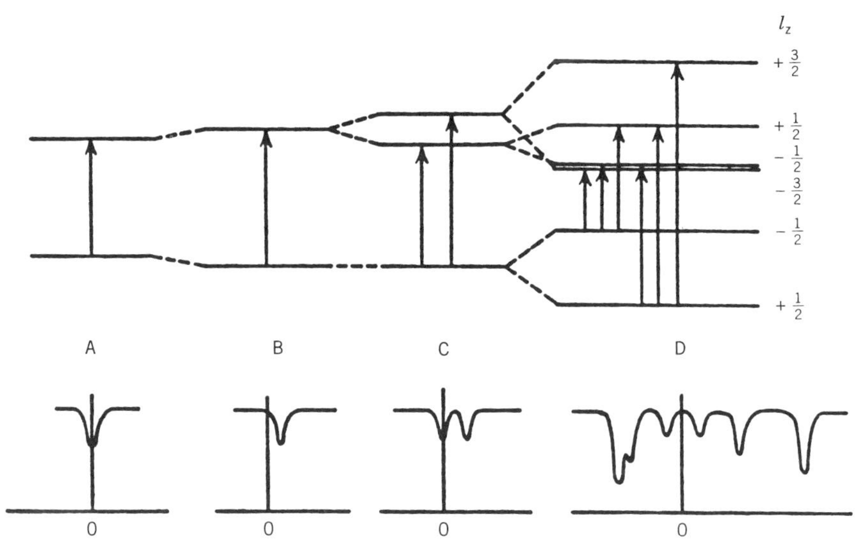

FIG. 8.24. Effect of absorber properties on observed Mössbauer spectra. (A) Unsplit nuclear levels, the same as in the source. (B) Reduced electron density in the absorber increases the separation of levels. (C) Electric field gradient splits the upper level into $\pm\frac{3}{2}$ and $\pm\frac{1}{2}$ components. This quadrupole splitting produces a two-line spectrum. (D) The presence of a magnetic field causes further splitting into two four levels, $\pm\frac{3}{2}$ and $\pm\frac{1}{2}$ for the excited nuclear state, and two levels, $\pm\frac{1}{2}$, for the ground state, producing a six-line spectrum. It is assumed in this example that the magnetic field is applied along the principal axis of the electric field gradient. [Reproduced with permission from G. Lang, *Quart. Rev. Biophys.* **3,** 1–60 (1970).]

Electric Quadrupole Splitting

Any nucleus with a spin $>\frac{1}{2}$ has a quadrupole moment that will interact with electronic charges that are not symmetrically distributed about it. The ground state of the ^{57}Fe nucleus has a spin of $\frac{1}{2}$ and so does not possess a quadrupole moment. However, its first excited state, 14.41 keV higher in energy, has a nuclear spin of $\frac{3}{2}$. If the electric field gradient at the nucleus is finite, then the energy level of the excited state is split in two, with the energy of the quadrupole splitting given by ΔE_Q. A two-line spectrum is the result (Fig. 8.24*c*). The value of ΔE_Q is not greatly affected by temperature and can be determined at 77 K, the boiling point of liquid nitrogen.

Magnetic Hyperfine Interaction

In the presence of a magnetic field the ^{57}Fe nuclear ground state is split in two ($+\frac{1}{2}$ and $-\frac{1}{2}$) and the excited state into four ($+\frac{3}{2}$, $+\frac{1}{2}$, $-\frac{1}{2}$, and $-\frac{3}{2}$). The selection rule for transitions is that the magnetic quantum number I_z can change by 0 or ±1 resulting in six possible transitions (Fig. 8.24*d*). A magnetic field at the Fe nucleus can be generated either by unpaired electrons surrounding the nucleus (internal field) or by the application of an external magnetic field or both. Due to electronic relaxation, the rate of which increases with increasing temperature, the internal field tends to average out at high temperatures. In general, magnetic hyperfine interaction is observed over a temperature range from the boiling point of liquid helium, 4.2 K, to about 20 K. The spin state of the iron and the zero field splitting can be established accurately in the presence of a strong applied field. Some of the magnetic parameters can be obtained by EPR, so the two spectroscopic techniques are complementary.

Superposition of Effects

In an actual Mössbauer spectrum, the chemical or isomer shift, the quadrupole splitting, and magnetic splitting, if they exist, will be superimposed, which results in a more complex spectrum. This is illustrated in Fig. 8.24*d* where a six-line spectrum is obtained.

Most frozen samples will contain randomly oriented molecules. However, application of the external field may polarize the unpaired electrons and thus orient the internal field even though the molecules in the frozen sample may be randomly oriented. The usual practice is to apply the external magnetic field parallel to the incident γ beam or perpendicular to it. Since the nuclear transition probabilities depend on the orientation of the magnetic field at the nucleus with respect to the γ beam, the absorption line intensities are affected by the orientation of the external field. We are now in a position to discuss some of the details of Mössbauer spectra of peroxidases.

8.4.3. Mössbauer Spectra of Horseradish Peroxidase

The first Mössbauer spectra of horseradish peroxidase were reported in 1969 (Fig. 8.25).[122] The heme was removed from the peroxidase, its iron was replaced with ^{57}Fe, and the enzyme was reconstituted. Compounds I and II were prepared, and frozen samples were studied at 77 K. The isomer shifts, δE,

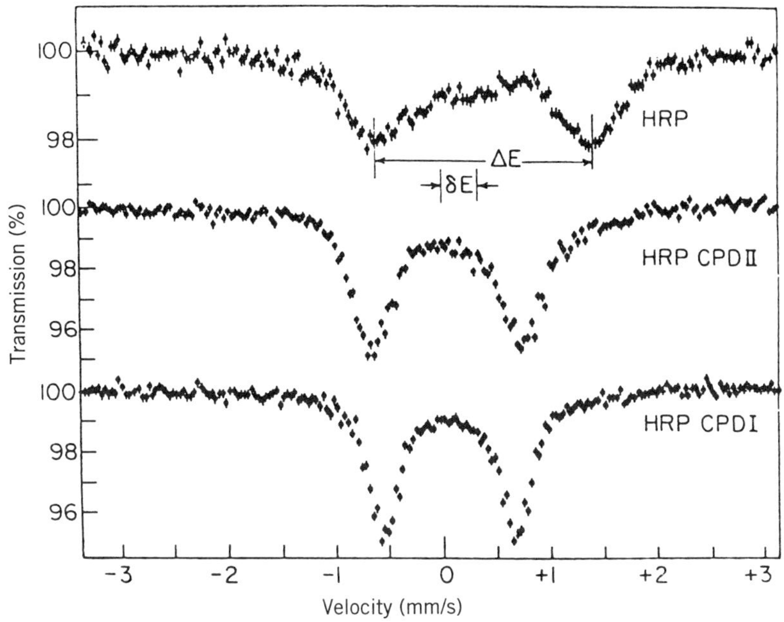

FIG. 8.25. Mössbauer spectra for resting horseradish peroxidase, compound II and compound I at 77 K. The isomer shift, δE, and quadrupole splitting, ΔE, for the resting enzyme are indicated. [Reproduced with permission from T. H. Moss, A. Ehrenberg, and A. J. Bearden, *Biochemistry* **8,** 4159–4162 (1969).]

relative to metallic iron were +0.25 for the resting enzyme and +0.03 and 0.00 for compounds II and I; quadrupole splittings, ΔE_Q, were 1.96 for resting enzyme and 1.36 and 1.20 for compounds II and I (all in units of mm s^{-1}). The smaller isomer shift for compound II relative to resting enzyme was compatible with the well-established idea that iron(IV) is present in compound II. The great similarity of the spectra of compounds I and II indicated that iron(IV) is also present in compound I. Therefore the additional oxidizing equivalent in compound I must be located elsewhere in the enzyme.

The detection of a weak EPR free radical signal on compound I[123] paved the way for further EPR work and an examination of the Mössbauer spectrum of compound I in a magnetic field (Fig. 8.26).[103] All of the following results were obtained with ^{57}Fe-enriched preparations. The results confirm the presence of iron(IV) in compound I, with an electron spin S of 1. Compound I has essentially the same Mössbauer parameters as compound II but the spin S weakly couples to a spin S' of $\frac{1}{2}$ for the free radical. A model based on this weak coupling satisfactorily accounts for the Mössbauer spectra at all temperatures and magnetic field strengths. This work was extended to horseradish peroxidase reconstituted with mesoheme, with the same conclusions.[124]

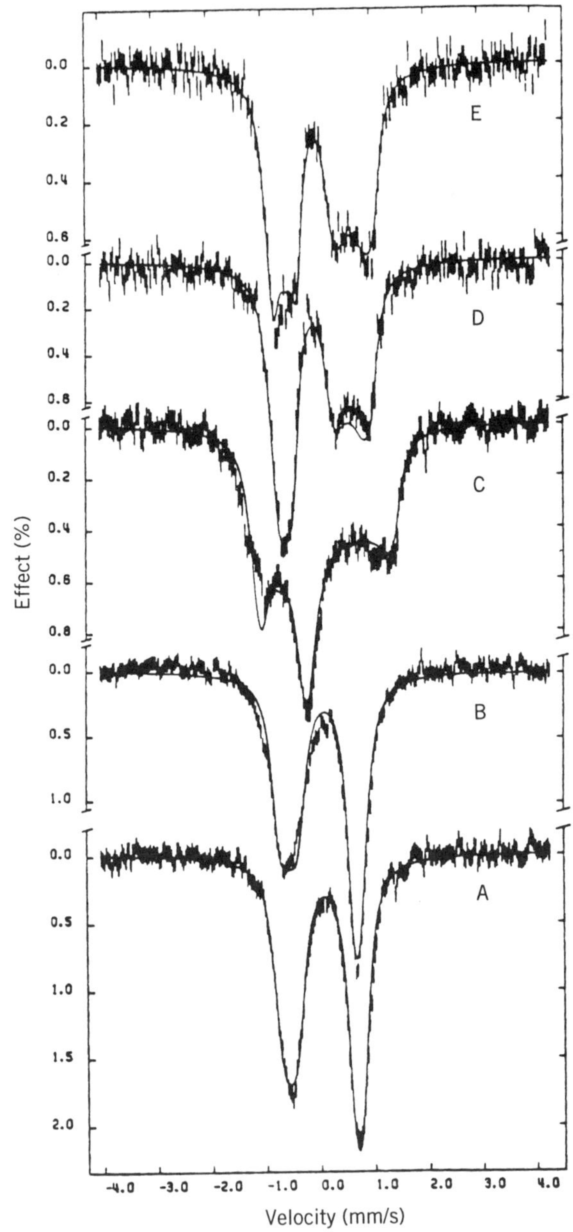

FIG. 8.26. Mössbauer spectra of horseradish peroxidase compound I under various conditions. Temperature, T, in Kelvin and magnetic flux density, H, in Tesla. (A) T = 1.5, H = 0.04 parallel to the γ beam. (B) Same conditions as (A) but H at right angles to the beam. (A) T = 4.2, H parallel = 5.5. (D) T = 26 and H parallel = 3.03. (E) T = 40 and H parallel = 3.03. [Reproduced from C. E. Schulz, P. W. Devaney, H. Winkler, P. G. Debrunner, N. Doan, R. Chiang, R. Rutter, and L. P. Hager, *FEBS Lett.* **103,** 102–105 (1979) with permission from Elsevier Science.]

In a more extensive study the Mössbauer spectra of native horseradish peroxidase and its compounds I, II, III, and X were examined,[103] the latter compound being prepared from native enzyme plus chlorite. In addition spectra were obtained for the benzhydroxamic acid complex with the native enzyme (Fig. 8.27). The reader is referred to this paper for the most detailed discussion of the Mössbauer spectra of horseradish peroxidase and its compounds, and also a comparison with other peroxidases. Because it was not possible to obtain a clean and reproducible EPR spectrum for frozen native horseradish

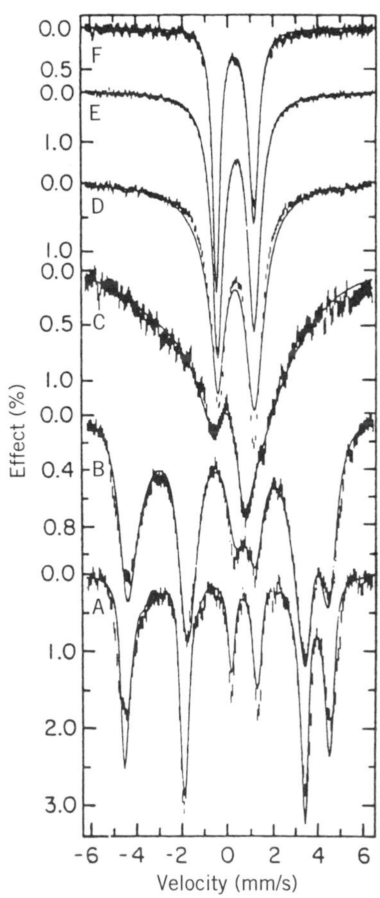

FIG. 8.27. Mössbauer spectra of the benzhydroxamic complex of resting horseradish peroxidase at (A) 4.2, (B) 10, (C) 20, (D) 36, (E) 60, (F) and 200 K. Magnetic flux density of 0.17 T applied at right angles to the beam (0.034 T at 60 K and 0.060 T at 200 K). [Reproduced with permission from C. E. Schulz, R. Rutter, J. T. Sage, P. G. Debrunner, and L. P. Hager, *Biochemistry* **23,** 4743–4754 (1984).]

peroxidase, its benzhydroxamic acid complex was studied instead. The spectrum is typical of a ferric high-spin species.

The Mössbauer spectrum of the fluoride complex of horseradish peroxidase is shown in Fig. 8.28. The data were fit assuming tetragonal symmetry, which implies that the histidine fifth ligand and fluoride sixth ligand are equivalent. The spectrum for compound III is shown in Fig. 8.29. The data indicate a diamagnetic $S = 0$ ground state.

Some of the results on horseradish peroxidase, and comparisons to other compounds I and II, are summarized in Table 8.1. The data indicate that all compounds I and II have $S = 1$ and $Fe^{IV}{=}O$ coordination. Chloroperoxidase is unique in its properties, to a large extent because of its thiolate fifth ligand. A needed valuable contribution is a discussion by a Mössbauer spectroscopist of the significance of differences in Mössbauer spectra for native horseradish peroxidase (five-coordinate at room temperature but six-coordinate at 4.2 K),

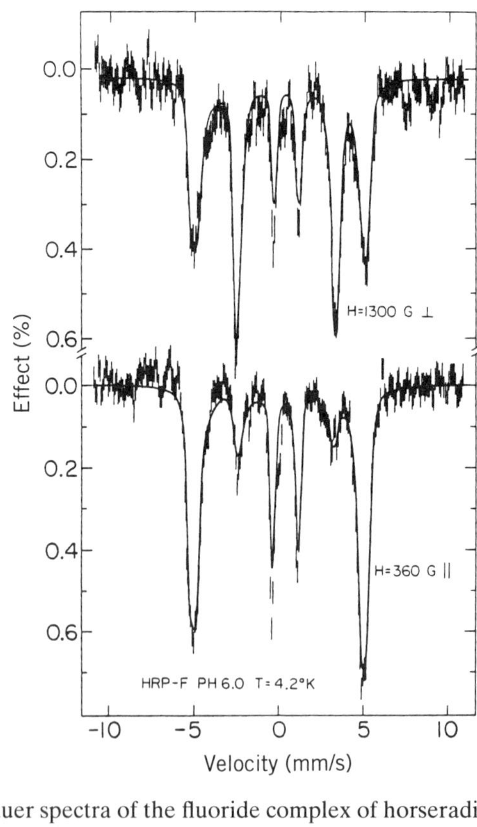

FIG. 8.28. Mössbauer spectra of the fluoride complex of horseradish peroxidase taken at 4.2 K with transverse and parallel magnetic fields. (Reproduced with permission from E. Münck in *The Porphyrins,* D. Dolphin, ed., Academic, New York, 1979, Vol. IVB, pp. 379–423.)

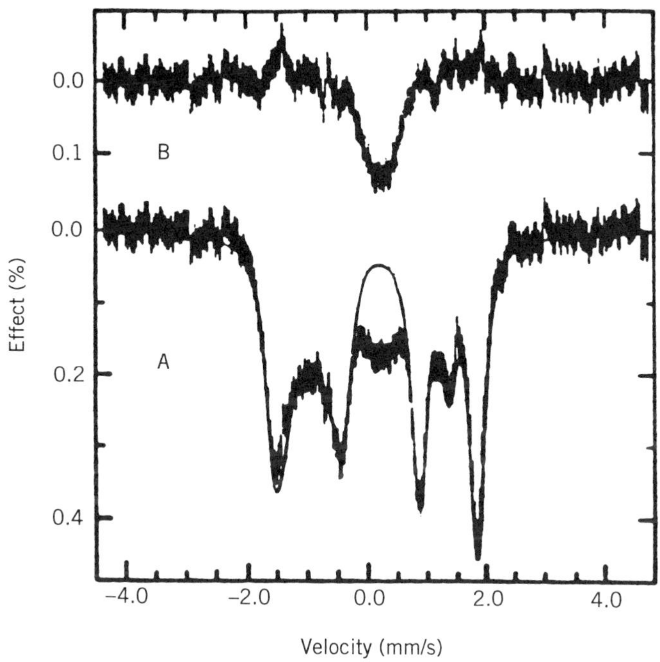

FIG. 8.29. (A) Mössbauer spectrum of compound III of horseradish peroxidase obtained at 4.2 K in a 4.03-T field parallel to the beam. The solid line is spectral simulation based upon a diamagnetic complex. Compound III was prepared by addition of hydrogen peroxide to compound II with some resultant oxidative damage. This is believed to account for the difference between the simulated and observed spectra, shown in (B). [Reproduced with permission from C. E. Schulz, R. Rutter, J. T. Sage, P. G. Debrunner, and L. P. Hager, *Biochemistry* **23,** 4743–4754 (1984).]

TABLE 8.1. Mössbauer Parameters for Various Peroxidase Species

Enzyme[a]	Temp., K	Quadrupole Splitting ΔE_Q, mm s^{-1}	Chemical Shift, δE, mm s^{-1}	References
HRP	77	1.96	0.25	122
HRP-BHA	4.2	1.60	0.40	98
HRP-F	4.2	0.8		121
HRP-I	4.2	1.25	0.08	98
HRP-I	77	1.20	0.00	122
HRP-II	4.2	1.61	0.03	98
HRP-II	77	1.36	0.03	122
Mb(Fe^{IV})	4.2	1.43	0.09	98
JRP-I	77	1.33	0.10	125
JRP-II	77	1.46	0.11	125
CcP-I	4.2	1.55	0.05	126
CPO-I	4.2	1.02	0.15	127
HRP-III	4.2	2.31	0.23	98

[a] Abbreviations: HRP, horseradish peroxidase; Mb, myoglobin; JRP, Japanese radish peroxidase; CcP, yeast cytochrome *c* peroxidase; CPO, chloroperoxidase; I, II, and III are compounds I, II and III; BHA, benzhydroxamic acid.

the benzhydroxamic acid complex, which is six-coordinate with water in the sixth position, and the six-coordinate fluoride complex.

8.5. THEORETICAL STUDIES

Because of the complexities of heme proteins, with the transition-metal ion at the center of the heme, there have been limited purely theoretical calculations, where the experimental instrument is a computer. Nevertheless, progress has been made. The interpretation of heme spectra was facilitated by use of iterative extended Hückel calculations, in work that has stood the test of time.[128–130] Calculated spin densities and quadrupole splitting on a heme model with an imidazole ligand indicated an Fe^{IV} ($S = 1$) porphyrin cation radical ($S = \frac{1}{2}$) in a a_{2u} ground state for the structure of compound I.[131] Charge iterative extended Hückel calculations on heme models, with and without an imidazole ligand, indicated an $Fe^{IV}{=}O$ structure for both compounds I and II, with the addition of a porphyrin π-cation radical for compound I. No commitment was made on whether the porphyrin cation radical had an a_{1u} or a_{2u} ground state. In the words of one of the contributors, "a cottage industry was born, to determine the relative merits of a_{1u} and a_{2u} electronic states."[132] Skipping to more recent results, an a_{1u} state was assigned to compound I of horseradish peroxidase and a_{2u} to compound I of catalase.[133] A conclusion of a density functional theory (DFT) investigation was that compound I might be an admixture of a_{1u} and a_{2u} ground states.[134]

Resting state models were characterized by *ab initio* calculations.[135] It was shown that spin-state changes could modulate the extent of water coordination. Theoretical predictions were made on the nature of the compound I precursor complex with hydrogen peroxide.[136,137]

There are many papers where theory and experiment are combined. Thus, resonance Raman and NMR spectra of oxidized hemes were used to interpret the nature of the electronic ground state.[138,139] Known structures and homology modeling were used to predict docking sites of substrates.[140,141]

The reactivities of a series of phenols with horseradish peroxidase oxidized compounds were interpreted using the Marcus equation for electron transfer.[142] Reactions of compounds I and II were found to be slowed in the presence of a magnetic field, a result interpreted in terms of the influence of the magnetic field on the reversible formation a triplet radical pair (two unpaired electrons) in the process of transferring the electron from a diamagnetic substrate to the enzyme.[143]

REFERENCES

1. J. A. Pople, W. G. Schneider, and H. J. Bernstein, *High Resolution NMR,* McGraw-Hill, New York, 1959.
2. K. Wüthrich, *NMR in Biological Research: Peptides and Proteins,* North-Holland, Amsterdam, 1976.
3. I. D. Campbell and R. A. Dwek, *Biological Spectroscopy,* Benjamin-Cummings, Menlo Park, CA, 1984, pp. 127–177.
4. W. R. Croasmun and R. M. K. Carlson, eds., *Two-Dimensional NMR Spectroscopy. Applications for Chemists and Biochemists,* 2nd ed., VCH, New York, 1994, pp. 915–926.
5. J. N. S. Evans, *Biomolecular NMR Spectroscopy,* Oxford University Press, Oxford, 1995.
6. H. Günther, *NMR Spectroscopy, 2nd ed.,* Wiley, New York, 1995.
7. J. Cavanagh, W. J. Fairbrother, A. G. Palmer III, and N. J. Skelton, *Protein NMR Spectroscopy. Principles and Practice,* Academic, New York, 1996.
8. K. Wüthrich in Ref. 2, p. 22.
9. G. N. La Mar, ed., *Nuclear Magnetic Resonance of Paramagnetic Molecules,* Kluwer Academic, Dordrecht, 1995.
10. A. Lanir and A. Schejter, *Biochem. Biophys. Res. Commun.* **62,** 199–203 (1975).
11. P. S. Burns, R. J. P. Williams, and P. E. Wright, *J. Chem. Soc. Chem. Comm.* 795–796 (1975).
12. I. Morishima, S. Ogawa, T. Inubushi, and T. Yonezawa, *FEBS Lett.* **80,** 177–181 (1977).
13. R. J. P. Williams, P. E. Wright, G. Mazza, and J. R. Ricard, *Biochim. Biophys. Acta* **412,** 127–147 (1975).
14. G. N. La Mar and J. S. de Ropp, *Biochem. Biophys. Res. Commun.* **90,** 36–41 (1979).

15. T. Iizuka, S. Ogawa, T. Inubushi, T. Yonezawa, and I. Morishima, *FEBS Lett.* **64,** 156–158 (1976).
16. I. Morishima and S. Ogawa, *J. Biol. Chem.* **254,** 2814–2820 (1979).
17. I. Morishima, S. Ogawa, and H. Yamada, *Biochemistry* **19,** 1569–1575 (1980).
18. Y. Shiro, M. Kurono, and I. Morishima, *J. Biol. Chem.* **261,** 9382–9390 (1986).
19. I. Morishima, M. Kurono, and Y. Shiro, *J. Biol. Chem.* **261,** 9391–9399 (1986).
20. I. Morishima and S. Ogawa, *J. Am. Chem. Soc.* **100,** 7125–7127 (1978).
21. I. Morishima and S. Ogawa, *Biochemistry* **17,** 4384–4388 (1978).
22. G. N. La Mar and J. S. de Ropp, *J. Am. Chem. Soc.* **102,** 395–397 (1980).
23. J. Sakurada, S. Takahashi, and T. Hosoya, *J. Biol. Chem.* **262,** 4007–4010 (1987).
24. I. Morishima, S. Ogawa, and T. Yonezawa, *Biochim. Biophys. Acta* **537,** 293–303 (1978).
25. S. Modi, D. V. Behere, and S. Mitra, *J. Biol. Chem.* **264,** 19677–19684 (1989).
26. G. N. La Mar, J. S. de Ropp, K. M. Smith, and K. C. Langry, *J. Biol. Chem.* **255,** 6646–6652 (1980).
27. G. N. La Mar, J. S. de Ropp, K. M. Smith, and K. C. Langry, *J. Biol. Chem.* **256,** 237–243 (1981).
28. V. Thanabal, J. S. de Ropp, and G. N. La Mar, *J. Am. Chem. Soc.* **109,** 265–272 (1987).
29. V. Thanabal, J. S. de Ropp, and G. N. La Mar, *J. Am. Chem. Soc.* **109,** 7516–7525 (1987).
30. J. S. de Ropp, L. P. Yu, and G.N. La Mar, *J. Biomol. NMR* **1,** 175–190 (1991).
31. N. C. Veitch, R. J. P. Williams, R. C. Bray, J. F. Burke, S. A. Sanders, R. N. F. Thorneley, and A. T. Smith, *Eur. J. Biochem.* **207,** 521–531 (1992).
32. N. C. Veitch and R. J. P. Williams, *Eur. J. Biochem.* **189,** 351–362 (1990).
33. J. S. de Ropp and G. N. La Mar, *J. Am. Chem. Soc.* **113,** 4348–4350 (1991).
34. V. Thanabal, J. S. de Ropp, and G. N. La Mar, *J. Am. Chem. Soc.* **110,** 3027–3035 (1988).
35. M. Sette, J. S. de Ropp, G. Hernández, and G. N. La Mar, *J. Am. Chem. Soc.* **115,** 5237–5245 (1993).
36. G. N. La Mar, Z. Chen, K. Vyas, and A. D. McPherson, *J. Am. Chem. Soc.* **117,** 411–419 (1995).
37. I. Bertini, C. Luchinat, and D. Tarchi, *Chem. Phys. Lett.* **203,** 445–449 (1993).
38. Z. Chen, J. S. de Ropp, G. Hernández, and G. N. La Mar, *J. Am. Chem. Soc.* **116,** 8772–8783 (1994).
39. G. N. La Mar, Z. Chen, and J. S. de Ropp in *Nuclear Magnetic Resonance of Paramagnetic Molecules*, G.N. La Mar, ed., Kluwer Academic, Amsterdam, 1995.
40. G. H. Loew, P. Du, and A. T. Smith, *Biochem. Soc. Trans.* **23,** 250–256 (1995).
41. G. N. La Mar, J. S. de Ropp, K. M. Smith, and K. C. Langry, *J. Am. Chem. Soc.* **106,** 4438–4444 (1984).
42. A. J. Sitter, J. R. Shifflett, and J. Terner, *J. Biol. Chem.* **263,** 13032–13038 (1988).
43. A. Feis, M. P. Marzocchi, M. Paoli, and G. Smulevich, *Biochemistry* **33,** 4577–4583 (1994).

44. N. Foote, P. M. A. Gadsby, M. J. Berry, C. Greenwood, and A. J. Thomson, *Biochem. J.* **246,** 659–658 (1987).
45. K.-G. Paul and P.-I. Ohlsson, *Acta Chem. Scand.* **B32,** 395–404 (1978).
46. J. Sakurada, S. Takahashi, and T. Hosoya, *J. Biol. Chem.* **261,** 9657–9662 (1986).
47. L. Banci, I. Bertini, T. Bini, M. Tien, and P. Turano, *Biochemistry* **32,** 5825–5831 (1993).
48. L. Banci, P. Carloni, and G. G. Savellini, *Biochemistry* **33,** 12356–12366 (1994).
49. N. C. Veitch, *Biochem. Soc. Trans.* **23,** 232–240 (1995).
50. R. K. Gupta, A. S. Mildvan, and G. R. Schonbaum, *Biochem. Biophys. Res. Commun.* **89,** 1334–1340 (1979).
51. G. Smulevich, A. M. English, A. R. Mantini, and M. P. Marzocchi, *Biochemistry* **30,** 772–779 (1991).
52. G. N. La Mar, G. Hernández, and J. S. de Ropp, *Biochemistry* **31,** 9158–9168 (1992).
53. N. C. Veitch and R. J. P. Williams, *Eur. J. Biochem.* **229,** 629–640 (1995).
54. N. C. Veitch, R. J. P. Williams, N. M. Bone, J. F. Burke, and A. T. Smith, *Eur. J. Biochem.* **233,** 650–658 (1995).
55. N. C. Veitch, D. J. Gilfoyle, C. G. White, and A. T. Smith in C. Obinger, U. Burner, R. Ebermann, C. Penel, and H. Greppin, eds., *Plant Peroxidases: Biochemistry and Physiology,* University of Geneva Press, Geneva, 1996, pp. 1–6.
56. N. C. Veitch, Y. Gao, A. T. Smith, and C. G. White, *Biochemistry* **36,** 14751–14761 (1997).
57. E. Gonzalez-Vergara, M. Meyer, and H. M. Goff, *Biochemistry* **24,** 6561–6567 (1985).
58. J. S. de Ropp, Z. Chen, and G. N. La Mar, *Biochemistry* **34,** 13477–13484 (1995).
59. N. C. Veitch, R. J. P. Williams, R. C. Bray, J. F. Burke, S. A. Sanders, R. N. F. Thorneley, and A. T. Smith, *Eur. J. Biochem.* **207,** 521–531 (1992).
60. S. Nagano, M. Tanaka, I. Ishimori, Y. Watanabe, and I. Morishima, *Biochemistry* **35,** 14251–14258 (1996).
61. M. Tanaka, K. Ishimori, and I. Morishima, *Biochem. Biophys. Res. Commun.* **227,** 393–399 (1996).
62. E. Zavoisky, *J. Phys. U.S.S.R.* **9,** 211–216 (1945).
63. M. Iwamura and N. Inamoto, *Bull. Chem. Soc. Jpn.* **40,** 702 (1967).
64. E. G. Janzen and J. I.-P. Liu, *J. Mag. Res.* **9,** 510–512 (1973).
65. E. G. Janzen, C. A. Evans, and E. R. Davis in *Organic Free Radicals,* W. A. Pryor, ed., ACS Symposium Series 69, American Chemical Society, Washington, DC, 1978, pp. 433–446.
66. I. Yamazaki, H. S. Mason, and L. Piette, *J. Biol. Chem.* **235,** 2444–2449 (1960).
67. I. Yamazaki and L. H. Piette, *Biochim. Biophys. Acta* **50,** 62–69 (1961).
68. I. Yamazaki and L. H. Piette, *Biochim. Biophys. Acta* **77,** 47–64 (1963).
69. B. W. Griffen, *Arch. Biochem. Biophys.* **190,** 850–853 (1978).
70. B. W. Griffen and P. L. Ting, *Biochemistry* **17,** 2206–2211 (1978).
71. B. Kalyanaraman and R. P. Mason, *Biochem. Biophys. Res. Commun.* **105,** 217–224 (1982).

72. P. D. Josephy, T. E. Eling, and R. P. Mason, *Mol. Pharmacol.* **23,** 766–770 (1983).
73. P. D. Josephy, T. E. Eling, and R. P. Mason, *J. Biol. Chem.* **257,** 3669–3675 (1982).
74. P. D. Josephy, T. E. Eling, and R. P. Mason, *Mol. Pharmacol.* **23,** 461–466 (1983).
75. L. S. Harman, C. Mottley, and R. P. Mason, *J. Biol. Chem.* **259,** 5606–5611 (1982).
76. V. Fischer and R. P. Mason, *Chem.-Biol. Interactions* **57,** 129–142 (1986).
77. V. Fischer, L. S. Harman, P. R. West, and R. P. Mason, *Chem.-Biol. Interactions* **60,** 115–127 (1986).
78. K. M. Morehouse, W. D. Flitter, and R. P. Mason, *FEBS Lett.* **222,** 246–250 (1987).
79. J. F. Curtis, M. F. Hughes, R. P. Mason, and T. E. Eling, *Carcinogenesis* **9,** 2015–2021 (1988).
80. C. Mottley and R. P. Mason, *Arch. Biochem. Biophys.* **267,** 681–689 (1988).
81. M. Valoti, H. J. Sipe, Jr., G. Sgaragli, and R. P. Mason, *Arch. Biochem. Biophys.* **269,** 423–432 (1989).
82. K. M. Morehouse, H. J. Sipe, Jr., and R. P. Mason, *Arch. Biochem. Biophys.* **273,** 158–164 (1989).
83. F. R. Gadelha, P. M. Hanna, R. P. Mason, and R. Docampo, *Chem.-Biol. Interactions* **85,** 35–48 (1992).
84. P. Kolodzieczyk, K. Reszka, and J. W. Lown, *Free Rad. Biol. Med.* **5,** 13–25 (1988).
85. V. Fischer and R. P. Mason, *J. Biol. Chem.* **259,** 10284–10288 (1984).
86. B. Kalyanaraman, E. G. Janzen, and R. P. Mason, *J. Biol. Chem.* **260,** 4003–4006 (1985).
87. L. S. Harman, D. K. Carver, J. Schreiber, and R. P. Mason, *J. Biol. Chem.* **261,** 1642–1648 (1986).
88. S. N. J. Moreno, K. Stolze, E. G. Janzen, and R. P. Mason, *Arch. Biochem. Biophys.* **265,** 267–271 (1988).
89. K. Stolze, S. N. J. Moreno, and R. P. Mason, *J. Inorg. Biochem.* **37,** 45–53 (1989).
90. J. Van der Zee, D. R. Duling, R. P. Mason, and T. E. Eling, *J. Biol. Chem.* **264,** 19828–19836 (1989).
91 D. N. R. Rao, V. Fischer, and R. P. Mason, *J. Biol. Chem.* **265,** 844–847 (1990).
92. Y. Morita and H. S. Mason, *J. Biol. Chem.* **240,** 2654–2659 (1965).
93. W. E. Blumberg, J. Peisach, B. A. Wittenberg, and J. B. Wittenberg, *J. Biol. Chem.* **243,** 1854–1862 (1968).
94. P. Douzou and F. Leterrier, *Biochim. Biophys. Acta* **220,** 338–340 (1970).
95. J. S. Leigh, M. M. Maltempo, P.-I. Ohlsson, and K.-G. Paul, *FEBS Lett.* **51,** 304–308 (1975).
96. M. Tamura and H. Hori, *Biochim. Biophys. Acta* **284,** 20–29 (1972).
97. M. M. Maltempo, P.-I. Ohlsson, K.-G. Paul, and A. Ehrenberg, *Biochemistry* **18,** 2935–2941 (1979).
98. C. E. Schulz, R. Rutter, J. T. Sage, P. G. Debrunner, and L. P. Hager, *Biochemistry* **23,** 4743–4754 (1984).
99. S. Vuk-Pavlovic and Y. Siderer, *Biochem. Biophys. Res. Commun.* **79,** 885–889 (1977).
100. K. Kobayashi, M. Tamura, K. Hayashi, H. Hori, and H. Morimoto, *J. Biol. Chem.* **255,** 2239–2242 (1980).

101. L. J. Young and L. M. Siegel, *Biochemistry* **27,** 2790–2800 (1988).

102. R. Aasa, T. Vänngård, and H. B. Dunford, *Biochim. Biophys. Acta* **391,** 259–264 (1975).

103. C. E. Schulz, P. W. Devaney, H. Winkler, P. G. Debrunner, N. Doan, R. Chiang, R. Rutter, and L. P. Hager, *FEBS Lett.* **103,** 102–105 (1979).

104. M. Chu, H. B. Dunford, and D. Job, *Biochem. Biophys. Res. Commun.* **74,** 159–163 (1977).

105. A. R. McIntosh and M. J. Stillman, *Biochem. J.* **167,** 31–37 (1977).

106. K. L. Moore, M. M. Moronne, and R. J. Melhorn, *Arch. Biochem. Biophys.* **299,** 47–56 (1992).

107. G. Feher, *Phys. Rev.* **103,** 500–501 (1956).

108. G. Feher, *Phys. Rev.* **103,** 834–835 (1956).

109. J. S. Hyde and A. H. Maki, *J. Chem. Phys.* **40,** 3117–3118 (1964).

110. J. S. Hyde, *J. Chem. Phys.* **43,** 1806–1818 (1965).

111. G. H. Rist and J. S. Hyde, *J. Chem. Phys.* **52,** 4633–4643(1970).

112. J. Fritz, R. Anderson, J. Fee, G. Palmer, R. H. Sands, J. C. M. Tibris, I. C. Gunsalus, W. H. Orme-Johnson, and H. Beinert, *Biochim. Biophys. Acta* **253,** 110–133 (1971).

113. G. Feher, *Electron Paramagnetic Resonance with Applications to Selected Problems in Biology,* Gordon and Breach, New York, 1970, pp. 13–22.

114. A. Abragam and B. Bleaney, *Electron Paramagnetic Resonance of Transition Ions,* Clarendon Press, Oxford, 1970, Chapter 4.

115. J. E. Wertz and J. R. Bolton, *Electron Spin Resonance. Elementary Theory and Practical Applications,* McGraw-Hill, New York, 1972, pp. 353–374.

116. P. B. Ayscough, *Electron Spin Resonance in Chemistry,* Methuen, London, 1967, pp. 130–135.

117. W. H. Orme-Johnson and R. H. Sands in *Iron-Sulfur Proteins,* Vol. II, W. Lovenberg, ed., Academic, New York, 1973, pp. 195–238.

118. T. D. Smith and J. R. Pilbrow in *Biological Magnetic Resonance,* Vol. 2, L. J. Berliner and J. Reuben, eds., Plenum, New York, 1980, pp. 85–168.

119. J. E. Roberts, B. M. Hoffman, R. Rutter, and L. P. Hager, *J. Biol. Chem.* **256,** 2118–2121 (1981).

120. G. Lang, *Quart. Rev. Biophys.* **3,** 1–60, (1970).

121. E. Münck in *The Porphyrins,* D. Dolphin, ed., Academic, New York, Vol. IVB, 1979, pp. 379–423.

122. T. H. Moss, A. Ehrenberg, and A. J. Bearden, *Biochemistry,* **8,** 4159–4162 (1969).

123. R. Aasa, T. Vänngård, and H. B. Dunford, *Biochim. Biophys. Acta* **391,** 259–264 (1975).

124. R. Rutter, M. Valentine, M. P. Hendrich, L. P. Hager, and P. G. Debrunner, *Biochemistry* **22,** 4769-4774 (1983).

125. T. Harami, Y. Maeda, Y. Morita, A. Trautwein, and U. Gonser, *J. Chem. Phys.* **67,** 1164–1169, (1977).

126. G. Lange, K. Spartalian, and T. Yonetani, *Biochim. Biophys. Acta* **451,** 250–258 (1976).

127. R. Rutter, Ph.D. Thesis, University of Illinois, Urbana-Champaign, 1982.
128. M. Zerner and M. Gouterman, *Theor. Chim. Acta* **4,** 44–63 (1960).
129. A. M. Schaffer, M. Gouterman, and E. R. Davidson, *Theor. Chim. Acta* **30,** 9–30 (1973).
130. M. Gouterman in *The Porphyrins,* Vol. III, D. Dolphin, ed., Academic, New York, 1978, pp. 1–165.
131. L. K. Hanson, C. K. Chang, M. S. Davis, and J. Fajer, *J. Am. Chem. Soc.* **103,** 663–670, (1981).
132. Personal communication, J. Fajer, Brookhaven National Laboratory, Upton, NY, January, 1997.
133. P. Du and G. H. Loew, *Biophys. J.* **68,** 69–80, (1995).
134. H. Kuramochi, L. Noodleman, and D. A. Case, *J. Am. Chem. Soc.* **119,** 11442–11451 (1997).
135. G. Loew and M. Dupuis, *J. Am. Chem. Soc.* **119,** 9848–9851 (1997).
136. G. Loew and M. Dupuis, *J. Am. Chem. Soc.* **118,** 10584–10587, (1996).
137. G. Loew and M. Dupuis, *J. Am. Chem. Soc.* **118,** 10588–10594 (1996).
138. K. Czarnecki, L. M. Proniewicz, H. Fujii, and J. R. Kincaid, *J. Am. Chem. Soc.* **118,** 4680–4685 (1996).
139. H. Fujii and K. Ichikawa, *Inorg. Chem.* **31,** 1110–1112, (1992).
140. G. H. Loew, P. Du, and A. T. Smith, *Biochem. Soc. Trans.* **23,** 250–256 (1995).
141. D. Zhao, D. Gilfoyle, A. T. Smith, and G. H. Loew, *Proteins: Structure, Function Genetics* **26,** 204–216, (1996).
142. L. K. Folkes and L. P. Candeias, *FEBS Lett.* **412,** 305–308 (1997).
143. M. B. Taraban, T. V. Lesina, M. A. Anderson, and C. B. Grissom, *J. Am. Chem. Soc.* **119,** 5768–5769 (1997).

CHAPTER 9

Yeast Cytochrome *c* Peroxidase. I: Properties and Reactions with Small Molecules

9.1. INTRODUCTION

Yeast cytochrome *c* peroxidase was discovered in 1940,[1] and preliminary investigations were conducted.[2] Little further work occurred until a simpler and better purification procedure using chromatographic techniques was published by Takashi Yonetani in 1965.[3] His work, which pioneered studies on the yeast enzyme, has continued into the 1990s. It includes comparisons of oxidized forms of the enzyme to those of horseradish peroxidase[4] and myoglobin,[5,6] measurements of room and low-temperature optical spectra,[7] and reconstitution with different prosthetic groups.[8–12]

The initial chromatographic preparation procedure of yeast cytochrome *c* peroxidase was followed rapidly by preparation of the crystalline enzyme.[13,14] Both improved and similar preparation procedures have since been published.[15–17] Crystallization was facilitated because the enzyme does not contain any carbohydrate.

The importance of yeast cytochrome *c* peroxidase as a model in establishing the plant peroxidase superfamilies is discussed in Chapter 2, and a discussion of its crystal structure as well as brief comparison of its properties to those of horseradish peroxidase in Chapter 3. Despite the voluminous literature on the enzyme, there is no agreement on several key questions concerning its mechanism of action. The question of the speed of establishment of equilibrium for two forms of compound II, or indeed whether the equilibrium exists, has not been settled. As far as I am aware, it is an original suggestion in Section 9.3.2 of this chapter that small substrates react with the iron(IV) site of both compounds I and II. As described in Chapter 10, the mechanisms of the reactions between the two oxidized forms of the enzyme and reduced cytochrome *c* are not clearly established.

The physiological role of cytochrome *c* peroxidase, as an electron donor to hydrogen peroxide and an electron acceptor from the iron(II) form of cytochrome *c*, appears not to be understood. Since the enzyme is isolated from aerobically grown baker's yeast, and there are several physiological pathways to reduce molecular oxygen, the enzyme may be present to remove the threat of oxidative damage from hydrogen peroxide. Alternatively, its role may be to oxidize cytochrome *c*.

9.2. OVERVIEW OF PROPERTIES

9.2.1. Native Enzyme

Yeast cytochrome *c* peroxidase contains one heme group bound to a single polypeptide chain.[3] The proximal ligand, in the fifth coordination position of the iron, is the imidazole ring of histidine 175. The molecular weight of the enzyme, determined from sedimentation and diffusion studies,[18] is 34,100. The isoelectric point of the enzyme occurs at pH 5.26.[19] The amino acid sequence is listed in Appendix A. Its circular dichroism spectrum has been recorded.[20] The enzyme is unstable outside the pH range of 4–8. Alkaline transitions have been studied,[21] as have tryptophan fluorescence,[22,23] the effect of pressure,[24] alkaline,[25] and thermal[26,27] denaturation. Hydrogen peroxide oxidizes three methionine residues in the apoenzyme, Met119, Met230, and Met231, to the corresponding sulfoxide derivatives.[28] Earlier work on the enzyme has been reviewed in detail.[29,30]

Coordination Number of Iron(III)

Confusion existed about the coordination number of the iron(III) of native yeast cytochrome *c* peroxidase (see Section 3.3.1). The correct answer is five, with the sixth coordination position vacant on the distal side of the heme. The first determination of the coordination number of five was obtained from a Raman spectral investigation in 1979.[31] However, more credence was given to the assertion, in the first crytallographic study, that water was present in the sixth coordination position, despite its long distance from the iron site.[32] The resonance Raman spectra of a single crystal of cytochrome *c* peroxidase confirmed that the heme iron(III) is five-coordinate high-spin.[33]

There are at least five factors that need to be discussed with regard to the coordination number of iron(III): (i) buffer effects, (ii) effect of nitrate, (iii) effect of aging, (iv) effect of pH, and (v) effect of temperature. Sometimes combined effects are observed. Some of the literature has been confusing because of lack of recognition of the existence of one or more effects.

Buffer Effects

A definitive study was conducted on the striking effect of acetate, often used as a buffer.[34] The effect of acetate on the optical spectrum of the enzyme is shown in Fig. 9.1. Acetate acts as a competitive inhibitor for binding or

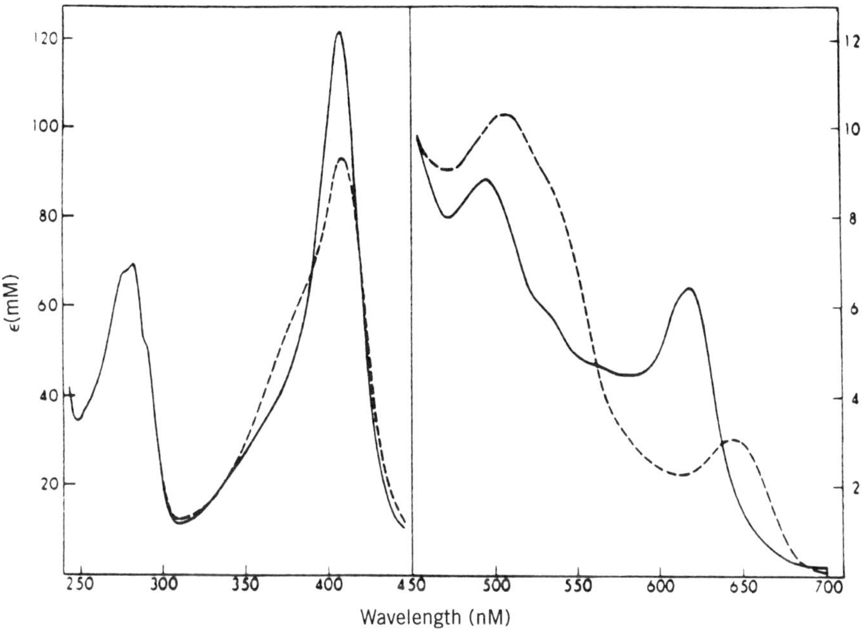

FIG. 9.1. Optical spectra of acetate-bound cytochrome c peroxidase—(0.1 M acetate, pH 5.3, 3°C), and of native enzyme ----- (0.2 M phosphate buffer, pH 8.1, 20°C). [Reproduced with permission from R. A. Mathews and J. B. Wittenberg, *J. Biol. Chem.* **254,** 5991–5996 (1979).]

reaction at the iron site since it binds in the sixth coordination position of the iron.

Effect of Nitrate

Nitrate was used as a supposedly inert electrolyte to maintain constant ionic strength in many of the kinetic studies of yeast cytochrome *c* peroxidase by the Erman group, and of horseradish peroxidase by the Dunford group. (See Section 4.5.2 on horseradish peroxidase). A pH-dependent effect of nitrate and possibly acetate on the rate of compound I formation of cytochrome *c* peroxidase is shown in Fig. 9.2. When nitrate and acetate are eliminated the pH-dependent effect disappears.

Effect of Aging

Unfortunately, some of the discussion of aging has been clouded because of the simultaneous effect of nitrate. The following descriptions apply to nitrate-free solutions. A resonance Raman investigation showed that cytochrome *c* peroxidase undergoes aging-induced processes (Fig. 9.3).[35] This was confirmed in a combined electronic spectral and electron paramagnetic resonance (EPR)

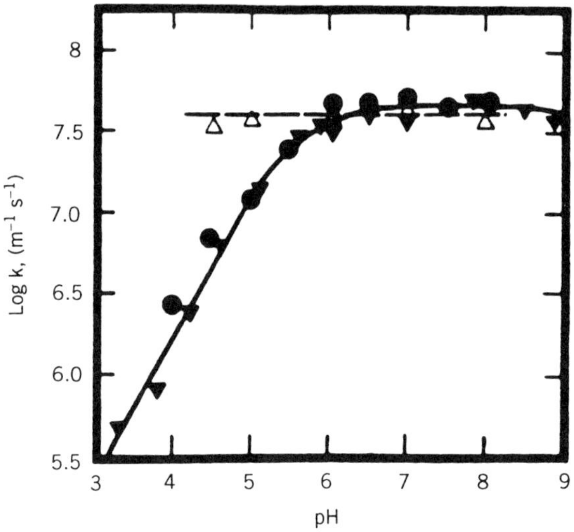

FIG. 9.2. Log of the rate constant for compound I formation from hydrogen peroxide and yeast cytochrome *c* peroxidase. Solid symbols: 0.01 M acetate (pH 4–5.5) or phosphate (pH 5.5–8) buffer with KNO_3 added to make the total ionic strength 0.1 M. The solid line was calculated assuming a pK_a of 5.5 for a group on the enzyme. Open triangles: enzyme in 0.1 M phosphate buffer. [Reproduced with permission from L. B. Vitello, M. Huang, and J. E. Erman, *Biochemistry* **29,** 4283–4288 (1990).]

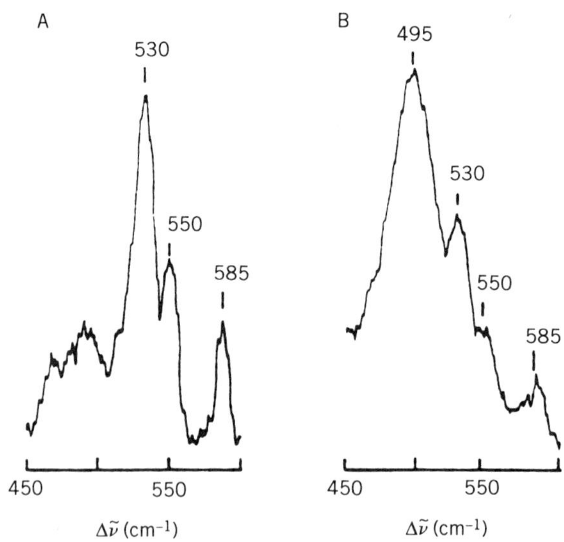

FIG. 9.3. Resonance Raman spectra of the CO adducts of ferrous cytochrome *c* peroxidase in 0.1 M phosphate buffer, pH 6. Fresh protein (A) was stored less than one week at 4°C. Aged protein (B) was stored at −10°C for one year. [Reproduced with permission from G. Smulevich, R. Evangelista-Kirkup, A. M. English, and T. G. Spiro, *Biochemistry* **25,** 4426–4430 (1986).]

study (Fig. 9.4).[36] It was recommended that zwitterionic buffers such as Mes, Mops, and Hepes be used, as well as phosphate, citrate, and succinate buffers. An effect of aging is to convert the iron(III) from five-coordinate high-spin to six-coordinate high-spin, with the most probable candidate for the sixth ligand being the nearby $H_2O^{(595)}$, detected in the crystal structure.

Detailed investigations were conducted on both freshly prepared and stored enzyme.[17,37] The iron(III) of enzyme freshly prepared by the Yonetani method is entirely five-coordinate, high-spin, between pH 4.3 and 7 with reinheitzahl (RZ) values between 1.25 and 1.30. However, storage of samples in solution for a week at 4°C caused changes. It was found that there was no difference in properties between freshly prepared enzyme and enzyme stored below −20°C in the crystalline state for up to 41 months. In a nuclear magnetic resonance (NMR) study it was concluded that the iron of the enzyme was six-coordinate, an indication of an aged sample.[38] When the iron is changed to six-coordinate, the RZ value actually increases. Therefore, the RZ value is not a good indication of enzyme viability.

A summary of this section, from some hard-earned lessons, is the following: Don't use nitrate to adjust the ionic strength, or acetate buffer; and only use enzyme samples obtained using preparation and storage protocols proven to give fully active enzyme.

9.2.2. Compound I

In accordance with all other peroxidase literature, the initial product of the reaction of yeast cytochrome *c* peroxidase with hydrogen peroxide is now usually called compound I,[39–41] a trend started in 1978. The earlier names of complex I and complex ES have been discarded because covalent attachment of hydrogen peroxide is not complex formation. Some investigators still like to use the name compound ES as an indication of the unique protein-free radical found in the oxidized yeast enzyme. However, compound ES is a contradiction; ES is a universal indication of a complex in the biochemical literature so the name literally means compound complex. The name yeast cytochrome *c* peroxidase compound I contains no ambiguity.

The role of the axial ligands in determining the properties of compound I, its formation and reactivity, have been examined on a comparative basis from the standpoint of X-ray crystallographic data,[42] site-directed mutagenesis,[43] ligand field parameters, spectroscopic data and reduction potentials,[44] model compounds,[45] molecular orbital calculations,[46] and calculated exchange interactions, based upon Mössbauer,[47] EPR,[48] and other spectroscopic data.

The optical spectra of native enzyme and compound I, and the difference spectrum, are shown in Fig. 9.5. (See also Fig. 3.2.) The native enzyme has an EPR signal characteristic of high-spin iron(III) that changes to an intense

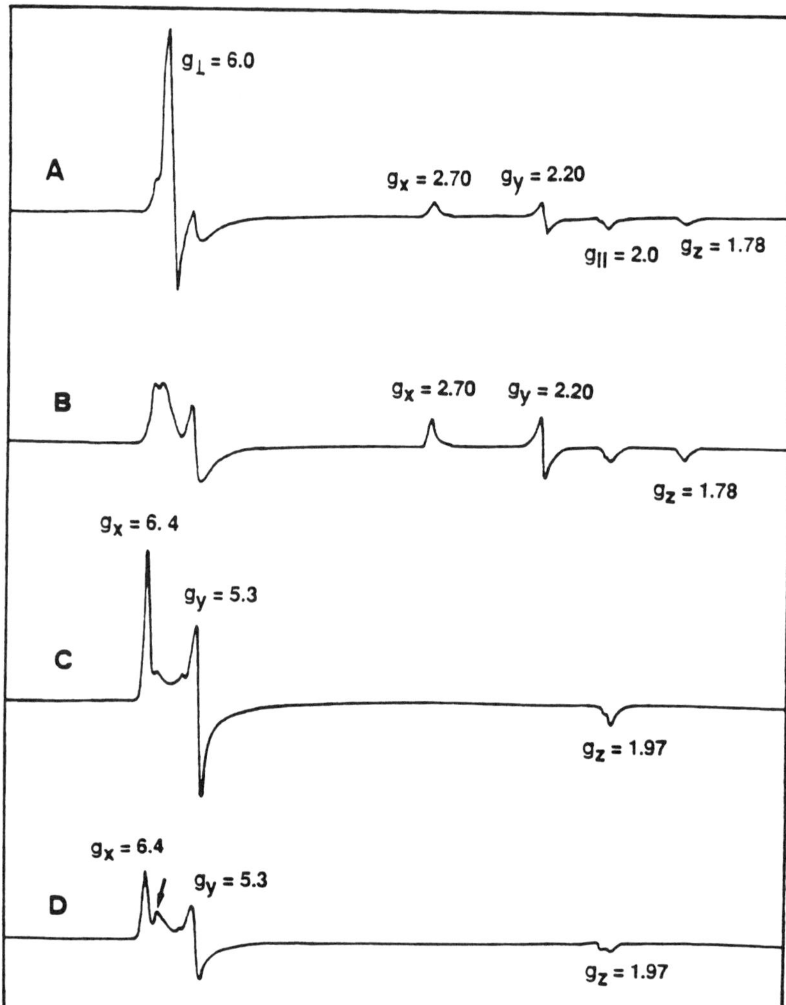

FIG. 9.4. X-band EPR spectra of cytochrome *c* peroxidase in 0.1 M Mes buffer, pH 7, at 10 K. (A, D) aged enzyme in the absence and presence of 67 vol % glycerol. (B, C) fresh enzyme in the absence and presence of 67 vol % glycerol. [Reproduced with permission from T. Yonetani and H. Anni, *J. Biol. Chem.* **262,** 9547–9554 (1987).]

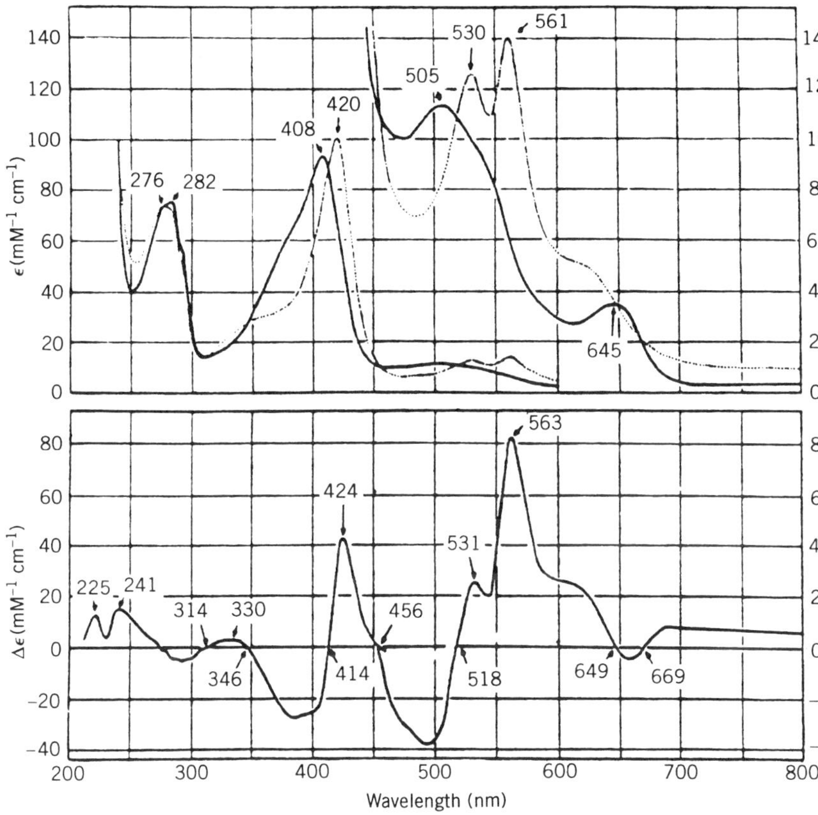

FIG. 9.5. Absolute and difference optical absorption spectra of native yeast cytochrome *c* peroxidase and its compound I at pH 7 and 20°C. (Reproduced with permission from T. Yonetani in *The Enzymes,* 3rd ed., P. D. Boyer, ed., Academic, New York, Vol. 13, 1976, pp. 345–361.)

EPR signal[49–51] centered on $g = 2$ upon compound I formation (Fig. 9.6). The signal is now known to be caused by an unpaired electron on Trp191.[52] Kinetic studies on the Trp191Phe mutant showed that compound I of the mutant had a significantly altered EPR signal,[53,54] and a ferryl porphyrin π-cation radical spectrum as found in horseradish peroxidase.[55] It was shown that the EPR line shape and ENDOR signals from compound I were only perturbed by mutations involving Trp191 or amino acid residues that were immediately adjacent.[56] The EPR signal is perturbed by a weak interaction between the free radical and the Fe=O group on the heme.[57] Electron nuclear double resonance (ENDOR) studies in which Trp191 was labeled in specific positions with ^{2}H, with ^{15}N at N1 of the indole ring, and with ^{13}C at C2 of the ring and at Cβ of the side chain proved conclusively that the free radical is located on Trp191 and exists as a cation radical.[58]

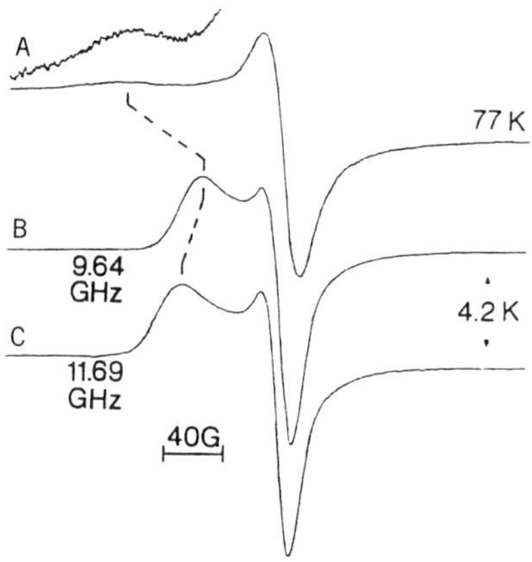

FIG. 9.6. EPR spectra of cytochrome *c* peroxidase compound I. (A) 77 K, 9.16 GHz; (B) 4.2 K, 9.64 GHz; and (C) 4.2 K, 11.69 GHz. Computer simulations indicate that the *g* values are (A) $g_{\|} = 2.09$, $g_{\perp} = 2.01$ at 77 K, and (B and C) $g_{\|} = 2.05$, $g_{\perp} = 2.01$ at 4.2 K. [Reproduced with permission from B. M. Hoffman, J. E. Roberts, C. H. Kang, and E. Margoliash, *J. Biol. Chem.* **256,** 6556–6564 (1981).]

If compound I, formed in the absence of exogenous donor, is allowed to decay, tyrosine crosslinks between residues 32 and 48 are formed.[59] The effect of excess hydrogen peroxide on tryptophan residues in the native enzyme and in Trp51Phe and Trp191Phe mutants has been examined.[60] The native enzyme and the initial complex with hydrogen peroxide have been studied by molecular dynamics simulations.[61]

Direct measurement of the potential of the formation of compound I/ reduction of hydrogen peroxide was made in a system where the native enzyme was adsorbed on a pyrolytic graphite "edge" (PGE) electrode and a value of +740 mV was obtained at pH 6.1, phosphate buffer, 4°C. Each enzyme molecule appeared to form a monolayer on the electrode.[62]

9.2.3. Complexities of Compound II

Because compound I of the yeast enzyme has its free radical on Trp191 instead of on the porphyrin ring, its optical spectrum looks like that of a conventional peroxidase compound II, dominated by the $Fe^{IV}{=}O$ porphyrin chromophore. This complicated the search for a compound II, since if the Trp191 radical of compound I is reduced first, little difference would be expected in the optical spectra when compound I is converted to compound II. If the $Fe^{IV}{=}O$ site of compound I is reduced first, then one would obtain a compound II optical spectrum that looks like that of native enzyme.

A great deal of valuable information was obtained in an elegant kinetic, EPR, and stoichiometric study.[63] It was confirmed that the enzyme and hydrogen peroxide react in a 1 : 1 ratio. It was shown that two equivalents of reducing agent are required to reduce compound I to the native enzyme and that the reductions occur in two sequential steps, the first fast and the second slow. Ferrocyanide was used as the reducing agent since it is known to be oxidized in a one-electron reaction to ferricyanide.

Two plausible mechanisms were examined for the reduction of compound I to native enzyme, an independent sites mechanism and an equilibrium mechanism (Fig. 9.7). In the independent sites mechanism, reduction of the $Fe^{IV}{=}O$ (rapid) and of the Trp191 radical (slow) would proceed independently. In the equilibrium mechanism for reduction, the rapid phase would correspond to formation of two forms of compound II, in rapid equilibrium with each other, followed by the slow phase in which the compound II is reduced to native

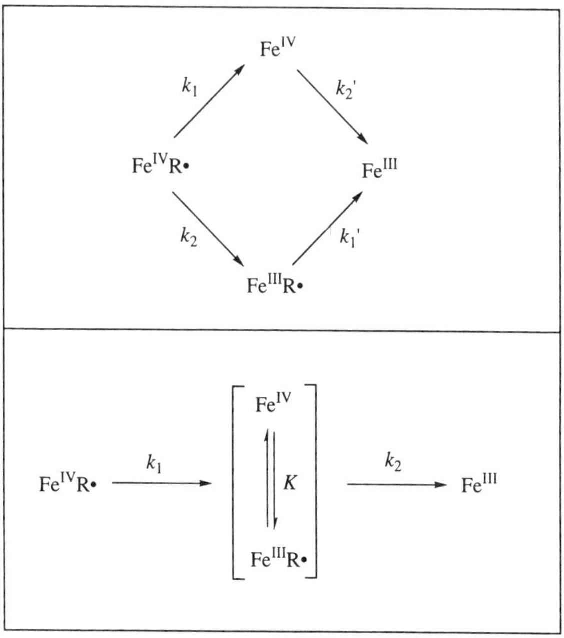

FIG. 9.7. Two mechanisms for the reduction of compound I to native enzyme. Top: the independent sites mechanism in which reduction of the radical site proceeds with rate constants k_1 and k_1', and reduction of iron(IV) with rate constants k_2 and k_2'. Bottom: the equilibrium mechanism in which the one-electron reduction of compound I produces an equilibrium mixture of the two forms of compound II, and this equilbrium mixture is reduced in a second one-electron step to native enzyme. [Reproduced with permission from A. F. W. Coulson, J. E. Erman, and T. Yonetani, *J. Biol. Chem.* **246,** 917–924 (1971).]

enzyme. An alternative hypothesis is that, under certain conditions, the transition between the two forms of compound II is slow compared to redox reactions of compounds I and II.

The equilibrium between the two forms of compound II, whatever its rate of attainment, can be represented as follows:

$$\begin{array}{ll} Fe^{IV} = O, \text{Trp191} \rightarrow & Fe^{III}, \text{Trp191}\bullet^{+} \\ \text{Compound II}_A & \leftarrow \text{Compound II}_B \end{array}$$

Compound II_A is the ferryl form, and compound II_B the radical form. Evidence for this equilibrium was obtained by trapping the Fe^{III}, Trp191$\bullet^+$ form with added fluoride, which can bind in the sixth coordination site of iron(III), but not in the blocked sixth site of $Fe^{IV}{=}O$. With the equilibrium shifted by added fluoride, a more intense EPR signal of the Trp191 $\bullet^+$ radical was obtained.[63]

The equilibrium between the two species, compound II_A and compound II_B, is pH-dependent, with the apparent pK_a near a value of 5.0.[63,64] At pH 8 compound II_A exists, and at lower pH an equilibrium mixture. No one has ever observed pure compound II_B since it cannot be formed in the absence of compound II_A. A caveat should perhaps be placed on the pK_a value of 5, since the value was obtained from experiments in which acetate buffer was used.

A further big step forward in studies of yeast cytochrome *c* peroxidase was attained when pure compound II_A was formed by photoreduction of the enzyme to its ferrous state followed by reaction with hydrogen peroxide at pH 7.[65] The reduction procedure involved addition of acetophenone/isopropyl alcohol followed by irradiation from a xenon–mercury arc lamp.[65] Previously no one had been able to isolate pure compound II in any form. In the previous studies, because of the comparable rates of reduction of compounds I and II, compound II would start to reduce before reduction of compound I would be complete.

At pH 5 an EPR signal indicated the presence of 10% of the enzyme in its free radical form, yielding a $g = 2$ signal.[65] The results were interpreted to mean that the $Fe^{IV}{=}O$ group did not oxidize the Trp191 radical, since at pH 5 a much higher yield of compound II_B might be expected. In other words, establishment of the equilibrium between compound II_A and compound II_B appeared to be blocked.

In order to explain the blocked reaction, it was postulated that electron transfer occurs on two occasions only, during the transient formation of either compound I or compound II.[65] However, the optical spectrum of the pH 5 preparation, reproduced in Fig. 9.8, has a high noise level, indicating the presence of only a small amount of the ferryl form, compound II_A. The pH 5 experiments were conducted in 0.2 M acetate buffer. At pH 5 there may be an equilibrium between three species: compound II_A, the acetate complex of native enzyme, and compound II_B, with or without bound acetate.

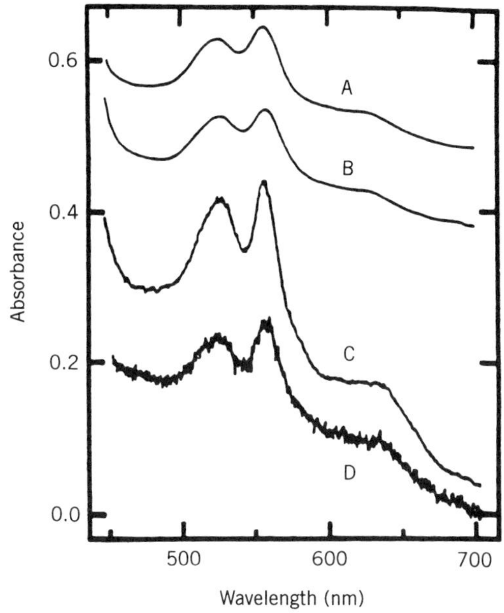

FIG. 9.8. Optical spectra of compound I and the ferryl form of compound II, compound II_A. Room temperature spectra of 10-μM samples of (A) compound I and (B) compound II_A, both in 0.1 M phosphate buffer, pH 7.0. Spectra at 77 K of (C) 1.2 mM compound I and (D) 0.82 mM compound II both in 0.2 M acetate buffer, pH 5.0. [Reproduced with permission from P. S. Ho, B. M. Hoffman, C. H. Kang, and E. Margoliash, *J. Biol. Chem.* **258,** 4356–4363 (1983).]

The spectrum of the actetate complex with native yeast cytochrome *c* peroxidase is fairly flat in the 550-nm region, and so would offer little interference to observation of the twin peaks of compound II_A. Furthermore the pH 5 spectra have broad shoulders in the 620–645 nm region, indicative of the presence of the native enzyme–acetate complex and possibly some native enzyme.[34] For further discussion of the effect of acetate see Section 9.2.3. The postulate of a blocked equilibrium between the two forms of compound II is not in accord with the observations on the effect of added fluoride on compound II.

A combined kinetic and EPR reexamination of the compound II_A–compound II_B equilibrium as a function of pH and in the absence of acetate would be of great value. The pK_a value for the equilibrium, whether the equilibrium is established spontaneously or requires a catalyst, and rate of interconversion of the two forms of compound II, could be firmly established.

The rate constant, 4×10^3 $M^{-1}s^{-1}$, obtained at pH 7 for reduction of compound II_A by ferrocyanide,[65] agrees well with the value obtained in the earlier study.[66]

The optical difference spectrum at pH 8 between compound I and compound II_A, which represents the perturbation of the $Fe^{IV}{=}O$ porphyrin spectrum by the free radical on Trp191, is shown in Fig. 9.9. This remarkably accurate difference spectrum was obtained by applying corrections for incomplete conversion and presence of some native enzyme. The differences in rates

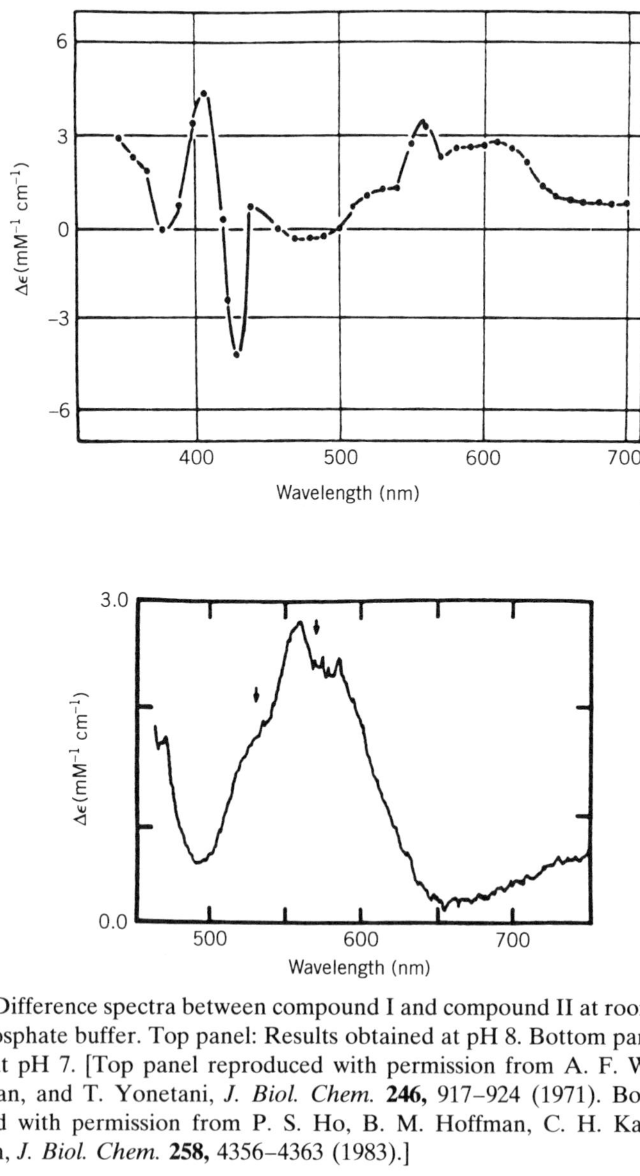

FIG. 9.9. Difference spectra between compound I and compound II at room temperature in phosphate buffer. Top panel: Results obtained at pH 8. Bottom panel: Results obtained at pH 7. [Top panel reproduced with permission from A. F. W. Coulson, J. E. Erman, and T. Yonetani, *J. Biol. Chem.* **246,** 917–924 (1971). Bottom panel reproduced with permission from P. S. Ho, B. M. Hoffman, C. H. Kang, and E. Margoliash, *J. Biol. Chem.* **258,** 4356–4363 (1983).]

between compound I and compound II reductions are too small, so that any titration of compound I with reducing substrate also reduces some compound II. The maximum difference, about $\pm 4\ mM^{-1}cm^{-1}$, occurs in the Soret region.[63] The difference spectrum in the visible region was measured directly in a later investigation,[65] also shown in Fig. 9.9.

The breakthrough in obtaining compound II_A in pure form,[65] was followed by a further investigation of the effect of fluoride on compound II_A at pH 7 in phosphate buffer. Titration of pure compound II_A with fluoride led to reduction of iron(IV) to iron(III) and transfer of the oxidizing equivalent to the free radical site,[67] now known to be Trp191. Thus, compound II_A was converted into the fluoride complex of compound II_B. The reaction may be represented as:

$$\underset{\text{Compound } II_A}{Fe^{IV} = O, \text{Trp191}} + F^- + 2H^+ \rightarrow \underset{\text{Compound } II_B - F^-}{Fe^{III}F^-, \text{Trp191}\bullet^+} + H_2O$$

One of the protons is an intrinsic part of compound II_A and the other must be added from solvent or buffer. A first-order rate constant of $0.115\ s^{-1}$ was observed for the reaction, which was independent of both enyme and fluoride ion concentration. The result was interpreted as a rate-controlling electron transfer from the radical site to iron(IV). However, if the fluoride occupies the sixth coordination position by a dissociative mechanism, the rate-controlling step could be the rate of formation and release of water. The ferryl oxygen of horseradish peroxidase is labile.[68] Whatever the mechanism of fluoride binding to the yeast enzyme, the results are in agreement with an equilibrium between compound II_A and compound II_B, greatly favored by the compound II_B form at pH 7, but shifted by the binding of fluoride. It would be of interest to determine whether the same result could be obtained by titration of compound II_A by a different ligand.

The intramolecular rate of reduction of the iron(IV) of compound I, by a ruthenium(II) derivative attached at His60,[69] was extremely slow, $3 \pm 1\ s^{-1}$.

9.3. KINETICS

9.3.1. Ligand Binding

Optical spectra of cyanide, azide, and fluoride complexes of native yeast cytochrome *c* peroxidase are shown in Fig. 9.10.

The most comprehensive stopped flow investigations of the kinetics of yeast cytochrome *c* peroxidase reactions have been conducted in the laboratory of James Erman, beginning in 1974.[70] Studies of fluoride and cyanide binding to the iron of the native yeast enzyme at low pH were complicated by the same problems encountered with horseradish peroxidase, caused by the presence of nitrate to maintain constant ionic strength and acetate buffer (Section

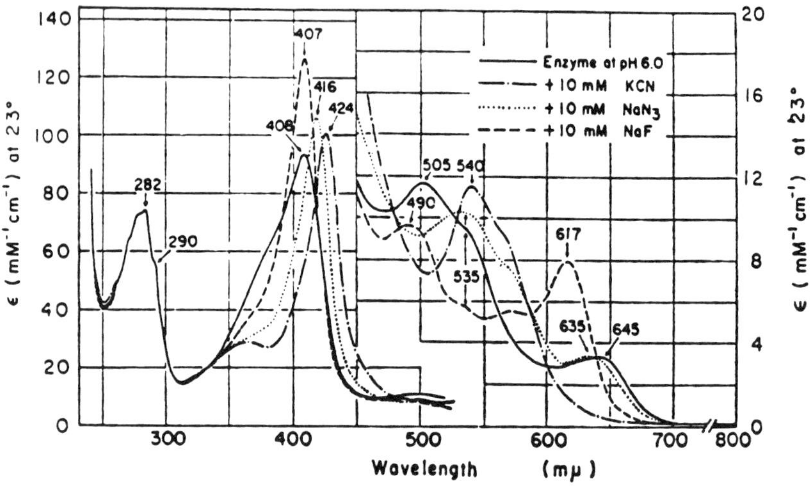

FIG. 9.10. Optical absorption spectra of yeast cytochrome *c* peroxidase and its complexes with cyanide, azide, and fluoride obtained at pH 6. (Reproduced with permission from T. Yonetani in *Advances in Enzymology,* F. F. Nord, ed., Wiley-Interscience, New York, 1970, Vol. 33, pp. 309–335.)

4.5.1)[71]. Results above pH 6 are free of complications caused by the two anions. The trend in pH-dependent results on the binding of fluoride are similar to those for horseradish peroxidase. The acid form of the ligand, HF, binds to a basic form of the enzyme. Resonance Raman studies on a series of peroxidases and mutants showed that fluoride binds to the distal arginine. The distal histidine also interacts with the bound fluoride, but more weakly, perhaps through an intervening water molecule.[72] Differences between horseradish and cytochrome *c* peroxidases are the following.

1. A pK_a value of 5.5 is indicated for an acid group in the active site of the yeast enzyme,[73] compared to ~2.5 for horseradish peroxidase, the latter value being obtained in the absence of nitrate. The question is whether the pK_a value is too high for the yeast enzyme because of the presence of 0.14 M nitrate, and the answer would appear to be yes. (See the discussion of the kinetics of compound I formation in Section 9.3.3).
2. The reaction is an order-of-magnitude faster. The pH-independent constant for binding of HF is 5.1×10^7 $M^{-1}s^{-1}$ for the yeast enzyme compared to 4.2×10^6 $M^{-1}s^{-1}$ for horseradish peroxidase.

The effect of ionic strength on fluoride binding kinetics was also investigated.[74] Cyanide binding occurs with a maximum rate constant of 1.1×10^5 $M^{-1}s^{-1}$, considerably slower than for fluoride. Again the acid form of the ligand, HCN, binds to the basic form of the enzyme.[71]

9.3.2. Kinetics of Ferrocyanide Oxidation

The kinetics of ferrocyanide oxidation, by yeast cytochrome *c* peroxidase,[66,75] offers both some marked contrasts and some similarities to the corresponding horseradish peroxidase reaction (Fig. 9.11).[76] First, I shall discuss results for the yeast enzyme.

Transient-State Kinetics for Cytochrome c Peroxidase

Because of the difficulty in distinguishing compound I from compound II_A and compound II_B from native enzyme, discussed above, reactions of compounds I and II were followed at the same wavelength, 424 nm, the maximum difference in absorbance between the two extremes in spectra, that of the native enzyme and of compound I. The two forms of compound II have spectra resembling one or the other of the two extremes. Biphasic, and occasionally triphasic, curves were obtained. The faster phase of a biphasic curve yielded the rate of reaction of ferrocyanide with compound I and the slower phase the rate with compound II. The two rates differ by a factor of only 3 over the entire accessible pH range. Apparent saturation kinetics were observed at high concentrations of ferrocyanide, but first-order rate constants were linearly proportional to ferrocyanide concentration at lower concentrations. The ionic strength dependence of rates was determined. Rate constants for both compounds I and II, extrapolated to zero ionic strength where there is no screening of electrostatic charge interaction, were correlated with total charge on the yeast enzyme.[77] Between pH 5 and 8 a fairly good straight-line fit with a slope of -1 for log of the rate constant versus pH is obtained, particularly for the compound II reaction with ferrocyanide. The results indicate that the mechanisms of the redox reactions are similar for compounds I and II. They are consistent with ferrocyanide having the same docking site on both compounds, and reducing the same site on the enzyme.

It is the ferryl ($Fe^{IV}{=}O$) forms of both compounds I and II that are initially reduced. Compound I is entirely in the $Fe^{IV}{=}O$, Trp191$\bullet^{+}$ form at all pHs, whereas compound II is predominantly in its $Fe^{IV}{=}O$, Trp191 form above pH 5, with amounts of its Fe^{III}, Trp191$\bullet^{+}$ form decreasing exponentially with increasing pH. The pH rate profile for compound II would be very different from that of compound I if reduction occurred at the radical site. Note the similarity of the ferrocyanide kinetics for both compounds I and II of the yeast enzyme with the compound II kinetics for horseradish peroxidase in Fig. 9.11.

Dicyano-bis(1,10 phenanthroline) iron(II) exhibited similar behavior to that of ferrocyanide in its reductions of compounds I and II.[78] The latter was extended to a temperature-dependent study of other substitution-inert iron(II)

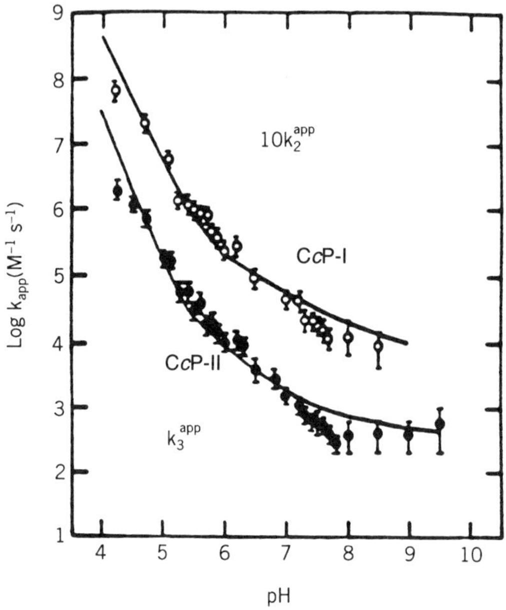

Log k_{app}(M^{-1} s^{-1})
$10k_2^{app}$
CcP-I
CcP-II
k_3^{app}
pH

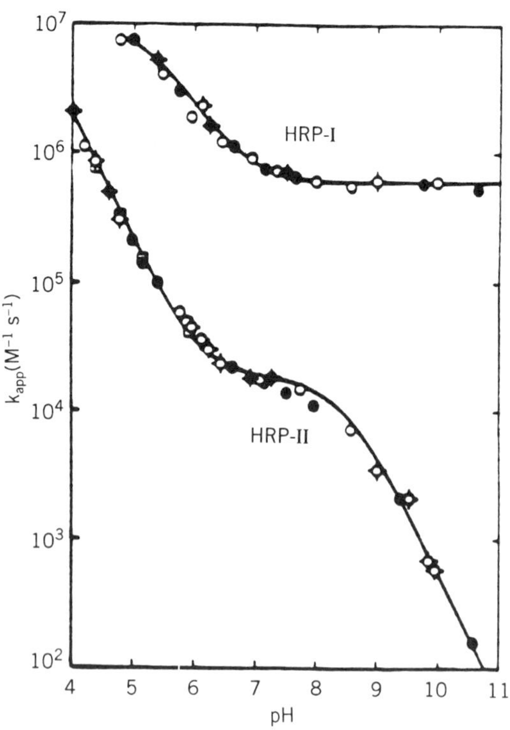

k_{app}(M^{-1} s^{-1})
HRP-I
HRP-II
pH

complexes: mono-, bis-, and tris(1,10-phenanthroline) and 2,2′-bipyridine complexes.[79] Marcus theory was applied to the results, the first time for peroxidase-catalyzed reactions. The reduction potential of the compound II–native enzyme couple was estimated to be 1.087 V.

Steady-State Kinetics for Cytochrome c Peroxidase

Sense can be made of apparently conflicting reports in the literature by considering the following two mechanisms for the reaction cycle of yeast cytochrome *c* peroxidase. The first mechanism is identical to that for other peroxidases, whereas the second involves reduction of the iron(IV) site of compound I, followed by interconversion of the two forms of compound II, followed by reduction of the iron(IV) site of compound II.

The conventional peroxidase mechanism, adapted to the terminology used here for the yeast enzyme, is the following. Proton balances are neglected and the reductant is indicated simply as an electron donor, e^-.

Conventional Mechanism

$$\underset{\text{CcP}}{\text{Fe}^{\text{III}}\text{, Trp}} + \text{H}_2\text{O}_2 \xrightarrow{k_1} \underset{\text{Compound I}}{\text{Fe}^{\text{IV}} = \text{O, Trp191}\bullet^+} + \text{H}_2\text{O} \tag{9.1}$$

$$\underset{\text{Compound I}}{\text{Fe}^{\text{IV}} = \text{O, Trp}\bullet^+} + e^- \xrightarrow{k_2} \underset{\text{Compound II}_\text{A}}{\text{Fe}^{\text{IV}} = \text{O, Trp}} \tag{9.2}$$

$$\underset{\text{Compound II}_\text{A}}{\text{Fe}^{\text{IV}} = \text{O, Trp}} + e^- \xrightarrow{k_3} \underset{\text{CcP}}{\text{Fe}^{\text{III}}\text{, Trp}} \tag{9.3}$$

The steady-state velocity, v, is given by:

$$\frac{v}{[\text{E}]_0} = \frac{1}{\dfrac{1}{k_1[\text{H}_2\text{O}_2]} + \dfrac{k_2 + k_3}{k_2 k_3\,[e^-]}} = \frac{1}{\dfrac{1}{k_1[\text{H}_2\text{O}_2]} + \dfrac{1}{k'\,[e^-]}} \tag{9.4}$$

where $[\text{E}]_0$ is the total enzyme concentration, $[e^-]$ the concentration of reducing substrate, and $k' = k_2k_3/(k_2 + k_3)$.

FIG. 9.11. pH dependences for the rates of oxidation of ferrocyanide by oxidized yeast cytochrome *c* peroxidase and horseradish peroxidase at 25°C. Top panel, yeast enzyme: open circles, compound I (C*c*P-I); closed circles, compound II (C*c*P-II). The compound I results are multiplied by a factor of 10 to offset them from the compound II results. Lower panel: horseradish enzyme. HRP-I, compound I; HRP-II, compound II. [Top panel reproduced with permission from H. C. Jordi and J. E. Erman, *Biochemistry* **13,** 3734–3741 (1974). Bottom panel reproduced with permission from M. L. Cotton and H. B. Dunford, *Can. J. Chem.* **51,** 582–587 (1973).]

The more complicated mechanism, apparently unique to yeast cytochrome c peroxidase under certain conditions, is the following.

$Fe^{IV}{=}O$ Reduction Mechanism

$$\underset{\text{CcP}}{\text{Fe}^{\text{III}}, \text{Trp}} + \text{H}_2\text{O}_2 \xrightarrow{k_1} \underset{\text{Compound I}}{\text{Fe}^{\text{IV}}{=}\text{O}, \text{Trp191}\bullet^+} + \text{H}_2\text{O} \tag{9.1}$$

$$\underset{\text{Compound I}}{\text{Fe}^{\text{IV}} = \text{O}, \text{Trp}\bullet^+} + e^- \xrightarrow{k_2} \underset{\text{Compound II}_\text{B}}{\text{Fe}^{\text{III}}, \text{Trp}\bullet^+} \tag{9.5}$$

$$\underset{\text{Compound II}_\text{B}}{\text{Fe}^{\text{III}}, \text{Trp}\bullet^+} \xrightarrow{k_\text{A}} \underset{\text{Compound II}_\text{A}}{\text{Fe}^{\text{IV}} = \text{O}, \text{Trp}} \tag{9.6}$$

$$\underset{\text{Compound II}_\text{A}}{\text{Fe}^{\text{IV}} = \text{O}, \text{Trp}} \xrightarrow{k_\text{B}} \underset{\text{Compound II}_\text{B}}{\text{Fe}^{\text{III}}, \text{Trp}\bullet^+} \tag{9.7}$$

$$\underset{\text{Compound II}_\text{A}}{\text{Fe}^{\text{IV}} = \text{O}, \text{Trp}} + e^- \xrightarrow{k_3} \underset{\text{CcP}}{\text{Fe}^{\text{III}}, \text{Trp}} \tag{9.8}$$

For the $Fe^{IV}{=}O$ mechanism the rate is given by:

$$\frac{v}{[\text{E}]_0} = \frac{1}{\dfrac{1}{k_1[\text{H}_2\text{O}_2]} + \left[\dfrac{k_\text{A}k_3 + k_\text{A}k_2 + k_\text{B}k_2}{k_\text{A}k_2k_3}\right]\dfrac{1}{[e^-]} + \dfrac{1}{k_\text{A}}} \tag{9.9}$$

$$\frac{v}{[\text{E}]_0} = \frac{1}{\dfrac{1}{k_1[\text{H}_2\text{O}_2]} + \dfrac{1}{k''[e^-]} + \dfrac{1}{k_\text{A}}} \tag{9.10}$$

where $k'' = k_\text{A}k_2k_3/(k_\text{A}k_3 + k_\text{A}k_2 + k_\text{B}k_2)$.

For the condition that a large concentration of hydrogen peroxide is used, Eq. (9.4) reduces to

$$\frac{v}{[\text{E}]_0} = k'\,[e^-] \tag{9.11}$$

and Eq. (9.10) to

$$\frac{v}{[\text{E}]_0} = \frac{1}{\dfrac{1}{k''[e^-]} + \dfrac{1}{k_\text{A}}} \tag{9.12}$$

The reciprocal of Eq. (9.12) is

$$\frac{[\mathrm{E}]_0}{v} = \frac{1}{k''\,[e^-]} + \frac{1}{k_\mathrm{A}} \tag{9.13}$$

Oxidation of Complexes of Different Charge

In an important and neglected steady-state study by Yandell and Yonetani, marked by its simplicity, reactions of ferrocyanide and positively charged cobalt and ruthenium complexes were studied over the pH range of 5–8 in 0.1 M KNO_3.[80] With sufficient excess of hydrogen peroxide so that compound I formation was not rate-limiting, the initial rate was given by Eq. (9.12) (Fig. 9.12). Thus, plots of $[\mathrm{E}]_0/v$ were linear [Eq. (13)], with the slopes equal to $1/k''$ and the intercept to $1/k_\mathrm{A}$. The values of k_A were independent of reducing substrate, going from ferrocyanide with a charge of −4 to the other complexes with charges of +2. These results are consistent with reduction of both compounds I and II occurring at the iron(IV) site, and the limiting rate for compound II reduction for large concentrations of reductant being the conversion of compound $\mathrm{II_B}$ ($\mathrm{Fe^{III}}$, Trp191•$^+$) into compound $\mathrm{II_A}$ ($\mathrm{Fe^{IV}{=}O}$, Trp191). The mean values of k_A for the three reductants at pH 6 and 7 were $(3.0 \pm 0.4) \times 10^2\ \mathrm{s^{-1}}$ and $51 \pm 4\ \mathrm{s^{-1}}$. Results using the $[\mathrm{Co(terpyridine)_2}]^{2+}$ ion gave k_A values of $(2 \pm 1) \times 10^3$ and $6.4 \pm 2\ \mathrm{s^{-1}}$ at pH 5.0 and 8.0. The results are not consistent with a conventional peroxidase mechanism, free radical site reduction followed by iron(IV) reduction. They are completely consistent with the $\mathrm{Fe^{IV}{=}O}$ reduction mechanism.

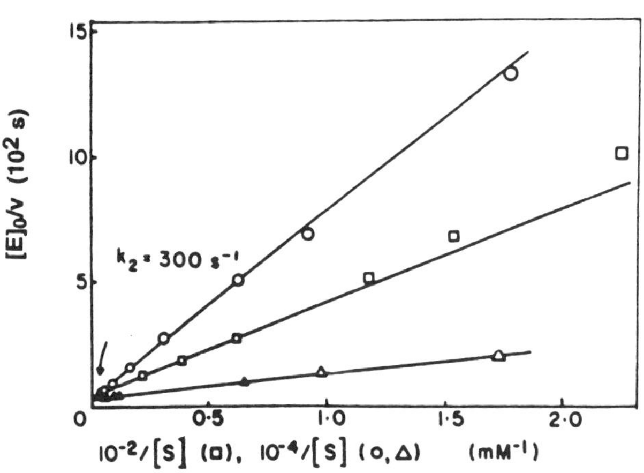

FIG. 9.12. Plots of $[\mathrm{E}]_0/v$ versus the reciprocal of reducing substrate concentration for reactions of cytochrome *c* peroxidase. (○) $[\mathrm{Co(terpyridine)_2}]^{2+}$, (□) $[\mathrm{Fe(CN)_6}]^{4-}$, (△) $[\mathrm{Ru(NH_3)_5pyridine}]^{2+}$. [Reproduced with permission from J. K. Yandell and T. Yonetani, *Biochim. Biophys. Acta* **748,** 263–270 (1983) with permission from Elsevier Science.]

At high pH, where the overall reactions are slower, and at high concentrations of the highly reactive substrates $[Co(terpyridine)_2]^{2+}$ and $[Ru(NH_3)_5pyridine]^{2+}$, the results fit a combination of Eqs. (9.11) and (9.12):

$$\frac{v}{[E]_0} = k'\,[e^-] + \frac{1}{\dfrac{1}{k''[e^-]} + \dfrac{1}{k_A}} \tag{9.14}$$

In the original paper, the $k'\,[e^-]$ term was attributed to reaction with a precursor of compound I. However, the values for k' were smaller than that for compound I formation. More likely, the $k'\,[e^-]$ in Eq. (9.14) indicates that at high pH the mechanism is starting to revert to that of a conventional peroxidase. Perhaps a second docking site for reducing substrate causes a switch in mechanism. This discussion may be relevant to the mechanism of reaction of yeast cytochrome *c* peroxidase with cytochrome *c,* discussed in Chapter 10.

An important point to remember in discussion of peroxidase mechanisms is that although binding of substrate is not observed in the kinetics of most reactions, binding must still occur. It simply means that subsequent reaction of the complex is faster than the binding process.

From the above discussion it is apparent that the yeast cytochrome *c* peroxidase cycle for reaction with hydrogen peroxide and small reducing substrates is native enzyme → compound I → compound II → native enzyme as it is with all the other peroxidases. However, major differences may occur in the site of reduction and in the behavior of compound II. For the yeast enzyme, it would appear that small reducing substrates react with the $Fe^{IV}{=}O$ site of both compounds I and II. Furthermore, the interconversion of the two forms of compound II plays an important role and may be rate-controlling. It cannot be assumed *a priori* that the two forms of compound II are in rapid equilibrium.

The above results were anticipated to the extent that reactions of small reductant molecules were assumed to occur near the δ-meso edge of the heme, whereas cytochrome *c* reacted near the γ-meso edge,[81,82] but the actual species being reduced was not specified.

Horseradish Peroxidase Plus Ferrocyanide

While the ionic strength dependence was not determined for ferrocyanide reacting with horseradish peroxidase compounds I and II, a qualitative correlation with total protein charge would appear possible for the compound II reaction, but definitely not for compound I. It would appear likely that there are different docking sites for ferrocyanide on horseradish peroxidase compounds I and II. In the horseradish peroxidase study,[76] several different oxidants were used to generate compound I, with no difference in the kinetics of compound I reduction. These experiments were performed because at that time (1973) arguments about compound I being an enzyme–peroxide complex ES, and therefore dependent upon the nature of the oxidant, had not been completely laid to rest.

9.3.3. Compound I Formation

The kinetics of yeast cytochrome *c* peroxidase compound I formation using hydrogen peroxide as oxidant were studied.[83,84] The effect of pH and ionic strength was examined.[85] The pH-independent rate constant is $(4.5 \pm 0.3) \times 10^7\ M^{-1}s^{-1}$, slightly more than twice as fast as the corresponding reaction for horseradish peroxidase. The same problems were encountered as with horseradish peroxidase in the presence of nitrate and acetate (Section 4.5.1). In a later study, using phosphate instead of actetate buffers and with no nitrate present, the rate of compound I formation was found to be pH-independent over the range 4.5–8.0,[86] a result confirmed in the Erman group (Fig. 9.2).[17] The effect of nitrate is to shift the apparent pK_a to higher values. The opposite effect, the apparent disappearance of the kinetic effect of the acid group upon removal of nitrate from the reaction system, means that the pK_a value is shifted to a value lower than the lowest pH of the study. Thus, the pK_a value would appear to be $\leq$4.0 for the native yeast enzyme in comparison to a value of ~2.5 on horseradish peroxidase.

Yeast cytochrome *c* peroxidase compound I formation kinetics were also studied as a function of solvent viscosity and temperature.[87] Viscosity was varied using various sucrose–water mixtures. The rate is independent of viscosity over a 15-fold range, showing that compound I formation is not diffusion controlled. At pH 7.0 in aqueous solutions of ionic strength 0.1 M, the enthalpy of activation $\Delta H\ddagger$ is 2.2 ± 1.0 kcal/mol and the entropy of activation $\Delta S\ddagger - 16.3 \pm 0.8$ e.u. (cal/deg mol). When a dimethyl ester of ferriprotoporphyrin IX was substituted for the natural heme, little change in the kinetics of compound I formation was observed, indicating that the propionate side chains of the heme are not directly involved in the mechanism.[88]

Several different oxidants were shown to produce compound I, strong evidence against an enzyme–substrate complex ES as the highest oxidized form of the enzyme.[89]

9.4. PROTON STOICHIOMETRY

The accepted mechanism for compound I formation does not involve proton uptake or release:

$$\underset{\text{Native CcP}}{Fe^{III}, \text{Trp191} + H_2O_2} \rightarrow \underset{\text{Compound I}}{Fe^{IV}{=}O, \text{Trp191}\bullet^{+}} + H_2O$$

Measurements of proton uptake in pH stat experiments showed that compound I formation is complicated by ionizations on both the native enzyme and compound I.[39]

Reduction of compound I to native enzyme via compound II by small molecules (not cytochrome *c*) can be discussed in terms of the following hypothetical scheme, which could help to explain the complicated results obtained in the pH stat experiments.[39]

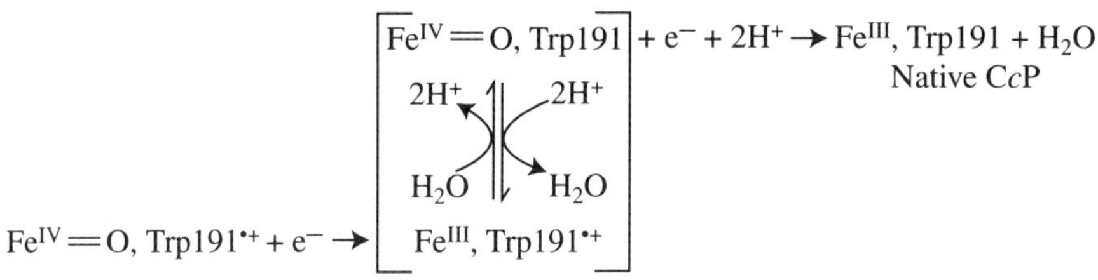

Compound I Compound II

The two forms of compound II are assumed to be in rapid equilibrium. At low pH where there is an appreciable amount of the free radical form of compound II, proton uptake could approach a value of 2 as compound I is reduced to compound II. At high pH, where the iron(IV) form of compound II is dominant, proton uptake could approach a value of 2 for reduction of compound II to native enzyme. Superimposed on these reactions are the pH-dependent equilibria of native enzyme, compound I, and each of the two forms of compound II.

9.5. SITE-DIRECTED MUTAGENESIS

The cloning of an enzyme may provide a better way to isolate the enzyme in high yield. One of the more difficult parts is to prove that there is no folding problem and a pure intact enzyme, identical with its native form, is obtained. Careful characterization using a battery of techniques is required. After that, site-directed mutagenesis becomes an almost routine matter.[90] Again careful work is required to identify whether any changes in enzyme properties have been introduced, and if so, precisely what changes.

Because yeast cytochrome *c* peroxidase contains no carbohydrate and because for many years it was the only peroxidase with a known crystal structure, it was especially attractive for site-directed mutagenesis experiments. The nuclear gene was cloned in 1982,[91,92] and site-directed mutagenesis experiments began shortly after.[93,94] Expression of the gene contained on a plasmid of *Saccharomyces cerevisiae* was accomplished. The first mutants, expressed in yeast, involved replacement of Met172.[94] The resultant proteins were purified, crystallized, and the spectroscopic and catalytic properties were examined. The conclusion was that Met172 is not directly involved in compound I formation.

A problem with the yeast system is that the enzyme is poorly expressed, complicating the isolation and purification. A better expression system was developed by inserting the plasmid into *Escherichia coli.* Initially, yields of more than 15 mg/L of cell culture were obtained, and these have now been increased to 1000 mg/L![95] The isolated enzyme was largely in its *apo* form but was readily converted to *holo* enzyme by addition of ferriprotoporphyrin IX. The first mutant produced in *E. coli* was by the substitution Trp51Phe and indicated that Trp51 contributed to the stability of compound I but was not essential for its formation.[96] This was followed by production of the Trp191Phe mutant,[53] which greatly affected the ability of the enzyme to oxidize cytochrome *c.*

The Asp235Ala mutant was prepared and isolated using an *E. coli* expression system. It was characterized using NMR spectroscopy and other techniques.[97] (For a discussion of the kinetics of the Asp235Asn mutant, see Section 9.5.1.) Many changes are induced by the alanine substitution, including a reduction in steady-state activity in the oxidation of cytochrome *c* by three orders of magnitude. The Asp235Ala mutant was also examined using XAS.[98] Rupture of the hydrogen bond between Asp235 and the proximal His175 causes an elongation of the iron to proximal imidazole nitrogen (Fe—Nε) bond from ~1.9 to ~2.1 Å.

9.5.1. Kinetic Studies on Mutants

The substitution Trp191Phe strongly affected the oxidation of ferrocyanide, one of the many pieces of evidence pointing to the major role of Trp191 in compound I formation.[53] Rate constants obtained with the parent cloned enzyme, compared to the wild type, were slightly smaller for compound I formation and slightly larger for the endogenous reduction of the ferryl heme.[99,100] The substitution His52Leu caused a decrease in the rate constant for compound I formation by a factor of 10^5. Thus, the distal His52 is a crucial residue for compound I formation.[101] For the substitution Arg48Leu the decrease was slightly greater than 10^2. The role of distal Arg48 appears to be to stabilize compound I.[100,102] The effect of the substitution Asp235Asn is to change the iron(III) from a five-coordinate species to six-coordinate with hydroxide ion in the sixth position. Despite this dramatic change in coordination, the rate constant for compound I formation is only decreased by a factor of 2. The aspartate residue is hydrogen-bonded to the proximal His175 residue. Several authors speculated about the influence of its negative charge on the character of the proximal histidine and in influencing the rate of compound I formation.[103–106] There may be offsetting influences in the Asp235Asn substitution.[100,107]

The kinetics of oxidation of a series of substituted anilines reacted with wild-type enzyme and two mutants, Trp51Phe and Trp51Ala were examined in detail.[108] Removal of the bulky distal trptophan residue resulted in enhanced rates of reaction. Rates were correlated with Brown–Okamoto σ^+ values. An important conclusion is that the mutants contain a more reactive ferryl species.

9.6. NUCLEAR MAGNETIC RESONANCE SPECTRA

Unique contributions to the understanding of yeast cytochrome *c* peroxidase have been made using NMR spectroscopy, mainly by James Satterlee and co-workers. In this section studies, which began in 1980,[109–111] on various forms of the enzyme by itself are described. For NMR studies on the enzyme–cytochrome *c* complex see Section 10.3.

The effect of compound I formation on the NMR spectrum is shown in Fig. 9.12. Cyanide complexes of the native enzyme were used to convert the iron(III) into a low-spin form. Axial histidyl resonances,[112] hindered heme vinyl rotations,[113] and exchange or nonexchange of protons with deuterium oxide solvent were studied.[114] Breakthroughs occurred first by use of selectively deuterated hemes, then in one-dimensional nuclear Overhauser experiments, followed by use of two-dimensional NMR.[115]

By use of various deuterated forms of ferriprotoporphyrin IX, the heme methyl groups were found to fall in the order $5 > 1 > 8 > 3$, going from low-field to high field.[38] In contrast those of metmyoglobin are in the order $8 > 5 > 3 > 1$, indicating different orientations of the heme groups. The study using different deuterated hemes was extended to cyanide and azide complexes.[116] The cyanide complex provides the best NMR model for compounds I and II. A rhombic perturbation was noted, caused partly by the orientation of the proximal histidine's imidazole plane and a specific interaction between pyrrole II and Trp51.

On the basis of nuclear Overhauser effects for cyanide-ligated cytochrome *c* peroxidase, the remaining unassigned downfield-shifted proton resonances were identified.[117] In addition to heme protons, resonances of His175, Thr180, and His52 were assigned. The study was extended to NMR NOESY[118] and COSY[119] experiments.

The effects of mutations have been studied by NMR. The mutation Asp235Asn disrupts the hydrogen bond to the proximal His175 ligand and converts the iron(III) into a low-spin form.[120] The Asn82Asp[121] and Asn82Ala[122] mutations affect the distal His52. The result of the aspartate substitution is to disrupt the hydrogen bonds of the histidine both to the residue in position 82 and to the cyanide in the sixth coordination position of the iron. The alanine substitution also disrupts hydrogen bonds and also causes doubling of some proton resonances.

A pH-dependent study on native yeast cytochrome *c* peroxidase indicated two acid–base interconversions. One with a pK_a of 4.9 was fast on the NMR time scale. The other of pK_a 5.8 was slow and appeared to correlate with kinetic observations. The observed values may be shifted from the true values since the study was conducted in 0.3 M potassium nitrate.[123]

Compound I was formed in deuterium oxide where the uncorrected pH meter reading was 6.9 (Fig. 9.13). The compound I was then titrated with ferrocyanide until native enzyme was regenerated. The heme methyl resonances of the native enzyme were obliterated upon compound I formation and were gradually restored during the ferrocyanide titration.[124]

Substitution of deuterohemin for ferriprotoporphyrin IX doubled some of the proton resonances.[113] The NMR spectra of the model compound, the cyanide complex of mono-γ-methylimidazole, were compared to the spectra of horseradish and yeast cytochrome *c* peroxidases. It was concluded that two proximal imidazole resonances could be detected and that the proximal imidazole of the peroxidases possessed considerable anionic character.[114] In a later study it was concluded that the proximal imidazole in the cyanide complex did not possess imidazolate character.[125] However, by use of the recombinant and mutant enzymes the imidazolate character of the proximal His175 was confirmed.[122] Heme methyl groups were assigned using selectively deuterium-labeled hemins.[38,118] A single proton resonance, that of the α-vinyl proton at heme position 4, was shown to indicate a pH-dependent rotation of the vinyl group, in rough accord with the enzyme activity.[126]

By use of the proton nuclear Overhauser effect, all remaining resolvable downfield-shifted resonances were assigned. In addition to heme protons, these included proton resonances of His175, Thr180, and His52.[127] Assigned resonances are shown in Fig. 9.14. From NOESY results partial assignments were made for Trp51 and Arg48.[128] A combination of COSY ane NOESY experiments were conducted.[129]

A review of proton NMR studies on yeast cytochrome *c* peroxidase, cytochromes *c* and *Glycera dibranchiata* monomeric hemoglobins has been published.[130] Studies have been extended to the recombinant enzyme and to mutants.[120] The Asn82Asp mutation of the cyanide complex eliminates the

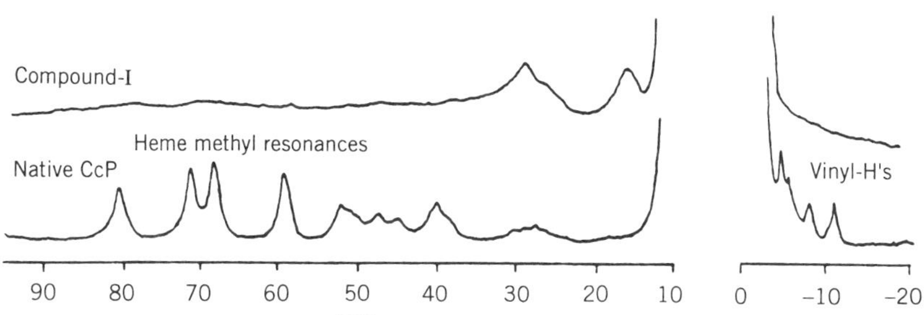

FIG. 9.13. Proton NMR spectra at 360 MHz of the native enzyme and compound I obtained at pH 7.4 and 21°C. [Reproduced with permission from J. D. Satterlee and J. E. Erman, *J. Biol. Chem.* **256,** 1091–1093 (1981).]

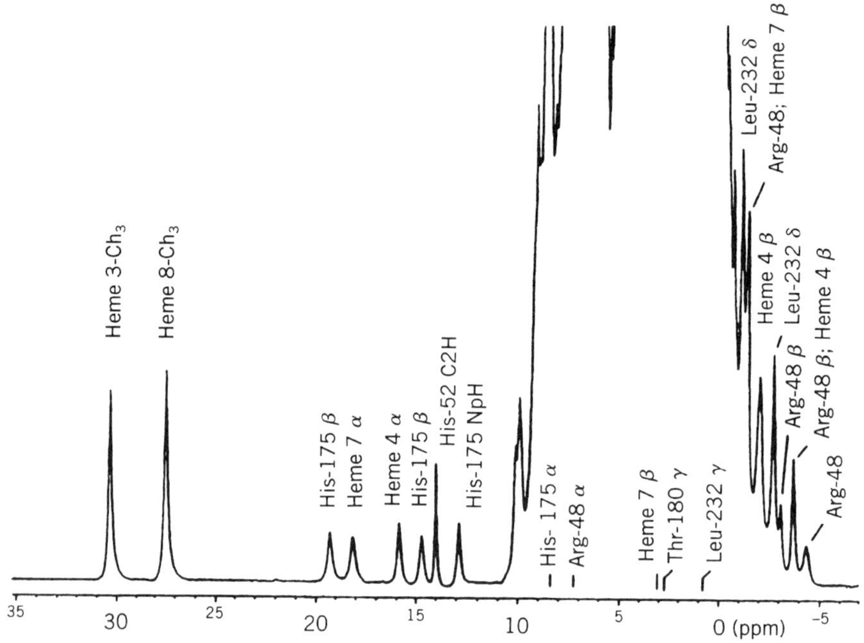

FIG. 9.14. Assignments of proton resonances taken at 500 MHz in the cyanide complex of cytochrome *c* peroxidase, obtained in deuterated buffer, pH 7.0 at 27°C. [Reproduced with permission from J. D. Satterlee, D. J. Russell and J. E. Erman, *Biochemistry* **30,** 9072–9077 (1991).]

His52 hydrogen bonds to Asn82 and to the bound cyanide.[131] Similar effects are observed in the Asn82Ala mutant. In addition 4 of the 24 resolved hyperfine shifted resonances are doubled.[132]

9.7. RESONANCE RAMAN SPECTRA

Resonance Raman spectroscopy played a key role in determining the correct coordination number of five for the iron in viable samples of yeast cytochrome *c* peroxidase (Section 9.2.1). In this section, other important contributions of resonance Raman spectroscopy are documented. Isotopically labeled hemes were used to assign bands of the recombinant enzyme, and one mutant, in which the Asp235Asn substitution was made.[133]

In a very important study of both the Fe^{II} and Fe^{III} forms of mutants and the cloned enzyme [CcP(MI)] from *E. coli,* relations among hydrogen-bonded networks were established, and information was obtained on coordination to the iron.[134] The iron(III) coordination number, normally five, is changed to six in the Trp51Phe mutant, showing that the hydrogen bond from Trp51 to the distal water inhibits its coordination to iron. The coordination number is also changed to six in the Asp235Asn mutant, showing that the Asp235–proximal His175 interaction helps prevent the iron from entering the heme plane, which would facilitate water coordination in the sixth position. The water coordination in the Asn235 mutant is sufficiently strong to convert the iron to its low spin-form. For a discussion of the Fe^{II} results see Section 9.8.

The effects of temperature and glycerol on the resonance Raman spectra of the native enzyme and the mutants Trp51Phe, Arg48Leu, Arg48Lys, Asp235Asn, and Trp191Phe at pH 7 were investigated.[135] At room temperature glycerol appears to weaken binding of ligands in the sixth coordination position and drives the equilbrium toward a five-coordinate high-spin state. Glycerol also retards conversion to a low-spin six-coordinate state at low temperature. Water is a likely candidate for the sixth ligand at low temperature. Studies were extended to single crystals of both native enzyme[33,35] and mutants,[136] on a variable temperature microscope stage. Measured depolarization ratios were accurately calibrated. The results, consistent with those observed in solution, indicate that as the temperature is lowered the heme iron gradually moves into the heme plane, where it binds water to change from a five-coordinate high-spin species to six-coordinate low-spin species.

Use of ^{54}Fe established the coexistence of two separate bands at 233 and 246 cm^{-1}, which arise from stretching of the iron–proximal histidine 175 ligand. The bands were assigned to tautomers of the His175 imidazole $N_{\delta}H$ proton and the Asp235 carboxylate side chain. In one tautomer the proton is on the imidazole and in the other it is transferred along the hydrogen bond to the carboxylate.[133] The effect of mutations of both distal[137] and proximal[138] histidines has been examined by resonance Raman spectroscopy. Mutations disrupt a hydrogen bonding network that extends from the distal to the proximal side.[139] Reviews of resonance Raman results on yeast cytochrome *c* peroxidase have been published.[127,139,140]

9.8. FERROUS ENZYME

9.8.1. Redox Potential of the Ferrous–Ferric Couple

The reduction potential of the ferric enzyme was determined as a function of pH over the range 4.5–8.0.[141] At pH 7 and 25°C the value is −194 mV relative to the standard hydrogen electrode. This value can be compared to −270 mV for ferric horseradish peroxidase[142] and −109 mV for turnip peroxidase isoenzyme P_7.[143]

9.8.2. Studies on the Enzyme and Its CO and O_2 Adducts

The status of the Fe^{II}–proximal histidine bond in the *E.-coli*-generated enzyme and several mutants was monitored by resonance Raman spectroscopy.[134] Its stretching frequency is ~240 cm^{-1} in the recombinant enzyme and all distal mutants because of the hydrogen bond from the His–175 Nδ to the carboxylate on Asp235. There are two components to the band at 233 and 246 cm^{-1}, which are suggested to arise from H-bonded and deprotonated imidazole, with a double-well potential. For the proximal Asn235 mutant the 240 cm^{-1} is missing and is replaced by one at 205 cm^{-1}, attributed to non-H-bonded imidazole.

Both resonance Raman and infrared spectra of CO adducts of native and mutant enzyme have been examined.[35,144] Alternative conformational states, one perpendicular to the heme plane, and the other tilted, were detected.[145] The kinetics of CO binding,[146] photodissociation,[147] and geminate recombination,[148] have been measured. The deuterium isotope effect on the heme-coordinated CO vibrational band has been determined.[149] The kinetics of dioxygen binding have been measured.[150] The dioxygen is reduced by the Trp191 site. Marcus theory was applied to the kinetics of ferricyanide reduction by ferro cytochrome *c* peroxidase.[151]

REFERENCES

1. A. M. Altschul, R. Abrams, and T. R. Hogness, *J. Biol. Chem.* **136,** 777–794 (1940).
2. R. Abrams, A. M. Altschul, and T. R. Hogness, *J. Biol. Chem.* **142,** 303–316 (1942).
3. T. Yonetani and G. S. Ray, *J. Biol. Chem.* **240,** 4503–4508 (1965).
4. T. Yonetani, *J. Biol. Chem.* **241,** 2562–2571 (1966).
5. T. Yonetani and H. Schleyer, *J. Biol. Chem.* **242,** 1974–1979 (1967).
6. T. Yonetani and T. Asakura, *J. Biol. Chem.* **244,** 4580–4588 (1969).
7. T. Yonetani, D. F. Wilson, and B. Seamonds, *J. Biol. Chem.* **241,** 5347–5352 (1966).
8. T. Yonetani and T. Asakura, *J. Biol. Chem.* **243,** 3996–3998 (1968).
9. T. Yonetani and T. Asakura, *J. Biol. Chem.* **243,** 4715–4721 (1968).
10. T. Asakura and T. Yonetani, *J. Biol. Chem.* **244,** 537–544 (1969).
11. T. Asakura and T. Yonetani, *J. Biol. Chem.* **244,** 4573–4579 (1969).
12. T. Asakura, H. R. Drott, and T. Yonetani, *J. Biol. Chem.* **244,** 6626–6631 (1969).
13. T. Yonetani, B. Chance, and S. Kajiwara, *J. Biol. Chem.* **241,** 2981–2986 (1966).
14. N. Ellfolk, *Acta Chem. Scand.* **21,** 175–181 (1967).
15. C. E. Nelson, E. V. Sitzman, C. H. Kang, and E. Margoliash, *Anal. Biochem.* **83,** 622–631 (1977).
16. A. M. English, M. Laberge, and M. Walsh, *Inorg. Chim. Acta,* **123,** 113–116 (1986).
17. L. B. Vitello, M. Huang, and J. E. Erman, *Biochemistry* **29,** 4283–4288 (1990).
18. N. Ellfolk, *Acta Chem. Scand.* **21,** 1921–1924 (1967).
19. T. Yonetani, J. Biol. Chem. **242,** 5008–5013 (1967).
20. G. Sievers, *Biochim. Biophys. Acta* **536,** 212–225 (1978).
21. B. K. Dhaliwal and J. E. Erman, *Biochim. Biophys. Acta* **827,** 174–182 (1985).
22. T. Fox, L. Ferreira-Rajabi, B. C. Hill, and A. M. English, *Biochemistry* **32,** 6938–6943 (1993).

23. T. Fox, G. Tsaprailis, and A. M. English, *Biochemistry* **33,** 186–191 (1994).
24. J. A. Kornblatt, A. M. English, and G. Hui Bon Hoa, *Eur. J. Biochem.* **139,** 39–43 (1986).
25. R. J. Dowe and J. E. Erman, *Biochim. Biophys. Acta* **827,** 183–189 (1985).
26. M. T. Gross and J. E. Erman, *Biochim. Biophys. Acta* **830,** 140–146 (1985).
27. G. C. Kresheck and J. E. Erman, *Biochemistry* **27,** 2490–2496 (1988).
28. K. Kim and J. E. Erman, *Biochim. Biophys. Acta* **954,** 95–107 (1988).
29. T. Yonetani in *Advances in Enzymology,* F. F. Nord, ed., Wiley-Interscience, New York, Vol. 33, 1970 pp. 309–335.
30. T. Yonetani in *The Enzymes,* P. D. Boyer, ed., 3rd ed., Academic, New York, Vol. 13, 1976, pp. 345–361.
31. G. Sievers, K. Österlund, and N. Ellfolk, *Biochim. Biophys. Acta* **581,** 1–14 (1979).
32. T. L. Poulos and J. Kraut, *J. Biol. Chem.* **255,** 8199–8205 (1980).
33. G. Smulevich, Y. Wang, S. L. Edwards, T. L. Poulos, A. M. English, and T. G. Spiro, *Biochemistry* **29,** 2586–2592 (1990).
34. R. A. Mathews and J. B. Wittenberg, *J. Biol. Chem.* **254,** 5991–5996 (1979).
35. G. Smulevich, R. Evangelista-Kirkup, A. M. English, and T. G. Spiro, *Biochemistry* **25,** 4426–4430 (1986).
36. T. Yonetani and H. Anni, *J. Biol. Chem.* **262,** 9547–9554 (1987).
37. S. Dasgupta, D. L. Rousseau, H. Anni, and T. Yonetani, *J. Biol. Chem.* **264,** 654–662 (1989).
38. J. D. Satterlee, J. E. Erman, G. N. La Mar, K. M. Smith, and K. C. Langry, *Biochim. Biophys. Acta* **743,** 246–255 (1983).
39. C. W. Conroy and J. E. Erman, *Biochim. Biophys. Acta* **527,** 370–378 (1978).
40. H. B. Dunford, *Adv. Inorg. Biochem.* **4,** 41–68 (1982).
41. H. R. Bosshard, H. Anni, and T. Yonetani in *Peroxidases in Chemistry and Biology,* J. Everse, K. E. Everse and M. B. Grisham, eds., CRC Press, Boca Raton, FL, Vol. II, 1991, pp. 51–84.
42. T. L. Poulos, *J. Biol. Inorg. Chem.* **1,** 356–359 (1996).
43. D. B. Goodin, *J. Biol. Inorg. Chem.* **1,** 360–363 (1996).
44. L. Banci, A. Rosato, and P. Turano, *J. Biol. Inorg. Chem.* **1,** 364–367 (1996).
45. Z. Gross, *J. Biol. Inorg. Chem.* **1,** 368–371 (1996).
46. I. M. C. M. Rietjens, A. M. Osman, C. Veeger, O. Zakharieva, J. Antony, M. Grodzicki, and A. X. Trautwein, *J. Biol. Inorg. Chem.* **1,** 372–376 (1996).
47. R. Weiss, D. Mandon, T. Wolter, A. X. Trautwein, M. Müther, E. Bill, A. Gold, K. Jayaraj, and J. Terner, *J. Biol. Inorg. Chem.* **1,** 377–383 (1996).
48. B. M. Hoffman, J. E. Roberts, C. H. Kang, and E. Margoliash, *J. Biol. Chem.* **256,** 6556–6564 (1981).
49. T. Yonetani and H. Schleyer, *J. Biol. Chem.* **241,** 3240–3243 (1966).
50. T. Yonetani and H. Schleyer, *J. Biol. Chem.* **242,** 3919–3925 (1967).
51. B. A. Wittenberg, L. Kampa, J. B. Wittenberg, W. E. Blumberg, and J. Peisach, *J. Biol. Chem.* **243,** 1863–1870 (1968).
52. M. Sivaraja, D. B. Goodin, M. Smith, and B. M. Hoffman, *Science* **245,** 738–740 (1989).

53. J. M. Mauro, L. A. Fishel, J. T. Hazzard, T. E. Meyer, G. Tollin, M. A. Cusanovich, and J. Kraut, *Biochemistry* **27,** 6243–6256 (1988).
54. C. P. Scholes, Y. Liu, L. A. Fishel, M. F. Farnum, J. M. Mauro, and J. Kraut, *Isr. J. Chem.* **29,** 85–92 (1989).
55. J. E. Erman, L. B. Vitello, J. M. Mauro, and J. Kraut, *Biochemistry* **28,** 7992–7995 (1989).
56. L. A. Fishel, M. F. Farnum, J. M. Mauro, M. A. Miller, J. Kraut, Y. Liu, X. Tan, and C. P. Scholes, *Biochemistry* **30,** 1986–1996 (1991).
57. A. L. P. Houseman, P. E. Doan, D. B. Goodin, and B. M. Hoffman, *Biochemistry* **32,** 4430–4443 (1993).
58. J. E. Huyett, P. E. Doan, R. Gurbiel, A. L. P. Houseman, M. Sivaraja, D. B. Goodin, and B. M. Hoffman, *J. Am. Chem. Soc.* **117,** 9033–9041 (1995).
59. B. D. Spangler and J. E. Erman, *Biochim. Biophys. Acta* **872,** 155–157 (1986).
60. G. Tsaprailis and A. M. English, *Can. J. Chem.* **74,** 2250–2257 (1996).
61. J. R. Collins, P. Du, and G. H. Loew, *Biochemistry* **31,** 11166–11174 (1992).
62. M. S. Mondal, H. A. Fuller, and F. A. Armstrong, *J. Am. Chem. Soc.* **118,** 263–264 (1996).
63. A. F. W. Coulson, J. E. Erman, and T. Yonetani, *J. Biol. Chem.* **246,** 917–924 (1971).
64. J. E. Erman, private communication, January 9, 1996.
65. P. S. Ho, B. M. Hoffman, C. H. Kang, and E. Margoliash, *J. Biol. Chem.* **258,** 4356–4363 (1983).
66. H. C. Jordi and J. E. Erman, *Biochemistry* **13,** 3734–3741 (1974).
67. P. S. Ho, B. M. Hoffman, N. Solomon, C. H. Kang, and E. Margoliash, *Biochemistry* **23,** 4122–4128 (1984).
68. S. Hashimoto, Y. Tatsuno, and T. Kitagawa, *Proc. Natl. Acad. Sci. U.S.A.* **83,** 2417–2421 (1986).
69. T. Fox, J. T. Hazzard, S. L. Edwards, A. M. English, T. L. Poulos, and G. Tollin, *J. Am. Chem. Soc.* **112,** 7426–7428 (1990).
70. J. E. Erman, *Biochemistry* **13,** 34–39 (1974).
71. J. E. Erman, *Biochemistry* **13,** 39–44 (1974).
72. F. Neri, D. Kok, M. A. Miller, and G. Smulevich, *Biochemistry* **36,** 8947–8953 (1997).
73. J. A. Shelnutt, J. D. Satterlee, and J. E. Erman, *J. Biol. Chem.* **258,** 2168–2173 (1983).
74. B. Lent, C. W. Conroy, and J. E. Erman, *Arch. Biochem. Biophys.* **177,** 56–61 (1976).
75. H. C. Jordi and J. E. Erman, *Biochemistry* **13,** 3741–3745 (1974).
76. M. L. Cotton and H. B. Dunford, *Can. J. Chem.* **51,** 582–587 (1973).
77. C. W. Conroy and J. E. Erman, *Biochim. Biophys. Acta* **537,** 396–405 (1978).
78. J. E. Erman, *Biochim. Biophys. Acta* **397,** 36–42 (1975).
79. W. L. Purcell and J. E. Erman, *J. Am. Chem. Soc.* **98,** 7033–7037 (1976).
80. J. K. Yandell and T. Yonetani, *Biochim. Biophys. Acta* **748,** 263–270 (1983).
81. G. D. DePillis, B. P. Shista, A. G. Mauk, and P. R. Ortiz de Montellano, *J. Biol. Chem.* **266,** 19334–19341 (1991).

82. P. R. Ortiz de Montellano, *Annu. Rev. Pharmacol. Toxicol.* **32,** 89–107 (1992).
83. J. E. Erman and T. Yonetani, *Biochim. Biophys. Acta* **393,** 343–349 (1975).
84. J. E. Erman and T. Yonetani, *Biochim. Biophys. Acta* **393,** 350–357 (1975).
85. S. Loo and J. E. Erman, *Biochemistry* **14,** 3467–3470 (1975).
86. C. Balny, H. S. Anni, and T. Yonetani, *FEBS Lett.* **221,** 349–354 (1987).
87. S. Loo and J. E. Erman, *Biochim. Biophys. Acata* **481,** 279–282 (1977).
88. R. J. Dowe and J. E. Erman, *J. Biol. Chem.* **257,** 2403–2405 (1982).
89. A. F. W. Coulson and T. Yonetani, *Biochemistry* **14,** 2389–2396 (1975).
90. M. Smith, *Annu. Rev. Genet.* **19,** 423–462 (1985).
91. S. Goltz, J. Kaput, and G. Blobel, *J. Biol. Chem.* **257,** 11186–11190 (1982).
92. Kaput, S. Goltz, and G. Blobel, *J. Biol. Chem.* **257,** 15054–15058 (1982).
93. G. J. Pielak, A. G. Mauk, and M. Smith, *Nature* **313,** 152–154 (1985).
94. D. B. Goodin, A. G. Mauk, and M. Smith, *Proc. Natl. Acad. Sci. U.S.A.* **83,** 1295–1299 (1986).
95. D. B. Goodin, Scripps Research Institute, personal communication, December 30, 1997.
96. L. A. Fishel, J. E. Villafranca, J. M. Mauro, and J. Kraut, *Biochemistry* **26,** 351–360 (1987).
97. J. C. Ferrer, P. Turano, L. Banci, I. Bertini, I. K. Morris, K. M. Smith, M. Smith, and A. G. Mauk, *Biochemistry* **33,** 7819–7829 (1994).
98. R. Sinclair, S. Hallam, M. Chen, B. Chance, and L. Powers, *Biochemistry* **35,** 15120–15128 (1996).
99. L. B. Vitello, J. E. Erman, J. M. Mauro, and J. Kraut, *Biochim. Biophys. Acta* **1038,** 90–97 (1990).
100. J. E. Erman, L. Vitello, M. A. Miller, and J. Kraut, *J. Am. Chem. Soc.* **114,** 6592–6593 (1992).
101. J. E. Erman, L. B. Vitello, M. A. Miller, A. Shaw, K. A. Brown, and J. Kraut, *Biochemistry* **32,** 9798–9806 (1993).
102. L. B. Vitello, J. E. Erman, M. A. Miller, J. Wang, and J. Kraut, *Biochemistry* **32,** 9807–9818 (1993).
103. T. L. Poulos, *Adv. Inorg. Biochem.* **7,** 1–36 (1987).
104. J. H. Dawson, *Science* **240,** 433–439 (1988).
105. H. B. Dunford in *Peroxidases in Chemistry and Biology,* J. Everse, K. E. Everse, and M. B. Grisham, eds., CRC Press, Boca Raton, FL, 1991, Vol. II, pp. 1–24.
106. D. B. Goodin and D. E. McCree, *Biochemistry* **32,** 3313–3324 (1993).
107. L. B. Vitello, J. E. Erman, M. A. Miller, J. M. Mauro, and J. Kraut, *Biochemistry* **31,** 11524–11535 (1992).
108. J. A. Roe and D. B. Goodin, *J. Biol. Chem.* **268,** 20037–20045 (1993).
109. J. D. Satterlee and J. E. Erman, *Arch. Biochem. Biophys.* **202,** 608–616 (1980).
110. J. D. Satterlee and J. E. Erman, *J. Biol. Chem.* **256,** 1091–1093 (1981).
111. J. D. Satterlee and J. E. Erman, *J. Am. Chem. Soc.* **103,** 199–200 (1981).
112. G. N. La Mar, J. S. de Ropp, V. P. Chapko, J. D. Satterlee, and J. E. Erman, *Biochim. Biophys. Acta* **708,** 317–325 (1982).

113. J. D. Satterlee and J. E. Erman, *J. Biol. Chem.* **258,** 1050–1056 (1983).
114. J. D. Satterlee and J. E. Erman, *Biochim. Biophys. Acta* **743,** 149–154 (1983).
115. J. D. Satterlee, S. Alam, Q. Yi, J. E. Erman, I. Constantinidis, D. J. Russell, and S. J. Moench in *Biological Magnetic Resonance,* L. J. Berliner and J. Reuben, eds., Plenum, New York, 1993, Vol. 12, pp. 275–298.
116. J. D. Satterlee, J. E. Erman, G. N. La Mar, K. M. Smith, and K. C. Langry, *J. Am. Chem. Soc.* **105,** 2099–2104 (1983).
117. J. D. Satterlee, J. E. Erman, and J. S. de Ropp, *J. Biol. Chem.* **262,** 11578–11583 (1987).
118. J. D. Satterlee and J. E. Erman, *Biochemistry* **30,** 4398–4405 (1991).
119. J. D. Satterlee and J. E. Erman, *Biochemistry* **30,** 9072–9077 (1991).
120. J. D. Satterlee, J. E. Erman, J. M. Mauro, and J. Kraut, *Biochemistry* **29,** 8797–8804 (1990).
121. J. D. Satterlee, S. L. Alam, J. M. Mauro, J. E. Erman, and T. L. Poulos, *Eur. J. Biochem.* **224,** 81–87 (1994).
122. S. L. Alam, J. D. Satterlee, J. M. Mauro, T. L. Poulos, and J. E. Erman, *Biochemistry* **34,** 15496–15503 (1995).
123. J. D. Satterlee and J. E. Erman, *Arch. Biochem. Biophys.* **202,** 608–616 (1980).
124. J. D. Satterlee and J. E. Erman, *J. Biol. Chem.* **256,** 1091–1093 (1981).
125. J. D. Satterlee and J. E. Erman, *Biochim. Biophys. Acta* **743,** 149–154 (1983).
126. J. D. Satterlee and J. E. Erman, *J. Biol. Chem.* **258,** 1050–1056 (1983).
127. J. D. Satterlee, J. E. Erman, and J. S. de Ropp, *J. Biol. Chem.* **262,** 11578–11583 (1987).
128. J. D. Satterlee and J. E. Erman, *Biochemistry* **30,** 4398–4405 (1991).
129. J. D. Satterlee, D. J. Russell, and J. E. Erman, *Biochemistry* **30,** 9072–9077 (1991).
130. J. D. Satterlee, S. Alam, Q. Yi, J. E. Erman, I. Constantinidis, D. J. Russell, and S. J. Moench in *NMR of Paramagnetic Molecules,* L. J. Berliner and J. Reuben, eds., Plenum, New York, 1983, pp. 275–298.
131. J. D. Satterlee, J. M. Mauro, J. E. Erman, and T. L. Poulos, *Eur. J. Biochem.* **224,** 81–87 (1994).
132. S. L. Alam, J. D. Satterlee, S. L. Alam, J. M. Mauro, T. L. Poulos, and J. E. Erman, *Biochemistry* **34,** 15496–15503 (1995).
133. G. Smulevich, S. Hu, K. R. Rodgers, D. B. Goodin, K. M. Smith, and T. G. Spiro, *Biospectroscopy* **2,** 365–376 (1996).
134. G. Smulevich, J. M. Mauro, L. A. Fishel, A. M. English, J. Kraut, and T. G. Spiro, *Biochemistry* **27,** 5477–5485 (1988).
135. G. Smulevich, A. R. Mantini, A. M. English, and J. M. Mauro, *Biochemistry* **28,** 5058–5064 (1989).
136. G. Smulevich, Y. Wang, J. M. Mauro, J. Wang, L. A. Fishel, J. Kraut, and T. G. Spiro, *Biochemistry* **29,** 7174–7180 (1990).
137. G. Smulevich, M. A. Miller, J. Kraut, and T. G. Spiro, *Biochemistry* **30,** 9546–9558 (1991).
138. G. Smulevich, F. Neri, O. Willemsen, K. Choudhury, M. P. Marzocchi, and T. L. Poulos, *Biochemistry* **34,** 13485–13490 (1995).

139. G. Smulevich in *Biomolecular Spectroscopy, Part A,* R. J. H. Clark and R. E. Hester, eds., Wiley, New York, 1993, pp. 163–193.

140. T. G. Spiro, G. Smulevich, and C. Su, *Biochemistry* **29,** 4497–4508 (1990).

141. C. W. Conroy, P. Tyma, P. H. Daum, and J. E. Erman, *Biochim. Biophys. Acta* **537,** 62–69 (1978).

142. H. A. Harbury, *J. Biol. Chem.* **225,** 1009–1023 (1957).

143. J. Ricard, G. Mazza, and R. J. P. Williams, *Eur. J. Biochem.* **28,** 566–578 (1972).

144. G. Smulevich, J. M. Mauro, L. A. Fishel, A. M. English, J. Kraut, and T. G. Spiro, *Biochemistry* **27,** 5486–5492 (1988).

145. G. Smulevich, R. Envangelista-Kirkup, A. English, and T. G. Spiro, *J. Mol. Struct.* **141,** 411–414 (1986).

146. A. M. English, G. McLendon, and K. Taylor, *J. Am. Chem. Soc.* **106,** 6448–6449 (1984).

147. G. Smulevich, S. Dasgupta, A. English, and T. G. Spiro, *Biochim. Biophys. Acta* **873,** 88–91 (1986).

148. A. B. Berinstain, A. M. English, B. C. Hill, and D. Sharma, *J. Am. Chem. Soc.* **112,** 9649–9651 (1990).

149. J. D. Satterlee and J. E. Erman, *J. Am. Chem. Soc.* **106,** 1139–1140 (1984).

150. M. A. Miller, D. Bandyopadhyay, J. M. Mauro, T. G. Traylor, and J. Kraut, *Biochemistry* **31,** 2789–2797 (1992).

151. E. Cheung and A. M. English, *Can. J. Chem.* **73,** 1181–1186 (1995).

CHAPTER 10

Yeast Cytochrome *c* Peroxidase. II: Interaction with Cytochrome *c*

10.1. INTRODUCTION

The reaction of oxidized yeast cytochrome *c* peroxidase with reduced cytochrome *c* appears at present to be the best model to examine the mechanism of electron transfer between two proteins. The crystal structures of both the enzyme and many cytochromes *c* are known, and crystal structures for some cytochrome *c* peroxidase–cytochrome *c* complexes were recently determined. The Marcus theory for electron transfer was developed and proven for small molecules,[1] and its applicability to long-range electron transfer can be tested. Furthermore, the controversy as to whether electron transfer occurs along well-defined pathways through π, σ and hydrogen bonds[2] or through "free space"[3] will eventually be resolved by studies on proteins such as cytochrome *c* peroxidase and cytochrome *c*, and their complexes. Results on the mechanism of electron transfer from cytochrome *c* to cytochrome *c* peroxidase, to 1996, have been well reviewed.[4]

The natural reductants for oxidized yeast cytochrome *c* peroxidase are the cytochromes *c* found in yeast. There are two of them, designated iso-1 cytochrome *c* and iso-2 cytochrome *c*. In many studies horse cytochrome *c* has been used instead of the natural reducing substrates.

Studies have been complicated by uncontrollable factors such as the difficulty in distinguishing the different oxidized forms of the enzyme, spectral interference because of the presence of two heme groups in the intermolecular complex, rates that are too fast to measure, and experimental parameters, measurable for one set of conditions, that disappear for another. Rates, and perhaps mechanism, are dependent upon ionic strength. Reduced cytochrome *c* can be autoxidized, and traces of hydrogen peroxide may be present unless scrupulous anaerobic conditions are used. Points of disagreement concern the answers to the following questions: Which redox center in compound I is reduced first, the oxyferryl heme or the Trp191 radical cation? Is the same cytochrome *c* binding site used for reduction of both the oxyferryl heme and the radical cation, or are different binding sites used? Where are these cytochrome *c* binding site(s) located? Does reduction by cytochrome *c* occur at the same site on both compounds I and II? If it does, which site? If it doesn't, which site is reduced first?

Points of agreement are the following: There are two binding sites on the enzyme for cytochrome *c*, one of high affinity and the other of low affinity. Compound I is a $Fe^{IV}{=}O$ trytophan 191 cation radical and compound II is an equilibrium mixture of $Fe^{IV}{=}O$ and an Fe^{III} trytophan 191 cation radical. The $Fe^{IV}{=}O$ form of compound II is favored at higher pH. Increased ionic strength slows the overall rate of reaction.

10.2. EQUILIBRIUM BINDING TO THE NATIVE ENZYME

A 1:1 complex of yeast cytochrome *c* peroxidase and cytochrome *c* was detected and isolated.[5] Kinetic, equilibrium, and titration studies were conducted on the complex, the source of the cytochrome *c* being horse heart.[6,7] Because cytochrome *c* is negatively charged and the enzyme positively charged at neutral pH, electrostatic interactions occur and were assumed to play an important role in the binding process. Increase in ionic strength weakens the complex as expected. At pH 6 the equilibrium association constant for the 1:1 complex, using horse heart cytochrome *c*, decreases from $(6.0 \pm 3.6) \times 10^6\ M^{-1}$ to $(2.2 \pm 1.9) \times 10^3\ M^{-1}$ as the ionic strength is increased from 0.01 to 0.2 M.[8] At lower ionic strength a second weaker binding of cytochrome *c* occurs.[9] The complex or complexes have been studied using a battery of techniques, including model studies, crosslinked complexes, sedimentation equilibrium,[10] calorimetry,[11] substitution of mesoporphyrin,[12] and zinc- or metal-free-porphyrins. Detection techniques include potentiometric titration,[13] fluorescence quenching,[14] absorption difference, nuclear magnetic resonance (NMR) spectroscopy, electron paramagnetic resonance (EPR) spectroscopy at room temperature in sol–gel glasses,[15] and gel filtration. A review of work up to 1991 has been published.[16]

Changes were detected in the resonance Raman spectra of the ferric enzyme upon binding of both horse and yeast cytochromes *c*.[17] Both cytochromes increase the percentage of six-coordinate low-spin iron(III) in the enzyme. The decrease in intensity of the 403 cm^{-1} band, characteristic of five-coordinate high-spin cytochrome *c* peroxidase, is apparent in Fig. 10.1. Upon binding of the yeast cytochrome, in contrast to that from horse, the heme vinyl and propionate side chains become more rigidly fixed in the protein matrix.

Titration calorimetry was used to study the interaction of horse ferricytochrome *c* with the wild-type yeast enzyme and six enzyme mutants that affect the docking site.[18] The conclusion was that the solution studies are consistent with crystal structure results defining the binding site. Mutations in the docking region on the enzyme decreased the binding constants by a factor of up to 4.

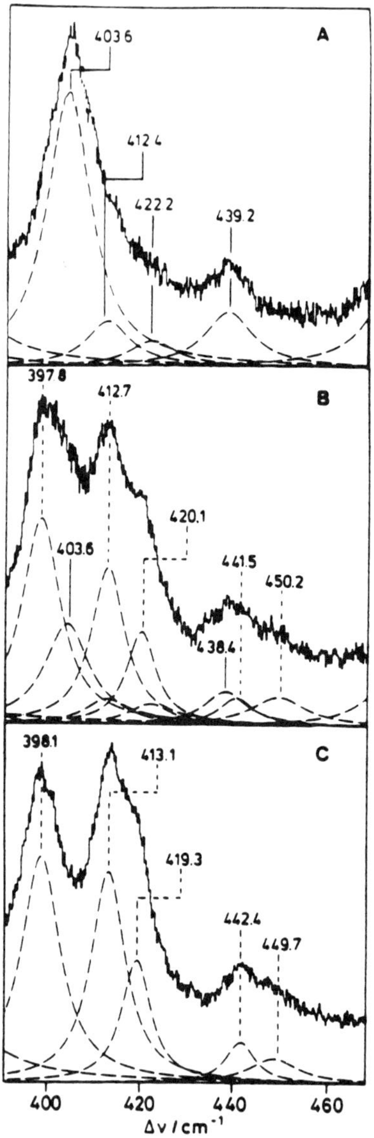

FIG. 10.1. Resonance Raman spectra of (A) cytochrome *c* peroxidase (CCP), (C) ferric yeast cytochrome *c* [Cy + (Y)], and (B) 1:1 complex of the two proteins [CCP − Cy + (Y) complex]. [Reproduced with permission from P. Hildebrandt, A. M. English, and G. Smulevich, *Biochemistry* **31,** 2384–2392 (1992).]

10.3. NUCLEAR MAGNETIC RESONANCE STUDIES

Nuclear magnetic resonance studies on the yeast cytochrome *c*–cytochrome *c* peroxidase complex began in 1987. A covalently linked complex using horse heart cytochrome *c* was prepared using carbodiimide[19] and examined by electrophoresis and proton NMR.[20,21] In a noncovalently linked 1:1 complex, again using horse cytochrome *c*, the proton NMR peaks of the heme 8-methyl and 3-methyl groups, single-proton resonances near 19 ppm and −4 ppm, and the methionine 80 methyl group of the cytochrome, were unambiguously shifted at low ionic strength.[22,23] Studies were extended to both covalently and noncovalently linked 1:1 complexes using horse, tuna, and yeast cytochromes *c*.[24] The proton NMR spectra of the natural substrates, yeast iso-1- and iso-2-cytochromes *c*, exhibited distinctly different proton NMR spectra in the 1:1 complexes with the enzyme from the unnatural substrates.

When the peroxidase was converted to its low-spin form by ligation with cyanide, resonance shifts on the enzyme, upon complexation with cytochrome *c*, were detected for the first time.[25] The latter results were extended using both one- and two-dimensional NMR with horse and yeast cytochromes,[26] as well as tuna.[27] Results obtained using yeast iso-1-cytochrome c are shown in Fig. 10.2. Exchange dynamics of the complexes with the yeast cytochromes were quantitated.[28] Differences in hydrogen–deuterium exchange on the backbone of cytochrome *c* upon binding to the peroxidase were determined.[29] The above NMR results have been summarized in reviews.[30,31]

An extensive NMR study was made on wild-type yeast iso-1 ferricytochrome when it is free in solution and noncovalently bound to yeast cytochrome *c* peroxidase. Both DEFT-NOESY and CLEAN-TOCSY experiments were performed. The results are in agreement with the X-ray results on the complex (Sections 3.5 and 10.7), which show that the iso-1 heme pyrrole ring B is most affected by complexation, and indicate by inference that the enzyme polarizes the structural and electronic properties of ferrocytochrome *c* to facilitate electron transfer to the heme B edge.[32]

10.4. STEADY-STATE KINETICS

Steady-state kinetics have been performed in several laboratories.[5,6,33] The steady-state reaction of hydrogen peroxide and ferrous cytochrome *c*, catalyzed by yeast cytochrome *c* peroxidase, was shown to exhibit non-Michaelis–Menten-type kinetics.[34–36] Here I discuss a later and more detailed study conducted using horse heart cytochrome *c* at pH 7, temperature of 25°C and ionic strength of 0.01 M.[37] Biphasic Eadie–Hofstee plots were obtained both as a function of hydrogen peroxide and ferrous cytochrome *c* concentrations (Fig. 10.3). Essentially the same results were obtained from the nonlinear least-squares analysis, whether two independently measured rate constants

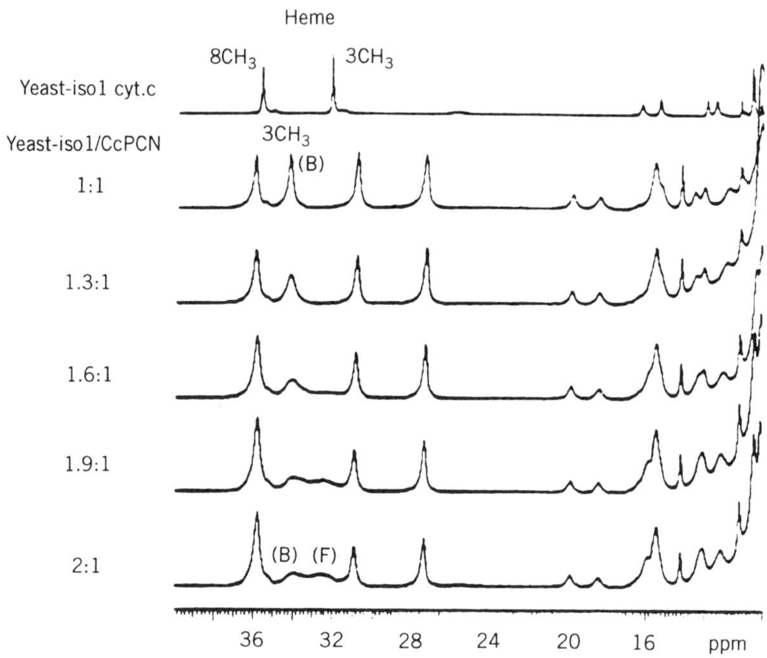

FIG. 10.2. Identification of bound (B) and free (F) ferric yeast iso-1-cytochrome *c* in its complexes with cytochrome *c* peroxidase. For a 1 : 1 ratio only the bound form exists. For amounts of cytochrome *c* in excess of the 1 : 1 ratio, the free form is detected. [Reproduced with permission from Q. Yi, J. D. Satterlee, and J. E. Erman, *Mag. Res. Chemistry* **31,** S53–S58 (1993).]

and a calculated diffusion-controlled rate constant were utilized or all parameters were allowed to vary until convergence to the best fit was obtained. The minimal mechanism that fit the experimental data was:

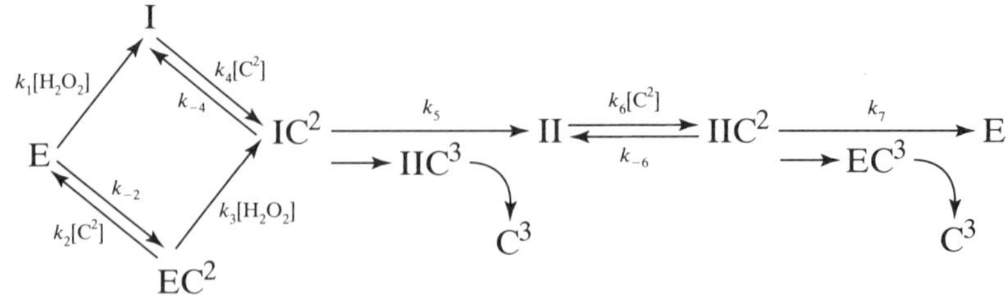

(10.1)

where E, I, and II are native enzyme and compounds I and II; C^2 and C^3 are reduced and oxidized cytochrome *c;* and combinations of these terms are enzyme–cytochrome *c* complexes. The values of both k_1 and k_3 were determined to be $4.4 \times 10^7\ M^{-1}s^{-1}$ from independent measurements, and k_2 was

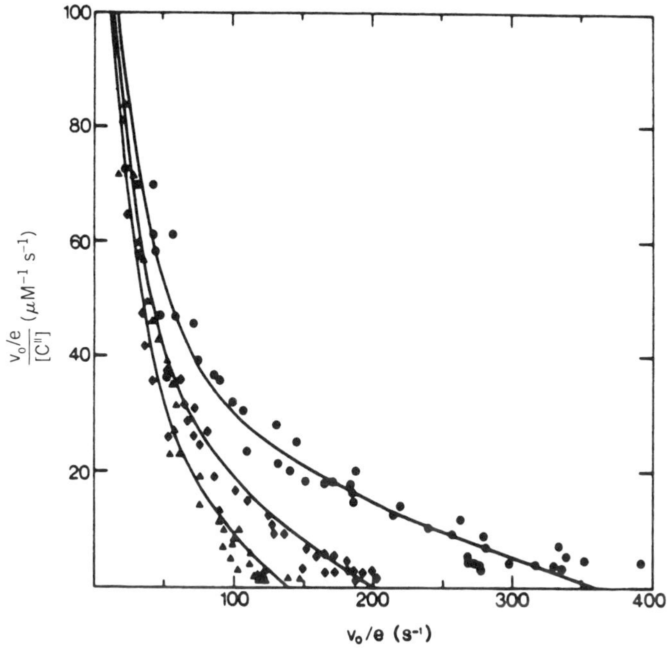

FIG. 10.3. Eadie–Hofstee plots of initial velocities as a function of ferrocytochrome *c* peroxidase concentration at constant hydrogen peroxide concentration. Ferrocytochrome *c* varied from 0.27 to 104 μM, and hydrogen peroxide was 5.0 (▲), 10.0 (◆), and 200 μM (●). The solid lines were calculated according to Eq. (10.1). [Reproduced with permission from D. S. Kang, and J. E. Erman, *J. Biol. Chem.* **257,** 12775–12779 (1982).]

calculated to be 6.4×10^{10} $M^{-1}s^{-1}$ from diffusion theory. Since the binding constant to the native enzyme is $(6.0 \pm 3.6) \times 10^6$ M^{-1}, then the value of k_{-2} can be calculated to be of the order of 10^4 s^{-1}. The value of k_{-2} obtained from the computer fit was in the range 5×10^5 to 10×10^5 s^{-1}.

There are undoubtedly shortcomings in an analysis of a mechanism in which there are at least seven unknown parameters. Nevertheless, the above scheme summarized a great deal of known information, placed certain restrictions on the mechanism, and provided a basis for further discussion.

During the interval 1988–1996 a consensus developed that the free radical site in compound I is reduced by cytochrome *c* at all ionic strengths. Some confusion existed in the literature about which site was being reduced first, caused in part because some experimental results were influenced by traces of hydrogen peroxide contamination. The confusion caused by the influence of hydrogen peroxide contamination now has been totally resolved, but some details appear to require clarification about the mechanism of compound II reduction. Some of the highlights of the pathway to the current state of knowledge, as best as I could summarize it in March 1998, are outlined in the remainder of this chapter.

The steady-state kinetics for the horse heart ferrous cytochrome *c* oxidation,[38] and the corresponding oxidation of yeast iso-1 cytochrome *c*,[39] were studied as a function of ionic strength (Fig. 10.4).[40] The equilibrium binding constants determined from the steady-state kinetics for the iso-1 cytochrome *c* agreed well with values obtained by other techniques. Maximum turnover rate at the high-affinity site increased from 2 to 860 s^{-1} as the ionic strength was increased from 0.01 to 0.20 M. Binding at the second low-affinity site also influences the rate. The total maximum turnover when both sites are occupied increases from 1100 to 2700 s^{-1} as the ionic strength is increased from 0.01 to 0.07 M.[39]

A comprehensive steady-state investigation as a function of ionic strength was made on enzyme mutants at pH 6.0. The two binding sites for cytochrome *c* were accommodated in the mechanism. Rapid intramolecular electron transfer occurs via the high-affinity binding site and is limited by how rapidly the oxidized cytochrome *c* dissociates from the complex. Binding at the low-affinity site increases the rate of dissociation at the high-affinity site, and hence increases the rate of reaction.[41] Simulations to determine unknown rate constants were completed using known rate constants from the literature. The reaction scheme for the high-affinity site is shown in Fig. 10.5. Superimposed on this scheme is a correction obtained by replacement of the one binding site by a two-binding-site model, which indirectly affects the rate. It is assumed that compound I formation is not affected by cytochrome *c* binding, and that the dissociation of cytochrome *c* is not affecting oxidation states of either enzyme or cytochrome *c*. Two electron-transfer reactions are labeled eta and

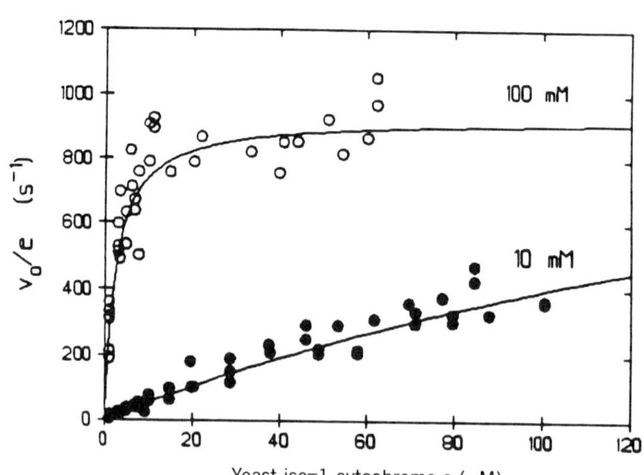

FIG. 10.4. Influence of ionic strength on the rate of oxidation of yeast iso-1 ferrous cytochrome *c* by cytochrome *c* peroxidase at pH 7.5, 25°C. [Reproduced with permission from J. E. Erman, D. S. Kang, K. L. Kim, F. E. Summers, A. L. Matthis, and L. B. Vitello, *Mol. Cryst. Liq. Cryst.* **194,** 253–258 (1991).]

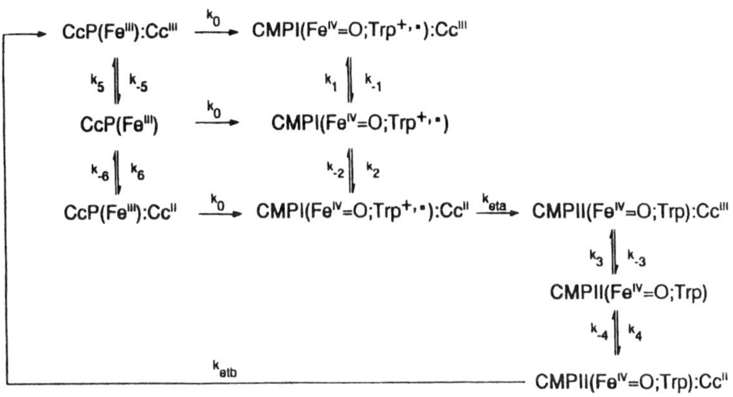

FIG. 10.5. Reaction cycle for yeast cytochrome *c* peroxidase showing the electron-transfer reactions occurring at the high-affinity site for cytochrome *c*. [Reproduced with permission from M. A. Miller, *Biochemistry* **35,** 15791–15799 (1996).]

etb. The values of k_{eta}, ranging from 2×10^6 s^{-1} at ionic strength 5 mM to 1.2×10^6 s^{-1} at ionic strength 105 mM, and k_{etb} ranging from 5000 to 3500 s^{-1} for the same change in ionic strength, were obtained using a cytochrome *c* with a ruthenium label at position 39.[42] The value of k_{-1} was obtained from the same ruthenium-labeled cytochrome *c*.[43] Using these kinetic parameters the reaction scheme shown in Fig. 10.5 provided excellent agreement with the steady-state measurements. It was assumed that reduction of compound II occurs at the $Fe^{IV}{=}O$ site by cytochrome *c* located at the high-affinity binding site. The ruthenium labeling experiments described in Section 10.6.2 show that the $Fe^{IV}{=}O$ reduction is mediated by the Trp191 radical, which is made possible since the two sites are in rapid equilibrium.

10.5. TRANSIENT-STATE KINETICS

The effect of cytochrome *c* binding upon the rates of small molecules reacting with yeast cytochrome *c* peroxidase has been examined. Although the rate of cyanide binding is reduced by nearly 90%, the rates of fluoride binding and hydrogen peroxide reacting are only slightly affected by the presence of enzyme-bound cytochrome *c*.[44]

The negatively charged complex between the enzyme and cytochrome *c* has an electrostatic effect on rates of reduction of both compound II and the ferric ion of cytochrome *c* by small molecules. Thus, rates may be enhanced or inhibited depending upon the charge on the small reductant. However, overall, there is little effect of the binding of cytochrome *c* on the accessibility to the heme of the enzyme by small molecules.[45]

The kinetic properties of a covalently linked complex between cytochrome *c* and the enzyme were examined.[46] Electron transfer from cytochrome *c* to to cytochrome *c* peroxidase was demonstrated across a bilayer lipid membrane, an observation that has important consequences for theories of the mechanism of electron transfer.[47]

A stopped flow investigation was made of the reduction of compounds I and II in the presence of horse heart ferrous cytochrome *c*.[48] Biphasic curves were measured at 417 nm, pH 7.5, ionic strength 10 mM, for the reaction starting with compound I and using excess ferrous cytochrome *c*. The fast phase corresponded to conversion of compound I to compound II. Ionic-strength-dependent kinetic studies were made using both horse[49] and yeast iso-1 cytochromes *c*.[50]

The reaction of excess yeast cytochrome *c* peroxidase compound I with reduced horse cytochrome *c* was studied by stopped flow kinetics at high ionic strength (0.1 M NaCl) to slow the reaction. The biphasic kinetics were consistent with fast reduction of the free radical site in compound I [$k = (1.3 \pm 0.2) \times 10^8\ M^{-1}s^{-1}$] and the subsequent reduction of the ferryl group of compound II had a rate constant of $(9 \pm 1) \times 10^6\ M^{-1}s^{-1}$. At lower ionic strengths the reaction rates increased.[51] Similar experiments with site-directed mutants of cytochrome *c* peroxidase led to the following conclusions:[52] The same interaction domain is used for the reactions of cytochrome *c* with the radical of compound I and with the oxyferryl form of compound II, and the interaction domain for the electron-transfer reactions in solution resembles the interaction domain of the crystalline enzyme–cytochrome *c* complex (Section 10.7).

When tryptophan 191 is replaced by phenylalanine, the initial compound I formed is an iron(IV) species with a transient porphyrin π-cation radical.[53] The rate of electron transfer from ferrous cytochrome *c* to compound I was reduced by a factor of at least 10^4.

10.6. RATE OF INTRACOMPLEX ELECTRON TRANSFER

Electron transfer between the two proteins has been described in terms of electron tunneling.[54] Attempts have been made to measure directly the intermolecular electron transfer rate within the complex. A zinc-substituted porphyrin was bound to yeast apocytochrome *c* peroxidase and was converted to its triplet state by flash photoexcitation. From analysis of the resultant kinetics, electron-transfer rates to oxidized cytochrome *c* and back to the porphyrin of the enzyme were determined.[55] The back transfer is analogous to the physiological oxidation of cytochrome *c* by cytochrome *c* peroxidase. The rate constant for oxidation of horse cytochrome *c* was $17 \pm 3\ s^{-1}$ and for yeast cytochrome *c*, $138 \pm 12\ s^{-1}$, both at 20°C, pH 7.0, and ionic strength 0.01 M.

The first direct measurement of the intracomplex rate constant for electron transfer was obtained under comparable conditions, pH 7.0, ionic strength

0.01 M, and 24°C, but was for electron transfer to the ferrous enzyme,[56] formed by the photoreduction of ferric cytochrome *c* peroxidase using acetophenone/isopropyl alcohol.[57] The measured rate constant was very small, $0.23 \pm 0.02\ s^{-1}$. This value may be suspect because measurements could not be started until 3 s after the termination of the photoreduction, to ensure a well-mixed solution.

At high ionic strength, up to 0.5 M, the rate-controlling step in the reaction between ferric cytochrome *c* and ferrous cytochrome *c* peroxidase appears to be a conformational change in the enzyme.[58,59] Fluorescence studies using porphyrin cytochrome c indicate no complex formation at high ionic strength.

10.6.1. Use of Semiquinone Reductants

An alternative approach has been developed. In a mixture of oxidized enzyme, oxidized cytochrome *c,* and a trace of flavin, laser flash photolysis can be used to produce semiquinones from the flavin. These in turn can reduce one of the oxidized proteins, and the subsequent electron transfer between the two proteins can be monitored in real time.[60]

$$\mathrm{Fl_{ox}} \xrightarrow{h\nu, k_1} \mathrm{FlH\bullet}$$
$$P_{1,\mathrm{ox}} + \mathrm{FlH\bullet} \xrightarrow{k_2} P_{1,\mathrm{red}} + \mathrm{Fl_{ox}} + \mathrm{H^+}$$
$$P_{1,\mathrm{red}} + P_{2,\mathrm{ox}} \xrightarrow{k_3} P_{1,\mathrm{ox}} + P_{2,\mathrm{red}}$$

Oxidized and reduced forms are indicated by the subscripts ox and red, Fl is flavin, FlH• the semiquinone radical, and P_1 and P_2 the protein redox partners. The possibility exists that both $P_{1,\mathrm{ox}}$ and $P_{2,\mathrm{ox}}$ could be reduced by the semiquinone radical. However, there is often a large difference in rate constants for the reduction of the two proteins by the semiquinone, and the rates can also be controlled by the concentrations of P_1 and P_2. It is often the lesser favorable reaction of FlH•, from the standpoint of reduction potentials, which occurs. Results obtained at low ionic strength indicated that it is the ferryl $Fe^{IV}{=}O$ site that is reduced in compound I, a result in disagreement with more recent results (Fig. 10.6).[61]

Values of k_3, the intracomplex electron-transfer rate constant between compound I and ferrous cytochrome *c,* are highly dependent upon ionic strength.[62] On the other hand, if the complex is covalently crosslinked the ionic strength dependence disappears.[63] Replacement of His181 by glycine only changed the electron-transfer rate by a factor of 2.[64] Further studies of surface amino acid replacements were conducted.[65]

10.6.2. Use of Ruthenium Labels

The use of ruthenium labels provides a different technique to study the kinetics of the electron-transfer reactions. By attachment of a photo-redox active

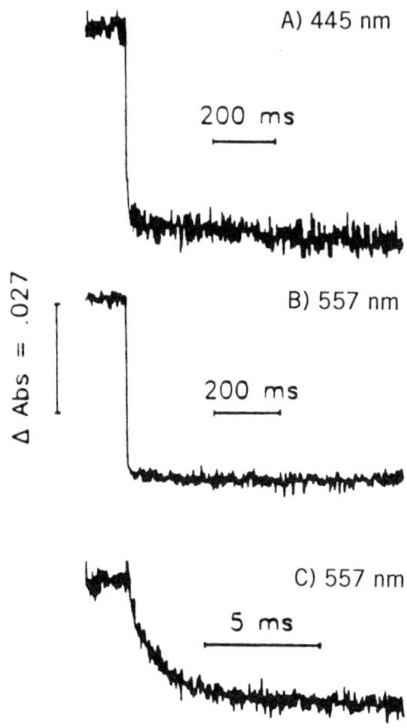

FIG. 10.6. Reduction of compound I by ferrous cytochrome *c,* generated from semiquinone radicals. The 557 nm is an isosbestic point between the two protein reactants, and 445 nm is as close to the heme peak as was allowed by flavin absorption. [Reproduced with permission from J. T. Hazzard and G. Tollin, *J. Am. Chem. Soc.* **113,** 8956–8957 (1991).]

ruthenium label, an iron or protein site can be reduced to the ferrous state, and the kinetics of the subsequent electron transfer to the partner in the intermolecular complex can be measured. Techniques were developed to covalently attach a tris(bipyridine)ruthenium complex to cysteine sulfhydryl,[66–68] lysine amino,[69–71] and histidine imidazole groups.[72] Attachment of the ruthenium label to cytochrome *c,* followed by flash photolysis using a dye laser resulted in the sequence of events within the cytochrome shown in Scheme 10.1:

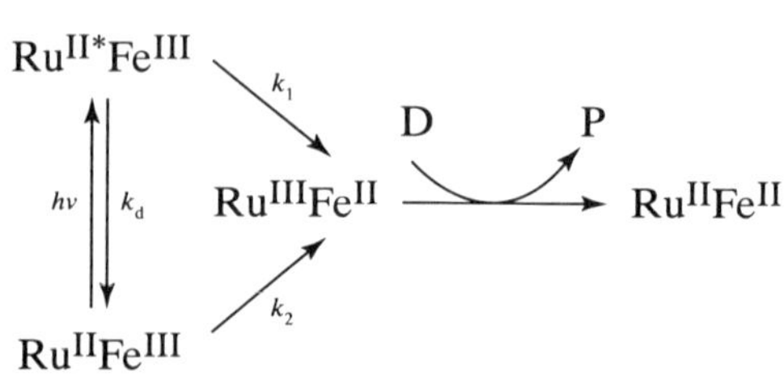

SCHEME 10.1.

The flash creates a charge-transfer complex between the ruthenium and its ligands, Ru^{II*}, which is a powerful reducing agent. The iron(III) of the heme *c* is reduced to iron(II). A donor molecule, D, an aniline or ethylenediaminetetraacetic acid (EDTA), is added to reduce Ru^{III} back to Ru^{II}, which ensures that reversion of $Ru^{III}\,Fe^{II}$ back to $Ru^{II}\,Fe^{III}$ does not occur. If conditions are used where these three steps occur rapidly enough, then the subsequent electron transfer from the iron(II) of the cytochrome *c* to the enzyme can be followed.

Alternatively, if the donor molecule can be omitted, the $Ru^{III}\,Fe^{II}$ form of cytochrome *c* is a strong oxidizing agent and may react rapidly with native resting cytochrome *c* peroxidase. This was demonstrated with some ruthenium labels attached to lysine residues of cytochrome *c,* and the interesting result was that Trp191 of the enzyme was oxidized to its free radical form.[73] Thus, the tryptophan free radical was obtained without the use of hydrogen peroxide.

A pH-dependent transient-state kinetic study was made of recombinant and mutant yeast cytochrome *c* peroxidase compounds I with native, mutant, and ruthenium-labeled horse cytochrome *c*.[74] The data could be accommodated by the following mechanism:

$$\text{Cmpd I }(\text{Fe}^{\text{IV}},\text{R}\bullet) + \text{cyt } c^{2+} \xrightarrow{k_1} \text{Cmpd II }(\text{Fe}^{\text{IV}},\text{R}) + \text{cyt } c^{3+}$$

$$\text{Cmpd II }(\text{Fe}^{\text{IV}},\text{R}) \overset{K}{\leftrightarrows} \text{Cmpd II }(\text{Fe}^{\text{III}},\text{R}\bullet)$$

$$\text{Cmpd II }(\text{Fe}^{\text{III}},\text{R}\bullet) + \text{cyt } c^{2+} \xrightarrow{k_2} \text{Native CcP }(\text{Fe}^{\text{III}},\text{R}) + \text{cyt } c^{3+}$$

At 100 mM ionic strength and pH 7.5 the value of k_1 was $1.3 \times 10^8\ \text{M}^{-1}\text{s}^{-1}$, and this value was virtually unchanged from pH 5 to 8. In contrast, the apparent rate constant for compound II reduction decreased 33-fold as the pH is increased from 5 to 8. This was explained in terms of the pH-dependent equilibrium between the two forms of compound II. None of the radical form of compound II, which is the reactive species with reduced cytochrome *c,* exists at high pH, but at lower pH the equilibrium shifts from the iron(IV) form of compound II to the free radical form. All of the data could be fit by assuming $k_1 = k_2 = 1.2 \times 10^8\ \text{M}^{-1}\text{s}^{-1}$ and the value of the equilibrium constant K changes from 1.2 at low pH to 0.005 at high pH. The value of K is dependent upon a single proton ionization with $pK_a = 5.7$. The apparent rate of compound II reduction is given by Kk_2. This mechanism implies that the same binding site is used by cytochrome *c* for the reductions of both compound I and compound II. Two pieces of evidence ruled against the alternative scheme in which cytochrome *c* has two different binding sites, one for reaction with the free radical site, and one for reaction with iron(IV). The value of K is the same whether one uses ferrocyanide or cytochrome *c* as reductant. Compound I of the Trp191Phe mutant is reduced at only 0.1% of the rate for the native compound I, even though the electronic properties of the iron(IV) form and the structure of the mutant appear to be similar to those of the native enzyme.[75,76] This comparatively early work remains as one of the pieces of

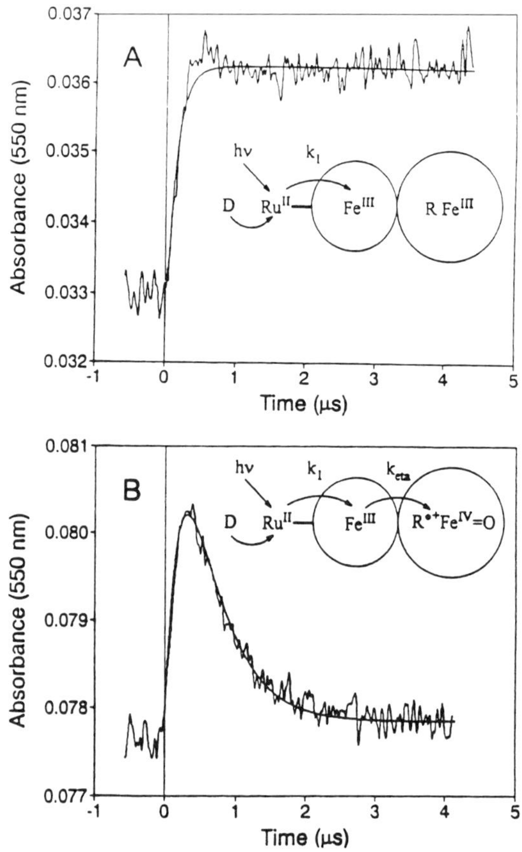

FIG. 10.7. Photoinduced electron transfer from ferrous Ru-39–cytochrome *c* to cytochrome *c* peroxidase compound I in the presence of a sacrificial electron donor. (A) Solution containing 11 μM Ru-39–C*c* and 12.5 μM C*c*P in 2 mM sodium phosphate, pH 7.0, and 2 mM aniline was excited with a 450-nm laser flash. The aniline rapidly reduced Ru^{III} and prevented the back electron-transfer reaction (k_2) from the C*c* iron site to Ru. The smooth line is the best fit to a single exponential equation with $(k_1 + k_d) = 6.5 \times 10^6\ s^{-1}$ and $k_2 = 0$ (rate constants defined in Scheme 10.1). (B) H_2O_2 (12.5 μM) was added to the solution in (A) to form compound I and the experiment was repeated. The smooth line is the best fit to a two-exponential equation with the same values of rate constants as in (A), and the rate constant for intracomplex electron transfer $k_{eta} = 2.0 \times 10^6\ s^{-1}$ for the second exponential. [Reproduced with permission from K. Wang, H. Mei, L. Geren, M. A. Miller, A. Saunders, X. Wang, J. L. Waldner, G. J. Pielak, B. Durham, and F. Millett, *Biochemistry* **35,** 15107–15119 (1996).]

supporting evidence for reaction at the free radical site in compound II. The classic problem with the mechanism of reaction with species in equilibrium is that one cannot tell which of the two species is reacting. (It might be both.) If the equilibrium is not maintained during the course of the reaction, then a different parameter is introduced into the rate analysis.

Both stopped flow and ruthenium-labeled experiments rule out direct roles of several aromatic residues in the reduction of the Trp191 radical[74] and support the mechanism proposed on the basis of the X-ray structure of the enzyme–protein complex.[77] Ruthenium labeling experiments are discussed in greater detail below. A role for Met230 in the intramolecular electron transfer of compound II was deduced from a study of the Met230Ile mutant.[78]

A combination of mutations with ruthenium labels provided further evidence that the Trp191 free radical is the initial site of reduction of both compounds I and II.[42,76] Results with the ruthenium label at position 39 of cytochrome *c* provided particularly clear-cut results, since the label is on the opposite side of the cytochrome from where the binding with the enzyme takes place. Evidence for photoinduced electron transfer from Ru-39–cytochrome *c* to the tryptophan 191 radical site of compound I is shown in Fig. 10.7. The rate constant for intracomplex electron transfer to the Trp191 radical of compound I k_{eta} is $2 \times 10^6\ s^{-1}$. The rate constant for the reduction of the $Fe^{IV}{=}O$ group of compound II is 5000 s^{-1} by Ru-39–cytochrome *c*, a process mediated by the Trp191 radical in rapid equilibrium with the $Fe^{IV}{=}O$ group. A further use of ruthenium labels was used to detect that the rate-limiting step in enzyme turnover is product dissociation at low ionic strength, and intracomplex electron transfer to reduce compound II at physiological ionic strength.[43]

In site-directed mutagenesis studies, in which cysteine residues were deleted in one position and added in another in the enzyme, differences in reactivities of the horse and yeast cytochrome *c*'s were detected, which were in accord with the known crystal structures of the peroxidase–cytochrome *c* complexes.[79] The studies show that the high-affinity binding site is utilized in both electron-transfer reactions, in accord with the Pelletier–Kraut crystal structure (next section).

10.7. CRYSTAL STRUCTURES FOR THE COMPLEX

A major breakthrough occurred when the crystal structure of the complex between yeast cytochrome *c* peroxidase and yeast iso-1 cytochrome *c* at physiological ionic strength of 150 mM was determined at 2.3 Å resolution.[77] The structure is different from that predicted by modeling studies, which placed heavier emphasis on electrostatic interactions.[80] The binding domain determined in the crystal structure includes hydrophobic interactions as well as electrostatic interactions between lysine amino groups on the cytochrome and carboxylate groups on the enzyme. A unique electron-transfer pathway was proposed on the basis of the shortest straight-line pathway between the two hemes in the complex. The pathway extends from a heme methyl group on the cytochrome through Ala193, Ala194, and Gly192 to the radical on Trp191 on the enzyme. The latter is in van der Waals contact with the enzyme heme group. The structure of the yeast enzyme–horse cytochrome *c* complex was also determined and found to be similar. Representations of the proposed electron-transfer pathway are shown in Section 3.5.

10.8. CONCLUSIONS

The oxidation of small molecules by oxidized forms of yeast cytochrome *c* peroxidase appears to be accomplished by the $Fe^{IV}{=}O$ group, as discussed in Chapter 9. The most reliable results to date show conclusively that the oxidation of reduced cytochrome *c,* on the other hand, is accomplished by the radical on Trp191 of compound I. The same oxidation site at Trp191 appears to be utilized to accomplish the reduction of compound II as well as compound I, facilitated by the equilibrium between $Fe^{IV}{=}O$, Trp and Fe^{III}, $Trp{\bullet}^{+}$ forms of compound II. The precise nature of the equilibrium between the two reaction sites of compound II, the speed with which equilibrium is attained, and the influence of the equilibrium on the reactivity of compound II at different pH values would appear to be areas of fruitful further research.

REFERENCES

1. R. A. Marcus, *J. Chem. Phys.* **24,** 966–989 (1956).
2. D. N. Beratan, J. N. Betts, and J. N. Onuchic, *Science* **252,** 1285–1288 (1991).
3. C. C. Moser, J. M. Keske, K. Warncke, R. S. Faird, and P. L. Dutton, *Nature* **325,** 796–802 (1992).
4. J. M. Nocek, J. S. Zhou, S. De Forest, S. Priyadarshy, D. N. Beratran, J. N. Onuchic, and B. M. Hoffman, *Chem. Rev.* **96,** 2459–2489 (1996).
5. P. Nicholls and E. Mochan, *Nature New Biology* **230,** 276–277 (1971).
6. P. Nicholls and E. Mochan, *Biochem. J.* **121,** 55–67 (1971).
7. E. Mochan and P. Nicholls, *Biochem. J.* **121,** 69–82 (1971).
8. J. E. Erman and L. B. Vitello, *J. Biol. Chem.* **255,** 6224–6227 (1980).
9. J. A. Kornblatt and A. M. English, *Eur. J. Biochem.* **155,** 505–511 (1986).
10. R. J. Dowe, L. B. Vitello, and J. E. Erman, *Arch. Biochem. Biophys.* **232,** 566–573 (1984).
11. G. C. Kresheck, L. B. Vitello, and J. E. Erman, *Biochemistry* **34,** 8398–8405 (1995).
12. H. Anni, J. M. Vanderkooi, K. A. Sharp, T. Yonetani, S. C. Hopkins, L. Herenyi, and J. Fidy, *Biochemistry* **33,** 3475–3486 (1994).
13. M. R. Mauk, J. C. Ferrer, and A. G. Mauk, *Biochemistry* **33,** 12609–12614 (1994).
14. L. B. Vitello and J. E. Erman, *Arch. Biochem. Biophys.* **258,** 621–629 (1987).

15. C. T. Lin, C. M. Catuara, J. E. Erman, K. C. Chen, S. F. Huang, W. J. Wang, and H. H. Wei, *J. Sol-Gel Sci. Technol.* **7,** 19–26 (1996).
16. H. R. Bosshard, H. Anni, and T. Yonetani in *Peroxidases in Chemistry and Biology, Vol. II,* J. Everse, K. E. Everse and M. B. Grisham, eds., CRC, Boca Raton, FL, 1991, pp. 51–84.
17. P. Hildebrandt, A. M. English, and G. Smulevich, *Biochemistry* **31,** 2384–2392 (1992).
18. J. E. Erman, G. C. Kresheck, L. B. Vitello, and M. A. Miller, *Biochemistry* **36,** 4054–4060 (1997).
19. B. Waldemeyer and H. R. Bosshard, *J. Biol. Chem.* **260,** 5184–5190 (1985).
20. S. J. Moench, J. D. Satterlee, and J. E. Erman, *Biochemistry* **26,** 3821–3826 (1987).
21. S. J. Moench, J. E. Erman, and J. D. Satterlee, *Int. J. Biochem.* **25,** 1335–1342 (1993).
22. J. D. Satterlee, S. J. Moench, and J. E. Erman, *Rec. Trav. Chim. Pays-Bas* **106,** 286 (1987).
23. J. D. Satterlee, S. J. Moench, and J. E. Erman, *Biochim. Biophys. Acta* **912,** 87–97 (1987).
24. S. J. Moench, S. Chroni, B.-H. Lou, J. E. Erman, and J. D. Satterlee, *Biochemistry* **31,** 3661–3670 (1992).
25. Q. Yi, J. E. Erman, and J. D. Satterlee, *J. Am. Chem. Soc.* **114,** 7907–7909 (1992).
26. Q. Yi, J. E. Erman, and J. D. Satterlee, *Biochemistry* **32,** 10988–10994 (1993).
27. Q. Yi, J. E. Erman, and J. D. Satterlee, *Mag. Res. Chem.* **31,** S53–S58 (1993).
28. Q. Yi, J. E. Erman, and J. D. Satterlee, *J. Am. Chem. Soc.* **116,** 1981–1987 (1994).
29. Q. Yi, J. E. Erman, and J. D. Satterlee, *Biochemistry* **33,** 12032–12041 (1994).
30. Q. Yi, S. Alam, Y. Ge, J. D. Satterlee, and J. E. Erman, *Techniques in Protein Chemistry IV,* Academic, New York, 1993, pp. 605–613.
31. J. E. Erman and J. D. Satterlee, *Adv. Biophys. Chem.* **5,** 141–178 (1995).
32. S. F. Sukits, J. E. Erman, and J. D. Satterlee, *Biochemistry* **36,** 5251–5259 (1997).
33. T. Yonetani and G. S. Ray, *J. Biol. Chem.* **241,** 700–706 (1966).
34. C. H. Kang, S. Ferguson-Miller, and E. Margoliash, *J. Biol. Chem.* **252,** 919–926 (1977).
35. C. H. Kang, D. L. Brautigan, N. Osheroff, and E. Margoliash, *J. Biol. Chem.* **253,** 6502–6510 (1978).
36. M. B. Smith and F. Millet, *Biochim. Biophys. Acta* **626,** 64–72 (1980).
37. D. S. Kang and J. E. Erman, *J. Biol. Chem.* **257,** 12775–12779 (1982).
38. K. L. Kim, D. S. Kang, L. B. Vitello, and J. E. Erman, *Biochemistry* **29,** 9150–9159 (1990).
39. A. L. Matthis and J. E. Erman, *Biochemistry* **34,** 9985–9990 (1995).
40. J. E. Erman, D. S. Kang, K. L. Kim, F. E. Summers, A. L. Matthis, and L. B. Vitello, *Mol. Cryst. Liq. Cryst.* **194,** 253–258 (1991).
41. M. A. Miller, *Biochemistry* **35,** 15791–15799 (1996).
42. K. Wang, H. Mei, L. Geren, M. A. Miller, A. Saunders, X. Wang, J. L. Waldner, G. J. Pielak, B. Durham, and F. Millett, *Biochemistry* **35,** 15107–15119 (1996).
43. H. Mei, K. Wang, S. McKee, X. Wang, J. L. Waldner, G. J. Pielak, B. Durham, and F. Millett, *Biochemistry* **35,** 15800–15806 (1996).

44. L. R. Hoth and J. E. Erman, *Biochim. Biophys. Acta* **788,** 151–153 (1984).
45. P. Cokic and J. E. Erman, *Biochim. Biophys. Acta* **913,** 257–271 (1987).
46. J. E. Erman, K. L. Kim, L. B. Vitello, S. J. Moench, and J. D. Satterlee, *Biochim. Biophys. Acta* **911,** 1–10 (1987).
47. Z. Salamon, L. B. Vitello, J. E. Erman, P. Butko, and H. T. Tien, *Bioelectrochem. Bioenerg.* **21,** 213–221 (1989).
48. F. E. Summers and J. E. Erman, *J. Biol. Chem.* **263,** 14267–14275 (1988).
49. M. R. Nuevo, H.-H. Chu, L. B. Vitello, and J. E. Erman, *J. Am. Chem. Soc.* **115,** 5873–5874 (1993).
50. A. L. Matthis, L. B. Vitello, and J. E. Erman, *Biochemistry* **34,** 9991–9999 (1995).
51. S. Hahm, L. Geren, B. Durham, and F. Millett, *J. Am. Chem. Soc.* **115,** 3372–3373 (1993).
52. M. A. Miller, R.-Q. Liu, S. Hahm, L. Geren, S. Hibdon, J. Kraut, B. Durham, and F. Millett, *Biochemistry* **33,** 8686–8693 (1994).
53. M. A. Miller, L. B. Vitello, and J. E. Erman, *Biochemistry* **34,** 12048–12058 (1995).
54. M. J. Potasek, *Science* **201,** 151–153 (1978).
55. P. S. Ho, C. Sutoris, N. Liang, E. Margoliash, and B. M. Hoffman, *J. Am. Chem. Soc.* **107,** 1070–1071 (1984).
56. E. Cheung, K. Taylor, J. A. Kornblatt, A. M. English, G. McLendon, and J. R. Miller, *Proc. Natl. Acad. Sci. U.S.A.* **83,** 1330–1333 (1986).
57. P. S. Ho, B. M. Hoffman, C. H. Kang, and E. Margoliash, *J. Biol. Chem.* **258,** 4356–4363 (1983).
58. E. Cheung and A. M. English, *Inorg. Chem.* **27,** 1078–1081 (1988).
59. E. Cheung and A. M. English, *Inorg. Chim. Acta* **201,** 243–246 (1992).
60. G. Tollin and J. T. Hazzard, *Arch. Biochem. Biophys.* **287,** 1–7 (1991).
61. J. T. Hazzard and G. Tollin, *J. Am. Chem. Soc.* **113,** 8956–8957 (1991).
62. J. T. Hazzard, T. L. Poulos, and G. Tollin, *Biochemistry* **26,** 2836–2848 (1987).
63. J. T. Hazzard, S. J. Moench, J. E. Erman, J. D. Satterlee, and G. Tollin, *Biochemistry* **27,** 2002–2008 (1988).
64. M. A. Miller, J. T. Hazzard, J. M. Mauro, S. L. Edwards, P. C. Simons, G. Tollin, and J. Kraut, *Biochemistry* **27,** 9081–9088 (1988).
65. A. F. Corin, R. A. Hake, G. McLendon, J. T. Hazzard, and G. Tollin, *Biochemistry* **32,** 2756–2762 (1993).
66. L. Geren, S. Hahm, B. Durham, and F. Millett, *Biochemistry* **30,** 9450–9457 (1991).
67. A. Willie, P. S. Stayton, S. G. Sligar, B. Durham, and F. Millett, *Biochemistry* **31,** 7237–7242 (1992).
68. L. M. Geren, J. R. Beasley, B. R. Fine, A. Saunders, S. Hibdon, G. J. Pielak, B. Durham, and F. Millett, *J. Biol. Chem.* **270,** 2466–2472 (1995).
69. L. P. Pan, B. Durham, J. Wolinska, and F. Millett, *Biochemistry* **27,** 7180–7184 (1988).
70. B. Durham, L. P. Pan, J. Long, and F. Millett, *Biochemistry* **28,** 8659–8665 (1989).
71. S. Hahm, B. Durham, and F. Millett, *Biochemistry* **31,** 3472–3477 (1992).
72. F. Millett, M. A. Miller, L. Geren, and B. Durham, *J. Bioenerg. Biomem.* **27,** 341–351 (1995).

73. R.-Q. Liu, S. Hahm, M. Miller, B. Durham, and F. Millett, *Biochemistry* **34,** 973–983 (1995).
74. S. Hahm, M. A. Miller, L. Geren, J. Kraut, B. Durham, and F. Millett, *Biochemistry* **33,** 1473–1480 (1994).
75. J. M. Mauro, L. A. Fishel, J. T. Hazzard, T. E. Meyer, G. Tollin, M. A. Cusanovich, and J. Kraut, *Biochemistry* **27,** 6243–6256 (1988).
76. J. Wang, J. M. Mauro, S. L. Edwards, S. J. Oatley, L. A. Fishel, V. A. Ashford, N. H. Xuong, and J. Kraut, *Biochemistry* **29,** 7160–7173 (1990).
77. H. Pelletier and J. Kraut, *Science* **258,** 1748–1755 (1992).
78. R.-Q. Liu, M. A. Miller, G. W. Han, S. Hahm, L. Geren, S. Hibdon, J. Kraut, B. Durham, and F. Millett, *Biochemistry* **33,** 8678–8685 (1994).
79. M. A. Miller, L. Geren, G. W. Han, A. Saunders, J. Beasley, G. J. Pielak, B. Durham, F. Millett, and J. Kraut, *Biochemistry* **35,** 667–673 (1996).
80. T. L. Poulos and B. C. Finzel, *Pept. Prot. Rev.* **4,** 115–171 (1984).

CHAPTER 11

Other Class I Peroxidases: Ascorbate Peroxidase and Bacterial Catalase-Peroxidases

Among the multitude of surprises in peroxidase research was the discovery of the high degree of homology between yeast cytochrome *c* peroxidase, plant ascorbate peroxidases, and the bacterial catalase-peroxidases, which led to the designation of class I peroxidases, those of prokaryotic origin (Section 3.2). The monumental amount of work on the yeast enzyme is described in Chapters 2, 3, 9, and 10. In this chapter other members of class I are described.

11.1. ASCORBATE PEROXIDASE

Identification of ascorbate peroxidase occurred in 1979, and work on highly purified enzyme began about 1989. Therefore, this enzyme is one of the more recent additions to the peroxidase family. Ascorbate peroxidases from three principal sources have been studied: from chloroplasts, from plant cytosol, and from root nodules with nitrogenase activity.

The first report was a study of spinach chloroplasts.[1] A heat-labile peroxidatic activity in washed lamellae from isolated chloroplasts was detected using 3,3′-diaminobenzidine and ascorbate as electron donors. A sharp pH optimum activity occurred between pH 7.5 and 8.0. A heat-stable extractable activity was attributed to nonenzymatic reactions. Soluble ascorbate peroxidase activity was detected in 20-day-old pea shoots and in spinach and wheat leaves. Use of the enzyme in a new assay for ascorbate was described.[2,3] The soluble enzyme is now known as cytosolic ascorbate peroxidase. Specific cytosolic ascorbate peroxidase activity was detected in the species *Euglena gracilis*.[4] Low peroxidase activity was detected in soybean root nodules. Ascorbate was not tested as a reductant; guaiacol was, and the latter substrate is now known to be a poor substrate for ascorbate peroxidase.[5] Later work established the soybean root nodule enzyme as an ascorbate peroxidase.[6,7] Ascorbate peroxidase was also detected in cyanobacteria.[8] Scavenging of photosynthetically generated hydrogen peroxide by ascorbate peroxidase was reported.[9]

The role of ascorbate peroxidase is now clear: Its purpose is to catalyze the removal of hydrogen peroxide by ascorbate in the following reaction.[10]

$$H_2O_2 + 2\ \text{Ascorbate} \rightarrow H_2O + 2\ \text{Monodehydroascorbate}$$

The structures of ascorbic acid, the free radical monodehydroascorbic acid, and the final oxidation product dehydroascorbic acid are shown in Fig. 11.1. The monodehydroascorbate is a fairly unreactive free radical, which partially explains its protective role against oxidative damage. Its formation may remove a much more reactive free radical. Monodehydroascorbate may be further oxidized to dehydroascorbate or reduced back to ascorbate. Two of the radicals can disproportionate to form one dehydroascorbate and reform one ascorbate.

Chloroplastic ascorbate peroxidase from tea leaves was purified to homogeneity[11] and partially sequenced.[12] Highly purified cytosolic enzymes were obtained from *E. gracilis*[4] and soybean root nodules.[7] Pea cytosolic ascorbate peroxidase was purified to homogeneity,[13] cloned, and the gene was characterized.[14,15] The deduced amino acid sequence of 249 residues is reproduced in Appendix A.

The recombinant pea cytosolic enzyme was crystallized and its crystal structure determined.[16] The enzyme is a noncovalently bound dimer. Interpretation of the X-ray data was complicated by the fact that there were two dimers per unit cell. The structure of yeast cytochrome *c* peroxidase was used as a starting model for the interpretation.

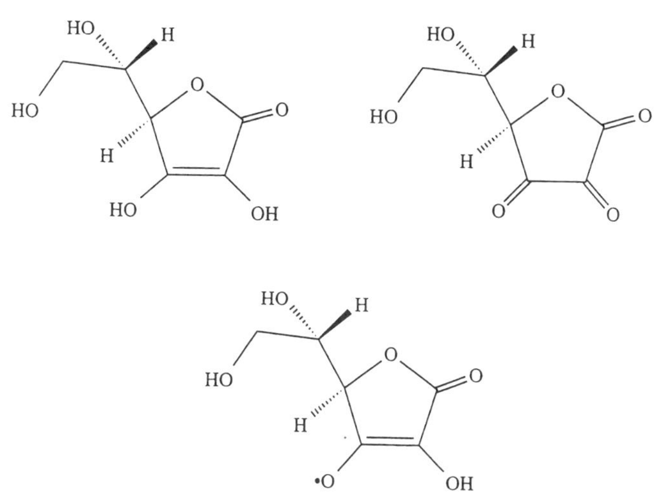

FIG. 11.1. Structure of ascorbic acid, dehydroascorbic acid, and the free radical monodehydroascorbic acid.

The amino acid sequence and crystal structure show that an amazing similarity exists between pea cytosolic ascorbate peroxidase and yeast cytochrome *c* peroxidase. Therefore, ascorbate peroxidase is properly classified as a type I plant peroxidase, one of prokaryotic origin.[17] A representation of the heme environment of ascorbate peroxidase is shown in Fig. 11.2. Ascorbate peroxidase has a tryptophan residue on the proximal side of the heme. The corresponding residue in the yeast enzyme is the site of an unpaired electron in compound I. Amazingly, compound I of ascorbate peroxidase contains a conventional porphyrin π-cation radical, as determined from the electron paramagnetic resonance (EPR) spectrum obtained using the recombinant enzyme (Fig. 11.3)[18] and from the optical spectrum using the wild-type enzyme (Fig. 11.4).[19] The apparent reason for the different compounds I is that ascorbate peroxidase has a metal ion binding site near the proximal trytophan (Fig. 11.5).[18] The site might be occupied by a calcium ion, but a potassium ion was favored because there is one carboxylate ligand to provide overall neutrality of charge. The presence of a nearby positive ion would make it more difficult for electron transfer to occur from the trytophan to the porphyrin; and from that standpoint the higher the charge on the metal ion, the better.

The rate constant for ascorbate peroxidase compound I formation using hydrogen peroxide is 8.0×10^7 $M^{-1}s^{-1}$ at pH 7.8 and 20°C. The rate of compound I formation decreases at lower pH with an inflection in the pH rate profile at pH 5. Therefore, ascorbate peroxidase compound I formation occurs in similar fashion to compound I formation in yeast cytochrome *c*

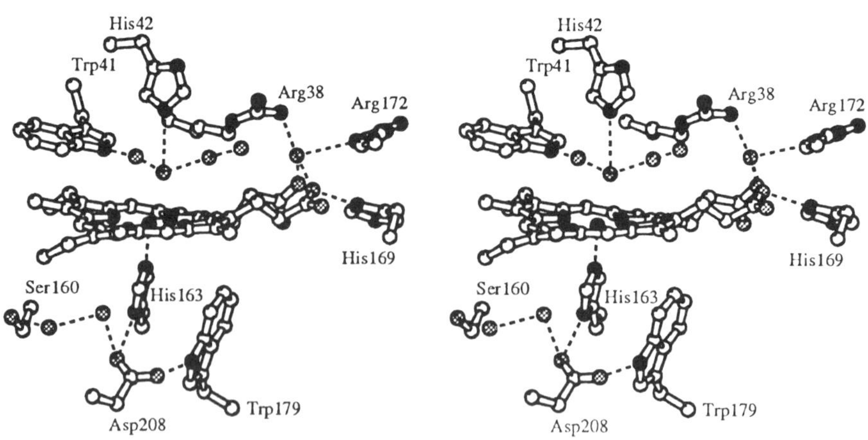

FIG. 11.2. Stereo representation of the heme environment of recombinant pea cytosolic ascorbate peroxidase. Carbon atoms are white, nitrogen atoms and iron black, and water oxygens gray. [Reproduced with permission from W. R. Patterson and T. L. Poulos, *Biochemistry* **34,** 4331–4341 (1995).]

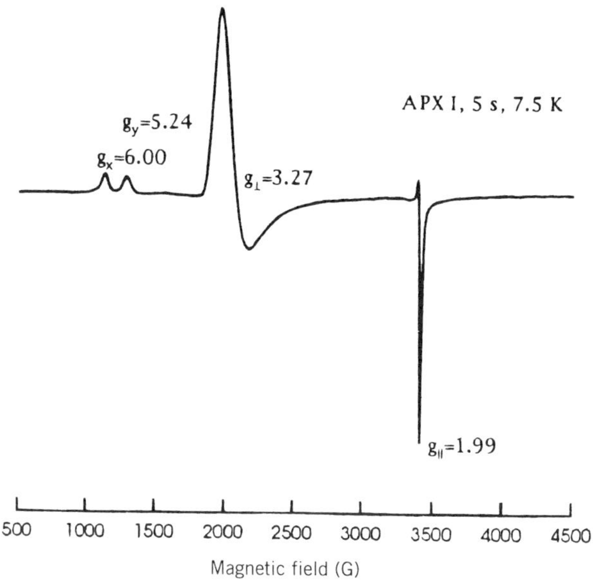

FIG. 11.3. EPR spectrum of ascorbate peroxidase compound I obtained at 7.5 K: 0.16 mM enzyme in 50 mM potassium phosphate, pH 7 and 1.1 equiv hydrogen peroxide, frozen 5 s after mixing. [Reproduced with permission from W. R. Patterson, T. L. Poulos and D. B. Goodin, *Biochemistry* **34,** 4342–4345 (1995).]

peroxidase and horseradish peroxidase, although somewhat faster, and the inflection point at pH 5 occurs at a higher pH.[19]

There is a marked difference in reactivities of the compounds I of ascorbate, cytochrome *c*, and horseradish peroxidases with ascorbate. The rate constant for ascorbate peroxidase compound I reacting with ascorbate is the same as the rate of compound I formation: $8.0 \times 10^7\ M^{-1}s^{-1}$ at pH 7.8 and 20°C. In contrast the yeast enzyme, so similar in structure, has almost no reactivity as indicated in steady-state experiments,[20] and horseradish peroxidase compound I has a maximum rate constant of $2.3 \times 10^5\ M^{-1}s^{-1}$ at pH 4.5.[21]

A comprehensive study was made of recombinant pea cytosolic ascorbate peroxidase and of ascorbate and cyanide binding to it.[22] By use of the pyridine hemochromogen test, the maximum molar absorptivity of the enzyme was determined to be $88 \pm 2\ mM^{-1}cm^{-1}$ at 403 nm. By comparison a value of $104\ mM^{-1}cm^{-1}$ at 419 nm was obtained for the cyanide complex. Also by comparison, maximum molar absorptivities of $92.5\ mM^{-1}cm^{-1}$ at 415 nm for compound II and $60\ mM^{-1}cm^{-1}$ at 405 nm for compound I can be calculated.[19] The latter value for compound I may be somewhat high, as judged by accurate spectra obtained for horseradish peroxidase compound I.

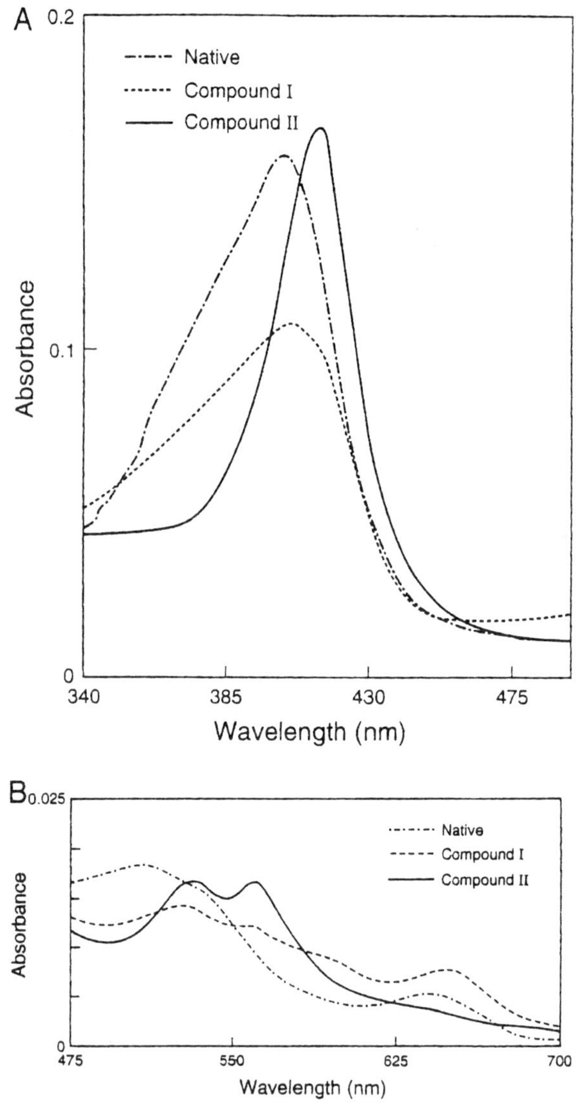

FIG. 11.4. Optical spectra of wild-type pea cytosolic ascorbate peroxidase and its compounds I and II obtained at pH 7.8 in phosphate buffer. Enzyme concentration 1 μM. Compound I was formed using 1 equiv hydrogen peroxide; compound II, 10 equiv. [Reproduced with permission from L. A. Marquez, M. Quitoriano, B. A. Zilinskas, and H. B. Dunford, *FEBS Lett.* **389,** 153–156 (1996) with permission from Elsevier Science.]

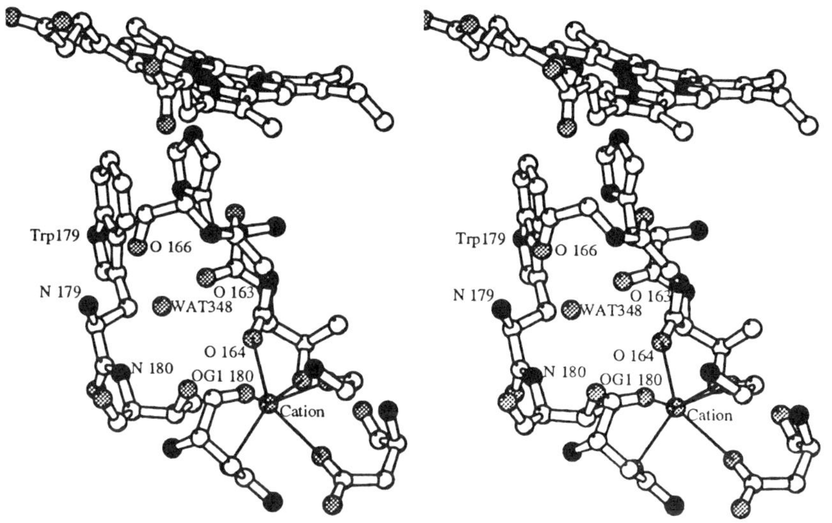

FIG. 11.5. Stereo representation of the ascorbate peroxidase cation binding site. Atom coloring as in Fig. 11.2, with the cation shown in gray. [Reproduced with permission from W. R. Patterson and T. L. Poulos, *Biochemistry* **34,** 4331–4341 (1995).]

From electrospray mass spectrometry a molecular weight of 27,194 ± 3 daltons was determined compared to 27,192.77 predicted by the DNA sequence.[22] A constant of 11.6 ± 0.4 μM was obtained for dissociation of ascorbate from the enzyme–cyanide complex. The results indicate 1 : 1 binding of ascorbate at a site different from the cyanide in the sixth coordination position of the iron. From the X-ray data, nuclear magnetic resonance (NMR) data and molecular modeling a suitable binding site for ascorbate were found near C20 of the porphyrin and close to alanine 134. Inactivation during turnover using phenylhydrazine indicated modification of C20, in agreement with the proposed ascorbate binding site (Fig. 11.6). It was noted that position 134 is occupied by proline in class III plant peroxidases, which could hinder ascorbate binding.

The reaction of salicylic acid (2-hydroxybenzoic acid) with thylakoid-bound chloroplastic ascorbate peroxidase isolated from tea leaves was studied by spectrophotometry, transient kinetics, and EPR spectroscopy.[23] The effect of ascorbic acid on the salicylic acid reaction was also studied. It was concluded that salicylic acid is a normal but slow reacting reducing substrate, and therefore not an effective inhibitor of the enzyme, as had been reported earlier.[24]

Ascorbate peroxidase occurs in the plant *Arabidopsis thalania.* Database analyses retrieved 52 expressed sequence tags coding for ascorbate peroxidase.[25] Analysis of sequence data for all known ascorbate peroxidases resulted in the following classification: five cytosolic ascorbate peroxidases, three of

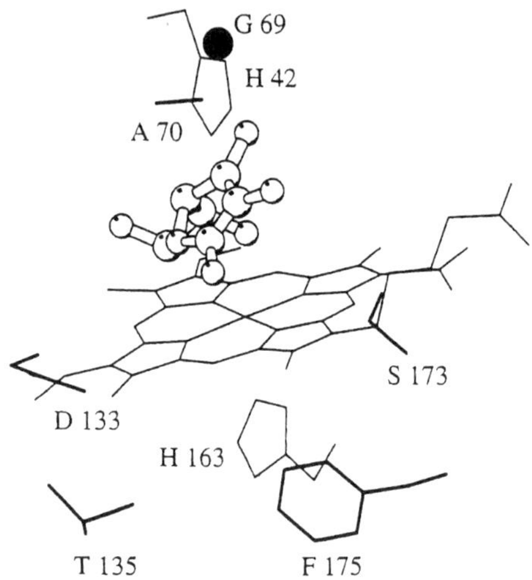

FIG. 11.6. Proposed ascorbate binding site on ascorbate peroxidase, based on molecular modeling and NMR and crystallographic constraints. [Reproduced with permission from A. P. Hill, S. Modi, M. J. Sutcliffe, D. D. Turner, D. J. Gilfoyle, A. T. Smith, B. M. Tam, and E. Lloyd, *Eur. J. Biochem.* **248,** 347–354 (1997).]

them membrane-bound, and two types of chloroplast ascorbate peroxidases. Key residues are conserved among the seven enzymes. A change from Phe175 to Trp175 may explain the higher ascorbate activity of the chloroplast enzymes. The proximal potassium ion binding site of the pea cytosolic enzyme appears to be preserved in *A. thaliana.*

The stimulating influence of ethylene on production of ascorbate peroxidase in soybean callus was examined.[26] The production facilitates elimination of excess hydrogen peroxide.

11.2. BACTERIAL CATALASE-PEROXIDASES

A new class of bacterial enzymes, the catalase-peroxidases, started to be identified beginning in 1979. These enzymes are different from the heme catalases, described in Chapter 18. The first detections were made in *Escherichia coli,* where two enzymes labeled hydroperoxidase I and hydroperoxidase II (HP-I and HP-II) were isolated.[27,28] Hydroperoxidase I, a constitutive enzyme, was found to be a tetramer of 337,000 molecular weight with only two hemes per tetramer. Hydroperoxidase II was not present in anaerobically grown *E. coli,* but was synthesized when electron transfer to oxygen was occurring. The gene for hydroperoxidase I, called *katG,* was identified and mutations caused by nitrosoguanidine were studied.[29] The nucleotide sequence of *katG* was obtained, which predicted a 726-amino-acid protein.[30] The amino acid sequence is reproduced in Appendix A. In the earlier literature sometimes the term hydroperoxidase is used, sometimes catalase, to describe the same enzyme.

The photosynthetic bacteria *Rhodopseudomonas capsulata* was shown to contain catalase-peroxidases similar to those of *E. coli* and their properties were compared.[31] The more cumbersome but more accurate term catalase-peroxidase was introduced[29] and used to describe this new group of enzymes.[31] The catalase-peroxidase genes from *Bacillus stearothermophilus*[32] and for hydroperoxidase II from *E. coli*[33] were also sequenced. These catalase-peroxidases were recognized as part of the class I family of plant peroxidases, those of prokaryotic origin.[34] They have little sequence homology with typical catalases, but high homology with yeast cytochrome *c* peroxidase and ascorbate peroxidase.

Whereas the peroxidases of plants, fungi, and yeast are monomeric and contain 290–350 amino acid residues, the subunit of bacterial catalase-peroxidases contain approximately 730 residues. The approximate doubling of molecular weight was shown to be caused by gene duplication.[34] It was shown that each half has high homology with yeast cytochrome *c* peroxidase and can be fit into the crystal structure of the yeast enzyme. The 10 helices of the yeast enzyme are also present in each half of the bacterial enzymes but are usually connected by longer surface loops. The C-terminal half of the subunit of the bacterial catalase-peroxidases do not bind heme. The catalase-peroxidases range in size from one[35] to four subunits. Some bacteria contain only one type of catalase[36]; *Klebsiella pneumoniae* contains three. One is a classic catalase, the second is a catalase-peroxidase, and the third a novel catalase with a chlorin-type heme.[37–39] The pH rate profiles for two of them are shown in Fig. 11.7. Wild-type *Rhodobacter capsulatus* contains a peroxidase and a catalase-peroxidase, but a mutant contains none of the latter enzyme.[40] *Bacillus subtilis* also contains three catalases.[41]

The catalase-peroxidases typically show a sharp maximum in their activity at pH 6.0–6.5, are not inhibited by 3-amino-1,2,4-triazole, and are inactivated by an ethanol–chloroform mixture, and their activity is half-destroyed by dialysis against 2 mM hydrogen peroxide in 20–30 min.[31] In contrast classic bacterial catalases have broad pH activity, from 5.5 to 10.0, are inactivated by aminotriazole, are unaffected by the same ethanol–chloroform mixture, and are not affected by the dialysis against hydrogen peroxide for periods up to one hour. Catalase-peroxidases are not activated by a temperature increase, classic catalases are.

Catalase-peroxidases may be important in antibiotic treatment of tuberculosis. Certain organisms that have reduced catalase-peroxidase activity are resistant to treatment by isoniazid.[42] Characterization was conducted in the isoniazid–active catalase-peroxidase from *Mycobacterium smegmatis*.[43] Manganese(II) was shown to bind to the catalase-peroxidase of *M. smegmatis*. Isoniazid was shown to reduce the resultant manganese(III), a reaction that

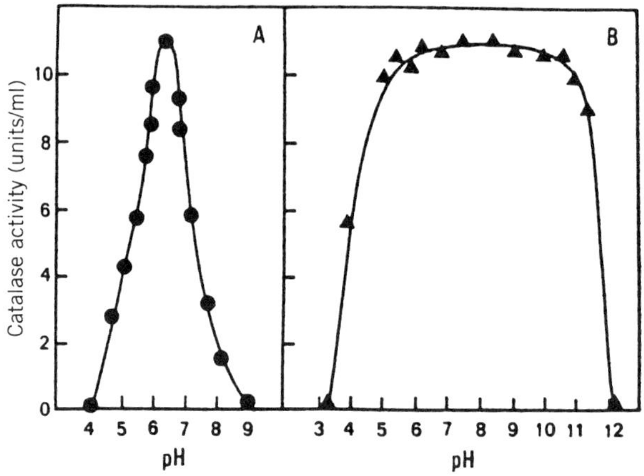

FIG. 11.7. Different pH rate profiles for two catalases from *K. pneumonia:* (A) the catalase-peroxidase and (B) the classic catalase. [Reproduced with permission from I. Goldberg and A. Hochman, *Arch. Biochem. Biophys.* **268,** 124–128 (1989).]

might convert the drug into an antibacterial agent.[44] Wild-type catalase-peroxidase from *Mycobacterium tuberculosis* as well as a specific mutant frequently found in isoniazid-resistant strains, were overexpressed in *E. coli.* The resultant yield permitted characterization, showing that they possess a proximal histidine. Both enzymes have high catalase activity and broad specifivity as peroxidases.[45,46]

REFERENCES

1. D. Groden and E. Beck, *Biochim. Biophys. Acta* **546,** 426–435 (1979).
2. G. J. Kelly and E. Latzko, *Naturwissenschaften* **66,** 617–618 (1979).
3. K.-P. Gerbling, G. J. Kelly, K.-H. Fischer, and E. Latzko, *J. Plant Physiol.* **115,** 59–67 (1984).
4. S. Shigeoka, Y. Nakano, and S. Kitaoka, *Arch. Biochem. Biophys.* **201,** 121–127 (1980).
5. A. Puppo, J. Rigaud, D. Job, J. Ricard, and B. Zeba, *Biochim. Biophys. Acta* **614,** 303–312 (1980).
6. D. A. Dalton, S. A. Russell, F. J. Hanus, G. A. Pascoe, and H. J. Evans, *Proc. Natl. Acad. Sci. U.S.A.* **83,** 3811–3815 (1986).
7. D. A. Dalton, F. J. Hanus, S. A. Russell, and H. J. Evans, *Plant Physiol.* **83,** 789–794 (1987).
8. E. Tel-Or, M. E. Huflejt, and L. Packer, *Arch. Biochem. Biophys.* **246,** 396–402 (1986).
9. Y. Nakano and K. Asada, *Plant Cell Physiol.* **22,** 867–880 (1981).

10. Y. Nakano and K. Asada, *Plant Cell Physiol.* **28,** 131–140 (1987).
11. G. X. Chen and K. Asada, *Plant Cell Physiol.* **30,** 987–998 (1989).
12. G. X. Chen, S. Sano, and K. Asada, *Plant Cell Physiol.* **33,** 109–116 (1992).
13. R. Mittler and B. A. Zilinskas, *Plant Physiol.* **97,** 962–968 (1991).
14. R. Mittler and B. A. Zilinskas, *FEBS Lett.* **289,** 257–259 (1991).
15. R. Mittler and B. A. Zilinskas, *J. Biol. Chem.* **267,** 21802–21807 (1992).
16. W. R. Patterson and T. L. Poulos, *Biochemistry* **34,** 4331–4341 (1995).
17. K. G. Welinder, *Curr. Opin. Struct. Biol.* **2,** 388–393 (1992).
18. W. R. Patterson, T. L. Poulos, and D. B. Goodin, *Biochemistry* **34,** 4342–4345 (1995).
19. L. A. Marquez, M. Quitoriano, B. A. Zilinskas, and H. B. Dunford, *FEBS Lett.* **389,** 153–156 (1996).
20. T. Yonetani and G. S. Ray, *J. Biol. Chem.* **240,** 4503–4508 (1965).
21. H. Zhang, M.Sc. Thesis, University of Alberta (1993).
22. A. P. Hill, S. Modi, M. J. Sutcliffe, D. D. Turner, D. J. Gilfoyle, A. T. Smith, B. M. Tam, and E. Lloyd, *Eur. J. Biochem.* **248,** 347–354 (1997).
23. M. Kvaratskhelia, S. J. George, and R. N. F. Thorneley, *J. Biol. Chem.* **272,** 20998–21001 (1997).
24. J. Durner and D. F. Klessig, *Proc. Natl. Acad. Sci. U.S.A.* **92,** 11312–11316 (1995).
25. H. M. Jespersen, I. V. H. Kjersgård, L. Østergaard, and K. G. Welinder, *Biochem. J.* **326,** 305–310 (1997).
26. S.-E. Oh, H. J. Lee, M. Y. Ha, M. Y. Lee, D. K. Kim, and S. Y. Kim in *Plant Peroxidases. Biochemistry and Physiology,* C. Obinger, U. Burner, R. Ebermann, C. Penel, and H. Greppin, eds., University of Geneva Press, Geneva, 1996, pp. 275–280.
27. A. Claiborne and I. Fridovich, *J. Biol. Chem.* **254,** 4245–4252 (1979).
28. A. Claiborne, D. P. Malinowski, and I. Fridovich, *J. Biol. Chem.* **254,** 11664–11668 (1979).
29. P. C. Loewen, B. L. Triggs, C. S. George, and B. E. Hrabarchuk, *J. Bacteriol.* **162,** 661–667 (1985).
30. B. L. Triggs-Raine, B. W. Doble, M. R. Mulvey, P. A. Sorby, and P. C. Loewen, *J. Bacteriol.* **170,** 4415–4419 (1988).
31. V. Nadler, I. Goldberg, and A. Hochman, *Biochim. Biophys. Acta* **882,** 234–241 (1986).
32. S. Loprasert, S. Negoro, and H. Okada, *J. Bacteriol.* **171,** 4871–4875 (1989).
33. I. von Ossowski, M. R. Mulvey, P. A. Leco, A. Borys, and P. C. Loewen, *J. Bacteriol.* **173,** 514–520 (1991).
34. K. G. Welinder, *Biochim. Biophys. Acta* **1080,** 215–220 (1991).
35. Y. Fukumori, T. Fujiwara, Y. Okada-Takahashi, Y. Mukohata, and T. Yamanaka, *J. Biochem.* **98,** 1055–1061 (1985).
36. J. J. Abrams and D. A. Webster, *Arch. Biochem. Biophys.* **279,** 54–59 (1990).
37. I. Goldberg and A. Hochman, *Arch. Biochem. Biophys.* **268,** 124–128 (1989).
38. I. Goldberg and A. Hochman, *Biochim. Biophys. Acta* **991,** 330–336 (1989).

39. A. Hochman and I. Goldberg, *Biochim. Biophys. Acta* **1077,** 299–307 (1991).
40. A. Hochman, A. Figueredo, and J. D. Wall, *J. Bacteriol.* **174,** 3386–3391 (1992).
41. P. C. Loewen and J. Switala, *J. Bacteriol.* **169,** 3601–3607 (1987).
42. B. Heym, P. M. Alzari, N. Honore, and S. T. Cole, *Mol. Microbiol.* **15,** 235–245 (1995).
43. J. A. Marcinkeviciene, R. S. Magliozzo, and J. S. Blanchard, *J. Biol. Chem.* **270,** 22290–22295 (1995).
44. R. S. Magliozzo and J. A. Marcinkeviciene, *J. Biol. Chem.* **272,** 8867–8870 (1997).
45. K. Johnsson, W. A. Froland, and P. G. Schultz, *J. Biol. Chem.* **272,** 2834–2840 (1997).
46. J. M. Nagy, A. E. G. Cass, and K. A. Brown, *J. Biol. Chem.* **272,** 31265–31271 (1997).

CHAPTER 12

Class II Peroxidases: Lignin, Manganese, and *Coprinus cinereus* Peroxidases

In this chapter the fungal peroxidases belonging to class II of the plant peroxidase superfamily are discussed in some detail, using three examples: lignin and manganese peroxidases, each named after one of its substrates, and *Coprinus cinereus* peroxidase, named after its source. A brief introduction was given in Chapter 3.

12.1. LIGNIN PEROXIDASE

12.1.1. Discovery and Properties

Lignin is second only to cellulose in abundance, and so plays a very important role in the carbon cycle. Its degradation is accomplished aerobically by a narrow array of microbes, which have their own hydrogen peroxide generating systems.[1] Lignin peroxidase from the white-rot Basidiomycete *Phanerochaete chrysosporium* was discovered independently by two groups in 1984, so that it is a newer addition to the plant peroxidase superfamily than ascorbate peroxidase and the catalase-peroxidases (Chapter 11). Lignin peroxidase was originally called a hydrogen-peroxide-dependent oxidase or a diarylpropane oxygenase by the group of Michael Gold[2,3] and ligninase by Kent Kirk and co-workers.[4] Oxygen was incorporated into some of the products of oxidation, but this is typical of many peroxidase reactions, in which a carbon-centered free radical is generated and is then scavenged by dioxygen in a nonenzymatic process. The name became lignin peroxidase when the similarities of the oxidized enzymatic compounds to those of horseradish peroxidase became apparent. In addition, Gold's group discovered manganese peroxidase (Section 12.2).[2]

Three isoenzyme forms, labeled I–III, were isolated, and shown to be glycoproteins.[5] The dominant form was isoenzyme II, which was readily isolated by chromatography. It is now known that there are at least eight isoenzymes of lignin peroxidase, and the amino acid sequences of two of them are reproduced in Appendix A.

Oxidized forms of the enzyme were characterized by their optical spectra.[6,7] The maximum molar absorptivity of the native enzyme at 407 nm was estimated by measuring the protein concentration of a solution of known absorbance, with the result a molar absorptivity of $1.33 \times 10^5\ M^{-1}cm^{-1}$ at 407 nm. The pyridine hemochrome absorption spectrum, reported in the same paper, established that there was one heme per protein molecule and that the heme was ferriprotoporphyrin IX.[3] The reinheitzahl (RZ) value (A_{407nm}/A_{280nm}) is 4.97. Optical spectral data are shown in Figs. 12.1–12.3 and Table 12.1.[8,9] At pH 6 the interconversions are slow enough that a standard spectrophotometer can be used.

Resonance Raman spectral results indicate that the iron(III) of lignin peroxidase is a high-spin five-coordinate species at room temperature and that its compound II has a low-spin hexacoordinate $Fe^{IV} = O$ structure.[10,11] The porphyrin π-cation radical was detected by electron paramagnetic resonance (EPR) spectroscopy, showing that lignin peroxidase has a conventional compound I.[12] The heme of lignin peroxidase is exceptionally sensitive to degradation by hydrogen peroxide, which hampered studies aimed at elucidating whether the heme edge was the site of electron-transfer reactions.[13]

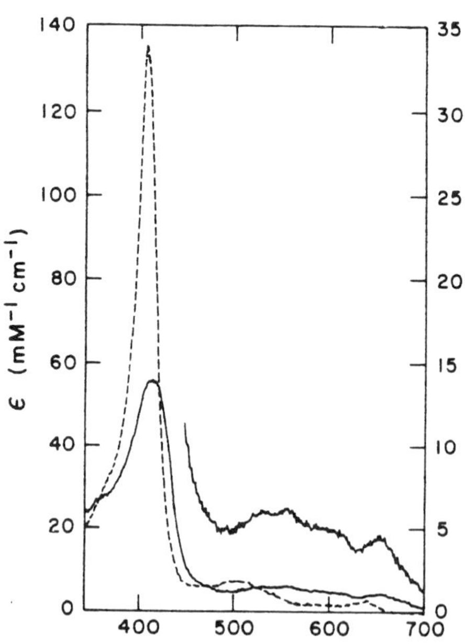

FIG. 12.1. Optical absorption spectra of native lignin peroxidase (dashed line) and compound I (solid lines) at pH 6. Compound I was obtained by mixing equimolar amounts of enzyme and hydrogen peroxide. [Reproduced with permission from V. Renganathan and M. H. Gold, *Biochemistry* **25,** 1626–1631 (1986).]

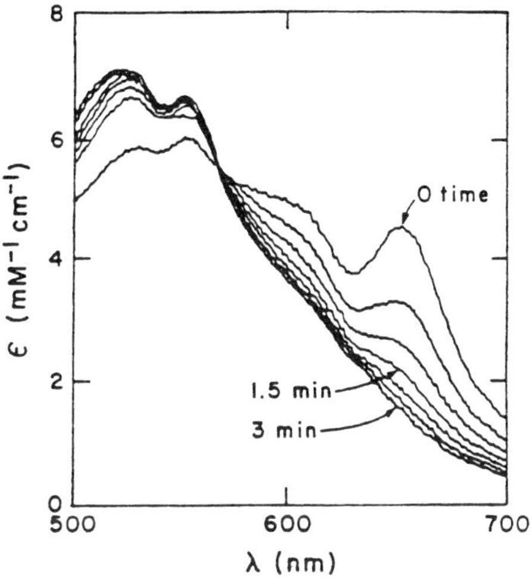

FIG. 12.2. Slow conversion of compound I of lignin peroxidase (9 μM) to compound II at pH 6. The isosbestic point occurs at 565 nm. [Reproduced with permission from V. Renganathan and M. H. Gold, *Biochemistry* **25,** 1626–1631 (1986).]

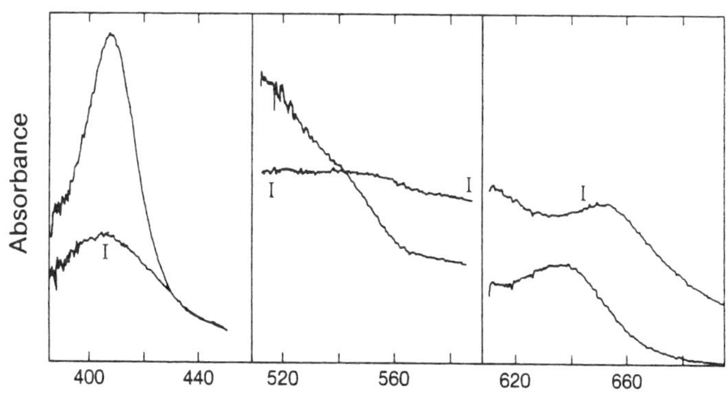

FIG. 12.3. Spectra obtained from the conversion of native lignin peroxidase to compound I in 0.2 s at pH 3.1. Concentrations after mixing: for the Soret region spectra, enzyme 1.5 μM, hydrogen peroxide 50 μM; for the visible spectra, 10 μM enzyme and 0.3 mM hydrogen peroxide. [Reproduced with permission from L. Marquez, H. Wariishi, H. B. Dunford, and M. H. Gold, *J. Biol. Chem.* **263,** 10549–10552 (1988).]

TABLE 12.1. Maximum Molar Absorptivities of Bands of Lignin Peroxidase and Its Compounds Obtained at pH 6.0 in Phosphate Buffer

Enzyme Species	Wavelength, nm		(ε_{max}, $mM^{-1}cm^{-1}$)		
Native	408 (133)	496 (8.1)			630 (2.7)
Ferrous	436 (93)		552 (11.0)		
Compound I	408 (55)		550	608	650 (4.5)
Compound II	420 (71)	525 (6.8)	556 (7.5)		
Compound III	419 (104)	543 (7.7)	578 (10.2)		

Reproduced with permission from V. Renganathan and M. H. Gold, *Biochemistry* **25**, 1626–1631 (1986).

There is substantial agreement in the literature over the rate of compound I formation from lignin peroxidase and hydrogen peroxide.[6,9,14] The rate constant, which is constant from pH 2 to 7.5, is $5.4 \times 10^5\ M^{-1}s^{-1}$ (Fig. 12.4). The activation energy, determined at pH 3.5, is 5.9 kcal/mol.[7] Nitrate ion had no effect at low pH,[7] which is in marked contrast to the behavior of horseradish peroxidase.

Compound II forms spontaneously from compound I, a process that is accelerated by excess hydrogen peroxide or *p*-cresol.[9] Compound III is readily formed from compound II by reaction with hydrogen peroxide.[15,16] Further reaction with hydrogen peroxide leads to a new compound, designated compound III*, and eventually to enzyme denaturation.

Lignin-degrading peroxidases have been found in other white-rot fungi. For example, both lignin and manganese peroxidases have been found and

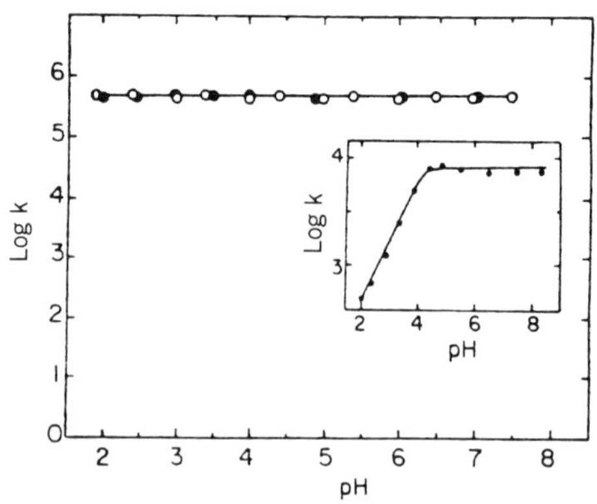

FIG. 12.4. pH independence of lignin peroxidase compound I formation using hydrogen peroxide. Open circles 0.001 M nitrate, closed circles 0.2 M nitrate. Inset shows the pH rate profile for compound I formation of horseradish peroxidase using *t*-butyl hydroperoxide for comparison. [Reproduced with permission from A. Andrawis, K. A. Johnson, and M. Tien, *J. Biol. Chem.* **263,** 1195–1198 (1988).]

partially characterized in the white-rot Basidiomycete *Trametes versicolor.*[17,18] However, studies have not been nearly as comprehensive as on the *P. chrysosporium* enzymes.

12.1.2. Veratryl Alcohol

Veratryl alcohol is a secondary metabolite of *P. chrysosporium,* found in cultures four days old or older.[19] It is a preferred substrate for lignin peroxidase. The principal, but not the only product, of veratryl alcohol oxidation is veratryl aldehyde (Fig. 12.5). In order to explain how the products were formed, it was correctly assumed that there was an initial one-electron oxidation to the veratryl alcohol cation radical.[20] Veratryl alcohol was shown to reduce compound I to compound II, and compound II to native enzyme, confirming one-electron redox reactions.[9] The rate of reaction of compound I with veratryl alcohol is linearly dependent on veratryl alcohol concentration, whereas compound II displays saturation kinetics (Fig. 12.6).[21] The rate constants for both compounds I and II reacting with veratryl alcohol display unusual pH dependencies, reacting much more rapidly at lower pH, with discontinuities at pH 3.5 and 4. The pH rate profile for compound I reacting with veratryl alcohol is shown in Fig. 12.7. Compound III also reacts readily

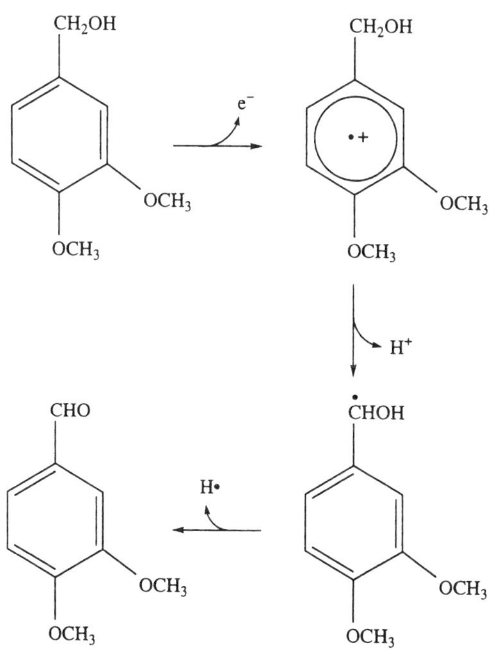

FIG. 12.5. Pathway for the conversion of veratryl alcohol to veratraldehyde, via its cation radical and a substituted benzylic radical.

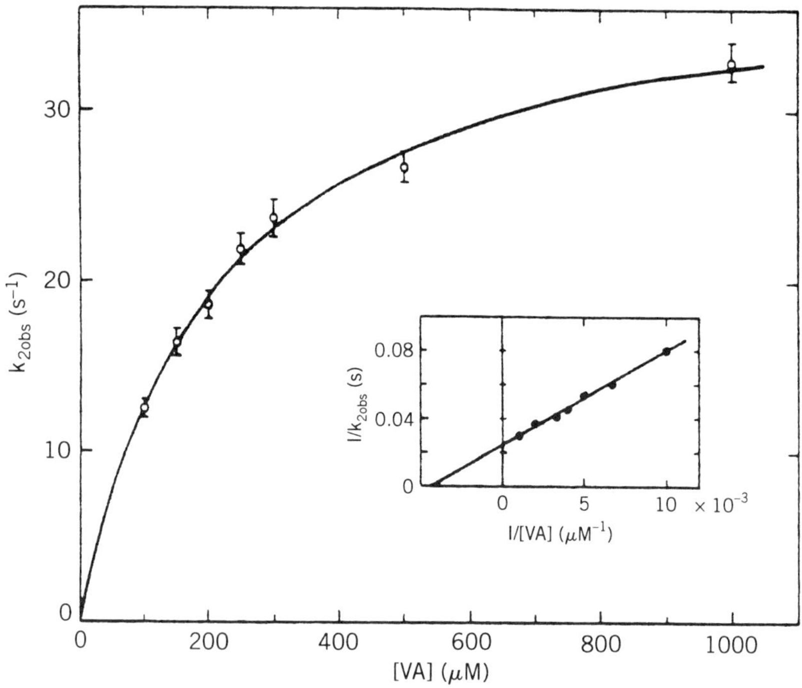

FIG. 12.6. Saturation kinetics displayed in the reaction of lignin peroxidase compound II with veratryl alcohol at pH 3.24. [Reproduced with permission from H. Wariishi, J. Huang, H. B. Dunford, and M. H. Gold, *J. Biol. Chem.* **266,** 20694–20699 (1991).]

with veratryl alcohol, suggesting that the alcohol plays a role in preventing enzyme denaturation. The reaction reduces compound III directly to native enzyme.[15]

One molecule of veratryl aldehyde product was formed per molecule of hydrogen peroxide reacted with lignin peroxidase,[4] and a half molar equivalent of veratryl alcohol was reported to reduce compound I to compound II, and also to reduce compound II to native enzyme.[8] How can these stoichiometric results be accommodated mechanistically?

An explanation is shown in Fig. 12.8 for the 2 : 1 stoichiometry in which two molecules of compound I are reduced to two molecules of compound II by one molecule of veratryl alcohol. The mechanism is based upon free release of veratryl alcohol cation radicals. A cation radical upon loss of its acid proton becomes a substituted benzyl radical. Disproportionation of a cation radical and its conjugate base radical yields a veratryl aldehyde molecule, and regenerates a veratryl alcohol molecule. Such a disproportionation would appear more likely than that between two positively charged cation radicals.

The two-electron reduction of compound I by excess veratryl alcohol could be explained by two one-electron reactions, the first of which produces the

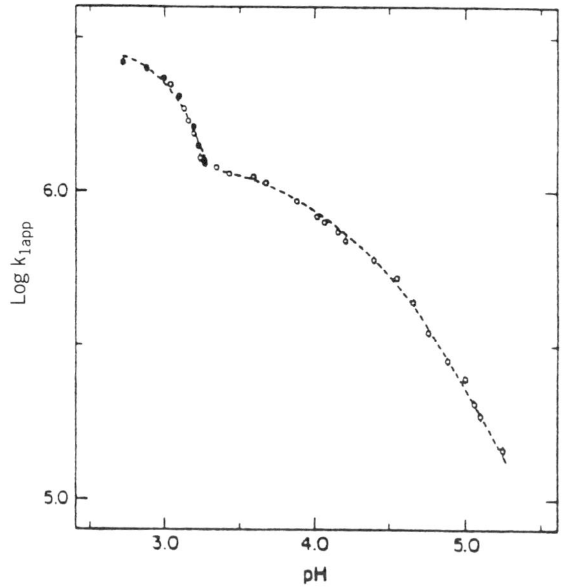

FIG. 12.7. pH rate profile for the reaction of lignin peroxidase compound I with veratryl alcohol. The log of the second-order rate constant is plotted versus pH. [Reproduced with permission from H. Wariishi, J. Huang, H. B. Dunford, and M. H. Gold, *J. Biol. Chem.* **266,** 20694–20699 (1991).]

cation radical and compound II. The cation radical or its conjugate base remains bound to compound II where it would be oxidized to veratrylaldehyde, and the native enzyme would be formed.[6] However, a more plausible mechanism, in my opinion, is again based upon free release of veratryl alcohol cation radicals from compounds I and II, as shown:

Compound I + Veratryl alcohol + H^+ $\longrightarrow$ Compound II + Cation radical

Compound II + Veratryl alcohol + H^+ $\longrightarrow$ Native LiP + Cation radical + H_2O

Cation radical $\longrightarrow$ Benzyl radical + H^+

Cation radical + Benzyl radical $\longrightarrow$ Veratryl alcohol + Veratryl aldehyde + H^+

Sum: Compound I + Veratryl alcohol = Native LiP + Veratryl aldehyde + H_2O

In the above scheme LiP is lignin peroxidase. The oxygen in the released water molecule comes from the Fe=O group of compound I.

Detection and Properties of the Veratryl Alcohol Cation Radical

The veratryl alcohol cation radical was detected spectrophotometrically in pulse radiolysis experiments.[22,23] A solution of Tl^+ ions was irradiated in the presence of nitrous oxide, which generated Tl^{2+} ions. These one-electron

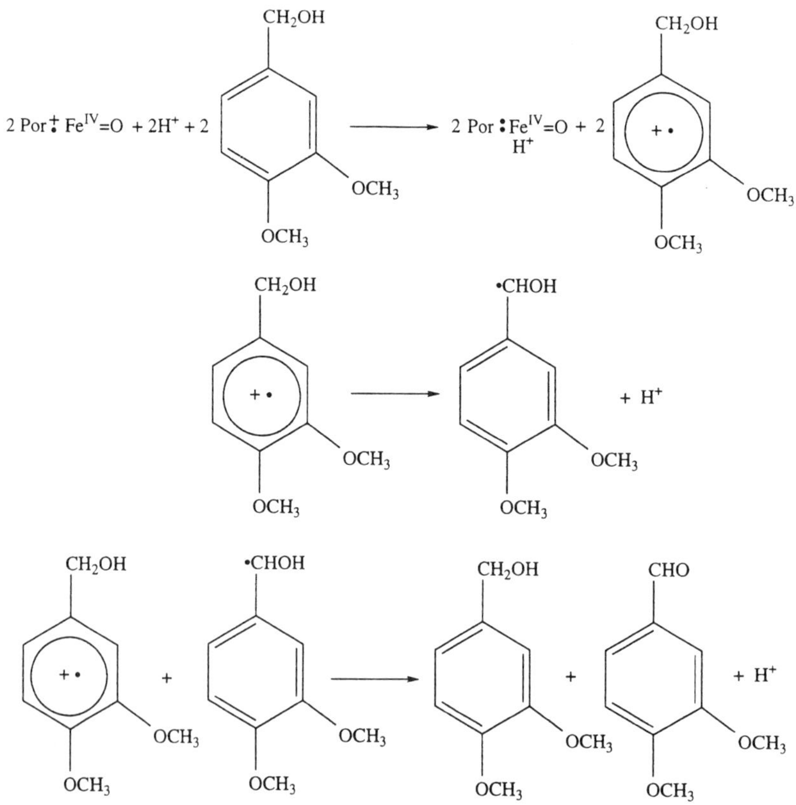

2 Compound I + 2 H^+ + 2 veratryl alcohols ⟶ 2 Compound II + 2 cation radicals

Cation radical ⟶ Benzyl radical + H^+

Cation radical + Benzyl radical ⟶ Veratryl alcohol + veratryl aldehyde + H^+

Sum: 2 Compound I + veratryl alcohol = 2 compound II + veratryl aldehyde

FIG. 12.8. Mechanistic explanation for the observed stoichiometry that two molecules of lignin peroxidase compound I can be reduced to two molecules of compound II by one molecule of veratryl alcohol, which produces one molecule of veratryl aldehyde. The representations show the porphyrin π-cation radical of compound I as $\text{Por}\bullet^{+}$ and the important proton in the distal region of compound II.

oxidants in turn oxidized veratryl alcohol to its cation radical. The spectrum has absorption peaks at 300 and 420 nm (Fig. 12.9), characteristic of benzyl cation radicals.[24,25] A first-order rate of decay for the absorbing species was obtained, $k = 17 \pm 2\ s^{-1}$, which defines a lifetime of 59 ± 8 ms, enough time for the cation radicals to diffuse 7 μm (700 Å). The decay involves a proton loss from a carbon atom to form an electrically neutral benzyl radical, as illustrated in Fig. 12.8. Carbon acids are typically strong acids, but they have a slow rate of dissociation,[26] a fact not always appreciated in the lignin peroxidase literature. The experiments described here were conducted anaerobically, but the same results were obtained in aerobic experiments where $SO_4\bullet^-$ was used as oxidant.[27] The cation radical lifetime was constant from pH 3 to 8. The 60-ms lifetime of the radicals precluded their detection by EPR.

In a series of three papers, it was claimed that the EPR spectra of the veratryl alcohol cation radical were obtained, which in the remainder of this paragraph I shall refer to as the cation radical. Properties of the cation radical were studied, both in the presence and absence of enzyme. First, the cation radical was generated in strong acid solution (10% nitric acid) using the one-electron oxidant cerium(IV) in a fast-flow system. Its absorption spectrum was measured in strong acid, and was entirely different from the spectrum in Fig. 12.9. The radical was found to be a strong acid (pK = −1.0), using the

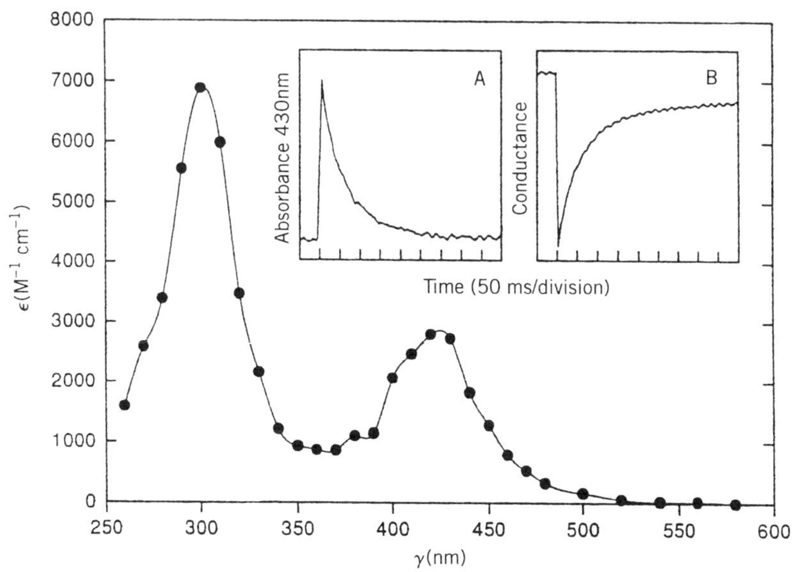

FIG. 12.9. Absorption spectrum of a species generated from veratryl alcohol 100 μs after pulses of radiation. Inset (A), decay of absorbance at 430 nm. Inset (B) time-dependent increase in conductance. [Reproduced with permission from L. P. Candeias and P. J. Harvey, *J. Biol. Chem.* **270,** 16745–16748 (1995).]

EPR g value, the spin concentration, and the visible absorption spectra as a function of pH. The strong acidity is a characteristic of cation radicals. The characterization of the radical properties facilitated its detection during turnover of lignin peroxidase.[28] Its decay rate was claimed to be fast ($k = 1.2 \times 10^3\ s^{-1}$). However, when generated enzymatically, its rate of decay is much slower ($k = 1.85\ s^{-1}$), indicating it is enzyme-bound. The room temperature EPR spectrum of the enzyme-generated cation radical was slightly broadened, in accordance with it being bound.[29] In low-temperature EPR results, the compound II–bound radical was detected. It was shown that a fraction of the bound radicals were EPR silent because they were coupled to the ferryl ion of compound II.[30] The neutral benzylic radical in bulk aerated aqueous solution would be rapidly scavenged by molecular oxygen, with the resultant peroxyl radical decaying to hydroperoxyl radical and veratryl aldehyde.[30] However, the electrically neutral benzylic radical must first be generated, and the rate of proton loss from the cation radical carbon acid is slow.

There are several problems with the EPR results described in the previous paragraph. First, the claimed lifetime of about 10^{-3} s is far too short for the radical ever to have been detected by EPR. The diffusion-controlled reaction of the cation radical with its conjugate base radical (Fig. 12.8) has been ignored. The pK_a value of -1.0 for the radical was obtained from a plot in which the changes occur over 0.3 pH units, whereas one would expect a change from 90 to 10% of the acid over two pH units! The method of measuring pH was not described. Finally, the hyperfine coupling constants appear to be unusual for a substituted benzyl radical,[25] and the hyperfine couplings at the 2 and 5 positions are in disagreement with the molecular orbital calculations.[31] The evidence, from sources other than EPR, is overwhelming for the existence of the veratryl alcohol cation radical. Clarification of the EPR results is required before the final story can be told.

Lignin-Peroxidase-Catalyzed Reactions of Veratryl Alcohol

Veratryl alcohol stimulated the oxidation of 4-methoxymandelic acid and anisyl alcohol by lignin peroxidase,[32] a result that is in contrast to experiments conducted in bulk solution.[22] The oxidations of anisyl alcohol,[33] and 4-methoxymandelic acid,[34] by lignin peroxidase in the presence of veratryl alcohol have been examined in detail. In the case of anisyl alcohol, it reacts with compound I but not with compound II. Veratryl alcohol reduces compound II to the native enzyme so the enzymatic cycle can be repeated. The substituted mandelic acid, on the other hand, is a slow reductant of compound I, but reacts rapidly in the presence of veratryl alcohol, which is an effective stimulant. It would appear that a veratryl alcohol radical is able to react with mandelic acid. Both anisyl alcohol and 4-methoxymandelic acid react to form anisaldehyde as principal products (Fig. 12.10). The two substrates, anisyl alcohol and 4-methoxymandelic acid, react on the enzyme by different mechanisms, which is quite typical of peroxidase behavior.

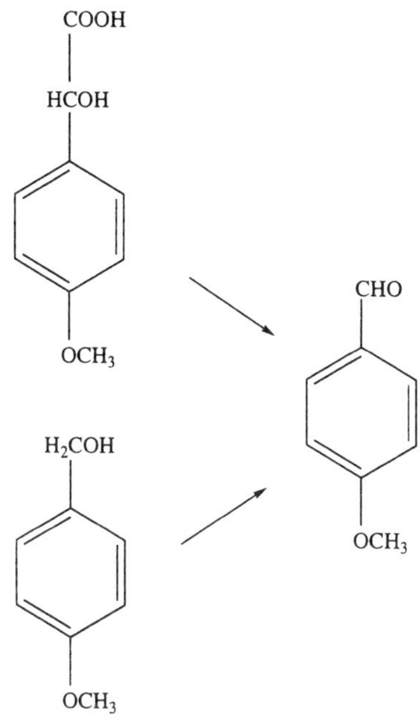

FIG. 12.10. 4-Methoxymandelic acid (top) and anisyl alcohol (bottom), both of which are oxidized to anisyl aldehyde (right) by lignin peroxidase.

In summary of this lengthy discussion of the role of veratryl alcohol, in the presence of recalcitrant substrates veratryl alcohol has been demonstrated to help catalysis by reducing lignin peroxidase compound II to native enzyme for recycling. This reduction helps prevent compound III formation and subsequent enzyme denaturation. If compound III is formed, veratryl alcohol can reduce it back to native enzyme. The compound II–veratryl alcohol cation radical complex can also oxidize a substrate that can approach closely enough. A direct role of veratryl alcohol in lignin degradation via its cation radical also appears to be likely, despite claims to the contrary. I find no reason to doubt that the veratryl alcohol cation radical in bulk solution has a lifetime of 60 ms, that it is immune to scavenging by molecular oxygen, and that the rate-controlling step in its decay is a pH-independent proton loss. Further work is in progress to clarify conflicting literature reports on the properties of the veratryl alcohol cation radical.

12.1.3. Reactions of Lignin Peroxidase with Lignin Models

Lignin precursors are shown in Fig. 12.11. Coumaryl alcohol, coniferyl alcohol, and sinapyl alcohol are related structures that are the building blocks of lignin, a phenyl propanoid polymer.[1] The ratio of the types of monomer units used in constructing lignins vary with the type of plant species.[35]

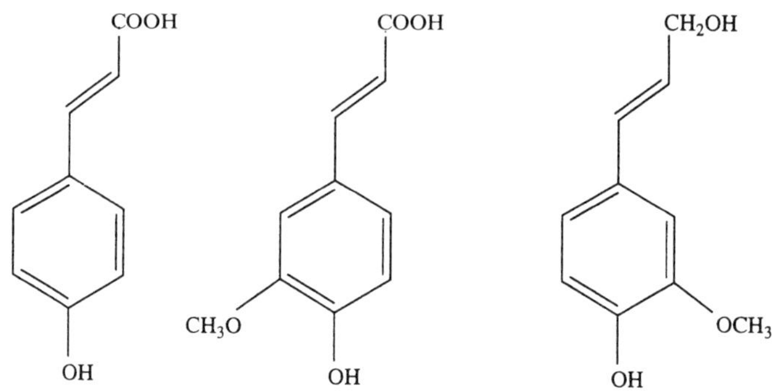

FIG. 12.11. Potential lignin precursors. From left to right: caffeic acid, ferulic acid, and coniferyl alcohol.

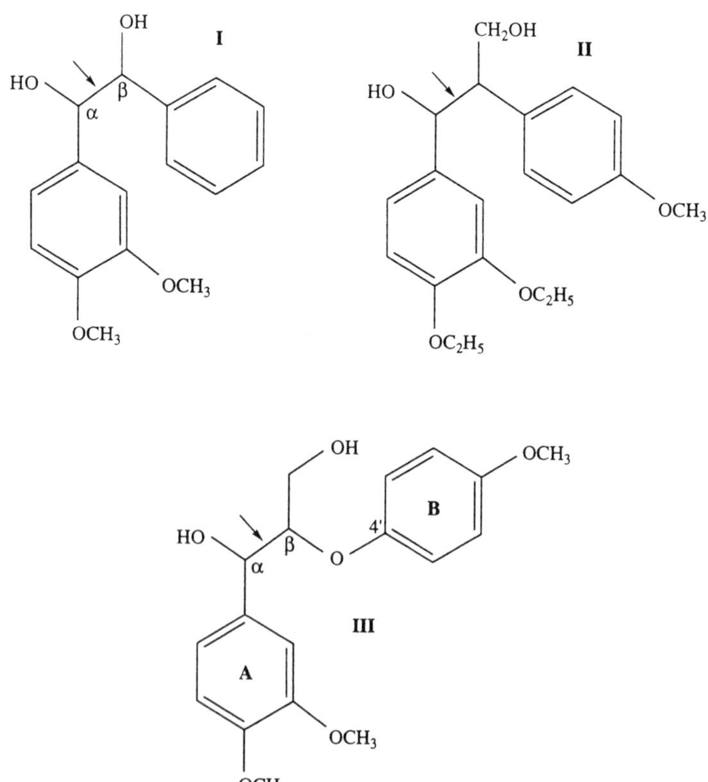

FIG. 12.12. Models of dimers of the lignin monomeric unit. Cleavage by lignin peroxidase usually occurs between C_α and C_β. The most accurate model is III, with the β-aryl ether linkage.

Lignin model dimers (Fig. 12.12, I and II) were shown to be cleaved between C_α and C_β, midway between the two aromatic rings, by lignin peroxidase.[36,37] A model more typical of lignin contains an arylglycerol β-aryl ether linkage (Fig.12.12, III), and again cleavage occurs between C_α and C_β.[38] The mechanism of cleavage was thought to involve inital oxidation of the substrate to a cation radical on ring A but on other models on ring B (Fig.12.12, III).[39] Aryl ether rearrangements were observed under anaerobic conditions[40] and ring cleavage in the presence of oxygen.[41] Cation radicals, formed by oxidation of methoxybenzenes by the enzyme, were detected by EPR (Fig.12.13).[42]

The role of methoxy substituents on the properties of lignin precursor radicals was examined. It was found that methoxy substituents increased the unpaired electron density on phenolic oxygens of precursor phenylpropanoid radicals, and this affects the nature of bond formation during polymerization.[43]

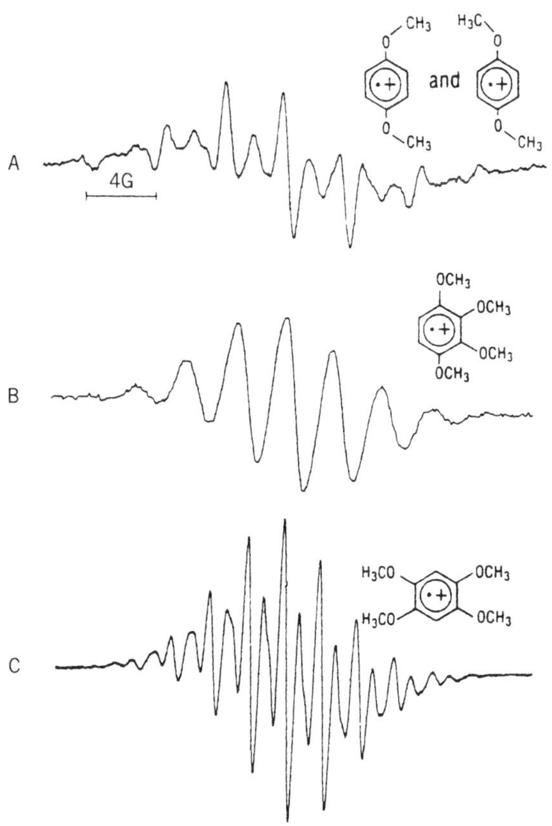

FIG. 12.13. EPR spectra of cation radicals obtained from reaction mixtures of lignin peroxidase, hydrogen peroxide, and methoxybenzenes. [Reproduced with permission from P. J. Kersten, M. Tien, B. Kalyanaraman, and T. K. Kirk, *J. Biol. Chem.* **260,** 2609–2612 (1985).]

Several other studies have been conducted on oxidation of model compounds by lignin peroxidase: 3,4-dimethoxybenzyl alcohol and its methyl ether,[44] dibenzo-*p*-dioxin,[45] guaiacol,[46] and chlorpromazine.[47] Unfortunately, these model compounds do not provide definitive answers for the mechanism of lignin degradation by lignin peroxidase.

12.1.4. Other Reactions

Lignin peroxidase was shown to oxidize bromide and iodide but not chloride.[48] Thus, its compound I is a stronger oxidizing agent than compound I of horseradish peroxidase, but not as strong as that of chloro- or myeloperoxidase.

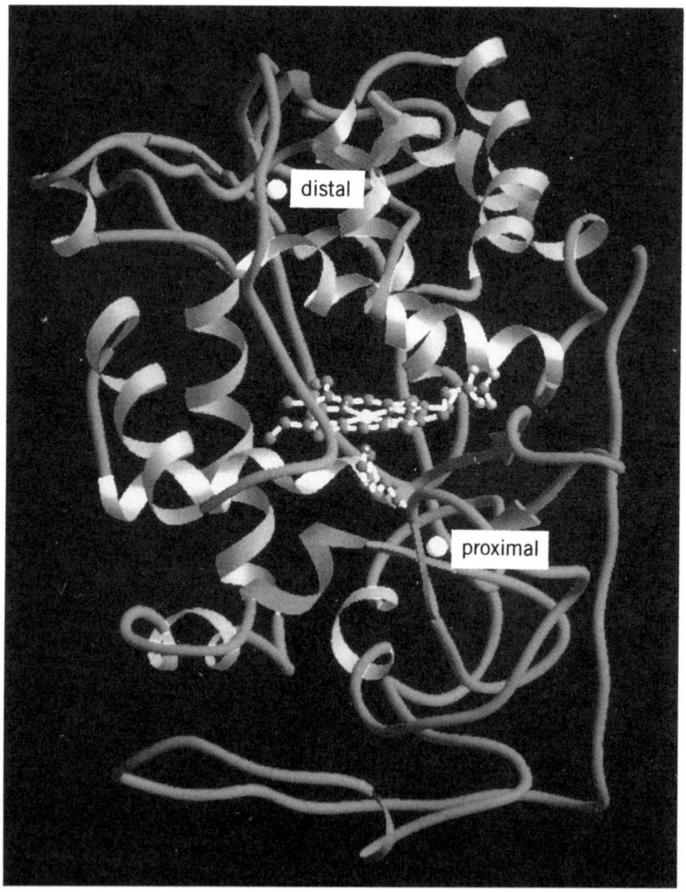

FIG. 12.14. Diagram of the lignin peroxidase structure showing the four disulfide bonds, the heme, proximal histidine ligand, and N and C terminal ends. See color insert. [Reproduced with permission from T. L. Poulos, S. L. Edwards, H. Wariishi, and M. H. Gold, *J. Biol. Chem.* **268,** 4429–4440 (1993).]

Lignin peroxidase and hydrogen peroxide only partially activate azide ion, causing subsequent heme destruction. The remainder of the enzyme formed an iron–azide complex.[49] Direct oxidation of manganese(II) by lignin peroxidase was demonstrated,[50] and it was shown that the reaction was not mediated by veratryl alcohol.[51]

12.1.5. Crystal Structure

The crystal structure of lignin peroxidase was obtained at 2.6 Å resolution[52] and refined to 2.0 Å.[53] The protein fold is similar to that of yeast cytochrome *c* peroxidase, with a proximal histidine residue hydrogen bonded to a buried aspartate, and with distal histidine and arginine residues. Lignin peroxidase belongs to class II, whereas the yeast enzyme belongs to class I in the plant peroxidase superfamily, so that beyond the similar protein fold and active site machinery, there are major differences. The tryptophans near the heme of the yeast enzyme are replaced by phenylalanines in lignin peroxidase. The approach to the distal side of the heme is much more constricted in lignin peroxidase. It contains two calcium ions compared to none in cytochrome *c* peroxidase. There are an extra 49 residues in lignin peroxidase at the C-terminal end (Fig. 12.14) (see p. 294 and color insert).

12.2. MANGANESE PEROXIDASE

12.2.1. Discovery and Properties

Manganese peroxidase from *P. chrysosporium* was discovered in 1984 by the group of Michael Gold.[2] Details of its purification were published.[54] The enzyme was found to be a monomeric glycoprotein with a molecular weight of 46,000, to contain one ferriprotoporphyrin IX and four calcium ions per protein molecule, to have an accessible active site to ligands such as cyanide and azide, and when reduced to bind carbon monoxide. Electronic absorption,[55] and EPR and resonance Raman spectral evidence,[56] indicated that the ferric heme iron of the native enzyme is high-spin pentacoordinate with histidine as the proximal fifth ligand. The enzyme exhibits typical native, compound I and compound II electronic spectra (Fig. 12.15).[57] The amino acid sequence of manganese peroxidase isoenzyme 1 is listed in Appendix A. Recombinant manganese peroxidase isoenzyme 4 has been expressed in *E. coli,* and the same difficulties in expressing horseradish peroxidase in *E. coli* were overcome in similar ways, by addition of heme, calcium ion, and oxidized glutathione. Highly active enzyme was isolated.[58,59] The rate of release of calcium ions during thermal inactivation has been measured.[60]

The rate constant for compound I formation using hydrogen peroxide is $(3.7 \pm 0.1) \times 10^6\ M^{-1}s^{-1}$, independent of the metal ion chelator present[61]; when *p*-nitroperbenzoic acid is used, the constant is $(5.9 \pm 0.1) \times 10^4\ M^{-1}s^{-1}$.

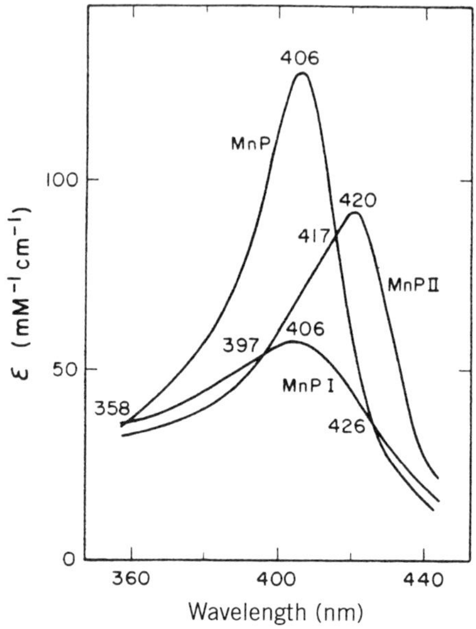

FIG. 12.15. Electronic absorption spectra of native manganese peroxidase (MnP) and its compound I (MnP I) and compound II (MnP II) obtained in 20 mM sodium lactate buffer, pH 4.5. Numbers indicate wavelengths of Soret peaks and isosbestic points. [Reproduced with permission from H. Wariishi, H. B. Dunford, I. D. MacDonald, and M. H. Gold, *J. Biol. Chem.* **264,** 3335–3340 (1989).]

The rate of compound I formation using hydrogen peroxide was found to be pH-independent over the range tested, from pH 3 to 8. Compound I reacted slowly with *p*-cresol: The rate constant is $(1.8 \pm 0.2) \times 10^3\ M^{-1}s^{-1}$.[57]

In the presence of hydrogen peroxide the enzyme oxidized manganese(II) to manganese(III) as detected in its pyrophosphate[54] or lactate complex.[55] Manganese ion was found to be a preferred substrate. In the presence of lactate at pH 4.5, Mn^{II} exhibits saturation kinetics in its reaction with compound I. Similar behavior was observed when lactate was replaced by citrate.[57] Saturation kinetics was also observed in the reaction of Mn^{II} with compound II. When complexed to lactate or other α-hydroxy acids, the manganese ion acted as a mediator for the oxidation of various dyes and lignin model compounds.[62–64] There can be thiol mediation of substrate oxidation in addition to that of manganese.[65]

Optimal enzyme activity involving Mn^{II} is a function of the metal chelator. Oxalate and malonate are the only chelators produced in significant amount by *P. chrysosporium.*[66] It was found that oxalate is present in 0.5 mM concentration in cultures in the organism, and at this concentration optimal enzyme activity is observed.[61,67,68] Higher concentrations of other chelators are required. The pH dependencies of the reactions of Mn^{II} with compound II in the presence of malonate and oxalate are shown in Fig. 12.16.

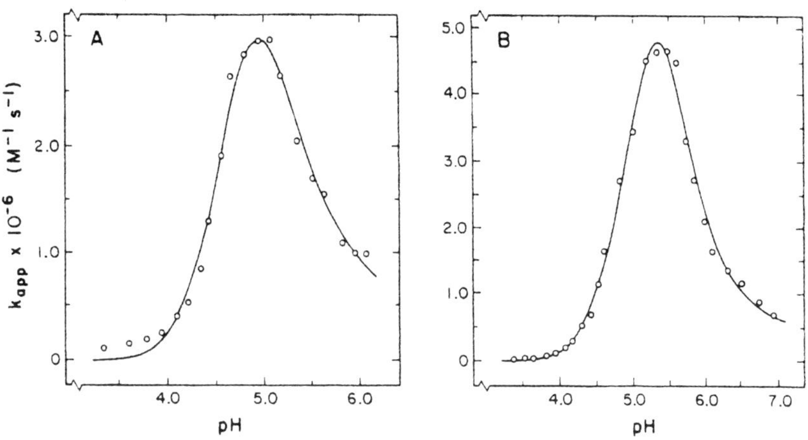

FIG. 12.16. pH dependence of the rate of reduction of manganese peroxidase compound II by Mn^{II}. The only differences in the curves are caused by the different pK_a values of the chelators: A, 2 mM oxalate and B, 20 mM malonate. [Reproduced with permission from K. Kishi, H. Wariishi, L. Marquez, H. B. Dunford, and M. H. Gold, *Biochemistry* **33,** 8694–8701 (1994).]

12.2.2. Characterization of the Active Site of Manganese Peroxidase

Reagents such as sodium azide in the presence of hydrogen peroxide, and substituted hydrazines, form adducts at the δ-meso edge of the heme of manganese peroxidase, indicating this may be the site of electron transfer for compound I reactions.[69] In this regard manganese peroxidase behaves in similar fashion to horseradish peroxidase and is shown to have a more open active site than lignin peroxidase. Site-directed mutants of manganese peroxidase isoenzyme 1 were created by expression of the modified gene in *P. chrysosporium.*[70] The purified mutant Asp179Asn displayed 1/200 of the rate of reduction of compound II by Mn^{II} compared to the wild-type compound II but did not affect the rate of reaction of *p*-cresol with compound II. The results indicated that Asp179 is a binding ligand of Mn^{II}. Further mutations Glu35Gln, Glu39Gln, and the double Glu35Gln-Asp179Asn also affected manganese binding, so that Glu35 and Glu39 are also probable ligands.[71]

12.2.3. Crystal Structure

The crystal structure of manganese peroxidase was obtained at 2.06 Å resolution.[72] The structures of the three fungal peroxidases discussed in this chapter are all very similar. Manganese peroxidase differs in having five, rather than four, disulfide bonds. Asp179, Glu35, Glu39, a heme propionate, and two water molecules were identified as the Mn^{II} binding ligands. An electron-transfer path along the sigma bonds of the propionate to the heme or iron was envisaged as the mechanism of Mn^{II} oxidation. A schematic representation

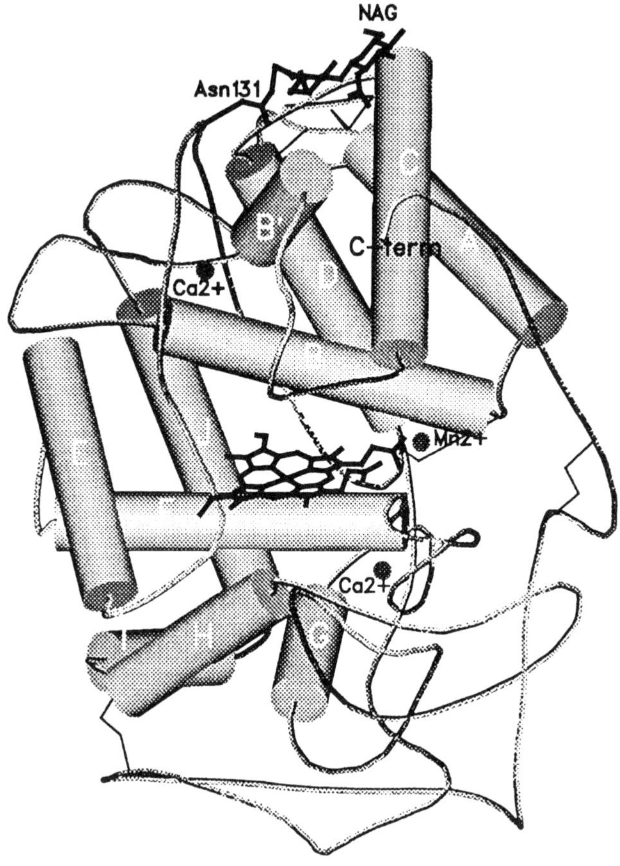

FIG. 12.17. Schematic representation of the complete manganese peroxidase polypeptide chain, showing the helices as cylinders, the heme, the calcium, and manganese binding sites and the two *N*-acetylglucosamine residues N-linked to Asn131. [Reproduced with permission from M. Sundaramoorthy, K. Kishi, M. H. Gold, and T. L. Poulos, *J. Biol. Chem.* **269,** 32759–32767 (1994).]

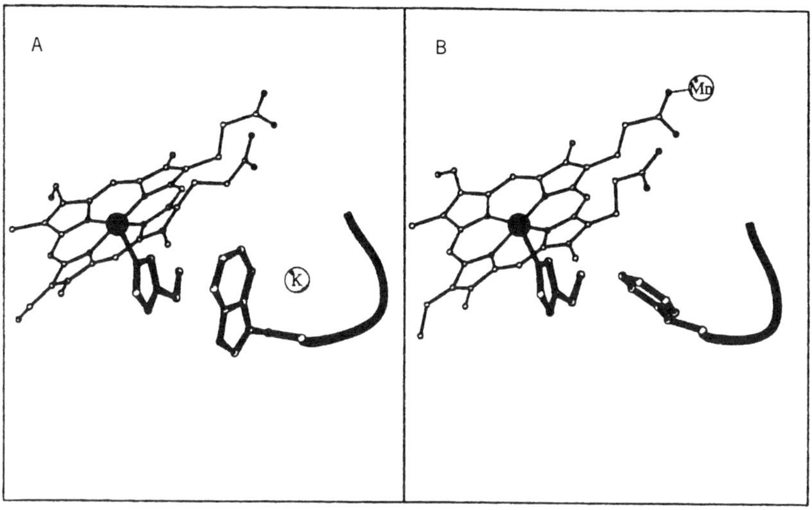

FIG. 12.18. Binding sites of K^+ in ascorbate peroxidase and Mn^{2+} in manganese peroxidase. The homologous section of the polypeptide chain for the K^+ binding site is also shown. (A) APX, (B) MnP. [Reproduced with permission from T. L. Poulos, W. R. Patterson, and M. Sundaramoorthy, *Biochem. Soc. Trans.* **23,** 228–232 (1995).]

of the structure is shown in Fig. 12.17. A comparison was made of the different binding sites of potassium ion in ascorbate peroxidase and manganese ion in manganese peroxidase (Fig. 12.18).[73]

12.3. *COPRINUS CINEREUS* (*ARTHROMYCES RAMOSUS*) PEROXIDASE

12.3.1. Discovery and Properties

The first reports on *C. cinereus* peroxidase appeared in the patent literature in 1986,[74,75] so that it is the most recent, of the three fungal peroxidases described in this chapter, that has been added to the peroxidase literature. The common name for the source is the black inkcap mushroom, from the Basidiomycete family of fungi. *Coprinus cinereus* peroxidase was proposed as an alternative for horseradish peroxidase for various commercial assays because of its stability and accessible active site. As discussed below there is an identical enzyme produced by the improperly classified species *A. ramosus* (Fungi imperfecti). Another name for *C. cinereus* is *Coprinus macrorhizus.* The properties of the two *Coprinus* peroxidases, *C. cinereus* and *C. macrorhizus,* and the one from *A. ramosus* were compared and found to be identical except for carbohydrate content. It was noted, however, that the latter varied from batch to batch of the same enzyme.[76,77] Thus, the post-translational modification of sugar moiety addition is not always reproducible. The amino acid sequences were determined from the cDNA for both the *C. cinereus* and the *A. ramosus* peroxidases and found to be identical.[78,79] A sequence for the *Arthromyces* enzyme, differing by one amino acid residue, an extra glycine near the N-terminal end, has also been published.[80] This difference did not exist in a commercial source of the latter enzyme. The extra glycine residue reported for the *A. ramosus* enzyme would be unlikely to affect its reactivity; however, it affects the numbering of residues. The sequence for the *Coprinus* enzyme is reproduced in Appendix A, and to the best of my knowledge this is also the correct sequence for *A. ramosus* peroxidase. I shall refer to *C. cinereus* peroxidase as much as possible, but use *A. ramosus* to describe important X-ray results, which have the latter name attached.

Coprinus cinereus peroxidase was found to be similar to horseradish peroxidase in many of its properties. The enzyme contains one ferriprotoporphyrin IX as its prosthetic group, it is a glycoprotein, and it forms conventional compounds I and II.[81] It is an acid protein, with a pI value of 3.5. Electronic absorption bands of its complexes were compared to those of other peroxidases.[82]

The kinetics and equilibria of cyanide binding were investigated as a function of pH, care being taken to avoid nitrate as a source of ionic strength and to avoid acetate buffer.[83] In general terms the results were similar to those obtained with other peroxidases: cyanide binds rapidly and very tightly, and the acid form of the ligand binds to a basic form of the enzyme. However, the pK_a value of the active site group in the *Coprinus* enzyme is 5.1 ± 0.2, much higher than in horseradish peroxidase. The kinetics of compound I formation exhibited similarities to cyanide binding, and again a pK_a value close to 5 was indicated.[84] The rate constant for compound I formation in the pH-independent region is $(7.1 \pm 0.1) \times 10^6\ M^{-1}s^{-1}$. Spectral characteristics of the native enzyme and compounds I, II, and III are listed in Table 12.2.

TABLE 12.2. Wavelengths (λ) of Absorption Peaks and Isosbestic Points (subscript *i*) and Corresponding Molar Absorptivities (ε) of Wild-Type *C. cinereus* Peroxidase (CiP) and Its Compounds I, II, and III (CiP-I, CiP-II, and CiP-III)

Enzyme Species	Absorption peaks λ, nm (ε, $mM^{-1}cm^{-1}$)	Isosbestic Points with CiP λ_i, nm (ε_i, $mM^{-1}cm^{-1}$)
CiP	405 (109), 504 (14), 645 (4)	
CiP-I	402 (63), 658 (6)	353 (39), 426 (43), 453 (13), 547 (8)
CiP-II	419 (92), 530 (11), 554 (10)	343 (31), 414 (90), 465 (12), 530 (11), 629 (3), 668 (2)
CiP-III	417 (105), 547 (14), 582 (10), 649 (2)	344 (32), 411 (103), 475 (13), 534 (11), 604 (3)

Reproduced with permission from M. B. Andersen, Y. Hsuanyu, K. G. Welinder, P. Schneider, and H. B. Dunford, *Acta Chem. Scand.* **45,** 1080–1086 (1991).

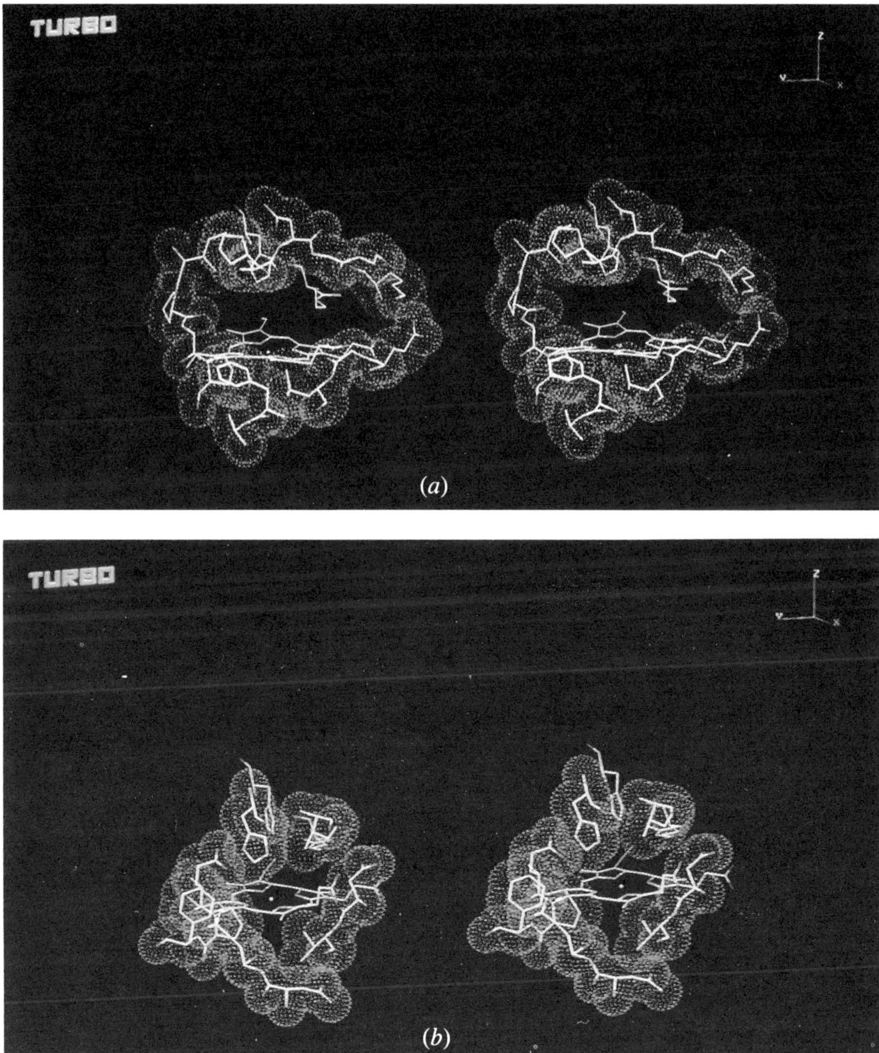

FIG. 12.19. Channels leading to the distal sides of the hemes of (*a*) *A. ramosus* and (*b*) lignin peroxidases. See color insert. For (*a*), residues Lys48, Pro90, Gly94, Ile152, Pro155, Glu189, Leu191, Phe 229, and Ala230 (corrected numbering) are involved. For (*b*), His82, Ile85, Glu146, Phe148, Asp183, Val184, Asn221, and Gln222. [Reproduced from N. Kunishima, K. Fukuyama, H. Matsubara, H. Hatanaka, Y. Shibano, and T. Amachi, *J. Mol. Biol.* **235,** 331–344 (1994) by permission of the publisher Academic Press Limited, London.]

Data for native enzyme and compounds I and III were obtained at pH 7.0, for compound II at pH 10.2. In addition to the data in Table 12.2, there are two isosbestic points between compounds I and II at 394 and 576 nm, with corresponding molar absorptivities of 60 and 7 $mM^{-1}cm^{-1}$. The observed spectra of the native enzyme changed at pH lower than 5, especially the Soret peak maximum, which was shifted toward the ultraviolet.

The folding and unfolding of *C. cinereus* peroxidase was studied as a function of pH, urea concentration, and temperature.[85] Monitoring of the heme spectrum and the rate of calcium ion uptake by EDTA provided accurate rate data. Such data are valuable for cloning and site-directed mutagenesis experiments in which the overexpressed enzyme forms inclusion bodies, which must be unscrambled to recover a viable product.

12.3.2. Crystal Structures

Preliminary data[86,87] and detailed structures were reported for the identical *C. cinereus*[88] and *A. ramosus*[89,90] peroxidases. The protein contains one heme group, two calcium ions, and two covalently bound sugar moieties. The only difference in helical topology between the *Coprinus* enzyme and yeast cytochrome *c* peroxidase is two additional helices, B″ and B′, between B and C in *Coprinus* peroxidase. Compared to the yeast peroxidase, both *C. cinereus* and

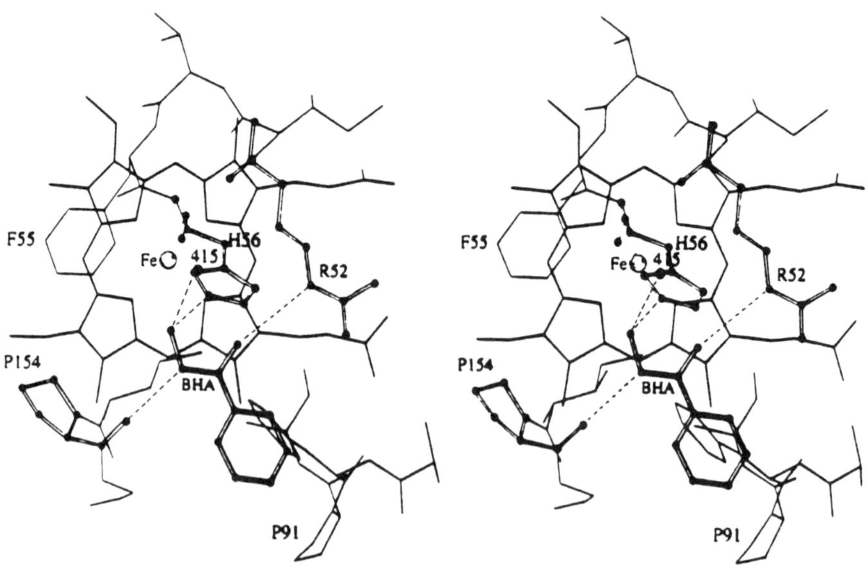

FIG. 12.20. Environment of benzhydroxamic acid bound to *A. ramosus* peroxidase shown by a stereo pair. Possible hydrogen bonds are shown by broken lines. [Reproduced from H. Itakura, Y. Oda, and K. Fukuyama, *FEBS Lett.* **412,** 107–110 (1997) with permission from Elsevier Science.]

lignin peroxidases have an additional 50 residues at the C-terminal end. The active site of *A. ramosus (C. cinereus)* peroxidase is much more accessible than in lignin peroxidase (Fig. 12.19) (see p. 301 and color insert), and different amino acid residues are involved.

At pH 4.5 the heme iron of *A. ramosus* peroxidase was shown to be five-coordinate in an X-ray study at 1.8 Å resolution,[90] and the binding mode

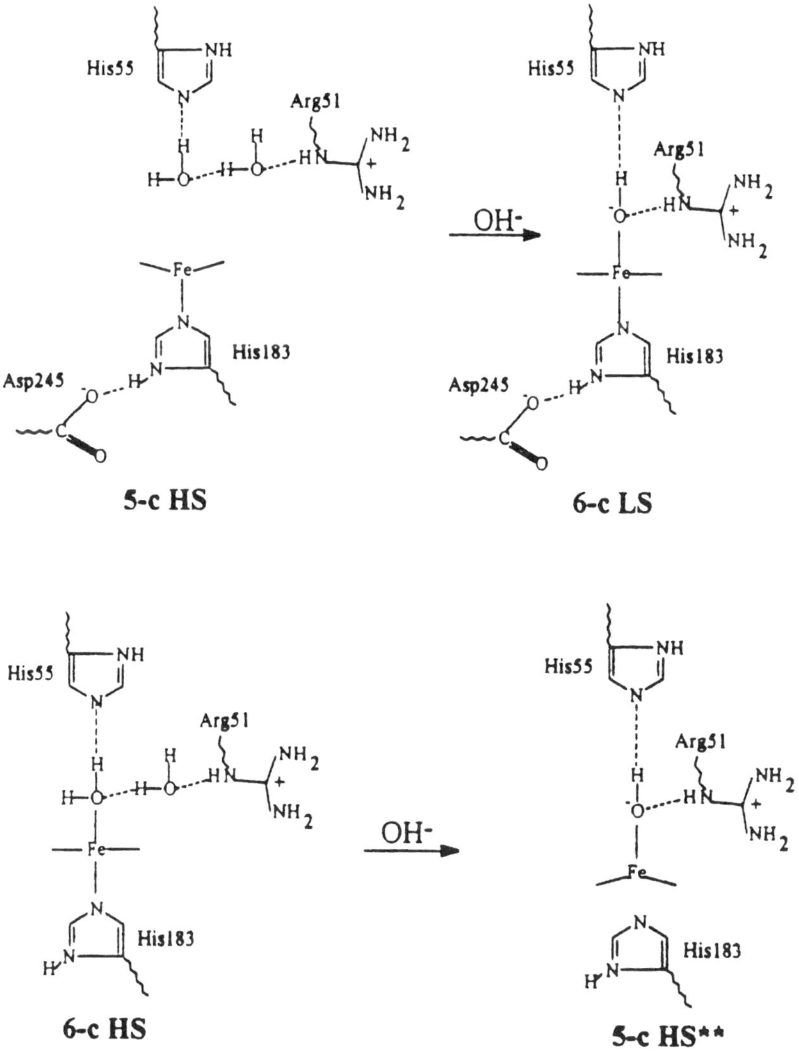

FIG. 12.21. Schematic representations of the neutral → alkaline transitions for wild-type *C. cinereus* peroxidase (top) and the Asp245Asn mutant (bottom). (Reproduced with permission from G. Smulevich, F. Neri, M. P. Marzocchi, and K. G. Welinder, *Biochemistry* **35,** 10576–10585 (1996).]

of benzhydroxamic acid was ascertained with certainty from a study with 1.6 Å resolution.[91] The functional groups of benzhydroxamic acid are held by hydrogen bonds to His56Nε, Arg52Nε, and Pro154O (Fig. 12.20). The sixth coordination position of the iron appeared to be occupied by a water molecule.

The binding of iodide to *A. ramosus* peroxidase was investigated by X-ray crystallographic analysis, ^{1}H and ^{127}I NMR spectroscopy and steady-state kinetics.[92] Iodide binds at the entrance to the distal cavity, farther from the heme iron than in horseradish peroxidase. The binding is influenced by a distal group with a pK_a of about 5.3, presumably the distal His56.

12.3.3. Resonance Raman and Nuclear Magnetic Resonance Studies on *Coprinus* Peroxidase

The versatility and confusing adaptability of peroxidases is well illustrated in resonance Raman spectral results. Although *C. cinereus* peroxidase is similar to the horseradish enzyme in its enzymatic properties, its resonance Raman spectra show it is much closer to yeast cytochrome *c* peroxidase in terms of heme cavity structure.[93] Marked changes are observed when Asp245, which is hydrogen-bonded to the proximal His183, is replaced with an asparagine. The wild-type enzyme changes from a five-coordinate to six-coordinate iron when the alkaline transition occurs, and the reverse occurs for the mutant. It is the proximal His183 that appears to vacate a ligand position in the mutant (Fig. 12.21).[94] Nuclear magnetic resonance studies show that there are no significant active site differences between the recombinant and wild-type enzymes.[77,95,96] Differences would be expected in the regions of carbohydrate binding since the recombinant enzyme has no sugar moieties. The NMR results on the Asp245Asn mutant are in agreement with the resonance Raman results.[97]

REFERENCES

1. T. K. Kirk and R. L. Farrell, *Ann. Rev. Microbiol.* **41,** 465–505 (1987).
2. M. Kuwahara, J. K. Glenn, M. A. Morgan, and M. H. Gold, *FEBS Lett.* **169,** 247–250 (1984).
3. M. H. Gold, M. Kuwahara, A. A. Chiu, and J. K. Glenn, *Arch. Biochem. Biophys.* **234,** 353–362 (1984).
4. M. Tien and T. K. Kirk, *Proc. Natl. Acad. Sci. U.S.A.* **81,** 2280–2284 (1984).
5. V. Renganathan, K. Miki, and M. H. Gold, *Arch. Biochem. Biophys.* **241,** 304–314 (1985).
6. M. Tien, T. K. Kirk, C. Bull, and J. A. Fee, *J. Biol. Chem.* **261,** 1687–1693 (1986).
7. A. Andrawis, K. A. Johnson, and M. Tien, *J. Biol. Chem.* **263,** 1195–1198 (1988).
8. V. Renganathan and M. H. Gold, *Biochemistry* **25,** 1626–1631 (1986).
9. L. Marquez, H. Wariishi, H. B. Dunford, and M. H. Gold, *J. Biol. Chem.* **263,** 10549–10552 (1988).

10. D. Kuila, M. Tien, J. A. Fee, and M. R. Ondrias, *Biochemistry* **24,** 3394–3397 (1985).
11. L. A. Andersson, V. Renganathan, T. M. Loehr, and M. H. Gold, *Biochemistry* **26,** 2258–2263 (1987).
12. A. Khindaria and S. D. Aust, *Biochemistry* **35,** 13107–13111 (1996).
13. G. D. DePillis, H. Wariishi, M. H. Gold, and P. R. Ortiz de Montellano, *Arch. Biochem. Biophys.* **280,** 217–223 (1990).
14. P. J. Harvey, J. M. Palmer, H. E. Schoemaker, H. L. Dekker, and R. Wever, *Biochim. Biophys. Acta* **994,** 59–63 (1989).
15. H. Wariishi and M. H. Gold, *FEBS Lett.* **243,** 165–168 (1989).
16. H. Wariishi, L. Marquez, H. B. Dunford, and M. H. Gold, *J. Biol. Chem.* **265,** 11137–11142 (1990).
17. T. Johansson and P. O. Nyman, *Arch. Biochem. Biophys.* **300,** 49–56 (1993).
18. T. Johansson, K. G. Welinder, and P. O. Nyman, *Arch. Biochem. Biophys.* **300,** 57–62 (1993).
19. T. K. Kirk in *Lignin Biodegredation: Microbiology, Chemistry and Potential Applications,* T. K. Kirk, T. Higuchi and H. Chang, eds., CRC, Boca Raton, 1980, Vol. II, pp. 51–63.
20. S. D. Haemmerli, H. E. Schoemaker, H. W. H. Schmidt, and M. S. A. Leisola, *FEBS Lett.* **220,** 149–154 (1987).
21. H. Wariishi, J. Huang, H. B. Dunford, and M. H. Gold, *J. Biol. Chem.* **266,** 20694–20699 (1991).
22. L. P. Candeias and P. J. Harvey, *J. Biol. Chem.* **270,** 16745–16748 (1995).
23. P. J. Harvey and L. P. Candeias, *Biochem. Soc. Trans.* **23,** 262–267 (1995).
24. P. O'Neill, S. Steenken, and D. Schulte-Frohlinde, *J. Phys. Chem.* **79,** 2773–2779 (1975).
25. R. W. Fessenden, Radiation Laboratory, Notre Dame University, Notre Dame, IN, U.S.A., personal communication, November 25, 1997.
26. M. Eigen, *Angew. Chemie, Intnl. Ed. (English)* **3,** 1–19 (1964).
27. L. P. Candeias, Gray Laboratory, Mount Vernon Hospital, Middlesex, UK, personal communication, December 5, 1997.
28. A. Khindaria, T. A. Grover, and S. D. Aust, *Biochemistry* **34,** 6020–6025 (1995).
29. A. Khindaria, I. Yamazaki, and S. D. Aust, *Biochemistry* **35,** 6418–6424 (1996).
30. A. Khindaria, G. Nie, and S. D. Aust, *Biochemistry* **36,** 14181–14185 (1997).
31. R. E. D. McClung, Department of Chemistry, University of Alberta, Edmonton, Canada, personal communication, November 27, 1997.
32. P. J. Harvey, H. E. Schoemaker, and J. M. Palmer, *FEBS Lett.* **195,** 242–246 (1986).
33. R. S. Koduri and M. Tien, *Biochemistry* **33,** 4225–4230 (1994).
34. M. Tien and D. Ma, *J. Biol. Chem.* **272,** 8912–8917 (1997).
35. I. A. Pearl, *The Chemistry of Lignin,* Marcel Dekker, New York, 1967, pp. 4–5.
36. K. E. Hammel, M. Tien, B. Kalyanaraman, and T. K. Kirk, *J. Biol. Chem.* **260,** 8348–8353 (1985).
37. V. Renganathan, K. Miki, and M. H. Gold, *Arch. Biochem. Biophys.* **246,** 155–161 (1986).
38. T. K. Kirk, M. Tien, P. J. Kersten, M. D. Mozuch, and B. Kalyanaraman, *Biochem. J.* **236,** 279–287 (1986).

39. K. Miki, V. Renganathan, and M. H. Gold, *Biochemistry* **25,** 4790–4796 (1986).
40. K. Miki, V. Renganathan, and M. H. Gold, *FEBS Lett.* **203,** 235–238 (1986).
41. K. Miki, V. Renganathan, M. B. Mayfield, and M. H. Gold, *FEBS Lett.* **210,** 199–203 (1987).
42. P. J. Kersten, M. Tien, B. Kalyanaraman, and T. K. Kirk, *J. Biol. Chem.* **260,** 2609–2612 (1985).
43. W. R. Russell, A. R. Forrester, A. Chesson, and M. J. Burkitt, *Arch. Biochem. Biophys.* **332,** 357–366 (1996).
44. H. W. H. Schmidt, S. D. Haemmerli, H. E. Schoemaker, and M. S. A. Leisola, *Biochemistry* **28,** 1776–1783 (1989).
45. D. K. Joshi and M. H. Gold, *Biochemistry* **33,** 10969–10976 (1994).
46. R. S. Koduri and M. Tien, *J. Biol. Chem.* **270,** 22254–22258 (1995).
47. D. C. Goodwin, S. D. Aust, and T. A. Grover, *Biochemistry* **34,** 5060–5065 (1995).
48. V. Renganathan, K. Miki, and M. H. Gold, *Biochemistry* **26,** 5127–5132 (1987).
49. M. Tatarko and J. A. Bumpus, *Arch. Biochem. Biophys.* **339,** 200–209 (1997).
50. A. Khindaria, D. P. Barr, and S. D. Aust, *Biochemistry* **34,** 7773–7779 (1995).
51. J. L. Popp, B. Kalyanaraman, and T. K. Kirk, *Biochemistry* **29,** 10475–10480 (1990).
52. S. L. Edwards, R. Raag, H. Wariishi, M. H. Gold, and T. L. Poulos, *Proc. Natl. Acad. Sci. U.S.A.* **90,** 750–754 (1993).
53. T. L. Poulos, S. L. Edwards, H. Wariishi, and M. H. Gold, *J. Biol. Chem.* **268,** 4429–4440 (1993).
54. J. K. Glenn and M. H. Gold, *Arch. Biochem. Biophys.* **242,** 329–341 (1985).
55. J. K. Glenn, L. Akileswaran, and M. H. Gold, *Arch. Biochem. Biophys.* **251,** 688–696 (1986).
56. Y. Mino, H. Wariishi, N. J. Blackburn, T. M. Loehr, and M. H. Gold, *J. Biol. Chem.* **263,** 7029–7036 (1988).
57. H. Wariishi, H. B. Dunford, I. D. MacDonald, and M. H. Gold, *J. Biol. Chem.* **264,** 3335–3340 (1989).
58. R. E. Whitwam, I. G. Gazarian, and M. Tien, *Biochem. Biophys. Res. Commun.* **216,** 1013–1017 (1995).
59. R. E. Whitwam and M. Tien, *Arch. Biochem. Biophys.* **333,** 439–446 (1996).
60. S. L. Timofeevski and S. D. Aust, *Arch. Biochem. Biophys.* **342,** 169–175 (1997).
61. K. Kishi, H. Wariishi, L. Marquez, H. B. Dunford, and M. H. Gold, *Biochemistry* **33,** 8694–8701 (1994).
62. H. Wariishi, K. Valli, and M. H. Gold, *Biochemistry* **28,** 6017–6023 (1989).
63. U. Tuor, H. Wariishi, H. E. Schoemaker, and M. H. Gold, *Biochemistry* **31,** 4986–4995 (1992).
64. H. Wariishi, K. Valli, and M. H. Gold, *Biochem. Biophys. Res. Commun.* **176,** 269–275 (1991).
65. H. Wariishi, K. Valli, V. Renganathan, and M. H. Gold, *J. Biol. Chem.* **264,** 14185–14191 (1989).
66. H. Wariishi, K. Valli, and M. H. Gold, *J. Biol. Chem.* **267,** 23688–23695 (1992).
67. I.-C. Kuan and M. Tien, *Proc. Natl. Acad. Sci. U.S.A.* **90,** 1242–1246 (1993).

68. I.-C. Kuan, K. A. Johnson, and M. Tien, *J. Biol. Chem.* **268,** 20064–20070 (1993).

69. R. Z. Harris, H. Wariishi, M. H. Gold, and P. R. Ortiz de Montellano, *J. Biol. Chem.* **266,** 8751–8758 (1991).

70. M. K.-v. Someren, K. Kishi, T. Lundell, and M. H. Gold, *Biochemistry* **34,** 10620–10627 (1995).

71. K. Kishi, M. K.-v. Someren, M. B. Mayfield, J. Sun, T. M. Loehr, and M. H. Gold, *Biochemistry* **35,** 8986–8994 (1996).

72. M. Sundaramoorthy, K. Kishi, M. H. Gold, and T. L. Poulos, *J. Biol. Chem.* **269,** 32759–32767 (1994).

73. T. L. Poulos, W. R. Patterson, and M. Sundaramoorthy, *Biochem. Soc. Trans.* **23,** 228–232 (1995).

74. Y. Shinmen, S. Asami, N. Amano, T. Amachi, and H. Yoshizumi in *Japanese Patent Provisional Publication,* Sho 61-104784, pp. 423–427 (1986).

75. T. Uchida and Y. Matsui in *Japanese Patent Provisional Publication,* Sho 61-128887, pp. 449–454 (1986).

76. M. Kjalke, M. B. Andersen, P. Schneider, B. Christensen, M. Schülein, and K. G. Welinder, *Biochim. Biophys. Acta* **1120,** 248–256 (1992).

77. P. Limongi, M. Kjalke, J. Vind, J. W. Tams, T. Johansson, and K. G. Welinder, *Eur. J. Biochem.* **227,** 270–276 (1995).

78. L. Baunsgaard, H. Dalbøge, G. Houen, E. M. Rasmussen, and K. G. Welinder, *Eur. J. Biochem.* **213,** 605–611 (1993).

79. A. K. Abelskov, A. T. Smith, C. B. Rasmussen, H. B. Dunford, and K. G. Welinder, *Biochemistry* **36,** 9453–9463 (1997).

80. H. Sawai-Hatanaka, T. Ashikari, Y. Tanaka, Y. Asada, T. Nakayama, H. Minakata, N. Kunishima, K. Fukuyama, H. Yamada, Y. Shibano, and T. Amachi, *Biosci. Biotech. Biochem.* **59,** 1221–1228 (1995).

81. Y. Morita, H. Yamashita, B. Mikami, H. Iwamoto, S. Aibara, M. Terada, and J. Minami, *J. Biochem.* **103,** 693–699 (1988).

82. G. S. Lukat, K. R. Rodgers, M. N. Jabro, and H. M. Goff, *Biochemistry* **28,** 3338–3345 (1989).

83. M. B. Andersen, Y. Hsuanyu, K. G. Welinder, P. Schneider, and H. B. Dunford, *Acta Chem. Scand.* **45,** 206–211 (1991).

84. M. B. Andersen, Y. Hsuanyu, K. G. Welinder, P. Schneider, and H. B. Dunford, *Acta Chem. Scand.* **45,** 1080–1086 (1991).

85. J. W. Tams and K. G. Welinder, *Biochemistry* **35,** 75573–7579 (1996).

86. N. Kunishima, K. Fukuyama, S. Wakabayashi, M. Sumida, M. Takaya, Y. Shibano, T. Amachi, and H. Matsubara, *Proteins: Struct. Funct. Genet.* **15,** 216–220 (1992).

87. J. F. W. Petersen, J. W. Tams, J. Vind, A. Svensson, H. Dalbøge, K. G. Welinder, and S. Larsen, *J. Mol. Biol.* **232,** 989–991 (1993).

88. J. F. W. Petersen, A. Kadziola, and S. Larsen, *FEBS Lett.* **339,** 291–296 (1994).

89. N. Kunishima, K. Fukuyama, H. Matsubara, H. Hatanaka, Y. Shibano, and T. Amachi, *J. Mol. Biol.* **235,** 331–344 (1994).

90. N. Kunishima, F. Amada, K. Fukuyama, M. Kawamoto, T. Masunaga, and H. Matsubara, *FEBS Lett.* **378,** 291–294 (1996).

91. H. Itakura, Y. Oda, and K. Fukuyama, *FEBS Lett.* **412,** 107–110 (1997).

92. K. Fukuyama, K. Sato, H. Itakura, S. Takahashi, and T. Hosoya, *J. Biol. Chem.* **272,** 5752–5756 (1997).

93. G. Smulevich, A. Feis, C. Focardi, J. Tams, and K. G. Welinder, *Biochemistry* **33,** 15425–15432 (1994).

94. G. Smulevich, F. Neri, M. P. Marzocchi, and K. G. Welinder, *Biochemistry* **35,** 10576–10585 (1996).

95. L. B. Dugad and H. M. Goff, *Biochim. Biophys. Acta* **1122,** 63–69 (1992).

96. N. C. Veitch, J. W. Tams, J. Vind, H. Dalbøge, and K. G. Welinder, *Eur. J. Biochem.* **222,** 909–918 (1994).

97. N. C. Veitch, Y. Gao, and K. G. Welinder, *Biochemistry* **35,** 14370–14380 (1996).

CHAPTER 13

Other Class III Plant Peroxidases: Peanut, Barley, Tobacco, and *Arabidopsis thaliana*

In this chapter, other examples of class III plant peroxidases are discussed, those belonging to the same family as horseradish peroxidase. Also discussed are the roles of peroxidases in growth, development, and wound healing, and in transgenic plants. The stereospecificity of lignin biosynthesis and cancer in plants are the concluding topics.

13.1. PEANUT PEROXIDASES

13.1.1. Discovery and Characterization

Work on peanut peroxidases and the career of Robert van Huystee are inextricably woven together. His initial research was on the effect of ionizing radiation on cell growth, for which he chose cultured peanut cells (*Arachis hypogaea* L.) from cotyledonary tissue. He isolated an anionic peroxidase from radioactively labeled culture[1] and later a cationic excreted peroxidase.[2,3] Both the anionic and cationic peroxidases were purified.[4] It was found that the cationic peroxidase contained 15% carbohydrate.[5] Tunicamycin was found to inhibit peanut peroxidase release, and from experiments with tritium-labeled mannose, to inhibit glycosylation at N-linked sites.[6] By use of ^{14}C-labeling, it was demonstrated that the peroxidase heme was derived from glutamic acid and was synthesized in mitochondria.[7] It was found that there is one heme unit per peanut peroxidase molecule, and the heme was identified as ferriprotoporphyrin IX.[8,9] Calcium ion was shown to be necessary for optimal enzyme activity,[10] and a co-dependency of calcium and porphyrin for an integrated molecular structure was demonstrated in a circular dichroism study.[11] Structural changes were observed by nuclear magnetic resonance (NMR) for the cationic peanut peroxidase with its two calcium ions removed, as it was titrated with calcium.[12]

The spectral and kinetic properties of the cationic peroxidase were studied.[13] Spectra of native peanut peroxidase and compounds II and III are shown in Fig. 13.1, and the Soret spectrum of compound I is shown in Fig. 13.2. The rate constant for compound I formation, obtained from the slope of Fig. 13.2*b*, is $(1.5 \pm 0.2) \times 10^7\ M^{-1}s^{-1}$, essentially the same as for horseradish peroxidase. From the pyridine hemochromogen spectrum, the molar absorptivity of peanut peroxidase was determined to be 112 $mM^{-1}cm^{-1}$ at 405 nm. Other Soret maximum absorbances: for compound I, 52 $mM^{-1}cm^{-1}$ at 399 nm; and for compound II, 101 $mM^{-1}cm^{-1}$ at 418 nm. Reinheitzahl (RZ) values (A_{405}/A_{280}) for purified cationic peanut peroxidase are of the order of 2.7. Spectra of cyanide, azide, and fluoride complexes were recorded, as were those of the alkaline form, the ferrous enzyme, and ferrous cyanide and carbon monoxide complexes.[13] Some rate constants for compound II reactions of both peanut and horseradish peroxidase are listed in Table 13.1. The similarities of peanut and horseradish peroxidases were also shown in a spectral and magnetic circular dichroism study.[14]

Further characterization indicated that there were two forms of the cationic peroxidase, perhaps differing only in the extent of glycosylation.[15] It was found that the rate of production of anionic and cationic peroxidases were essentially the same, although it is the cationic form that is predominantly excreted into the cell medium.[16]

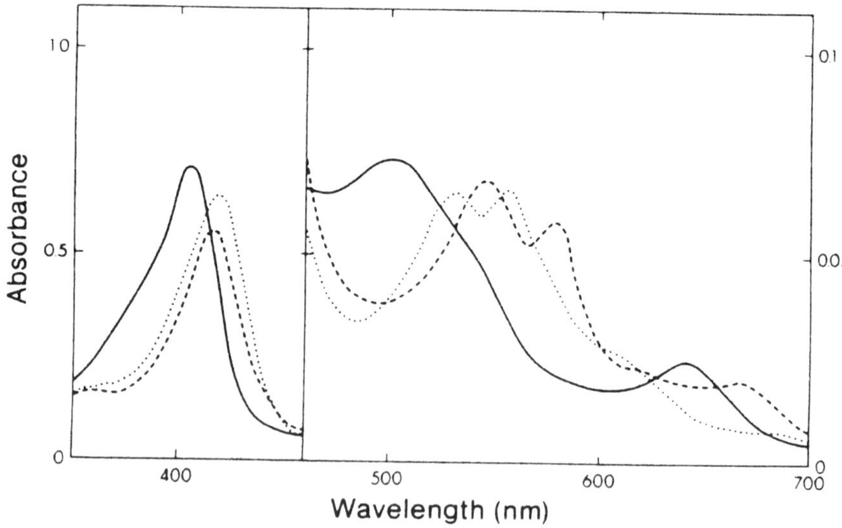

FIG. 13.1. Spectra of native cationic peanut peroxidase (——), compound II (.....), and compound III (----) obtained at pH 8 and 25°C. Compound II was formed by adding two equivalents of hydrogen peroxide and one of ferrocyanide to native enzyme, and compound III by adding three equivalents of hydrogen peroxide to compound II. [Reproduced with permission from A.-M. Lambeir, H. B. Dunford, R. B. van Huystee, and J. Lobarzewski, *Can. J. Biochem. Cell Biol.* **63,** 1086–1092 (1985).]

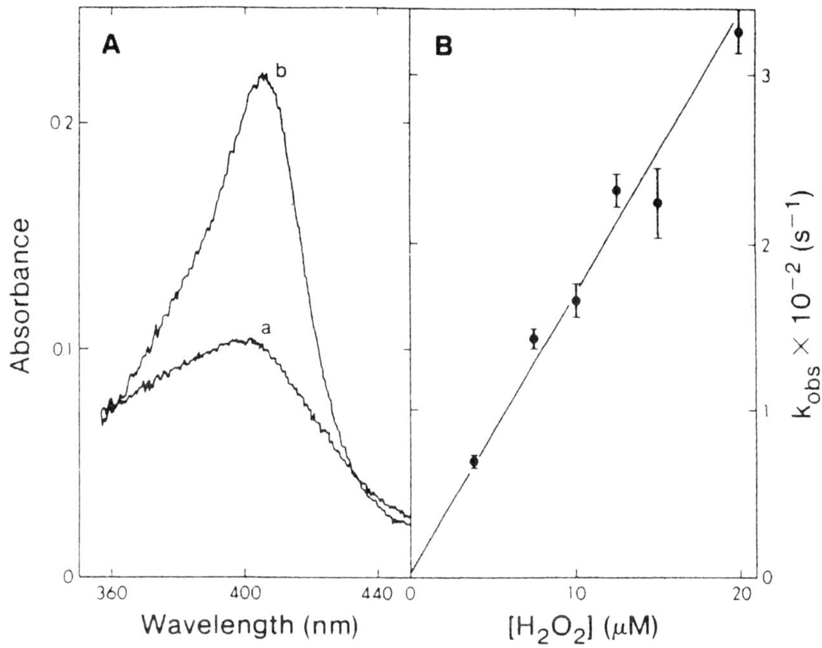

FIG. 13.2. (A) Soret spectrum of peanut peroxidase compound I at pH 7 and 25°C (*a*) obtained by 140 ms after mixing 15 μM hydrogen peroxide with 2 μM enzyme. The native enzyme spectrum (*b*) is shown for comparison. (B) Plot of pseudo-first-order rate constant versus concentration of hydrogen peroxide. [Reproduced with permission from A.-M. Lambeir, H. B. Dunford, R. B. van Huystee, and J. Lobarzewski, *Can. J. Biochem. Cell Biol.* **63,** 1086–1092 (1985).]

TABLE 13.1. Second-Order Rate Constants ($M^{-1}s^{-1}$) for Reactions of Peroxidase Compounds II at pH 7

Substrate	Peanut Compound II	Horseradish Compound II
L-Tyrosine	16.0 ± 0.2	(2 ± 1) × 10^3
Syringic acid	(1.02 ± 0.1) × 10^2	(6.8 ± 0.7) × 10^2
4-Hydroxybenzaldehyde	(1.57 ± 0.06) × 10^2	—
Ferrocyanide	(4.21 ± 0.02) × 10^2	1.3 × 10^4
Vanillin	(1.42 ± 0.04) × 10^5	(1.3 ± 0.1) × 10^6
Eugenol	(2.6 ± 0.1) × 10^5	(1.4 ± 0.1) × 10^6

Reproduced with permission from A.-M. Lambeir, H. B. Dunford, R. B. van Huystee, and J. Lobarzewski, *Can. J. Biochem. Cell Biol.* **63,** 1086–1092 (1985).

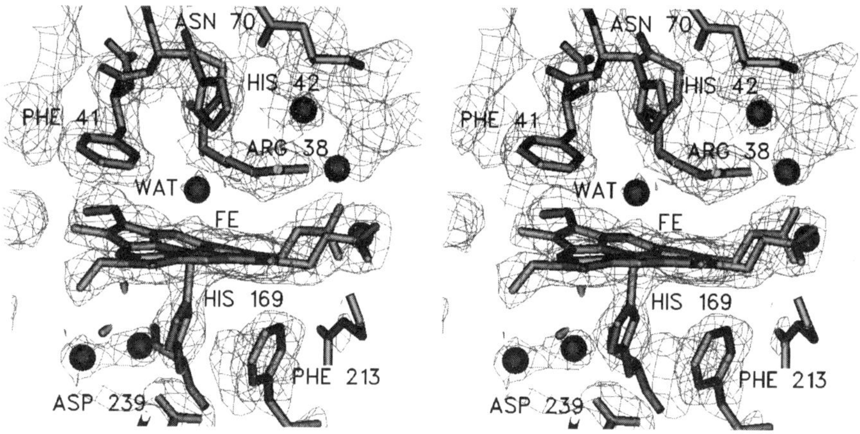

FIG. 13.3 Heme environment of peanut peroxidase. See color insert. [Reproduced with permission from D. J. Schuller, N. Ban, R. B. van Huystee, A. McPherson, and T. L. Poulos, *Structure* **4,** 311–321 (1996).]

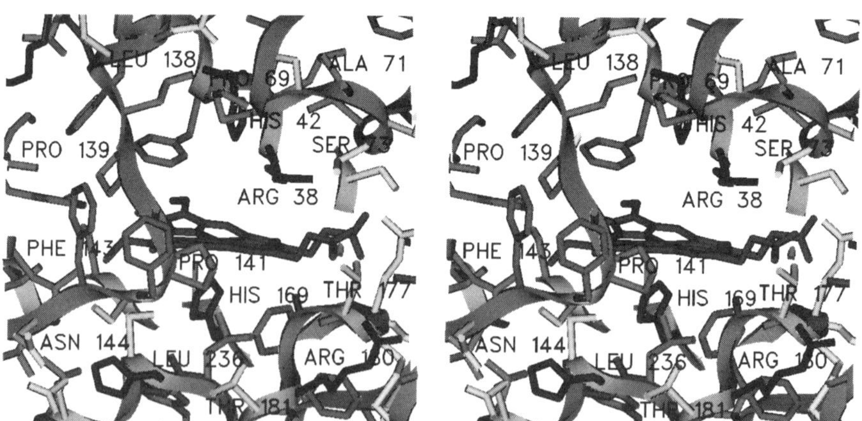

FIG. 13.4. The substrate access channel to the heme of peanut peroxidase. See color insert. [Reproduced with permission from D. J. Schuller, N. Ban, R. B. van Huystee, A. McPherson, and T. L. Poulos, *Structure* **4,** 311–321 (1996).]

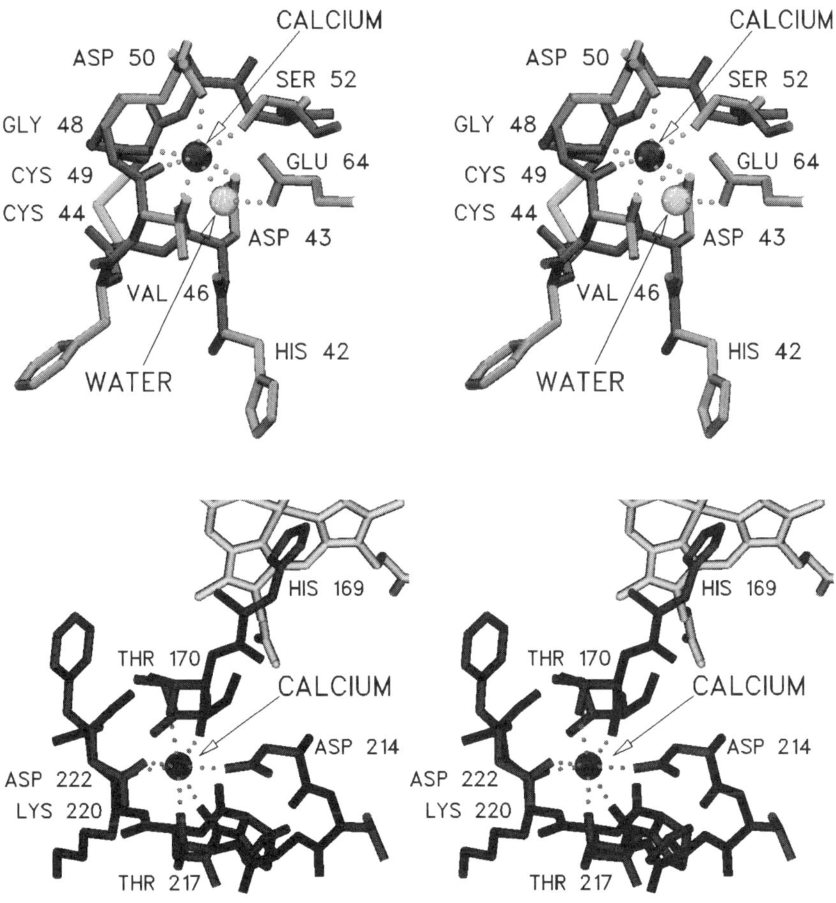

FIG. 13.5. Two calcium ion binding sites in peanut peroxidase. [Reproduced with permission from D. J. Schuller, N. Ban, R. B. van Huystee, A. McPherson, and T. L. Poulos, *Structure* **4,** 311–321 (1996).]

Ten monoclonal antibodies were produced against the cationic peanut peroxidase and screened by enzyme-linked immunosorbent assay (ELISA).[17] Eight were found to be of the immunoglobulin (Ig)G_1 subclass and two of IgA subclass. All antibodies cross reacted weakly with the anionic peanut peroxidase. Two monoclonal antibodies for the anionic peroxidase were developed.[18,19] Attention was then turned to the glycosylation of the peanut peroxidases. Four glycan chains of varying length were isolated from the cationic peroxidase but only one from the anionic enzyme.[20] Differences in the extent of glycosylation of the cationic peroxidase were detected,[21] and a role of glycans in maintaining the structural integrity of the enzymes was reported.[22] Robert van Huystee has reviewed his work on peanut peroxidases,[23–25] and a plant peroxidase newsletter was dedicated to him on the occasion of his retirement.[26]

13.1.2. Sequencing and Crystal Structure

Two cationic peanut peroxidases were cloned and their amino acid sequences determined.[27] These sequences are reproduced in Appendix A. The crystal structure of the major isoenzyme was determined, the first for any class III peroxidase.[28] Subsequently, the peanut peroxidase crystal structure was used to help solve the structure for horseradish peroxidase.

Peanut peroxidase lacks the short B′ helix found in all class I and II peroxidases. It possesses three additional helices not found in the class I and II structures: a small D′ between helices D and E, and F′ and F″ between helices F and G. The key distal residues, His42, Arg38, Phe41, and Asn70, of horseradish peroxidase are conserved with the same numbering in peanut peroxidase. His169 is the proximal ligand for peanut peroxidase. The heme environment of peanut peroxidase is shown in Fig. 13.3 (see p. 312 and color insert), and the substrate access channel to the heme is shown in Fig. 13.4 (see p. 312 and color insert). The binding sites for the two calcium ions are the same in both peanut and horseradish peroxidase (Fig. 13.5).

13.2. BARLEY PEROXIDASE

Barley peroxidase has been shown to have properties that, so far, are unique to class III peroxidases. The dominant isoenzyme, barley peroxidase 1 (commonly designated as BP1), has a messenger RNA that is expressed only in the barley seed endosperm, and only 15–20 days after flowering.[29] It has less than 50% sequence identity with other plant peroxidases, and it has a C-terminal propeptide that targets it to vacuoles.[30] The amino acid sequence is listed in Appendix A.

The reactivity, or more descriptively the lack of reactivity of barley peroxidase 1, came as a shock. The enzyme showed very little reactivity with hydrogen peroxide above pH 5 and exhibited saturation kinetics at lower pH.[31] An

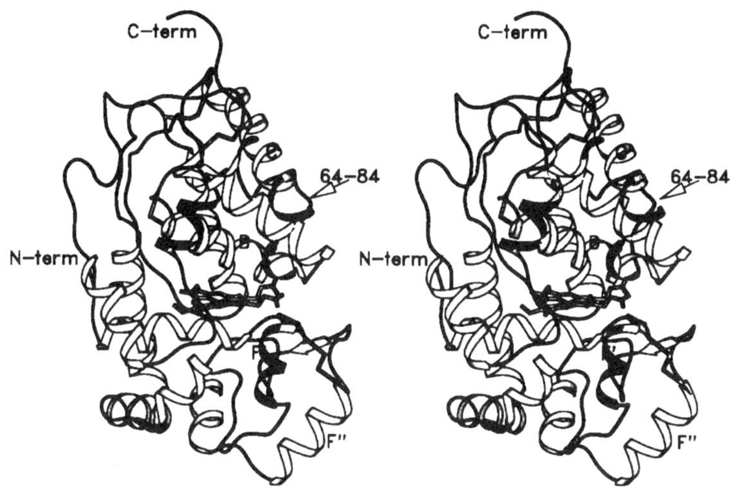

FIG. 13.6. Comparison of structures of peroxidases from barley and horseradish. Regions of the barley peroxidase structure with the largest deviations from that of horseradish are colored magenta. Residues 64–84 in the loop between helices B and C are severely disrupted in the barley enzyme. [Reproduced with permission from A. Henriksen, K. G. Welinder, and M. Gajhede, *J. Biol. Chem.* **273,** 2241–2248 (1998).]

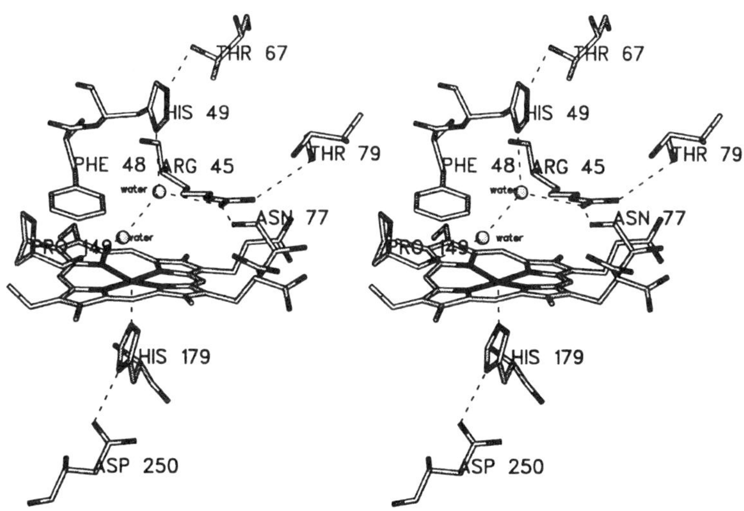

FIG. 13.7. Active site region of inactive barley peroxidase. [Reproduced with permission from A. Henriksen, K. G. Welinder, and M. Gajhede, *J. Biol. Chem.* **273,** 2241–2248 (1998).]

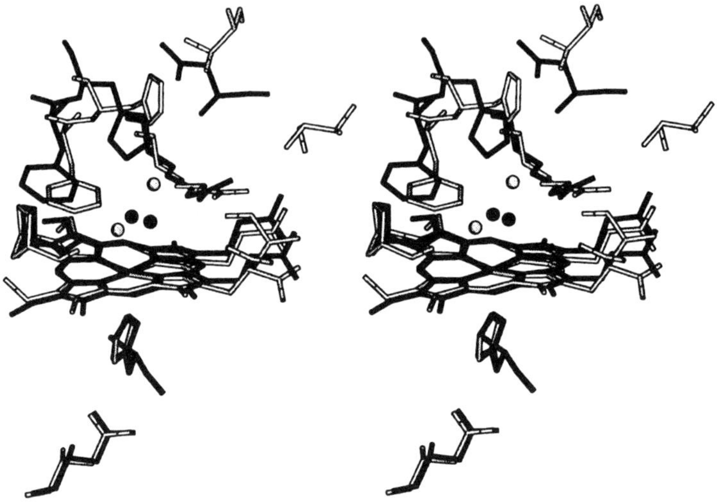

FIG. 13.8. Superposition of the active sites of barley (light) and horseradish (dark) peroxidases. [Reproduced with permission from A. Henriksen, K. G. Welinder, and M. Gajhede, *J. Biol. Chem.* **273,** 2241–2248 (1998).]

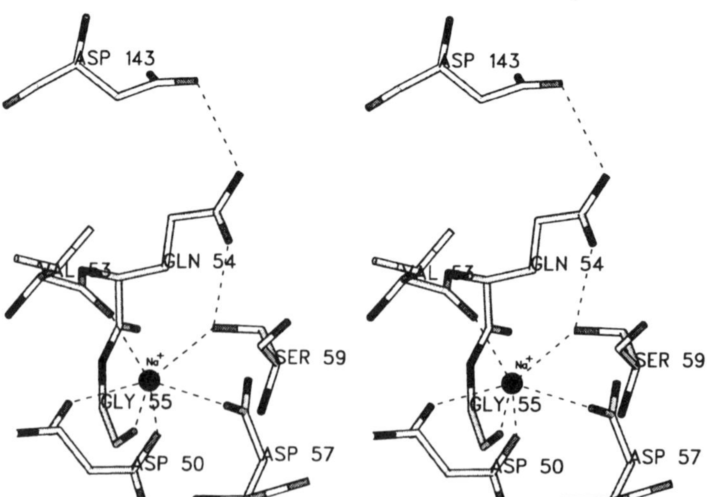

FIG. 13.9. Distal cation binding site of barley peroxidase. [Reproduced with permission from A. Henriksen, K. G. Welinder, and M. Gajhede, *J. Biol. Chem.* **273,** 2241–2248 (1998).]

investigation of reactions with naturally occurring reducing substrates ferulic acid, caffeic acid, and coniferyl alcohol was interpreted in terms of rate enhancement by these substrates, but at the time the stimulating effect of calcium ion, also present in the reaction mixture, was not fully appreciated.[32] The crucial factors in controlling the reactivity are now known to be pH and calcium ion concentration.[33] The enzyme as isolated is inactive at neutral pH. A decrease in pH to less than 3.7 converts the enzyme into a form that reacts slowly with hydrogen peroxide. The slowly reacting form appears to have a restricted active site. Calcium ion binding, with a dissociation constant of 4 mM, fully activates the enzyme. It then exhibits a pH-independent reaction with hydrogen peroxide, tested over the range pH 3.3–5.0. The fully active form binds the usual ligands for peroxidase iron(III): fluoride, azide, cyanide, and (weakly) chloride. In the presence of hydrogen peroxide and the reducing substrates mentioned above, the fully active enzyme exhibits reaction rates typical of other plant peroxidases. Large-scale prodution of barley peroxidase 1 has been accomplished.[34]

The crystal structure of the inactive form of barley peroxidase 1 has been published.[35] The crystals were obtained at pH 5.5, 7.5, and 8.5. The enzyme cannot form compound I above pH 5. Enzyme activity is restored at lower pH. The structure has overall similarity to those of peanut and horseradish peroxidases. The causes of the lack of reactivity at neutral pH appear to be the following: disruption of a hydrogen bond between the distal histidine and an arginine, wrong orientation of the distal histidine, substitution of sodium ion for calcium at the distal binding site, and a dramatic structural reaarangement to the loop region between helices B and C. Calcium ion binding at low pH restores the activity.

The structures of the class III peroxidases, those of barley and horseradish, are compared in Fig. 13.6, and the active site of barley peroxidase is shown in Fig. 13.7. A unique hydrogen bond from Thr67 to the distal histidine is seen. Superposition of the horseradish and barley enzyme active sites is shown in Fig. 13.8. The distal histidine of barley peroxidase is displaced too far from the iron to be catalytically active. Coordination to the distal sodium ion of inactive barley peroxidase is shown in Fig. 13.9. The sodium is replaced by calcium in the active form.

13.3. TOBACCO PEROXIDASES

There are at least 12 isoenzymes of tobacco peroxidases. Cationic peroxidases (pI 8–11), located in the central vacuole, have an indole-3-acetic acid activity in the absence of hydrogen peroxide.[36–38] Moderately anionic tobacco peroxidases (pI 4.5–6.5) are highly expressed in wounded stem tissue and may have a role in wound healing or suberization.[39] The highly anionic peroxidases (pI 3.5–4.0) are best understood. They are associated with the cell wall, catalyze the polymerization of cinnamyl alcohols *in vitro,* and function in lignification.[40,41] The highly anionic peroxidases have been purified from tobacco leaves.[42]

The role of the tobacco isoperoxidases to growth, development, and physiological state has been investigated.[43,44] Peroxidase activity was found to be an indissociable factor from auxin and polyamine metabolisms in the induction of rooting and flowering.[45–47]

The properties described above have been briefly reviewed in studies that describe the cloning and sequencing of the anionic tobacco peroxidase A.[48,49] The deduced amino acid sequence is reproduced in Appendix A. Different tissues were examined for their peroxidase isoenzyme distribution, and changes induced by wounding and tobacco mosaic virus infection were examined.[50] Tobacco plants were transformed with a chimeric tobacco anionic peroxidase gene that caused all tissues in the plants to produce high levels of the peroxidase. The subsequent effect of wounding was examined.[51,52] A nearly twofold greater synthesis of lignin was observed, and plant browning was correlated with peroxidase activity. The studies on transgenic plants were extended to transformed tomato plants into which the tobacco anionic peroxidase gene was introduced.[53,54]

The anionic peroxidase was isolated and purified from leaves of transgenic tobacco plants. An order-of-magnitude better yield was obtained compared to that from wild-type plants.[55] The high yield permitted studies of the tobacco peroxidase, which was shown to be similar to horseradish peroxidase A, the anionic enzyme. It was considerably different from horseradish peroxidase C, the neutral form of horseradish peroxidase. Reaction of tobacco anionic peroxidase with indole-3-acetic acid was also studied,[56] a topic discussed in Section 6.1.5.

13.4. *ARABIDOPSIS THALIANA* PEROXIDASES

In a European-Union-backed project, systematic screening for all of cDNAs encoding plant peroxidases was undertaken. The plant species *Arabidopsis thaliana* was chosen, which belongs to the Crucifereae family, a plant family that also contains horseradish peroxidase. Approximately 30 clones were identified as peroxidases and many of them have been sequenced.[57] A predominant anionic peroxidase with pI of 3.5 was purified and partially sequenced. The cDNA clone was completely sequenced and shown to encode a 30-residue signal peptide and a 305-residue mature protein. The sequence of the *Arabidopsis* enzyme is 95% identical to the anionic horseradish peroxidase isoenzyme A2; the latter has never been successfully cloned. The new enzyme was designated as *A. thaliana* A2.[58] Two novel new peroxidases were discovered in *A. thaliana,* and their many entries in the database of expressed sequence tags indicates they have essential roles. They lack the asparagine residue linked to the distal histidine and possess an N-glycosylation site right at the entrance to the heme channel.[59] The next step in the European Union program is heterologous expression of all of the isoenzymes in bacoluvirus systems, so large-scale preparations and purifications are possible.

13.5. OTHER CLASS III PLANT PEROXIDASES

Amino acid sequences are listed in Appendix A for three other plant peroxidases, tomato peroxidase, turnip peroxidase isoenzyme 7, and wheat peroxidase. Wheat peroxidase is subject to calcium ion control,[60] as is barley peroxidase. In addition valuable contributions have been made to the peroxidase literature from work on horseradish peroxidase isoenzyme A[61] and Japanese radish peroxidase.[62–64]

13.6. STEREOSPECIFIC LIGNIN BIOSYNTHESIS

A role for peroxidases in lignin biosynthesis is universally accepted. However, *in vitro* experiments only produce racemic products. Therefore, peroxidases and other oxidizing enzymes cannot control regio- and stereoselectivitiy by themselves. In an exciting development, a 78-kilodalton protein has been isolated, which in the presence of one-electron oxidants, can stereoselectively couple phenoxyl radicals.[65]

13.7. CANCER IN PLANTS

A 1996 paper entitled "Why don't plants get cancer?"[66] stimulated a strong response.[67] Plant cancerous cells cannot circulate throughout the organism, but that does not preclude their formation. The biochemical characteristics of cancerous plant cells were listed as hyperhydricity, deficiency of tetrapyrolle compounds (peroxidases), permanent oxidative stress, low ethylene production, and accumulation of polyamines. Programmed cell death was also a possible characteristic. Hyperhydricity in plants cultured in vitro was hypothesized to be caused from a burst in ethylene controlled by the peroxidase–indoleacetic acid system.[68] An examination was made of the role of oxidative stress in cancer initiation, progression, and maintenance.[69] The effect of erythromycin on catalase and peroxidase levels in normal and habituated (independence to hormones) sugarbeet calli was investigated.[70] Characteristics that define plant cancer cells at the end of a neoplastic progression were summarized.[71]

REFERENCES

1. R. B. van Huystee, *Can. J. Bot.* **54,** 876–880 (1976).
2. D. Stephan and R. B. van Huystee, *Can. J. Biochem.* **58,** 715–719 (1980).

3. D. Stephan and R. B. van Huystee, *Zeit. Plazenphysiol.* **101,** 313–321 (1981).
4. R. N. Chibbar and R. B. van Huystee, *Plant Physiol.* **75,** 956–958 (1984).
5. R. B. van Huystee and B. A. Maldonado, *Physiol. Plant.* **56,** 88–92 (1982).
6. K. Ravi, C. Hu, P. S. Reddi, and R. B. van Huystee, *J. Expt. Bot.* **37,** 1708–1715 (1986).
7. R. N. Chibbar and R. B. van Huystee, *Phytochemistry* **22,** 1721–1723 (1983).
8. R. N. Chibbar, R. Cella, and R. B. van Huystee, *Can. J. Biochem. Cell Biol.* **62,** 1046–1050 (1984).
9. R. N. Chibbar and R. B. van Huystee, *Phytochemistry* **25,** 585–587 (1986).
10. C. Hu, D. Lee, R. N. Chibbar, and R. B. van Huystee, *Physiol. Plant.* **70,** 99–102 (1987).
11. M. J. Rodríguez, M. J. Stillman, and R. B. van Huystee, *Biochem. Biophys. Res. Commun.* **194,** 326–332 (1993).
12. K. R. Barber, M. J. Rodríguez Marañón, G. S. Shaw, and R. B. van Huystee, *Eur. J. Biochem.* **232,** 825–833 (1995).
13. A.-M. Lambeir, H. B. Dunford, R. B. van Huystee, and J. Lobarzewski, *Can. J. Biochem. Cell Biol.* **63,** 1086–1092 (1985).
14. M. J. Rodríguez Marañón, D. Mercier, R. B. van Huystee, and M. J. Stillman, *Biochem. J.* **301,** 335–341 (1994).
15. J. P. O'Donnell, L. Wan, and R. B. van Huystee, *Biochem. Cell Biol.* **70,** 166–169 (1992).
16. C. Hu, M. Krol, and R. B. van Huystee, *Plant Cell Tissue Organ Culture* **22,** 65–70 (1990).
17. C. Hu, D. Carbonera, and R. B. van Huystee, *Plant Physiol.* **85,** 299–303 (1987).
18. Y. Xu, C. Hu, and R. B. van Huystee, *J. Expt. Bot.* **41,** 1479–1488 (1990).
19. Y. Xu and R. B. van Huystee, *J. Expt. Bot.* **42,** 935–945 (1991).
20. R. B. van Huystee, P. A. Sesto, and J. P. O'Donnell, *Plant Physiol. Biochem.* **30,** 147–152 (1992).
21. L. Wan and R. B. van Huystee, *Biochem. Biophys. Res. Commun.* **194,** 1398–1405 (1993).
22. R. B. van Huystee and L. Wan, *C. R. Acad. Sci. Paris/Sciences de la vie,* **317,** 789–794 (1994).
23. R. B. van Huystee, *Ann. Rev. Plant Physiol.* **38,** 205–219 (1987).
24. R. B. van Huystee in *Peroxidases in Chemistry and Biology,* J. Everse, K. E. Everse and M. B. Grisham, eds., CRC Press, Boca Rotan, 1991, Vol. II, pp. 155–170.
25. M. J. Rodríguez Marañón and R. B. van Huystee, *Phytochemistry* **37,** 1217–1225 (1994).
26. C. Penel, Th. Gaspar, and H. Greppin, eds., *Plant Peroxidase Newsletter,* University of Geneva Press, No. 10, October 1997.
27. D. Buffard, C. Breda, R. B. van Huystee, O. Asemota, M. Pierre, D. B. D. Ha, and R. Esnault, *Proc. Natl. Acad. Sci. U.S.A.* **87,** 8874–8878 (1990).
28. D. J. Schuller, N. Ban, R. B. van Huystee, A. McPherson, and T. L. Poulos, *Structure* **4,** 311–321 (1996).
29. S. K. Rasmussen, K. G. Welinder, and J. Hejgaard, *Plant Mol. Biol.* **16,** 317–327 (1991).

30. A. Johansson, S. K. Rasmussen, J. E. Harthill, and K. G. Welinder, *Plant Mol. Biol.* **18,** 1151–1161 (1992).
31. C. B. Rasmussen, M. Bakovic, K. G. Welinder, and H. B. Dunford, *FEBS Lett.* **321,** 102–105 (1993).
32. C. B. Rasmussen, H. B. Dunford, and K. G. Welinder, *Biochemistry* **34,** 4022–4029 (1995).
33. C. B. Rasmussen, A. N. P. Hiner, A. T. Smith, and K. G. Welinder, *J. Biol. Chem.* **273,** 2232–2240 (1998).
34. C. B. Rasmussen, A. Henriksen, A. K. Abelskov, R. B. Jensen, S. K. Rasmussen, J. Hejgaard, and K. G. Welinder, *Physiol. Plant.* **100,** 102–110 (1997).
35. A. Henriksen, K. G. Welinder, and M. Gajhede, *J. Biol. Chem.* **273,** 2241–2248 (1998).
36. M. Mäder, J. Ungemach, and P. Schloss, *Planta* **147,** 467–470 (1980).
37. M. Mäder in *Molecular and Physiological Aspects of Plant Peroxidases,* H. Greppin, C. Penel, and Th. Gaspar, eds., University of Geneva Press, Geneva, 1986, pp. 247–260.
38. H. J. Grambow and B. Langenbeck-Schwich, *Planta* **157,** 131–137 (1983).
39. K. E. Espelie, V. R. Franceschi, and P. E. Kolattukudy, *Plant Physiol.* **81,** 487–492 (1986).
40. M. Mäder, A. Nessel, and M. Bopp, *Z. Pflanzenphsiol.* **82,** 247–260 (1977).
41. D. T. A. Lamport in Ref. 37, pp. 199–208.
42. M. Mäder, *Z. Pflanzenphsiol.* **96,** 283–296 (1980).
43. T. H. Trinh, Th. Gaspar, K. Tran Thanh Van, and J. L. Marcotte, *Physiol. Plant.* **53,** 153–157 (1981).
44. Th. Gaspar, C. Penel, F. J. Castillo, and H. Greppin, *Physiol. Plant.* **64,** 418–423 (1985).
45. Th. Gaspar, C. Kevers, J. F. Hausman, J. Y. Berthon, and V. Ripetti, *Agronomie* **12,** 757–765 (1992).
46. Th. Gaspar, C. Kevers, J. F. Hausman, J. Y. Berthon, and V. Ripetti in *Physiology, Growth and Development of Plants in Culture,* P. J. Lumsden, J. R. Nicholas, and W. J. Davies, eds., Kluwer Academic, The Netherlands, 1994, pp. 289–298.
47. Th. Gaspar, C. Kevers, J. F. Hausman, C. Penel, L. Jouve, J. Martin-Tanguy, M. Aribaud, and H. Greppin in *Plant Peroxidases. Biochemistry and Physiology,* C. Obinger, U. Burner, R. Ebermann, C. Penel, and H. Greppin, eds., University of Geneva Press, Geneva, 1996, pp. 226–234.
48. L. M. Lagrimini, W. Burkhart, M. Moyer, and S. Rothstein, *Proc. Natl. Acad. Sci. U.S.A.* **84,** 7542–7546 (1987).
49. F. Diaz-De-Leon, K. L. Klotz, and L. M. Lagrimini, *Plant Physiol.* **101,** 1117–1118 (1993).
50. L. M. Lagrimini and S. Rothstein, *Plant Physiol.* **84,** 438–442 (1987).
51. L. M. Lagrimini, S. Bradford, and S. Rothstein, *Plant Cell* **2,** 7–18 (1990).
52. L. M. Lagrimini, *Plant Physiol.* **96,** 577–583 (1991).
53. L. M. Lagrimini, J. Vaughn, J. Finer, K. Klotz, and P. Rubaihayo, *J. Am. Hortic. Sci.* **117,** 1012–1016 (1992).

54. L. M. Lagrimini, J. Vaughn, W. A. Erb, and S. A. Miller, *Hort. Sci.* **28,** 218–221 (1993).

55. I. G. Gazaryan and L. M. Lagrimini, *Phytochemistry* **41,** 1029–1034 (1996).

56. I. G. Gazaryan and L. M. Lagrimini, *Phytochemistry* **42,** 1271–1278 (1996).

57. P. Simon, N. Capelli, J. Flach, S. Overney, M. Tognolli, C. Penel, and H. Greppin in *Plant Peroxidases. Biochemistry and Physiology,* C. Obinger, U. Burner, R. Ebermann, C. Penel and H. Greppin, eds., University of Geneva Press, Geneva, 1996, pp. 179–183.

58. L. Østergaard, A. K. Abelskov, O. Mattson, and K. G. Welinder, *FEBS Lett.* **398,** 243–247 (1996).

59. I. V. H. Kjaersgård, H. M. Jespersen, S. K. Rasmussen, and K. G. Welinder, *Plant Mol. Biol.* **33,** 699–708 (1997).

60. D. A. Converso and M. E. Fernández, *Arch. Biochem. Biophys.* **333,** 59–65 (1996).

61. I. Yamazaki and R. Nakajima in Ref. 37, pp. 71–84.

62. Y. Morita and H. S. Mason, *J. Biol. Chem.* **240,** 2654–2659 (1965).

63. Y. Maeda, T. Higashimura, and Y. Morita, *Biochem. Biophys. Res. Commun.* **29,** 362–367 (1967).

64. K. Hamaguchi, K. Ikeda, C. Yoshida, and Y. Morita, *J. Biochem.* **66,** 191–201 (1969).

65. L. B. Davin, H.-B. Wang, A. L. Crowell, D. L. Bedgar, D. M. Martin, S. Sarkanen, and N. G. Lewis, *Science* **275,** 362–366 (1997).

66. J. Doonan and T. Hunt, *Nature* **380,** 481–482 (1996).

67. Th. Gaspar, *Plant Physiol. Biochem.* **36,** 203–204 (1998).

68. C. Kevers, M. Coumans, M.-F. Coumans-Gillès, and Th. Gaspar, *Physiol. Plant.* **61,** 69–74 (1984).

69. Th. Gaspar, *Plant Tissue Culture Biotech.* **1,** 127–136 (1995).

70. B. Bisbis, C. Kevers, C. Hualt, J. P. Billard, and Th. Gaspar, *Plant Growth Regul.* **23,** 147–152 (1997).

71. Th. Gaspar, B. Bisbis, C. Kevers, C. Penel, H. Greppin, F. Le Dily, J. P. Billard, C. Hualt, F. Garnier, M. Rideau, and J. M. Foidart, *Plant Growth Regul.* **24,** 135–144 (1998).

CHAPTER 14

Chloroperoxidase and *Pseudomonas* Cytochrome *c* Peroxidase

In this chapter two disparate peroxidases, which do not fit into the Welinder classification, are described. Their unusual structures and properties add to the fascinating peroxidase story.

14.1. CHLOROPEROXIDASE

14.1.1. Discovery and Properties

The story of chloroperoxidase begins with a single report in 1940. No further work was published until 1959, the beginning of the intensive pioneering work conducted in the laboratory of Lowell Hager. An enzyme or enzyme system in *Caldariomyces fumago* was shown to convert β-ketoadipic acid to δ-chlorolevulinic acid.[1,2] Ring closure converted the latter to 2-chloro-1,3-diketocyclopentane. Further chlorination produced the dichloro derivative, caldariomycin (Fig. 14.1).[3] The enzyme system was found to contain glucose oxidase as a source of hydrogen peroxide, catalase to destroy excess hydrogen peroxide, and the chlorinating enzyme, called chloroperoxidase.[4–6]

Work now began in earnest to purify and characterize chloroperoxidase.[7] The enzyme was shown to be monomeric, with one ferriprotoporphyrin IX per molecule. Chloroperoxidase was found to contain 25–30% carbohydrate. Broadline nuclear magnetic resonance (NMR) results indicated that halide ions did not bind directly to the iron.[8] However, later experiments showed that chloride binds weakly in the sixth coordination site.[9] Nonspecific binding of chloride also occurs at low pH where the enzyme contains many positive charges.

Chloroperoxidase can oxidize chloride, bromide, and iodide,[10] as can myeloperoxidase. In contrast, lactoperoxidase can oxidize bromide and iodide, and horseradish peroxidase only iodide. An assay procedure was developed based on the conversion of monochlorodimedone to dichlorodimedone. Analysis of products in the chlorination of anisole by chloroperoxidase provided evidence that the oxidation of chloride by the enzyme was a two-electron reaction, producing chlorine in a +1 oxidation state.[11]

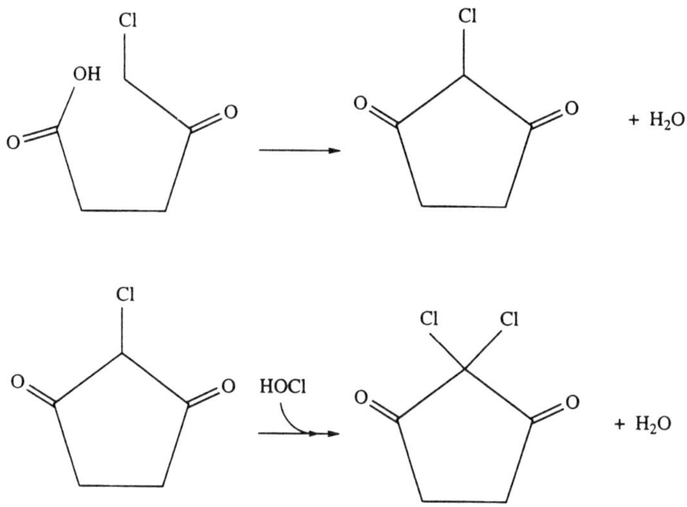

FIG. 14.1. Conversion of δ-chlorolevulinic acid to monochlorodimedone (top). Conversion of monochlorodimedone to dichlorodimedone (bottom).

Reactions of chloroperoxidase were classified into four categories. First, chlorination reactions had acid pH optima, and utilized three substrates: hydrogen peroxide or other hydroperoxides, chloride ion, and a wide range of compounds susceptible to chlorination. Second, some substrates such as uric acid also require both hydrogen peroxide and chloride, but the final product is oxidized, not chlorinated. Such substrates also have an acid pH optimum. It would appear that they may form a chlorinated intermediate. Third, in the absence of chloride, classic peroxidase reactions were observed, with pH optima in the range 4–7, dependent upon the organic substrate. Fourth, in the absence of chloride and organic substrate, the prime reaction of catalases is observed, the disproportionation of hydrogen peroxide to dioxygen and water. At the time of these observations, it was not realized that chloroperoxidase had properties in common with cytochrome P450. Thus, compared to other enzymes in the plant peroxidase superfamily, chloroperoxidase is capable of three unique reactions, chlorinations, the catalatic reaction,[12,13] and P450-type reactions.

Chloroperoxidase was shown to form conventional compounds I and II, which contain two and one oxidizing equivalents in excess of that in the resting ferric enzyme.[14] The determination of an accurate spectrum for compound I posed a problem because of the catalatic reaction. Replacement of hydrogen peroxide by an organic hydroperoxide helped to minimize the catalatic reaction. Methyl hydroperoxide was an improvement and peracetic acid was even better.[15] The spectra of native chloroperoxidase and its compound I are shown in Fig. 14.2. The compound I Soret spectrum has about half of the molar absorptivity of the native enzyme, which is typical for a peroxidase, but the spectrum is very unusual in that the Soret maximum is shifted dramatically toward the ultraviolet. The molar absorptivity (ε) of the Soret maximum for the native enzyme, shown in Fig. 14.2, is 85 $mM^{-1}cm^{-1}$ at 400 nm. This should be increased to 91 $mM^{-1}cm^{-1}$ at 399 nm.[16] The compound I Soret maximum occurs at 367 nm with ε = 52 $mM^{-1}cm^{-1}$ (corrected). Barely distinguishable visible region compound I peaks occur at 545 and 610 nm, and a pronounced peak at 688 nm (ε = 12.5 $mM^{-1}cm^{-1}$, corrected). Isosbestic points between native enzyme and compound I occur at 368, 444, 476, and 573 nm.

Use was made of the stoichiometry of the reaction of *m*-chloroperbenzoic acid with native enzyme to show that one atom of oxygen is incorporated directly from the acid into the enzyme structure; and, by analogy with other peroxidases, this must form a ferryl ($Fe^{IV}{=}O$) structure and a porphyrin π-cation radical. By use of double-labeling ^{18}O experiments using both hydrogen peroxide and *m*-chloroperbenzoic acid, the catalatic reaction was shown to occur by a nonscrambling mechanism as shown in Fig. 14.3.[17,18]

Reactions of chlorite and chlorine dioxide with chloro- and horseradish peroxidase were studied.[19,20] Incorporation of ^{36}Cl directly into substrate without exchange with excess chloride provided evidence for a halogenating intermediate of the form $Fe{-}O{-}X^{+}$.

The rate constants for compound I formation (k_1) and the catalatic reaction (k_2) were measured for three substrates at pH 4.7.

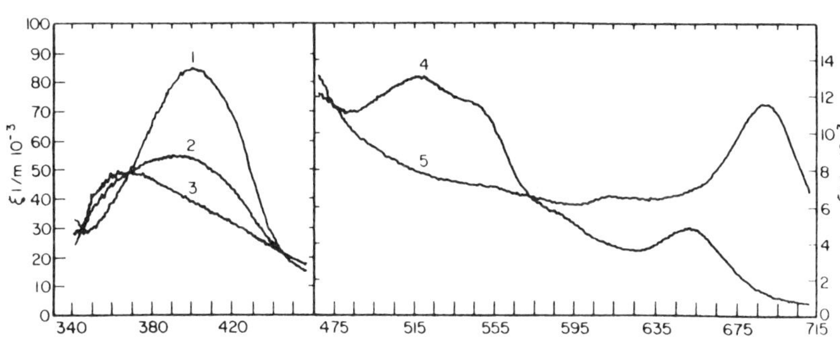

FIG. 14.2. Optical spectra of native chloroperoxidase with purity number (reinheitzahl, RZ) of 1.40–1.44, and of compound I in the Soret and visible regions. Spectra labeled 1 and 4 are of the native enzyme, and 3 and 5 of compound I obtained 22 ms after reaction of excess peracetic acid with native enzyme. The spectrum 2 shows partial conversion to compound I approximately 1 ms after mixing. Citrate buffer of pH ~ 3.5 was used. The ordinates are molar absorptivities ($mM^{-1}cm^{-1}$), obtained using a 1-cm pathlength. [Reproduced with permission from M. M. Palcic, R. Rutter, T. Araiso, L. P. Hager, and H. B. Dunford, *Biochem. Biophys. Res. Commun.* **94,** 1123–1127 (1980).]

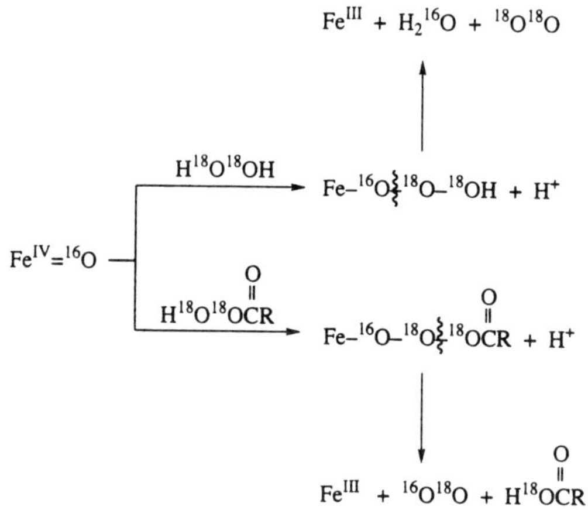

FIG. 14.3. Schematic representation of the catalase reaction of chloroperoxidase using ^{18}O double-labeling experiments. Starting with compound I, represented as $Fe^{IV}{=}O$, three-oxygen atom arrays are formed, which cleave differently for hydrogen peroxide compared to *m*-chloroperbenzoic acid. [Reproduced with permission from L. P. Hager, D. L. Doubek, R. M. Silverstein, J. H. Hargis, and J. C. Martin, *J. Am. Chem. Soc.* **94,** 4364–4366 (1972).]

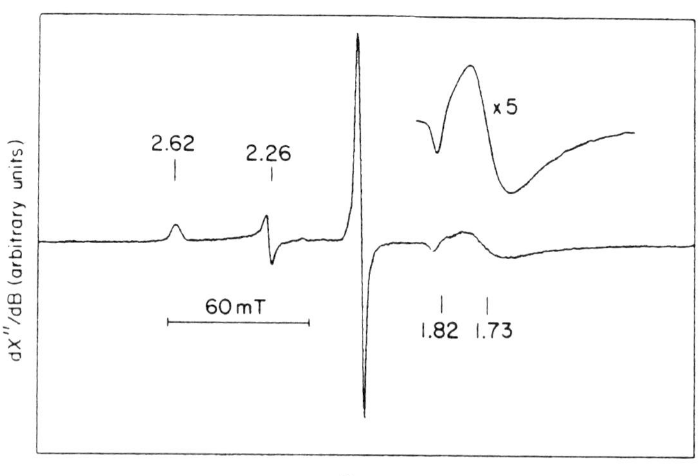

FIG. 14.4. EPR spectrum of chloroperoxidase compound I (55%) at 3.5 K. The derivative spectrum of compound I consists of the $g_{\perp} = 1.73$ and $g_{\parallel} = 2$ signals. The g = 1.82, 2.26, and 2.62 signals are from the 45% native enzyme. [Reproduced with permission from R. Rutter, L. P. Hager, H. Dhonau, M. Hendrich, M. Valentine, and P. Debrunner, *Biochemistry* **23,** 6809–6816 (1984).]

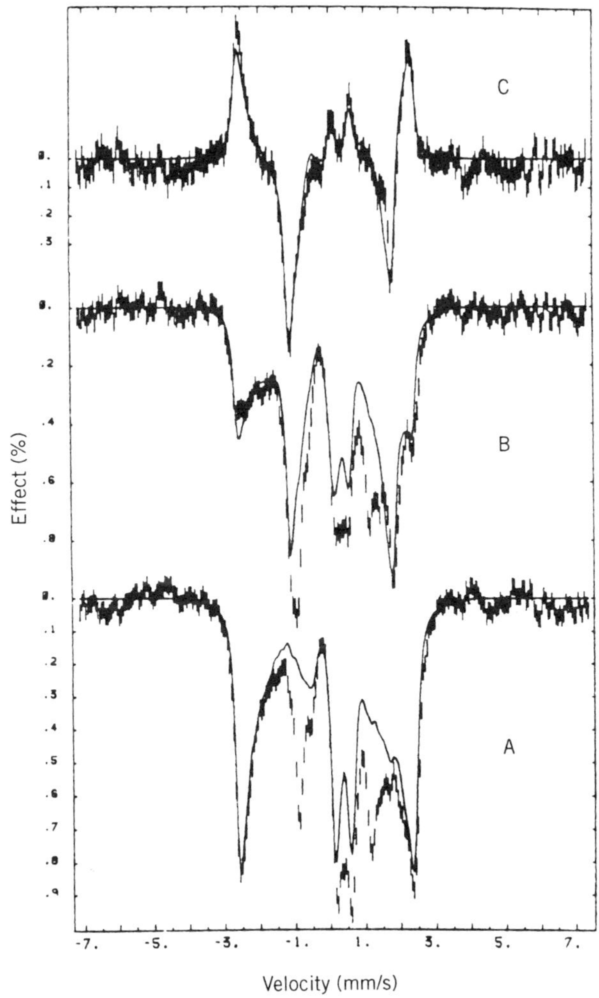

FIG. 14.5. Mössbauer spectra of chloroperoxidase compound I obtained at 4.2 K. Spectra from the native enzyme were subtracted. (A) Spectrum parallel to the direction of the gamma-ray beam, (B) perpendicular, (C) = (B)–(A). The solid lines are calculated from the spin simulation. [Reproduced with permission from R. Rutter, L. P. Hager, H. Dhonau, M. Hendrich, M. Valentine, and P. Debrunner, *Biochemistry* **23,** 6809–6816 (1984).]

$$\mathrm{Fe^{III} + ROOH \xrightarrow{k_1} Fe^{IV}{=}O + ROH}$$

$$\mathrm{Fe^{IV}{=}O + ROOH \xrightarrow{k_2} Fe^{III} + O_2 + ROH}$$

Fe^{III} and $Fe^{IV}{=}O$ are schematic representations of native enzyme and compound I. The values of k_1 for hydrogen peroxide, methyl hydroperoxide, and peracetic acid were 2.4×10^6, 4.7×10^5, and 3.8×10^6 $M^{-1}s^{-1}$. For k_2 the corresponding values are 3.4×10^5, 2.9×10^4 $M^{-1}s^{-1}$, and zero. In other words, a 100% yield of compound I can be obtained using peracetic acid.[21]

Electron paramagnetic resonance (EPR) and Mössbauer spectra were obtained of compound I (Figs. 14.4 and 14.5).[22,23] The results could be explained by a spin-coupling model between an $S = 1$ ferryl ion and an $S' = \frac{1}{2}$ porphyrin radical.

A major source of the unusual properties of chloroperoxidase is the proximal ligand. Evidence was accumulated that a thiolate ligand occupied the fifth coordination position of the heme iron. When the enzyme was reduced and its carbon monoxide complex was formed, the Soret band was shifted to 446 nm (Fig. 14.6), in very similar fashion to cytochrome P450, which is known to

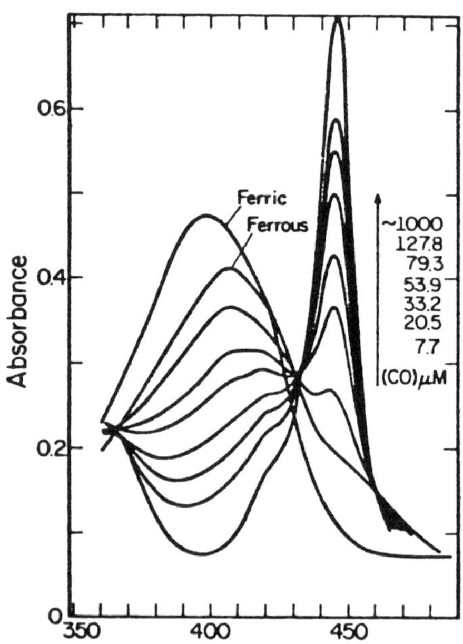

FIG. 14.6. Formation of the ferrous–carbon monoxide complex of chloroperoxidase as a function of carbon monoxide concentration. The ferric enzyme was first reduced using reduced methyl viologen. [Reproduced with permission from R. Makino, R. Chiang, and L. P. Hager, *Biochemistry* **15,** 4748–4754 (1976).]

have a thiolate ligand.[24,25] Data from Mössbauer spectra,[26,27] magnetic circular dichroism,[28] resonance Raman spectra,[29,30] extended X-ray absorption fine structure (EXAFS) studies,[31] EPR spectra,[32] hyperporphyrin spectra obtained upon thiol binding,[33] ligand binding studies that were compared to cytochrome P450 results,[34] and chemical modification studies[35] were all consistent with a thiolate fifth ligand in chloroperoxidase.[36]

The reduction potential for the ferric-ferrous couple displayed a discontinuity at pH 4.7 caused by a change to the ferrous enzyme. The potential shifted from +150 mV at pH 2.7 to −140 mV at pH 6.9. A comparison of potentials of chloro- and horseradish peroxidases as a function of pH is shown in Fig. 14.7.[25] By use of ^{60}Co gamma rays, chloroperoxidase was reduced to its ferrous state at 77 K and at room temperature.[37] A new low-spin form of the ferrous enzyme was produced at 77 K.

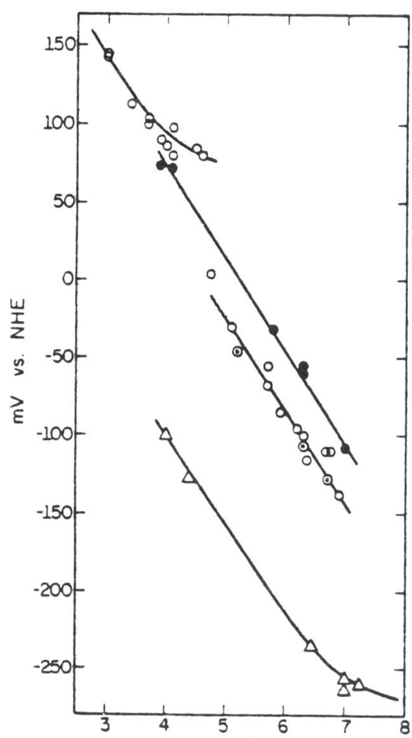

FIG. 14.7. Effect of pH on the redox potential of chloroperoxidase. Potential versus the normal hydrogen electrode is plotted as a function of pH. Open circles, chloroperoxidase; dotted open circles, duplicate experiments. Open triangles, horseradish peroxidase; solid circles indigo disulfonate, shown for comparison. [Reproduced with permission from R. Makino, R. Chiang, and L. P. Hager, *Biochemistry* **15,** 4748–4754 (1976).]

Chloroperoxidase is stable and active from pH 3 or lower to pH 7. At pH 7 a sharp, irreversible transition occurs to a first alkaline form on a time scale slow enough to observe by conventional spectrophotometry. If the pH is jumped from below pH 7 to above pH 10, the conversion to the first alkaline form is practically instantaneous on the stopped flow time scale. At pH 11 a further transition to a second alkaline form occurs, which may be the irreversible dissociation of the heme.[38] The alkaline transitions of both ferric and ferrous chloroperoxidase were characterized by electronic absorption, magnetic circular dichroism (MCD) and EPR spectroscopy.[39]

The ferrous form of chloroperoxidase was prepared by dithionite reduction, and compound III by reaction with oxygen.[40,41] The time course for the conversion of the ferrous enzyme to compound III, and the decay of compound III to the native enzyme, is shown in Fig. 14.8. The rate constant for oxygen binding to ferrous chloroperoxidase is pH-independent from pH 4 to 6, with a value of $5.5 \times 10^5\ M^{-1}s^{-1}$, comparable to cytochrome P450 rates, but an order-of-magnitude faster than oxygen binding to horseradish peroxidase.[40] The nitric oxide and alkyl isocyanide complexes were also formed from the ferrous enzyme, and MCD spectra were obtained.[41] Compound III was also formed by reaction of compound II with excess hydrogen peroxide.[42] The spectra of compounds II and III in both the Soret and visible regions are

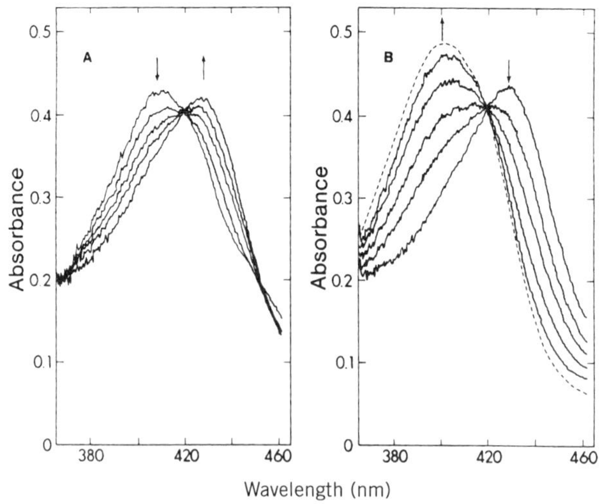

FIG. 14.8. Spectral changes after mixing dithionite-reduced chloroperoxidase with air-saturated buffer at pH 4. (A) Conversion of ferrous enzyme to compound III. Traces were recorded 0, 6, 12, and 26 ms after mixing. (B) Decay of compound III to ferric enzyme. Traces recorded 0.04, 0.3, 0.75, 2.25, and 8.25 s after mixing. The dashed line is the spectrum of the native enzyme. [Reproduced with permission from A.-M. Lambeir and H. B. Dunford, *Eur. J. Biochem.* **147,** 93–96 (1985).]

shown in Fig. 14.9. Bands of compound II occur at 438, 542, and 571 nm, and of compound III at 432, 555, and 586 nm.

14.1.2. Amino Acid Sequence and X-Ray Structure

The *c*DNA sequence of the chloroperoxidase gene was determined,[43,44] and subsequent modifications were made to the amino acid sequence on the basis of the crystal structure.[45] The determined sequence contained 20 residues that are the signal peptide at the N-terminal end. A 59-residue sequence at the C-terminal end could not be fit into the crystallographic data, which indicated there was no room for the sequence. Finally, a glutamine residue at the N-terminal end, excluded from the mature protein sequence, was found to be present in the crystal structure as the cyclized pyroglutamic acid residue. This was designated as residue number 0, in order to preserve the original numbering of the mature protein. Thus, the final secondary structure of mature chloroperoxidase consists of 299 amino acid residues, numbered from 0 to 298. This sequence is reproduced in Appendix A.

Cys29 is coordinated to the heme iron of chloroperoxidase. The Cys29 environment is shown in Fig. 14.10 (see p. 332 and color insert). Although

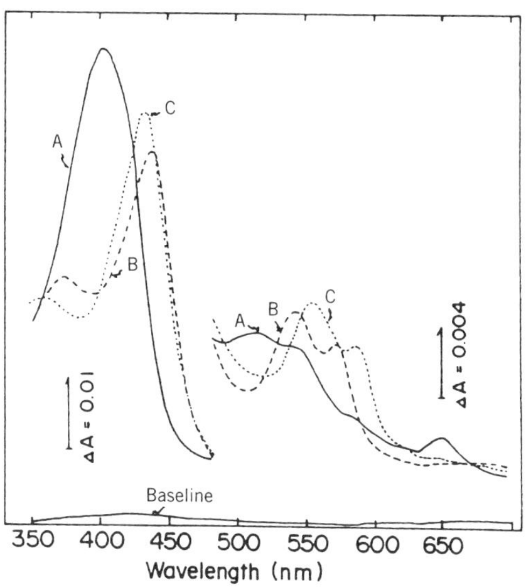

FIG. 14.9. Absorption spectra of chloroperoxidase and its compounds II and III obtained at pH 5.6. The size of the absorbance scales is indicated by ΔA values. (A) Spectrum of native enzyme, (B) of compound II prepared using hydrogen peroxide and ascorbate, and (C) of compound III formed by reacting compound II with excess hydrogen peroxide. [Reproduced with permission from R. Nakajima, I. Yamazaki, and B. W. Griffin, *Biochem. Biophys. Res. Commun.* **128,** 1–6 (1985).]

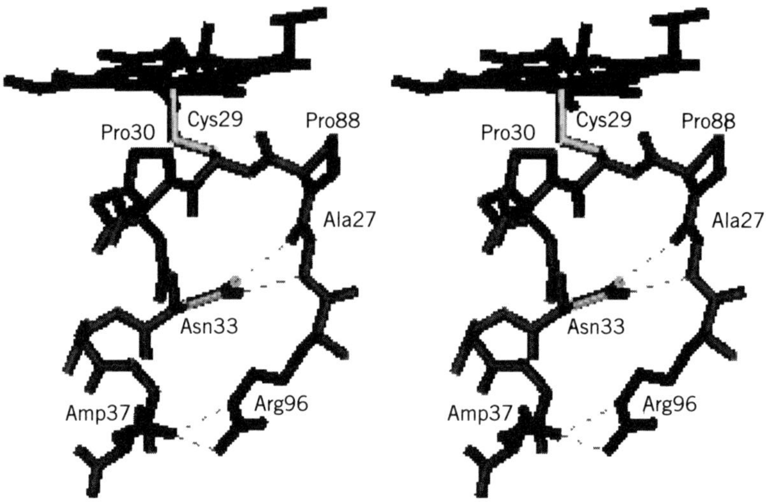

FIG. 14.10. Stereo diagram showing the ligation of Cys29 to the heme iron of chloroperoxidase and hydrogen bonds (dashed lines), which help to stabilize cysteine–ligand loop. See color insert. [Reproduced with permission from M. Sundaramoorthy, J. Terner, and T. L. Poulos, *Structure* **3,** 1367–1377 (1995).]

chloroperoxidase has a thiolate fifth ligand as does cytochrome P450, its structure is considerably different. The main part of the structural fold of chloroperoxidase consists of eight α-helices, labeled A–H. The helices on the proximal side lie parallel to the heme in most peroxidases and cytochrome P450; in chloroperoxidase heme A is perpendicular to the heme. In chloroperoxidase proximal heme A is near the N terminus; in other peroxidases and cytochrome P450 the proximal helices are closest to the C terminii.

Helix F contains the distal residues of chloroperoxidase. Other than the absence of a sixth ligand, the distal region has little similarity to other peroxidases. Instead of a distal histidine and arginine, as found in other plant peroxidases, Glu183 is closest to the heme iron. His105, hydrogen bonded to Glu183, is 3.5 Å away from a heme propionate, which is a ligand of manganese(II)! The role of the manganese ion is unknown. The positioning of helices A and F relative to the heme is shown in Fig. 14.11 (see p. 333 and color insert), and the ligands of the manganous ion in Fig. 14.12 electron nuclear double resonance (ENDOR) experiments at 2 K indicate that at this low temperature water is coordinated to the heme iron of chloroperoxidase.[46]

The crystals used for data analysis were of such high quality that many of the carbohydrate attachments to chloroperoxidase could be clearly defined. The data are for isoenzyme A, which is more heavily glycosylated than isoenzyme B (see Section 14.1.3). Fourteen glycosylation sites with a total of 21

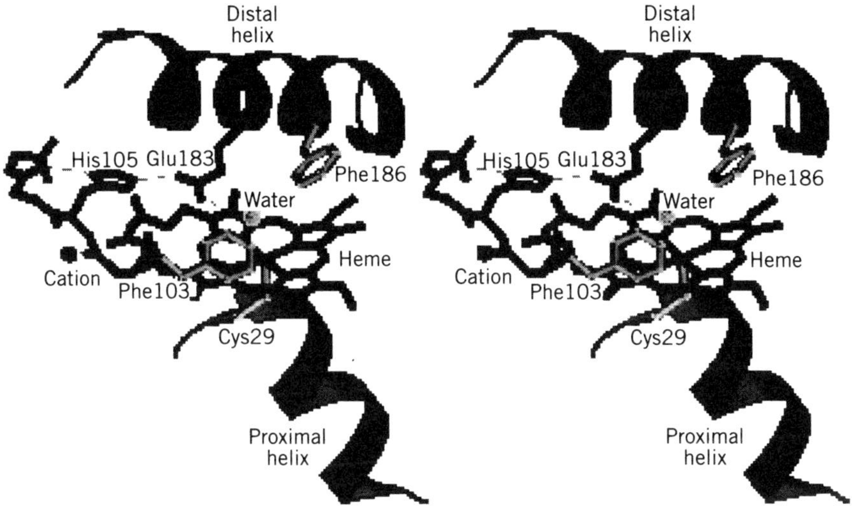

FIG. 14.11. Relation of the proximal (A) and distal (F) helices to the heme of chloroperoxidase. The water in the distal cavity is too far removed to be a ligand. See color insert. [Reproduced with permission from M. Sundaramoorthy, J. Terner, and T. L. Poulos, *Structure* **3,** 1367–1377 (1995).]

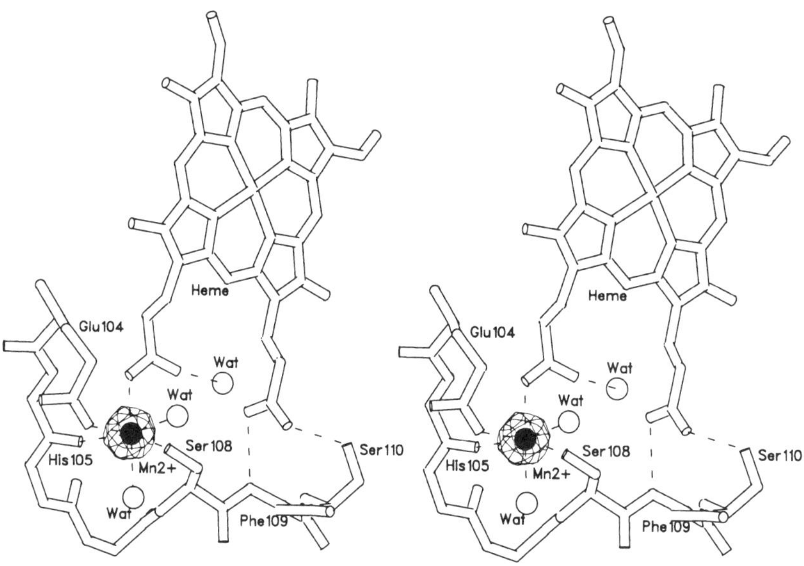

FIG. 14.12. Ligands and the environment of Mn^{II} in chloroperoxidase. [Reproduced with permission from M. Sundaramoorthy, J. Terner, and T. L. Poulos, *Structure* **3,** 1367–1377 (1995).]

sugar molecules were detected. There are three N-linked glycosyl chains, attached to asparagine residues 12, 93, and 216 with a minimum of two *N*-acetylglucosamine residues at each site. All three asparagines are in sequences of the form Asn-X-Ser-Thr. Eleven O-linked glycosylation sites were found on serine and threonine sites, all of which are near the C-terminal end. The serine sites are residues 239, 241, 242, 248, and 251; and the threonine sites, are 238, 250, 252, 275, 283, and 293. Most of the O-bound sugars are single mannose units.

The determination of the crystal structure of chloroperoxidase was a major triumph, both in obtaining satisfactory crystals[47] and in the data analysis for such a unique enzyme. Crystals of chloroperoxidase obtained earlier were not satisfactory for X-ray analysis.[7,48] The accurate static structure provides a scaffolding from which determinations of mechanisms can proceed with greater confidence for all the various reactions of chloroperoxidase.

14.1.3. Production and Isoenzymes of Chloroperoxidase

Improvements have been made to the original method of obtaining chloroperoxidase in which a glucose–malt extract medium was used to grow *Caldariomyces fumago*.[7] Removal of the malt and replacement of glucose with fructose lead to faster production with reduced pigmentation in the medium.[49] Further improvements in production were made by using a fructose–phytone,[50] instead of a fructose–salts, medium.[51,52] Two isoenzymes, A and B, were reported, with molecular weights estimated to be 46,000 and 40,000.[53] Later work showed that there are multiple forms, differing only in carbohydrate content.[50] Batch, semicontinuous, and continuous-flow methods of chloroperoxidase production were developed, so that large quantities of enzyme can be obtained.[54–56] A colorimetric assay for immobilized chloroperoxidase was developed using *N,N,N′,N′*-tetramethyl-*p*-phenylenediamine.[57]

14.1.4. Reactions of Chloroperoxidase

The very fast reversible binding of hydrazoic acid to chloroperoxidase was studied using an iodine laser temperature jump apparatus.[58] The inflection in the "on" rate corresponds exactly to the pK_a value of hydrazoic acid, showing it is the acid, not azide, ion, which is binding. The invariant "off" rate as a function of pH shows that the proton, required for the release of hydrazoic acid, is obtained from the enzyme, not the bulk solution. Therefore, both components of the hydrazoic acid, its proton, and the azide ion are attached to the enzyme upon binding and released upon dissociation (Fig. 14.13). The site of proton binding is undoubtedly a distal amino acid residue, most likely the catalytically active imidazole side chain of histidine.

By extrapolation from the influence of nitrate on the rate of cyanide binding, a pK_a value of 3.1 ± 0.3 was determined for an active site ionization on the

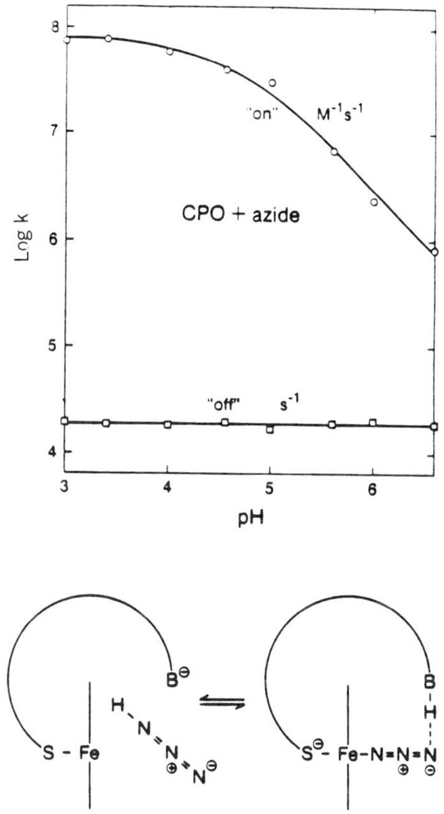

FIG. 14.13. pH rate profiles for the binding (on) and dissociation (off) of azide with chloroperoxidase. A schematic representation of the reversible process is shown. [Reproduced with permission from J. F. Holzwarth, F. Meyer, M. Pickard, and H. B. Dunford, *Biochemistry* **27,** 6628–6633 (1988).]

native enzyme.[59] The influence of nitrate on cyanide binding is shown in Fig. 14.14. The interesting effect of nitrate on horseradish peroxidase compound I formation is discussed in detail in Section 4.5.2.

One of the classic reactions of a peroxidase, oxidation of phenols, was studied.[60] The reactions were suggested as a method of removal of phenolics from waste waters, as had been suggested for horseradish peroxidase. An aromatic amine, 4-chloroaniline, was found to be oxidized to 4-chloronitrosobenzene by hydrogen peroxide in the presence of chloroperoxidase. The same reaction proceeded even faster in the presence of chloride, despite extensive ring chlorination.[61,62] Indole was shown to be oxidized almost exclusively to oxindole by hydrogen peroxide and chloroperoxidase.[63] Chloroperoxidase was shown to catalyze chlorination of steroids.[64] Thiols and disulfides were oxidized

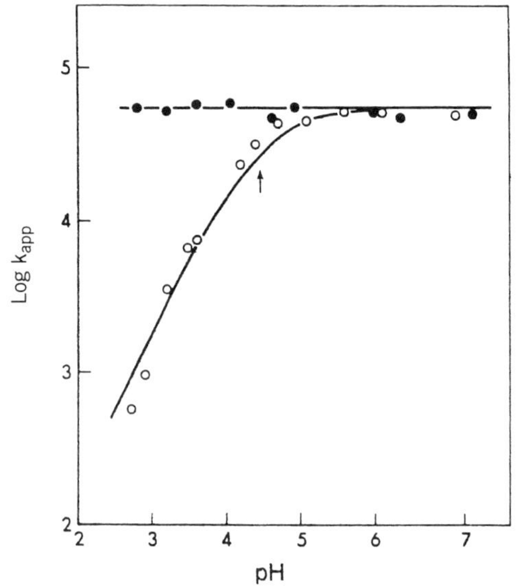

FIG. 14.14. pH profile for the rate of cyanide binding to chloroperoxidase at 25°C and ionic strength 0.11 M. Open circles indicate experiments in the presence of 0.06 M potassium nitrate; closed circles no nitrate present. [Reproduced with permission from A.-M. Lambeir, H. B. Dunford, and M. A. Pickard, *J. Inorg. Biochem.* **19,** 291–300 (1983) with permission from Elsevier Science.]

to sulfenyl chlorides by chloroperoxidase.[65] Chloride was shown to be incorporated into NADH both by hypochlorous acid and a chloroperoxidase-catalyzed reaction.[66]

Mechanism of Chlorination

The kinetics of the chlorination of monochlorodimedone to the dichloro compound has been exhaustively studied,[9,67] and the most plausible mechanism is shown in Fig. 14.15. A transient chlorinating intermediate, formed by addition of a chloride ion to the ferryl oxygen atom of chloroperoxidase compound I, is responsible for the chlorination reaction. The relevant rate constants at pH 3.2 are as follows: k_1 for compound I formation from peracetic acid, 2.3×10^6 $M^{-1}s^{-1}$; k_2 for chloride addition to compound I, 9×10^4 $M^{-1}s^{-1}$; and k_3 for the chlorination reaction, 2.4×10^7 $M^{-1}s^{-1}$. Therefore, the rate constant for reaction of the chlorinating intermediate with monochlorodimedone is more than two orders of magnitude faster than for its formation. The results obtained on monochlorodimedone indicate a binding site for the substrate that is sufficiently close to the iron center for reaction to occur. If a poorly oriented or nonbinding substrate is used, then release of free hypochlorous acid would be expected to occur:

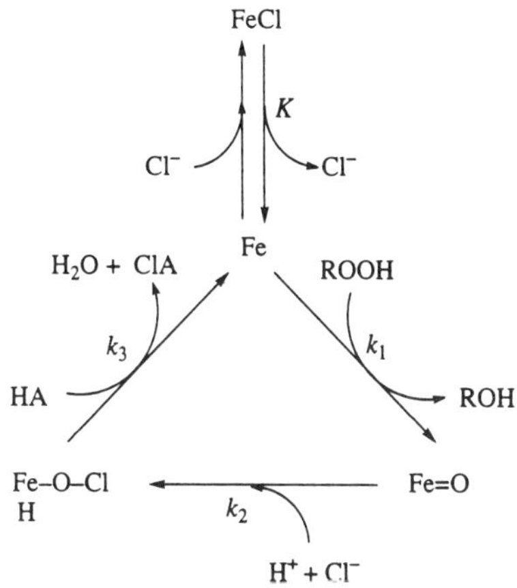

FIG. 14.15. Mechanism of chlorination of monochlorodimedone (HA) by chloroperoxidase in the presence of peracetic acid (ROOH) and chloride ion. Weak binding of chloride to the iron site occurs. Fe represents native enzyme, Fe=O compound I, and HFe—O—Cl the chlorinating intermediate. [Based upon, with permission, from A.-M. Lambeir and H. B. Dunford, *J. Biol. Chem.* **258,** 13558–13563 (1983).]

$$\underset{\text{H}}{\text{Fe}}{-}\text{O}{-}\text{Cl} \rightarrow \text{Fe} + \text{HOCl}$$

Generation of molecular chlorine might occur by the third-order reaction[68]:

$$\text{HOCl} + \text{Cl}^- + \text{H}^+ \rightarrow \text{Cl}_2 + \text{H}_2\text{O}$$

which has a rate constant of $1.80 \times 10^4\ \text{M}^{-2}\text{s}^{-1}$. To convert this to a second-order rate constant, at pH 3 the hydrogen ion concentration is 10^{-3} M, so the rate constant becomes $18\ \text{M}^{-1}\text{s}^{-1}$, indicative of a very slow reaction. Chlorine production might also occur by the reaction

$$\underset{\text{H}}{\text{Fe}}{-}\text{O}{-}\text{Cl} + \text{Cl}^- + \text{H}^+ \rightarrow \text{Fe} + \text{Cl}_2 + \text{H}_2\text{O}$$

Rate constants for the dissociation of either HOCl or Cl_2 from the active site of chloroperoxidase are unknown.

The available evidence does not support a radical mechanism for chlorination of monochlorodimedone.[69] In the absence of enzyme, HOCl reacts with monochlorodimedone with $k = 1.6 \times 10^6\ \text{M}^{-1}\text{s}^{-1}$. If chloride is also present, then generation of Cl_2 occurs in the nonenzymatic reaction, which sets off a chain reaction involving chlorine atoms.

In summary, an enzyme-assisted chlorination occurs if the substrate binds appropriately.[70–72] Otherwise the enzyme-bound hypochlorous acid dissociates, and HOCl is the chlorinating reagent. There appears to be little possibility that Cl_2 might play a role.

Mechanism of N*-Deamination Reactions*

The first report of a chloroperoxidase-catalyzed *N*-deamination reaction, similar to that of cytochrome P450, was the conversion of *N,N*-dimethylaniline to *N*-methylaniline and formaldehyde.[73] Chloroperoxidase did not catalyze the demethylation of *N,N*-dimethylaniline-*N*-oxide, indicating that the latter was not an intermediate. The comparative peroxidase and P450 dealkylation mechanisms have been extensively studied.[74,75] The most plausible chloroperoxidase mechanism appears to be that shown in Fig. 14.16. Loss of an electron from the amine produces an aminium cation radical. Disproportionation of cation radicals generates an iminium radical that hydrolyzes to formaldehyde and the deaminated product. The one-electron transfer that initiates the deamination could occur to both compounds I and II. Antipyrine was shown to be both chlorinated and deaminated by chloroperoxidase.[76]

Mechanisms of Reactions of Alkenes, Alkynes, and Cyclopropanes

Alkenes are converted to α,β-halohydrins, alkynes to mono- and di-halogenated ketones, and cyclopropanes to α,γ-chlorohydrins by haloperoxidases, hydrogen peroxide, and chloride ion.[77] The same products are formed when hypohalous acid is the halogenating reagent, indicating that hypohalous acid is the enzyme-generated reagent. The reactions with the unsaturated compounds followed Markovnikov's rule, with the halogen going to the carbon with the most hydrogens and the hydroxyl group forming on carbon best able to stabilize a positive charge. Little steroselectivity was reported in the chlorination of 2-methyl-4-propyl-cyclopentane-1,3-dione.[78]

In the absence of chloride, chloroperoxidase was shown to catalyze epoxidations, as does cytochrome P450.[79,80] By continuous addition of low levels

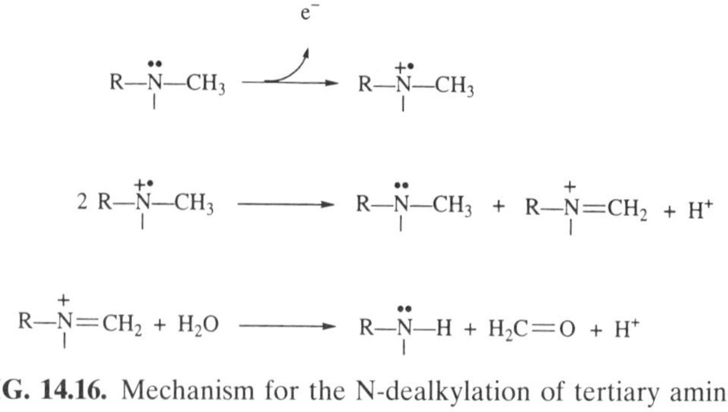

FIG. 14.16. Mechanism for the N-dealkylation of tertiary amines.

of hydrogen peroxide, which helped to prevent enzyme inactivation, stereoselective epoxidation of alkenes was achieved.[81–83] Even better results were obtained by replacing hydrogen peroxide by *t*-butylhydroperoxide.[84] Allylic and benzylic hydroxylations were demonstrated.[85] When allylbenzene is epoxidized, reversible covalent modification of a pyrolle ring of the enzyme's porphyrin occurs.[86,87]

Oxidation of Sulfides

Methionine was reported to yield a 50:50 mixture of the diastereo isomers using chloride.[78,83] However, with improved experiments, and excluding chloride from the reaction mixture, asymmetric oxidations of sulfides catalyzed by chloroperoxidase were demonstrated.[88,89]

14.2. *PSEUDOMONAS AERUGINOSA* CYTOCHROME *c* PEROXIDASE

A cytochrome *c* peroxidase was isolated and partially purified in the prechromatographic era from the organism called *Pseudomonas fluorescens* by the group of Nathan Kaplan.[90–92] Kaplan donated some of the same strain of organism, later identified as *Pseudomonas aeruginosa,*[93] to Nils Ellfolk and co-workers, who embarked on a comprehensive study of the enzyme. The enzyme was purified to homogeneity,[94] and its size and shape were investigated by ultracentrifugation.[95] Larger scale purifications starting with 125 g of cells were achieved.[96] It was established that the enzyme consisted of a single polypeptide chain with two heme *c* groups covalently attached.[97] The molecular weight was determined to be 40,000 ± 1500 by polyacrylamide gel electrophoresis in the presence of sodium dodecyl sulfate. Like yeast cytochrome *c* peroxidase, the enzyme contains no carbohydrate. Proteinase molecules are present in the organism so that it was found necessary to carry out the isolation procedure as rapidly as possible at low temperature.[98]

P. aeruginosa cytochrome *c* peroxidase uses hydrogen peroxide to produce oxidized forms. Its oxidized forms can reduce both cytochrome *c* and azurin, a copper-containing protein. When isolated, both heme groups contain ferric ion. The two heme groups can be totally reduced to ferrous using sodium dithionite.[99] A series of steady-state kinetic studies was undertaken, starting with fully reduced enzyme in the presence of reducing substrate, with the reactions initiated by adding hydrogen peroxide.[100–102] Presence of reducing substrate eliminated a lag period.[103] The two heme *c* groups were found to be inequivalent, with one that contains a high-spin iron being the target for hydrogen peroxide. The other iron is in a low-spin form.[104,105]

The kinetic studies were extended to rapid-scan stopped flow measurements. It was found that the totally reduced enzyme reacted slowly with hydrogen peroxide. However, the half-reduced form, containing one ferric and one ferrous heme, was highly reactive.[106,107] The conversion of the sluggish ferric–ferric form to the active ferric–ferrous form could be accomplished using either ferrocyanide or reduced azurin as the reducing agent.

The kinetics and equilibria of the reduction of *P. aeruginosa* cytochrome *c* peroxidase by azurin was investigated.[108] Difference spectra between the totally ferric form and successive additions of azurin, sufficient to form the half-reduced enzyme, are shown in Fig. 14.17. Biphasic kinetics were observed in the reduction of the ferric-ferric form to ferric-ferrous at pH 6.0, which were explained in terms of the scheme shown in Fig. 14.18. The extent of the fast, compared to the slow, phase was dependent upon azurin concentration, indicating an equilibrium between two forms of the enzyme that was affected by the rate of the reaction. The scheme shown in Fig. 14.18 was confirmed in a study conducted over the pH range of 5.5–8.0.[109]

Electron paramagnetic resonance spectral studies at 15 K indicated that both ferric hemes had six-coordinate iron: imidazole nitrogen–heme iron–methionine sulfur in one case, and imidazole nitrogen–heme iron–imidazole

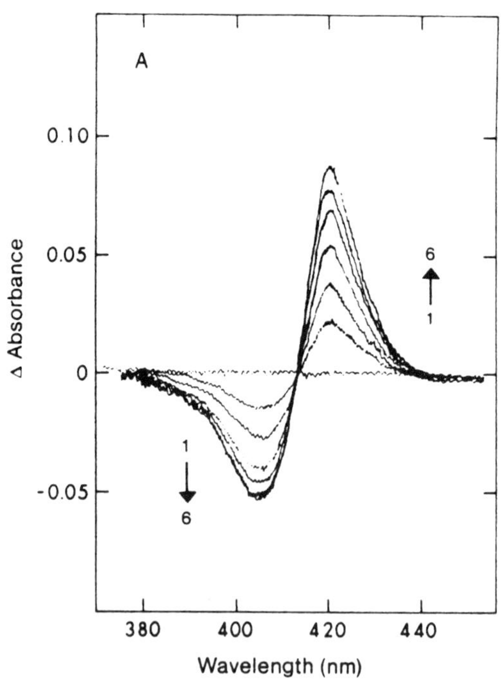

FIG. 14.17. Difference spectra of 0.9 μM *Pseudomonas* cytochrome *c* peroxidase as it is reduced from two Fe^{III} hemes to one Fe^{III} and one Fe^{II} heme by successive additions of azurin containing Cu^{I}. The concentration of azurin was incresed from 0.2 to 8.5 μM in six increments. [Reproduced with permission from M. Rönnberg, T. Araiso, N. Ellfolk, and H. B. Dunford, *J. Biol. Chem.* **256,** 2471–2474 (1981).]

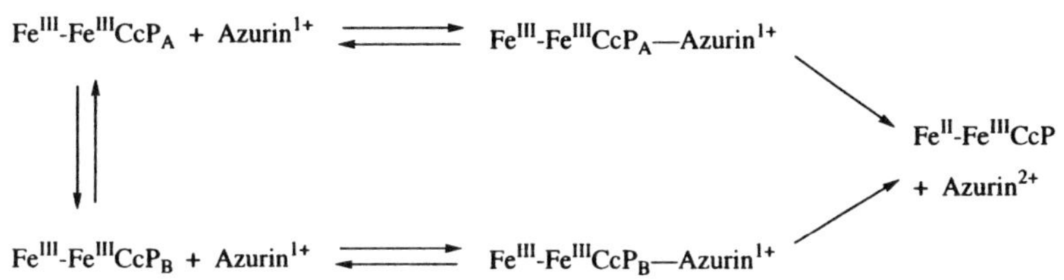

FIG. 14.18. Mechanism of reduction of two forms of *Pseudomonas* cytochrome *c* peroxidase by azurin. The two forms of the $Fe^{III}-Fe^{III}$ enzyme interconvert at a rate comparable to the rate of their reduction. [Based upon, with permission, M. Rönnberg, T. Araiso, N. Ellfolk, and H. B. Dunford, *J. Biol. Chem.* **256,** 2471–2474 (1981).]

nitrogen in the other. The two hemes were shown to interact. Reduction of one iron to its ferrous state induced a change in the EPR signal from the other heme.[110] Optical spectra at room temperature indicated the presence of a high-spin form. The reduction potentials of the two hemes were found to be +320 and −330 mV.[111] Compound I was found to be in a ferric-ferryl state and compound II in a ferric-ferric state. The latter is different from the inactive ferric-ferric form of the isolated enzyme. One would suspect that there would be differences in coordination to heme iron. The reaction cycle of *Pseudomonas* cytochrome *c* peroxidase is shown in Fig. 14.19.[112,113]

Once the enzyme is in its half-reduced form, it has a high-spin five-coordinate ferrous heme, readily accessible for reaction with hydrogen peroxide. By analogy with horseradish peroxidase, it would be expected to react to form a ferryl $Fe^{IV}{=}O$ group. Thus, compound I of *Pseudomonas* cytochrome *c* peroxidase would have $Fe^{IV}{=}O$ in one heme and Fe^{III} in the other. The rate constant for compound I formation using hydrogen peroxide as oxidant is $5.7 \times 10^7\ M^{-1}s^{-1}$. There is no absorption band in the 650-nm region, excluding a porphyrin π-cation structure.[112] The next step, reduction by reduced azurin,

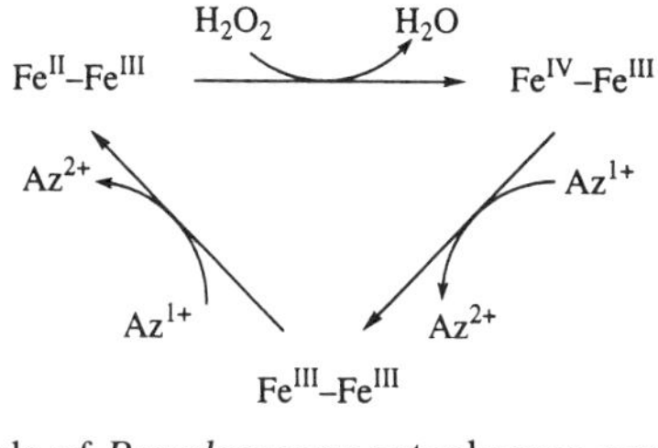

FIG. 14.19. Reaction cycle of *Pseudomonas* cytochrome *c* peroxidase. $Fe^{II}-Fe^{III}$ represents the active, half-reduced form of the enzyme, $Fe^{IV}-Fe^{III}$ is compound I, and $Fe^{III}-Fe^{III}$ compound II. Az^{+1} is reduced azurin.

produces the $Fe^{III}-Fe^{III}$ unique form of compound II, with a rate constant $> 2.5 \times 10^6\ M^{-1}s^{-1}$. A second reduction by azurin returns the enzyme to the start of its cycle, in a $Fe^{II}-Fe^{III}$ form, with a rate constant of $1.9 \times 10^6\ M^{-1}s^{-1}$. The azurin transfers its electron to the high potential heme, and the electron is then relayed to the low potential heme.[114] Use of azurin facilitated transient-state kinetic studies since the azurin did not interfere with spectrophotometric observation of the enzyme transitions. Cytochrome c_{551} is another natural substrate for the enzyme.

Cyanide binding kinetics to the half-reduced enzyme indicated an active site pK_a value of 3.9 ± 0.1.[115] The fully reduced enzyme binds a range of ligands, and, upon reduction, the methionine ligand of the high potential site becomes photosensitive.[116]

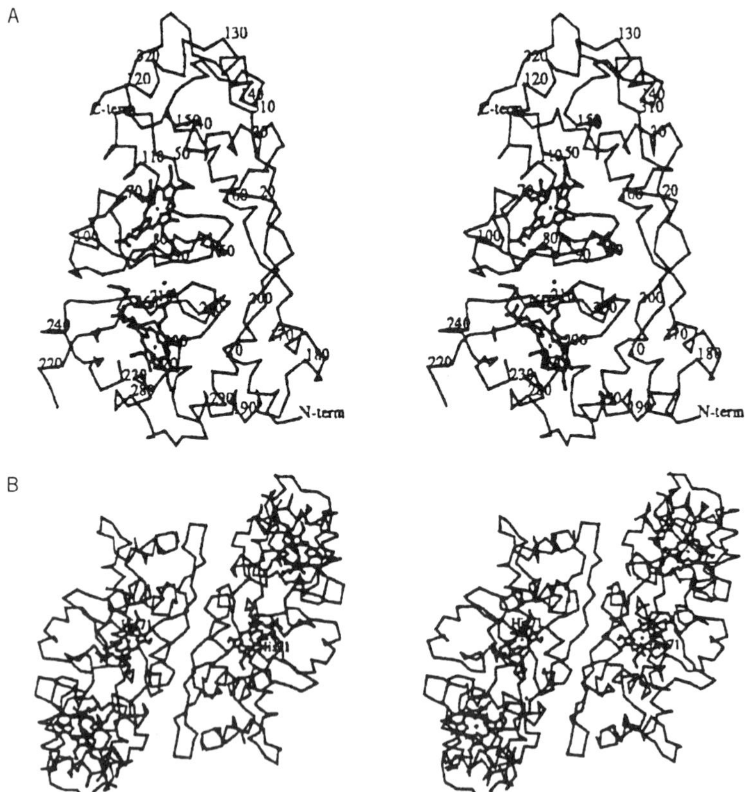

FIG. 14.20. Structure of the monomer and dimer of *Pseudomonas* cytochrome *c* peroxidase. (A) The monomer structure with every 10th α-carbon numbered. (B) The crystallographic dimer looking down the twofold axis. [Reproduced with permission from V. Fülöp, C. J. Ridout, C. Greenwood, and J. Hadju, *Structure* **3,** 1225–1233 (1995).]

The amino acid sequence was determined from the gene nucleotide sequence[117] and differed from an earlier sequence determination based upon cyanogen bromide fragments in which a 20-amino-acid hydrophobic region was missed.[118] The correct 323-amino acid sequence is reproduced in Appendix A. Some structural and functional features of the enzyme were predicted from the sequence.[119] The X-ray structure of the resting ferric-ferric form of the enzyme has been determined (Fig. 14.20).[120–122]

The molecule consists of two domains, each containing one of the heme *c*'s. The interface between the two domains was found, surprisingly, to contain a calcium ion. Both domains have marked similarities to cytochromes *c* but are somewhat larger in size. The N-terminal domain contains the low-potential heme. In the ferric-ferric state the low potential heme is six-coordinate with axial histidine residues 55 and 71. His71 is believed to leave its coordination site to permit entry of hydrogen peroxide when the enzyme is half-reduced. The high-potential heme is in the C-terminal domain. It is coordinated to Cys197 and Cys200. His201 and Met275 are ligands. The two hemes, perpendicular to each other, are located in hydrophobic pockets. The distance between the two irons is 20.9 Å. *Pseudomonas* cytochrome *c* peroxidase appears to be a product of an early evolutionary pathway in which this primitive organism learned to make use of cytochrome *c* structures to evolve a peroxidase that has no free radical intermediate in its cycle.

The *P. aeruginosa* enzyme has received by far the most attention among the diheme peroxidases. Other diheme peroxidases have been found in *Pseudomonas denitrificans,*[123] *Pseudomonas stutzeri,*[124] *Paracoccus denitrificans,*[125,126] and *Nitrosomonas europaea.*[127]

REFERENCES

1. P. W. Clutterbuck, S. L. Mukhopadhyay, A. E. Oxford, and H. Raistrick, *Biochem. J.* **34,** 664–677 (1940).
2. P. D. Shaw and L. P. Hager, *J. Am. Chem. Soc.* **81,** 1011–1012 (1959).
3. P. D. Shaw, J. R. Beckwith, and L. P. Hager, *J. Biol. Chem.* **234,** 2560–2564 (1959).
4. P. D. Shaw and L. P. Hager, *J. Am. Chem. Soc.* **81,** 6527–6528 (1959).
5. P. D. Shaw and L. P. Hager, *J. Biol. Chem.* **234,** 2565–2569 (1959).
6. P. D. Shaw and L. P. Hager, *J. Biol. Chem.* **236,** 1626–1630 (1961).
7. D. R. Morris and L. P. Hager, *J. Biol. Chem.* **241,** 1763–1768 (1966).
8. G. E. Krejcarek, R. G. Bryant, R. J. Smith, and L. P. Hager, *Biochemistry* **15,** 2508–2511 (1976).
9. A.-M. Lambeir and H. B. Dunford, *J. Biol. Chem.* **258,** 13558–13563 (1983).
10. L. P. Hager, D. R. Morris, F. S. Brown, and H. Eberwein, *J. Biol. Chem.* **241,** 1769–1777 (1966).
11. F. S. Brown and L. P. Hager, *J. Am. Chem. Soc.* **89,** 719–720 (1967).

12. J. A. Thomas, D. R. Morris, and L. P. Hager, *J. Biol. Chem.* **245,** 3129–3134 (1970).
13. W. Sun, T. A. Kadima, M. A. Pickard, and H. B. Dunford, *Biochem. Cell Biol.* **72,** 321–331 (1994).
14. J. A. Thomas, D. R. Morris, and L. P. Hager, *J. Biol. Chem.* **245,** 3135–3142 (1970).
15. M. M. Palcic, R. Rutter, T. Araiso, L. P. Hager, and H. B. Dunford, *Biochem. Biophys. Res. Commun.* **94,** 1123–1127 (1980).
16. P. F. Hollenberg and L. P. Hager, *J. Biol. Chem.* **248,** 2630–2633 (1973).
17. L. P. Hager, D. L. Doubek, R. M. Silverstein, J. H. Hargis, and J. C. Martin, *J. Am. Chem. Soc.* **94,** 4364–4366 (1972).
18. L. P. Hager, D. L. Doubek, R. M. Silverstein, T. T. Lee, J. A. Thomas, J. H. Hargis, and J. C. Martin in *Oxidases and Related Redox Systems,* T. E. King, H. S. Mason, and M. Morrison, eds., University Park Press, Baltimore, 1973, pp. 311–332.
19. P. F. Hollenberg, T. Rand-Meir, and L. P. Hager, *J. Biol. Chem.* **249,** 5816–5825 (1974).
20. S. Shahangian and L. P. Hager, *J. Biol. Chem.* **256,** 6034–6040 (1981).
21. T. Araiso, R. Rutter, M. M. Palcic, L. P. Hager, and H. B. Dunford, *Can. J. Biochem.* **59,** 233–236 (1981).
22. R. Rutter and L. P. Hager, *J. Biol. Chem.* **257,** 7958–7961 (1982).
23. R. Rutter, L. P. Hager, H. Dhonau, M. Hendrich, M. Valentine, and P. Debrunner, *Biochemistry* **23,** 6809–6816 (1984).
24. P. F. Hollenberg and L. P. Hager, *J. Biol. Chem.* **248,** 2630–2633 (1973).
25. R. Makino, R. Chiang, and L. P. Hager, *Biochemistry* **15,** 4748–4754 (1976).
26. P. M. Champion, E. Münck, P. G. Debrunner, P. F. Hollenberg, and L. P. Hager, *Biochemistry* **12,** 426–434 (1973).
27. P. M. Champion, R. Chiang, E. Münck, P. G. Debrunner, and L. P. Hager, *Biochemistry* **14,** 4159–4166 (1975).
28. J. H. Dawson, J. R. Trudell, G. Barth, R. E. Linder, E. Bunnenberg, C. Djerassi, R. Chiang, and L. P. Hager, *J. Am. Chem. Soc.* **98,** 3709–3710 (1976).
29. P. M. Champion, R. D. Remba, R. Chiang, D. B. Fitchen, and L. P. Hager, *Biochim. Biophys. Acta* **446,** 486–492 (1976).
30. R. D. Remba, P. M. Champion, D. B. Fitchen, R. Chiang, and L. P. Hager, *Biochemistry* **18,** 2280–2290 (1979).
31. S. P. Cramer, J. H. Dawson, K. O. Hodgson, and L. P. Hager, *J. Am. Chem. Soc.* **100,** 7282–7290 (1978).
32. P. F. Hollenberg, L. P. Hager, W. E. Blumberg, and J. Peisach, *J. Biol. Chem.* **255,** 4801–4807 (1980).
33. M. Sono, J. H. Dawson, and L. P. Hager, *J. Biol. Chem.* **259,** 13209–13216 (1984).
34. M. Sono, J. H. Dawson, K. Hall, and L. P. Hager, *Biochemistry* **25,** 347–356 (1986).
35. S. R. Blanke and L. P. Hager, *J. Biol. Chem.* **263,** 18739–18743 (1988).
36. J. H. Dawson and M. Sono, *Chem. Rev.* **87,** 1255–1276 (1987).
37. D. Metodiewa and H. B. Dunford, *Biochem. Biophys. Res. Commun.* **168,** 1311–1317 (1990).

38. A.-M. Lambeir and H. B. Dunford, *Arch. Biochem. Biophys.* **220,** 549–556 (1983).
39. S. R. Blanke, S. A. Martinis, S. G. Sligar, L. P. Hager, J. J. Rux, and J. H. Dawson, *Biochemistry* **35,** 14537–14543 (1996).
40. A.-M. Lambeir and H. B. Dunford, *Eur. J. Biochem.* **147,** 93–96 (1985).
41. M. Sono, K. M. Eble, J. H. Dawson, and L. P. Hager, *J. Biol. Chem.* **260,** 15530–15535 (1985).
42. R. Nakajima, I. Yamazaki, and B. W. Griffin, *Biochem. Biophys. Res. Commun.* **128,** 1–6 (1985).
43. G.-H. Fang, M. J. Axley, M. Nuell, and L. P. Hager, *Nucleic Acids Res.* **14,** 8061–8071 (1986).
44. M. J. Nuell, G.-H. Fang, M. J. Axley, P. Kenigsberg, and L. P. Hager, *J. Bacteriol.* **170,** 1007–1011 (1988).
45. M. Sundaramoorthy, J. Terner, and T. L. Poulos, *Structure* **3,** 1367–1377 (1995).
46. Y.-C. Fann, N. C. Gerber, P. A. Osmulski, L. P. Hager, S. G. Sligar, and B. M. Hoffman, *J. Am. Chem. Soc.* **116,** 5989–5990 (1994).
47. M. Sundaramoorthy, J. M. Mauro, A. M. Sullivan, J. Terner, and T. L. Poulos, *Acta Cryst. D* **51,** 842–844 (1995).
48. B. Rubin, J. vanMiddlesworth, K. Thomas, and L. Hager, *J. Biol. Chem.* **257,** 7768–7769 (1982).
49. M. A. Pickard, *Can. J. Microbiol.* **27,** 1298–1305 (1981).
50. A. Hashimoto and M. A. Pickard, *J. Gen. Microbiol.* **130,** 2051–2058 (1984).
51. M. A. Pickard and A. Hashimoto, *Can. J. Microbiol.* **28,** 1382–1388 (1982).
52. M. A. Pickard and A. Hashimoto, *Can. J. Microbiol.* **34,** 998–1002 (1988).
53. A. S. W. Sae and B. A. Cunningham, *Phytochemistry* **18,** 1785–1787 (1979).
54. R. D. Carmichael, A. Jones, and M. A. Pickard, *Appl. Environ. Microbiol.* **51,** 276–280 (1986).
55. R. D. Carmichael and M. A. Pickard, *Appl. Environ. Microbiol.* **55,** 17–20 (1989).
56. S. R. Blanke, S. Yi, and L. P. Hager, *Biotech. Lett.* **11,** 769–774 (1989).
57. T. A. Kadima and M. A. Pickard, *Can. J. Microbiol.* **36,** 302–304 (1990).
58. J. F. Holzwarth, F. Meyer, M. Pickard, and H. B. Dunford, *Biochemistry* **27,** 6628–6633 (1988).
59. A.-M. Lambeir, H. B. Dunford, and M. A. Pickard, *J. Inorg. Biochem.* **19,** 291–300 (1983).
60. R. Carmichael, P. M. Fedoruk, and M. A. Pickard, *Biotech. Lett.* **7,** 289–294 (1985).
61. M. D. Corbett, B. R. Chipko, and D. G. Baden, *Biochem. J.* **175,** 353–360 (1978).
62. M. D. Corbett, B. R. Chipko, and A. O. Batchelor, *Biochem. J.* **187,** 893–903 (1980).
63. M. D. Corbett and B. R. Chipko, *Biochem. J.* **183,** 269–276 (1979).
64. S. L. Neidleman, P. A. Diassi, B. Junta, R. N. Palmere, and S. C. Pan, *Tetrahed. Lett.* **44,** 5337–5342 (1966).
65. R. M. Silverstein and L. P. Hager, *Biochemistry* **13,** 5069–5073 (1974).
66. B. W. Griffin and R. Haddox, *Arch. Biochem. Biophys.* **239,** 305–309 (1985).
67. H. B. Dunford, A.-M. Lambeir, M. A. Kashem, and M. Pickard, *Arch. Biochem. Biophys.* **252,** 292–302 (1987).

68. M. Eigen and K. Kustin, *J. Am. Chem. Soc.* **84,** 1355–1361 (1962).
69. B. W. Griffin and P. L. Ashley, *Arch. Biochem. Biophys.* **233,** 188–196 (1984).
70. R. D. Libby, J. A. Thomas, L. W. Kaiser, and L. P. Hager, *J. Biol. Chem.* **257,** 5030–5037 (1982).
71. R. D. Libby, A. L. Shedd, K. A. Phipps, T. M. Beachy, and S. M. Gerstberger, *J. Biol. Chem.* **267,** 1769–1775 (1992).
72. R. D. Libby, T. M. Beachy, and A. K. Phipps, *J. Biol. Chem.* **271,** 21820–21827 (1996).
73. G. L. Kedderis, D. R. Koop, and P. F. Hollenberg, *J. Biol. Chem.* **255,** 10174–10182 (1980).
74. O. Okazaki and F. P. Guengerich, *J. Biol. Chem.* **268,** 1546–1552 (1993).
75. F. P. Guengerich, O. Okazaki, Y. Seto, and T. L. MacDonald, *Xenobiotica* **25,** 689–709 (1995).
76. P. L. Ashley and B. W. Griffin, *Arch. Biochem. Biophys.* **210,** 167–178 (1981).
77. J. Geigert, S. L. Neidleman, and D. J. Dalietos, *J. Biol. Chem.* **258,** 2273–2277 (1983).
78. K. Ramakrishnan, M. E. Oppenhuizen, S. Saunders, and J. Fisher, *Biochemistry* **22,** 3271–3277 (1983).
79. M. B. McCarthy and R. E. White, *J. Biol. Chem.* **258,** 9153–9158 (1983).
80. G. L. Kedderis and P. F. Hollenberg, *Arch. Biochem. Biophys.* **233,** 315–321 (1984).
81. E. J. Allain, L. P. Hager, L. Deng, and E. N. Jacobsen, *J. Am. Chem. Soc.* **115,** 4415–4416 (1993).
82. A. F. Dexter, F. J. Lakner, R. A. Campbell, and L. P. Hager, *J. Am. Chem. Soc.* **117,** 6412–6413 (1995).
83. F. J. Lakner and L. P. Hager, *J. Org. Chem.* **61,** 3923–3925 (1996).
84. F. J. Lakner, K. P. Cain, and L. P. Hager, *J. Am. Chem. Soc.* **119,** 443–444 (1997).
85. A. Zaks and D. R. Dodds, *J. Am. Chem. Soc.* **117,** 10419–10424 (1995).
86. A. F. Dexter and L. P. Hager, *J. Am. Chem. Soc.* **117,** 817–818 (1995).
87. H.-I. Lee, A. F. Dexter, Y.-C. Fann, F. J. Lakner, L. P. Hager, and B. M. Hoffman, *J. Am. Chem. Soc.* **119,** 4059–4069 (1997).
88. S. Collonna, N. Gaggero, L. Casella, G. Carrea, and P. Pasta, *Tetrahedron: Asymmetry* **3,** 95–106 (1992).
89. H. Fu, H. Kondo, Y. Ichikawa, G. C. Look, and C.-H. Wong, *J. Org. Chem.* **57,** 7265–7270 (1992).
90. H. M. Lenhoff and N. O. Kaplan, *Nature* **172,** 730–731 (1953).
91. H. M. Lenhoff and N. O. Kaplan, *J. Biol. Chem.* **220,** 967–982 (1956).
92. H. M. Lenhoff, D. J. D. Nicholas, and N. O. Kaplan, *J. Biol. Chem.* **220,** 983–995 (1956).
93. P. T. Cohen and N. O. Kaplan, *J. Biol. Chem.* **245,** 2825–2836 (1970).
94. N. Ellfolk and R. Soininen, *Acta Chem. Scand.* **24,** 2126–2136 (1970).
95. N. Ellfolk and R. Soininen, *Acta Chem. Scand.* **25,** 1535–1540 (1971).
96. R. Soininen, *Acta Chem. Scand.* **26,** 2535–2537 (1972).

97. R. Soininen, N. Ellfolk, and N. Kalkkinen, *Acta Chem. Scand.* **27,** 1106–1107 (1973).
98. R. Soininen and N. Ellfolk, *Acta Chem. Scand.* **B29,** 134–136 (1975).
99. M. Ellfolk, Ph.D. Thesis, Department of Biochemistry, University of Helsinki, 1979.
100. N. Ellfolk, M. Rönnberg, and R. Soininen, *Acta Chem. Scand.* **27,** 2171–2178 (1973).
101. M. Rönnberg and N. Ellfolk, *Acta Chem. Scand.* **B29,** 719–727 (1975).
102. M. Rönnberg, *Acta Chem. Scand.* **B30,** 721–726 (1976).
103. M. Rönnberg and N. Ellfolk, *Biochim. Biophys. Acta* **504,** 60–66 (1978).
104. M. Rönnberg, N. Ellfolk, and R. Soininen, *Biochim. Biophys. Acta* **578,** 392–400 (1979).
105. M. Rönnberg and N. Ellfolk, *Biochim. Biophys. Acta* **581,** 325–333 (1979).
106. T. Araiso, M. Rönnberg, H. B. Dunford, and N. Ellfolk, *FEBS Lett.* **118,** 99–102 (1980).
107. M. Rönnberg, T. Araiso, N. Ellfolk, and H. B. Dunford, *Arch. Biochem. Biophys.* **207,** 197–204 (1981).
108. M. Rönnberg, T. Araiso, N. Ellfolk, and H. B. Dunford, *J. Biol. Chem.* **256,** 2471–2474 (1981).
109. T. Brittain and C. Greenwood, *J. Inorg. Biochem.* **48,** 71–77 (1992).
110. R. Aasa, N. Ellfolk, M. Rönnberg, and T. Vänngård, *Biochim. Biophys. Acta* **670,** 170–175 (1981).
111. N. Ellfolk, M. Rönnberg, R. Aasa, L.-E. Andréasson, and T. Vänngård, *Biochim. Biophys. Acta* **743,** 23–30 (1983).
112. M. Rönnberg, A.-M. Lambeir, N. Ellfolk, and H. B. Dunford, *Arch. Biochem. Biophys.* **236,** 714–719 (1985).
113. N. Foote, R. Turner, T. Brittain, and C. Greenwood, *Biochem. J.* **283,** 839–843 (1992).
114. C. Greenwood, N. Foote, P. M. A. Gadsby, and A. J. Thomson, *Chim. Scr.* **28A,** 79–84 (1988).
115. M. Rönnberg, A.-M. Lambeir, N. Ellfolk, and H. B. Dunford, *Biochim. Biophys. Acta* **828,** 67–72 (1985).
116. C. Greenwood and Q. H. Gibson, *J. Biol. Chem.* **264,** 119022–19027 (1989).
117. C. J. Ridout, R. James, and C. Greenwood, *FEBS Lett.* **365,** 152–154 (1995).
118. M. Rönnberg, N. Kalkkinen, and N. Ellfolk, *FEBS Lett.* **250,** 175–178 (1989).
119. N. Ellfolk, M. Rönnberg, and K. Österlund, *Biochim. Biophys. Acta* **1080,** 68–77 (1991).
120. V. Fülöp, R. Little, A. Thompson, C. Greenwood, and J. Hadju, *J. Mol. Biol.* **232,** 1208–1210 (1993).
121. V. Fülöp, R. Little, A. Thompson, C. Greenwood, and J. Hadju, *Structure* **3,** 1225–1233 (1995).
122. V. Fülöp in *Plant Peroxidases. Biochemistry and Physiology,* C. Obinger, U. Burner, R. Ebermann, C. Penel, and H. Greppin, eds., University of Geneva Press, Geneva, 1996, pp. 45–51.

123. A. F. W. Coulson and R. I. C. Oliver, *Biochem. J.* **181,** 159–169 (1979).
124. J. Villalain, J. Moura, M. Liu, W. J. Payne, J. LeGall, A. V. Xavier, and J. J. Moura, *Eur. J. Biochem.* **141,** 305–312 (1984).
125. C. F. Goodhew, I. B. H. Wilson, D. J. B. Hunter, and G. W. Pettigrew, *Biochem. J.* **271,** 707–712 (1990).
126. S. Prazeres, I. Moura, R. Gilmour, G. Pettigrew, N. Ravi, and B. H. Huynh in *Nuclear Magnetic Resonance of Paramagnetic Molecules,* G. N. La Mar, ed., Kluwer Academic, Netherlands, 1995, pp. 141–163.
127. D. M. Arciero and A. B. Hooper, *J. Biol. Chem.* **269,** 11878–11886 (1994).

CHAPTER 15

Myeloperoxidase and Eosinophil Peroxidase: Phagocytosis and Microbial Killing

15.1. DISCOVERY OF MYELOPEROXIDASE

Oxidation of guaiacol by pus was demonstrated in the nineteenth century and by leukocytes early in the twentieth century. The molecules responsible for the reaction were not clearly identified until 1941. Myeloperoxidase was isolated by Kjell Agner from the sputum of tuberculosis patients.[1] Because the enzyme was green in color, he originally called it verdoperoxidase, in order to distinguish it from other peroxidases known at that time. The name verdoperoxidase was dropped when Hugo Theorell discovered that lactoperoxidase was also green in color,[2] and was changed to myeloperoxidase because the enzyme is found in the granules of myelocytes, precursors of neutrophils. Agner's description of the isolation and purification also contains a fascinating review of the relevant biochemistry and physiology, beginning in the nineteenth century. He also isolated the enzyme from leukocytes of a patient with myeloid leukemia.[3]

In order to appreciate the sources of animal peroxidases and their role in defence mechanisms, the nonbiological/medical researcher needs a glossary of cell classification. *Leukocytes,* or white blood cells, can be subdivided into cells that contain a large number of granules, *granulocytes,* and those that contain a small number, originally thought to be nongranular, the *agranulocytes.*

The granulocytes were subdivided on the basis of their response to histochemical staining into three categories: *neutrophils, eosinophils,* and *basophils.* A property of granulocytes, particularly neutrophils, is that they possess an irregularly shaped nucleus, one with many lobes. For this reason neutrophils are also called *polymorphonuclear (PMN) leukocytes.* There is a comprehensive review in the literature on neutrophils up to 1977,[4] and a review of all phagocytic cells up to 1988.[5] When the agranulocytes are immature cells, they are called *monocytes.* As mature cells they are called *macrophages* (big eaters).

Phagocytes are leukocytes capable of *phagocytosis,* the ingestion of foreign cells. The major types of phagocytes are the neutrophils, eosinophils, and

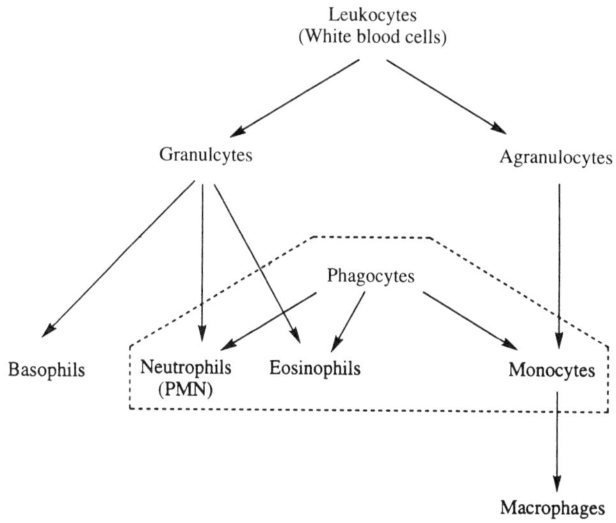

FIG. 15.1. Classification of various types of leukocytes. PMN is the acronym for polymorphonuclear leukocyte.

monocytes. The macrophages, which develop from the monocytes, are also phagocytes but to a much lesser degree. Thus, the phagocytes encompass members from both the granulocytes and agranulocytes. These relations are summarized in Fig. 15.1.

Available evidence now indicates that the myeloperoxidase found in neutrophils and monocytes is identical, but that found in eosinophils is distinctly different. The latter is called eosinophil peroxidase. The peroxidase found in basophils also appears to be different from that in neutrophils.[6] A green spleen heme protein was detected and isolated and proven to be identical to the myeloperoxidase from neutrophils.[7–10] The source of the spleen peroxidase is apparently from dead neutrophils, which would explain why identical peroxidases are found in both sources. The neutrophil is the dominant type of white blood cell, which explains why work on myeloperoxidase has dominated studies of blood peroxidases.

The remainder of this chapter is devoted to a description of neutrophil or spleen myeloperoxidase and its reactions, to the less well known eosinophil peroxidase, to the phagocytic process and to microbial killing.

15.2. PREPARATION AND PROPERTIES OF MYELOPEROXIDASE

The isolation of myeloperoxidase from pus[11,12] and leukocytes has been improved,[13–17] and beef spleens have proven to be an excellent alternative source.

Several investigations have shown that the leukocyte and spleen enzymes are identical.[7,8,18,19]

Three bands were observed after polyacrylamide gel electrophoresis of the human enzyme, a major band and two minor ones. One homogeneous band was observed by ultracentrifugation.[14] Although the existence of the three different species, myeloperoxidases I, II, and III, is beyond doubt (Fig. 15.2),[16,20,21] there appears to be no clear evidence that they are genuine isoenzymes with distinct genes. Electron paramagnetic resonance (EPR) spectra for the three species are identical (Fig. 15.3).[22] Heterogeneous mRNAs coding for myeloperoxidase are formed from a single gene. However, relative amounts have not been determined, and it is not certain they are responsible for the multiple forms of myeloperoxidase.[23] Differences may be caused by different extents of glycosylation. Myeloperoxidase I exhibits some differences in reactivity toward iodide, compared to myeloperoxidases II and III,[16] but in other assays it shows the same reactivity.[20] In subsequent discussion, myeloperoxidase will be treated as a single enzymatic species, in accordance with the majority of the literature in which results on pooled samples are reported.

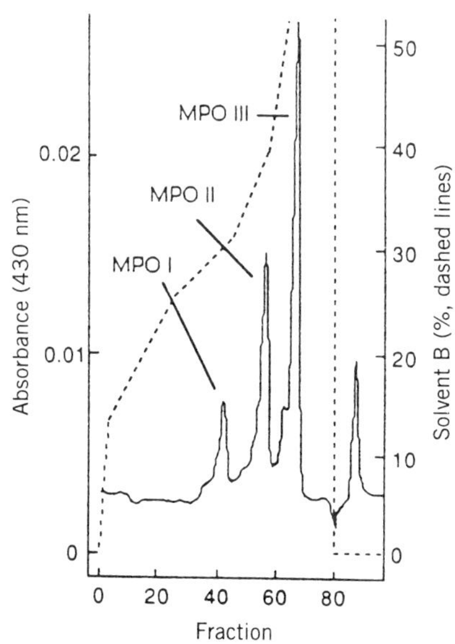

FIG. 15.2. Typical pattern of isomers isolated from neutrophil granules of a single patient using a salt gradient for intermediate-pressure liquid chromatography. Solvent A was 0.02 M sodium acetate, pH 4.5. Solvent B the same but with 2.0 M sodium chloride. [Reproduced with permission from K. T. Miyasaki, J.-P. Song, and A. R. K. Murthy, *Anal. Biochem.* **193,** 38–44 (1991).]

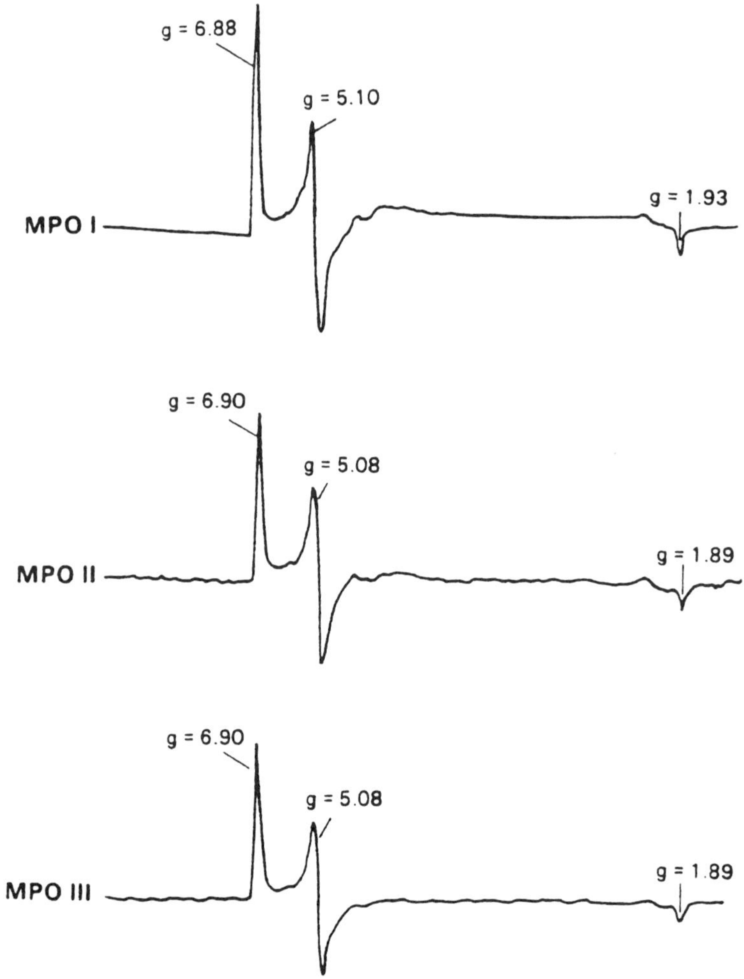

FIG. 15.3. EPR spectra for myeloperoxidase resolved into its three isoenzymatic forms. [Reproduced with permission from J. Wright, N. Bastian, T. A. Davis, C. Zuo, S. Yoshimoto, W. H. Orme-Johnson, and A. I. Tauber, *Blood* **75,** 238–241 (1990).]

The native enzyme was shown to have a total molecular weight of 150,000 with one atom of heme iron per half-molecule[24] and one of calcium.[25] The timing and role of heme insertion on processing of the enzyme has been examined.[26] The enzyme also contains 3–4% carbohydrate,[12] which consists of mannose, glucose, and *N*-acetylglucosamine.[14] The enzyme was cleaved in half by reduction of one connecting disulfide bridge. Each hemimyeloperoxidase consists of a heavy and light chain, molecular weights 57,500 and 14,000.[27,28] A schematic representation of the two halves of the native enzyme is shown in Fig. 15.4. Myeloperoxidase is denaturated by guanidine hydrochloride and can be partially renatured under carefully controlled conditions.[29]

The hemi-myeloperoxidase appears to be identical in its activity to the holoenzyme.[30,31] Why then does nature make a dimeric enzyme instead of monomeric? The answer may be that the dimer is more resistant to secretion. The dimeric structure would help to confine myeloperoxidase in the vacuole of a neutrophil, which is the site of microbial killing.

Discrepancies exist in the literature on the absolute molar absorptivities of myeloperoxidase. The secondary standard used for determinations of all peroxidase concentrations is often the molar absorptivity of the Soret peak of the native enzyme at 430 nm. Values of ε_{430} of 89 and 95 $mM^{-1}cm^{-1}$ per heme have been reported for canine[11] and porcine myeloperoxidase.[32] The value for ε_{430} of 89 $mM^{-1}cm^{-1}$ was 25% higher than the first report on the human enzyme.[1] Different authors have assumed that a literature value of ε_{430} for one source of myeloperoxidase is applicable to myeloperoxidase from a different species.[33,34] Because of the unknown nature of the heme and its covalent attachment, the usual primary standard of the pyridine hemochromogen spectrum was replaced by iron determination in order to calibrate the optical absorbance spectra of myeloperoxidase in absolute terms.

The most accurate Soret spectrum of compound I was obtained in 1994[35] and is shown in Fig. 15.5. The spontaneous conversion of compound I to compound II is shown in Fig. 15.6. Some spectral data are recorded in Table 15.1. The isosbestic point between native enzyme and compound I occurs at 456 nm and between compounds I and II at 429 nm. The large peak of the native enzyme in the visible region, centered at 571 nm, is responsible for the green color of myeloperoxidase.

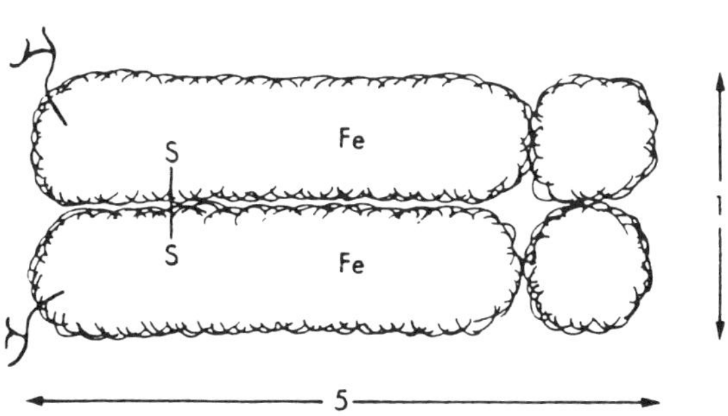

FIG. 15.4. Schematic representation of the gross structure of myeloperoxidase. [Reproduced with permission from P. C. Andrews and N. I. Krinsky, *J. Biol. Chem.* **256,** 4211–4218 (1981).]

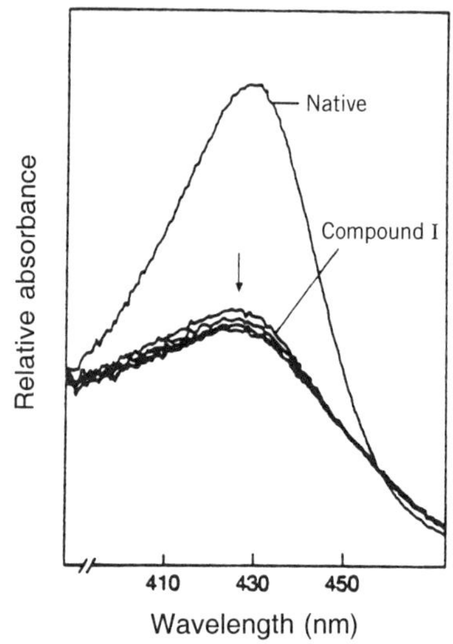

FIG. 15.5. Conversion of 1 μM beef spleen myeloperoxidase into compound I using a 20-fold excess of hydrogen peroxide. Spectra were taken 4–7 ms after flow had stopped. [Reproduced with permission from L. A. Marquez, J. T. Huang, and H. B. Dunford, *Biochemistry* **33,** 1447–1454 (1994).]

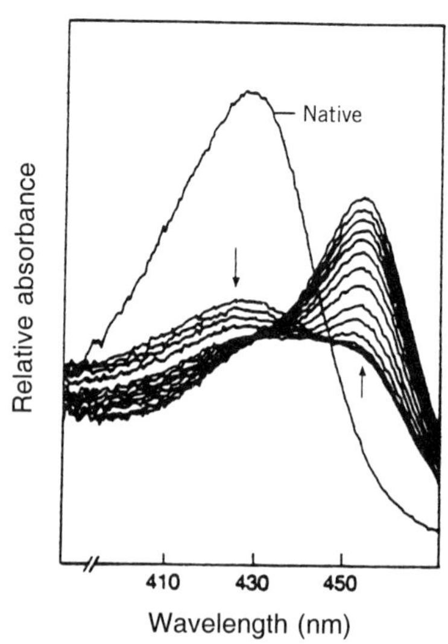

FIG. 15.6. Spontaneous conversion of compound I to compound II, using a 40-fold excess of hydrogen peroxide. Spectra were recorded starting 3 ms after flow had stopped, then in 1-ms intervals. [Reproduced with permission from L. A. Marquez, J. T. Huang, and H. B. Dunford, *Biochemistry* **33,** 1447–1454 (1994).]

TABLE 15.1 Optical Spectral Data for Myeloperoxidases, Expressed per Heme Group

Myeloperoxidase Species	λ_{max}, nm	ε_{max}, $mM^{-1}cm^{-1}$
Canine native enzyme[a]	430	95[b]
	496	Shoulder
	571	15.9
	620	8.9
	689	5.4
Beef spleen compound I[c]	428	47
Canine compound I[a]	572	9.7
Human compound II[d,e]	455	85
	628	18.0
Beef spleen compound II[f]	454	87
Human compound III[d]	450	78
	625	33.5

[a] Ref. 33. A peak reported at 625 nm for compound I is probably caused by a trace of compound II.
[b] Ref. 32.
[c] Ref. 35.
[d] Ref. 34. Values based on ε_{430} for the native enzyme of 89 $mM^{-1}cm^{-1}$.
[e] Values applicable ≤pH 8.0.
[f] Ref. 36.

15.3. CLONING OF MYELOPEROXIDASE, SITE-DIRECTED MUTAGENESIS

The gene for human myeloperoxidase has been cloned, characterized,[37–41] and expressed in a baculovirus system.[42] The deduced amino acid sequence is reproduced in Appendix A. An 84-kilodalton recombinant myeloperoxidase heavy chain containing a heme has been expressed in Chineses hamster ovary cells, and its spectral and enzymatic properties have been documented.[43,44] The promoter for human myeloperoxidase has been identified and characterized.[45] Regulation of the myeloperoxidase enhancer binding protein has been studied.[46] The high-capacity calcium-binding protein calreticulin was shown to act as a molecular chaperone for myeloperoxidase. Calreticulin does not interact with the apo-myeloperoxidase until it is glycosylated. Then it maintains the myeloperoxidase in a conformation suitable for heme insertion.[47]

Site-directed mutants of the 84-kilodalton heavy chain of myeloperoxidase have been prepared. The terminology to describe the results includes the amino acid residue number for the holoenzyme first and for the heavy chain second. Thus, His261(95) was replaced with Ala, a result that helped confirm His261 as the proximal heme ligand. Similarly, Asp260(94), Glu408(242), and Met409(243) were replaced by Asn, Gln, and Glu, respectively. All four residues are necessary for maintaining the correct conformation of the active site and contribute to the unique spectral properties of the heme.[48] The

Glu408(242)Gln mutation was intensively studied, using EPR and resonance Raman spectroscopies, and activity studies. The mutant cannot produce hypochlorous acid. The Glu residue is responsible for many spectral properties, previously attributed to other causes.[49]

15.4. CRYSTAL STRUCTURE

The crystal structure of canine myeloperoxidase to 3 Å resolution was published in 1992.[50] A schematic representation of the half-structure is shown in the introduction to X-ray crystallography (Section 3.3, Fig. 3.4) as a typical mammalian peroxidase. In addition to the disulfide linkage between the two hemi-myeloperoxidases, each hemi-myeloperoxidase possesses an additional

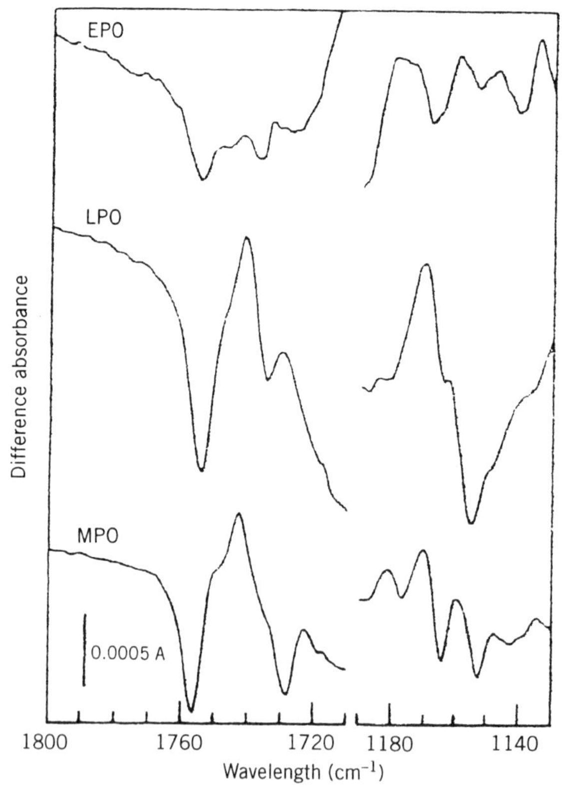

FIG. 15.7. Fourier transform infrared difference spectra between ferric peroxidases and ferrous peroxidases formed by photoreduction. EPO, eosinophil peroxidase; LPO, lactoperoxidase; and MPO, myeloperoxidase. [Reproduced with permission from I. M. Kooter, A. J. Pierik, M. Merkx, B. A. Averill, N. Moguilevsky, A. Bollen, and R. Wever, *J. Am. Chem. Soc.* **119,** 11542–11543 (1997).]

five disulfide bridges in its large polypeptide and one in its small. The structure is dominated by 19 α-helices. The proximal ligand is His336, which is hydrogen-bonded to Asn421. Distal His95 and Arg239 appear to be positioned to participate in the catalytic mechanism.

15.4.1. Nature of the Heme and Coordination Number of the Iron

The green color of native myeloperoxidase provided an early lead that the heme chromophoric group is not ferriprotoporphyrin IX. Since the heme is covalently bound, its structure could not be determined by the usual chemical techniques. A variety of spectroscopic techniques were applied, with limited success in terms of positive structural identification.[51,52] Similarities to the structure of chlorins[53–55] and the heme of lactoperoxidase[56,57] were noted.

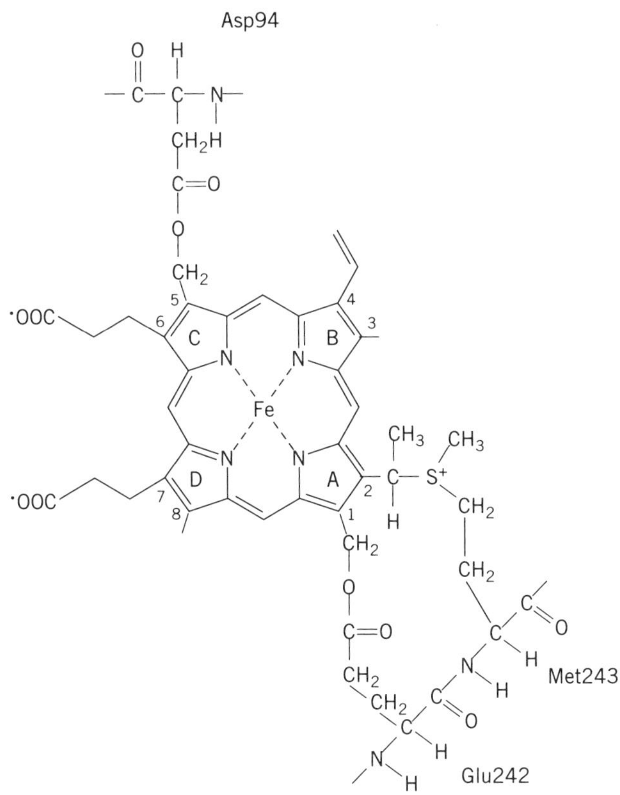

FIG. 15.8. Heme structure of myeloperoxidase. [Reproduced with permission from I. M. Kooter, A. J. Pierik, M. Merkx, B. A. Averill, N. Moguilevsky, A. Bollen, and R. Wever, *J. Am. Chem. Soc.* **119,** 11542–11543 (1997).]

Studies of the autolytic cleavage product of myeloperoxidase led to the conclusion that the heme was covalently bound via a sulfonium bond to Met409,[58] and that the heme is a derivative of ferriprotoporphyrin IX.

A crystal structure of human myeloperoxidase appears to have been at high enough resolution (2.28 Å) to define the heme structure. Modified methyl groups were found to form ester linkages with Glu242 and Asp94.[59] A covalent bond between a vinyl side chain of the heme and Met243 was confirmed. However, the details the sulfur atom linkage needed clarification. It was originally reported as a vinyl sulfonium ion[60] but later changed to sulfur atom α-carbon linkage by analogy with the thioether linkage of cytochrome *c*.[61] Fourier transform infrared (FTIR) difference spectra were used to confirm the ester linkages (Fig. 15.7).[61] The approach to the distal heme cavity is shown in Fig. 15.8, and the revised myeloperoxidase heme structure in Fig. 15.9.

The coordination number of the heme iron of myeloperoxidase has not received the intensive scrutiny applied to horseradish and cytochrome *c* peroxidases. Neveretheless, there are similarities in the literature descriptions. The iron was reported to be six-coordinate with water as the sixth ligand.[62] The refined crystal structure of the heme site indicates that although there is water in the distal cavity, it is not coordinated to the iron.[59]

The crystal structure of the human myeloperoxidase–salicylhydroxamic acid inhibitor complex has been determined at 2.3 Å resolution.[63] On the basis of the similarity of the inhibitor to perbenzoic acids, a mechanism for compound I formation, similar to that for the plant peroxidases was proposed.

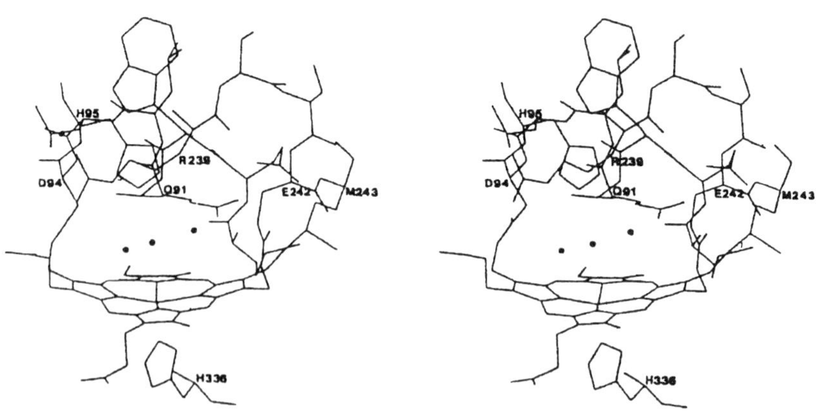

FIG. 15.9. View from the entrance to the distal heme pocket of myeloperoxidase. His336 provides the proximal ligand. Covalent linkages from the heme to Asp94, Glu242, and Met243 are shown. His95 and Arg239 are believed to play catalytic roles, while that of Gln91 is unknown. Also shown are three water molecules in the distal pocket. [Reproduced with permission from R. Fenna, J. Zeng, and C. Davey, *Arch. Biochem. Biophys.* **316,** 635–642 (1995).]

15.5. ASSAYS FOR MYELOPEROXIDASE

Myeloperoxidase is one of two heme enzymes with a compound I capable of oxidizing chloride ion. (The other is chloroperoxidase.) Thus, myeloperoxidase has both peroxidatic and chlorinating activity. These activities have different mechanisms and different pH maxima so that one should choose an assay procedure appropriate for the problem of interest.

An assay for peroxidase activity at pH 5.4 in phosphate-buffered saline utilizes a nonmutagenic derivative of benzidine, 3,3′,5,5′-tetramethyl benzidine (TMB) to produce a color measured at 655 nm. The reaction, conducted at 37°C, was quenched by sodium acetate buffer (pH 3.0) and the absorbance was measured after 3 min.[64] This assay procedure was reexamined. The presence of saline means chloride ion is present, a competitor for the peroxidase reaction of compound I. Saline solution was eliminated, and the initial rate of absorbance change at 652 nm and 20°C was monitored.[65] A linear increase in absorbance as a function of time is an indicator of a preparation containing no interfering reactants and that the colored product, which has a finite lifetime, is stable for the time of the assay. The colored product is a charge-transfer complex. The kinetics of formation of the charge-transfer complex were elucidated.

A peroxidase assay using guaiacol is also widely used. It is based on the assumption that guaiacol is oxidized exclusively to tetraguaiacol.[66] The chemiluminescence of singlet oxygen, produced from the the reaction of hypobromous acid with hydrogen peroxide, was used as a myeloperoxidase assay. Eosinophil contamination interfered with its use to detect myeloperoxidase in neutrophils.[67] A highly sensitive light emission assay was developed using a luciferin derivative.[68]

Methods of assaying for chlorination activity have been reviewed. Five methods were examined: use of a hydrogen peroxide electrode and assays that produce taurine chloramine, or utilize reactions of monochlorodimedone or ascorbate or TMB. Of the five methods, use of a hydrogen peroxide electrode was the recommended choice.[69]

15.6. REACTIONS OF NATIVE MYELOPEROXIDASE

15.6.1. Cyanide Binding

The kinetics of cyanide binding to myeloperoxidase from human leukocytes,[70] bovine spleen,[71] and canine pus[31] has been studied with similar results. A pK_a value of approximately 4.0 is indicated for an active site acid group, with HCN binding to the basic form of the enzyme. At neutral pH the association rate constant is in the range $1\text{–}4 \times 10^6$ $M^{-1}s^{-1}$, faster by at least an order of magnitude than with peroxidases from the plant superfamily. The equilibrium dissociation constant is 5×10^{-7} M, indicating a dissociation rate constant of

$\sim$1–2 s^{-1}. Thus, cyanide binds very tightly. In all of these studies the data are reported per heme group, and, as far as can be ascertained, the two halves of myeloperoxidase behave in identical fashion. The dissociable group on the enzyme involved in ligand binding has a ΔH value of about 8 kcal/mol and was assigned to the imidazole group of a distal histidine.[71]

The resonance Raman spectra of the cyanide complex of myeloperoxidase differ from those of horseradish peroxidase and model compounds.[72] The order of the Fe—C stretching and Fe—CN bending modes is reversed in myeloperoxidase, which was explained in terms of a bent axial ligand structure for the myeloperoxidase complex. The cyanide complex has also been examined by magnetic circular dichroism, circular dichroism, and EPR spectroscopies.[73] The EPR spectra of myeloperoxidase and its cyanide complex are compared in Fig. 15.10. Both the visible (Fig. 15.11) and near-infrared spectra were also reported. The dominant feature is the expected change from predominantly high-spin iron in the ferric enzyme, to low-spin when complexed with cyanide, but there appears to be more than one binding mode of cyanide.

15.6.2. Halide Ion Binding

Halides were shown to bind directly to the heme iron.[62,74,75] Whereas the acidic form of cyanide, HCN, binds to the basic form of the enzyme, the reverse occurs with chloride.[71] Chloride ion binds to the acidic form of the enzyme at a very rapid rate (k = 1.5 × 10^7 $M^{-1}s^{-1}$). The pK_a value for

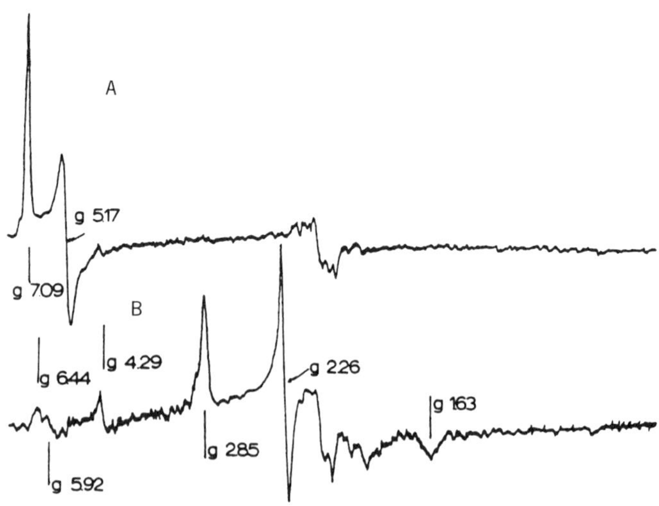

FIG. 15.10. EPR spectra of (A) native myeloperoxidase and its (B) cyanide complex obtained at 10 K. [Reproduced with permission from D. G. Eglington, D. Barber, A. J. Thomson, C. Greenwood, and A. W. Segal, *Biochim. Biophys. Acta* **703,** 187–195 (1982) with permission from Elsevier Science.]

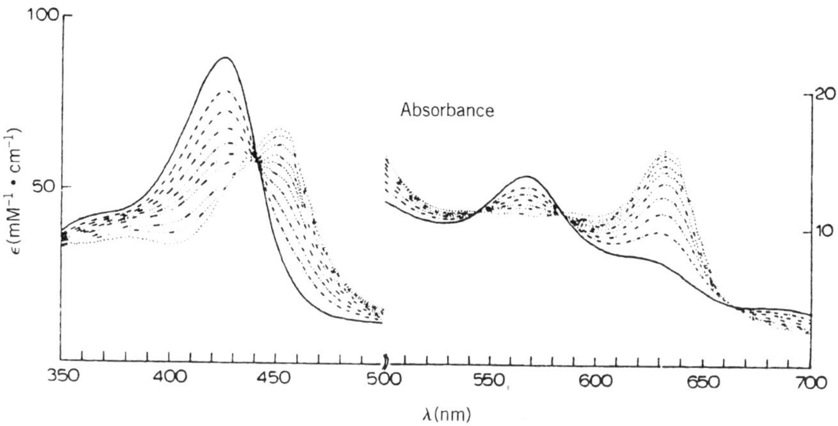

FIG. 15.11. Soret and visible optical spectra of myeloperoxidase titrated with cyanide at pH 7 and 22°C. Solid line, native enzyme. Additions of cyanide range from 3.35 μM to 9 mM. Protein concentration changes from 14.1 to 13.4 μM. [Reproduced with permission from D. G. Eglington, D. Barber, A. J. Thomson, C. Greenwood, and A. W. Segal, *Biochim. Biophys. Acta* **703,** 187–195 (1982) with permission from Elsevier Science.]

the enzyme group involved in chloride binding is 4.0, the same as for cyanide binding. Thus, chloride binds most tightly at low pH (Fig. 15.12), and it can displace cyanide. The effect of chloride on cyanide binding is shown in Fig. 15.13,[70] and of chloride on the reduction potential of the ferric-ferrous couple in Fig. 15.14.[75]

15.6.3. Nitrite Binding

Nitrite binding to myeloperoxidase was detected by both optical and EPR spectroscopy. The enzyme is converted from ferric high-spin to low-spin by nitrite.[76] Because of the large spectral changes, one would expect that nitrite binds directly to the iron(III).

15.6.4. Aromatic Substrate Binding

Aromatic substrate molecule binding was examined by EPR and by model building based on the 3 Å resolution structure of canine myeloperoxidase.[77] Phenol, *p*-cresol, resorcinol, and 4-aminosalicylate generate a high-spin EPR signal slightly different from that of the native enzyme. In contrast, salicylhydroxamic acid and benzhydroxamic acid generate a low-spin signal that is strongly temperature-dependent. All substrates bind in a hydrophobic pocket on the distal side of the heme, and the hydroxamic side chains appeared to interact directly with the iron. Later the X-ray structure of the salicylhydroxamic acid complex was determined,[63] and its binding near the heme is shown in Fig. 15.15.

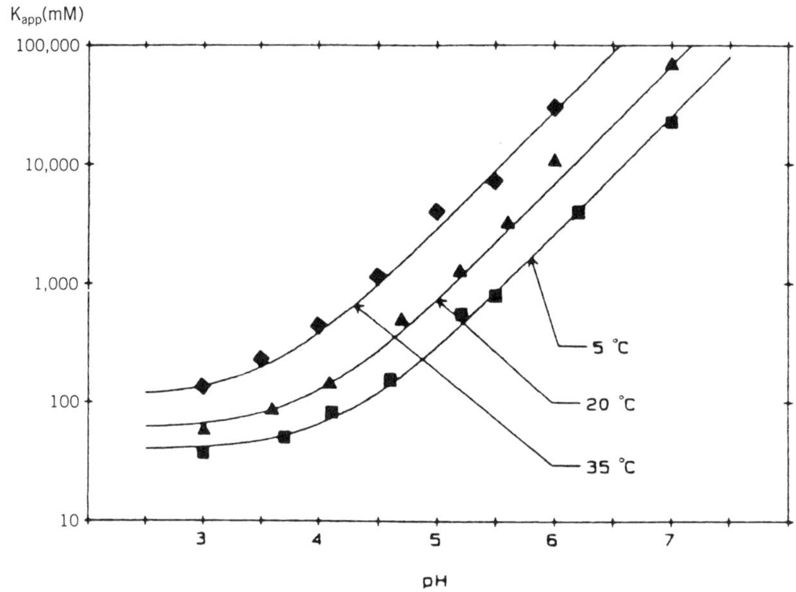

FIG. 15.12. pH dependence of the apparent dissociation constant of chloride ion from its complex with myeloperoxidase at 5, 20, and 35°C. [Reproduced with permission from M. Ikeda-Saito, *Biochemistry* **26,** 4344–4349 (1987).]

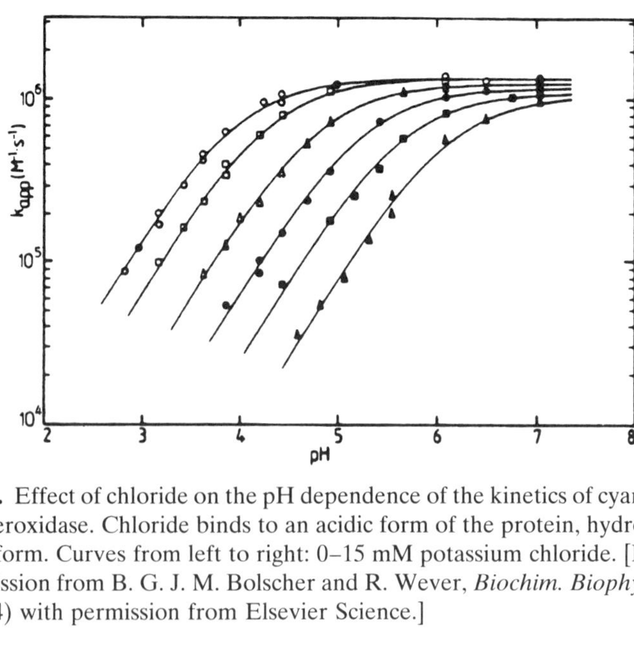

FIG. 15.13. Effect of chloride on the pH dependence of the kinetics of cyanide binding to myeloperoxidase. Chloride binds to an acidic form of the protein, hydrocyanic acid to a basic form. Curves from left to right: 0–15 mM potassium chloride. [Reproduced with permission from B. G. J. M. Bolscher and R. Wever, *Biochim. Biophys. Acta* **788,** 1–10 (1984) with permission from Elsevier Science.]

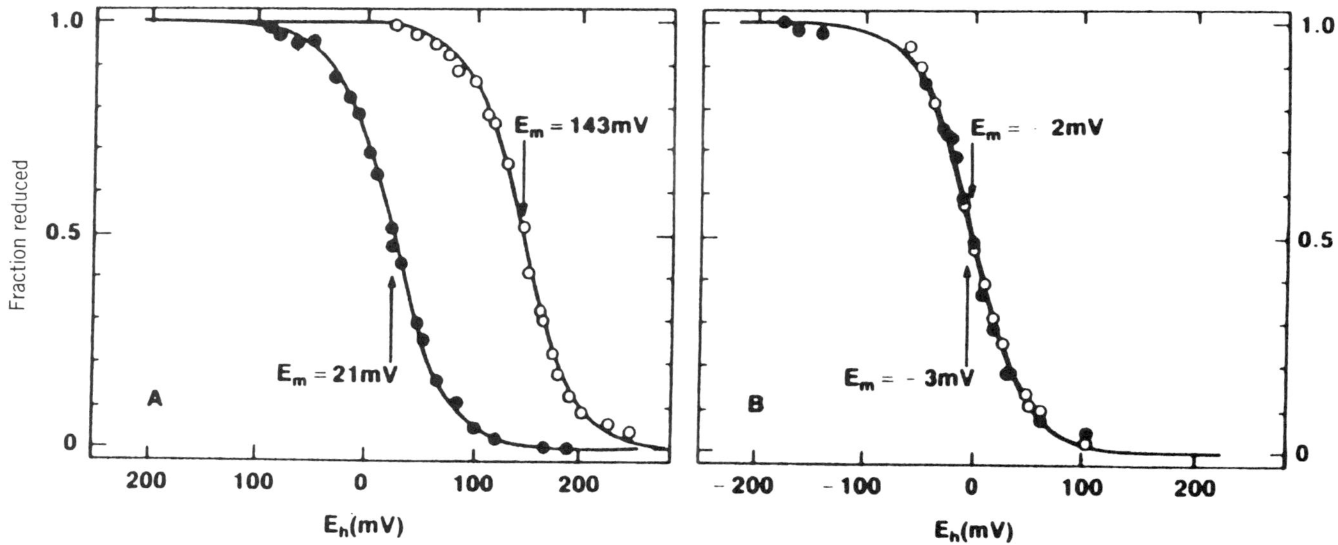

FIG. 15.14. Effect of chloride ion on the reduction potential of the ferric-ferrous couple for myeloperoxidase. Panel A, pH 4.3; panel B, pH 7.9. Open and closed circles correspond to the absence and presence of 0.1 M sodium chloride. [Reproduced with permission from M. Ikeda-Saito and R. C. Prince, *J. Biol. Chem.* **260,** 8301–8305 (1985).]

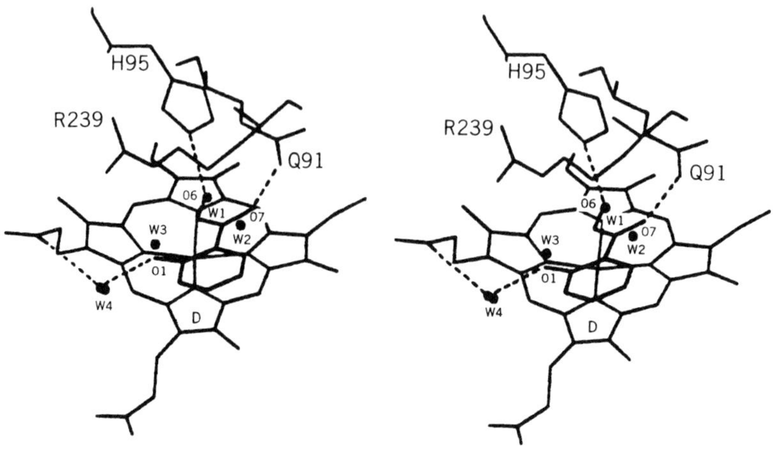

FIG. 15.15. Model for salicylhydroxamic acid bound to human myeloperoxidase. The positions of the three oxygen atoms of the acid are close to those formerly occupied by the three displaced water molecules (W1, W2, and W3). [Reproduced with permission from C. A. Davey and R. E. Fenna, *Biochemistry* **35,** 10967–10973 (1996).]

15.7. TRANSIENT-STATE KINETICS OF MYELOPEROXIDASE REACTIONS

15.7.1. Kinetics of Formation of Compound I

There are two papers that elucidate the unusual mechanism for compound I formation from myeloperoxidase. In the first, studies were performed on the human enzyme,[70] and in the second, on the bovine spleen enzyme.[35] Similar results were obtained. The main features are contained in the following mechanism:

$$\text{MPO} + \text{H}_2\text{O}_2 \underset{k_{-1}}{\overset{k_1}{\rightleftarrows}} \text{MPO-I} + \text{H}_2\text{O} \tag{15.1}$$

$$\text{MPO-I} + \text{H}_2\text{O}_2 \xrightarrow{k_2} \text{MPO-II} + \text{HO}_2\bullet \tag{15.2}$$

A large excess of hydrogen peroxide is required to complete the formation of compound I because of the reversibility of the reaction. Thus, compound I of myeloperoxidase is able to oxidize water to hydrogen peroxide. An examination of the relevant redox potentials shows that they are favorable for reversibility in compound I formation.[78] A report that myeloperoxidase has significant catalatic activity is erroneous.[79]

It is virtually impossible to obtain a 100% yield of myeloperoxidase compound I, because if a large excess of hydrogen peroxide is used in an attempt to drive the equilibrium for compound I formation to near completion, the excess hydrogen peroxide will commence to reduce the newly formed compound I to compound II. The highest yield of compound I appears to be that described in Section 15.2.[35]

The value of k_1 is $(2.3 \pm 0.1) \times 10^7\ M^{-1}s^{-1}$ for human myeloperoxidase[70] and $(1.8 \pm 0.1) \times 10^7\ M^{-1}s^{-1}$ for the beef spleen enzyme.[35] The value of k_{-1} is $58 \pm 4\ s^{-1}$ and so the value of the equilibrium constant $K = k_{-1}/k_1$ is $3.2 \pm 0.3\ \mu M$. The same value of K is obtained for the human enzyme from a corrected Scatchard plot. The value of k_2 is $(8.1 \pm 0.1) \times 10^4\ M^{-1}s^{-1}$ for the beef spleen enzyme.[35] A much smaller value was estimated from a steady-state study of the human enzyme.[80] When ethyl hydroperoxide is reacted with human myeloperoxidase, k_1 is $(2.8 \pm 0.3) \times 10^5\ M^{-1}s^{-1}$.[70]

There are further complications. Compound III formation occurs slowly by two different pathways. In one route it is formed from compound II and hydrogen peroxide, with a rate constant of $50\ M^{-1}s^{-1}$. It is also formed from superoxide, which must be reacting with native enzyme.[80] Superoxide also affects the rate of reduction of compound I to compound II either directly or indirectly.[35] In the steady state a mixture of compound II and compound III exists.[80] For further discussion of the role of superoxide, see Section 15.12.

For organic hydroperoxides, the analog of $HO_2\bullet$ formation is the $ROO\bullet$ radical. Formation of such radicals was demonstrated using both neutrophils and purified myeloperoxidase.[81] In addition to peroxyl radicals, alkoxyl and alkyl radicals were detected. It was suggested that formation of these radicals may be responsible for the cytotoxicity of organic hydroperoxides.

15.7.2. Reaction with Hypochlorous Acid

Myeloperoxidase reacts rapidly with hypochlorous acid to form compound I.

$$MPO + HOCl \rightarrow MPO\text{-}I + Cl^- + H^+ \tag{15.3}$$

Rate constants of $(2.3 \pm 0.4) \times 10^6\ M^{-1}s^{-1}$ and $2 \times 10^7\ M^{-1}s^{-1}$ were measured at pH 9.2 and 8.3. At pH 7.2 the reaction is too fast to measure on a stopped flow apparatus.[82] A competitive reaction is the nonenzymatic reaction of hypochlorous acid with chloride ion:

$$HOCl + Cl^- \rightarrow Cl_2 + OH^- \tag{15.4}$$

There are interesting implications of the above pair of reactions. If compound I formation is dominant [reaction (15.3)], then myeloperoxidase is hampered in producing a chlorinating reagent. As soon as hypochlorous acid is produced, it reacts with native enzyme to reform compound I and regenerate chloride ion. If reaction (15.4) is dominant, then the chlorinating reagent is Cl_2.

However, there is an alternative possibility. Hypochlorous acid may not be released from the enzyme. Rather, an enzyme-bound form of hypochlorous acid may be the chlorinating reagent. The latter possibility is discussed in Section 15.11.2.

15.7.3. Reactions of and with Peroxynitrite

The discovery of the endothelium-derived relaxing factor (EDRF) to be nitric oxide,[83,84] followed by the discovery and isolation of nitric oxide synthase,[85,86] caused great excitement. Nitric oxide is also synthesized in neutrophils and macrophages. Perhaps as a result of the presence of superoxide in stimulated neutrophils, peroxynitrite is formed.

$$\bullet NO + O_2\bullet^- \rightarrow OONO^-$$

Peroxynitirite is a strong oxidizing agent. Its reaction with myeloperoxidase, lactoperoxidase, and horseradish peroxidase was examined.[87] With horseradish peroxidase compound I formation was observed, but with the other two peroxidases only compound II. It is not known whether a one-electron redox reaction is occurring with myelo- and lactoperoxidase or whether their compounds I are very unstable in the presence of peroxynitrite.

15.7.4. Reactions of Compound I

When a suitable second reducing substrate for compound I of myeloperoxidase is introduced into the system, then the reversibility of compound I formation and the reduction of compound I by hydrogen peroxide become relatively unimportant. By proper choice of concentrations the dominant reactions are compound I formation from hydrogen peroxide and compound I reduction by the second substrate. In a conventional stopped flow apparatus, myeloperoxidase in one syringe is mixed with hydrogen peroxide and the second substrate in the second syringe. A pre–steady state is observed in which compound I formation is incomplete before reduction by the second substrate is started. In a sequential mixing stopped flow, the first two syringes contain myeloperoxidase and hydrogen peroxide. Conditions can be adjusted so compound I formation goes to near completion before reaction with the second substrate is started. Valid quantitative results can sometimes be obtained from the conventional stopped flow experiments, but much more readily from sequential mixing results.

Reaction of Compound I with Chloride Ion

When myeloperoxidase is reacted with a mixture of hydrogen peroxide and chloride ion, hypochlorous acid is produced,[88] in a two-electron oxidation typical of peroxidase reactions with halide ions. Only myelo- and chloroperoxidase compounds I possess sufficient potential for chloride ion oxidation. The two-step mechanism for hypochlorous acid production would be reaction (15.1) for compound I formation followed by

$$\text{MPO-I} + \text{Cl}^- \xrightarrow{\text{H}^+} \text{MPO} + \text{HOCl} \tag{15.5}$$

The existence of reaction (15.5) has never been directly proven, although it provides the simplest explanation for formation of hypochlorous acid. A single sequential mixing experiment, in which compound I was reacted with chloride ion, has been reported. A rate constant of $(4.7 \pm 0.1) \times 10^6\ M^{-1}s^{-1}$ at pH 7.4 was obtained.[89] This is comparable to the rate constant obtained from steady-state measurements, not for liberation of free hypochlorous acid, but for formation of enzyme-bound hypochlorous acid at pH 4.7, discussed in Section 15.11.2. Clearly more work is required.

Reactions with Tyrosine and Dityrosine

One of the few rate constants reported so far for reactions of myeloperoxidase compound I is from sequential mixing stopped flow experiments on the oxidation of tyrosine and *o,o′*-dityrosine.[89] The one-electron process for the oxidation of tyrosine has a rate constant of $(7.7 \pm 0.1) \times 10^5\ M^{-1}s^{-1}$, and for dityrosine $(1.12 \pm 0.01) \times 10^5\ M^{-1}s^{-1}$, both at pH 7.4. The pH rate profile is shown in Fig. 15.16. Dityrosine is highly fluorescent, whereas tyrosine is not. The fluorescence has been the basis for an assay for lipoprotein damage and oxidative stress. Because dityrosine is nearly as reactive as tyrosine with

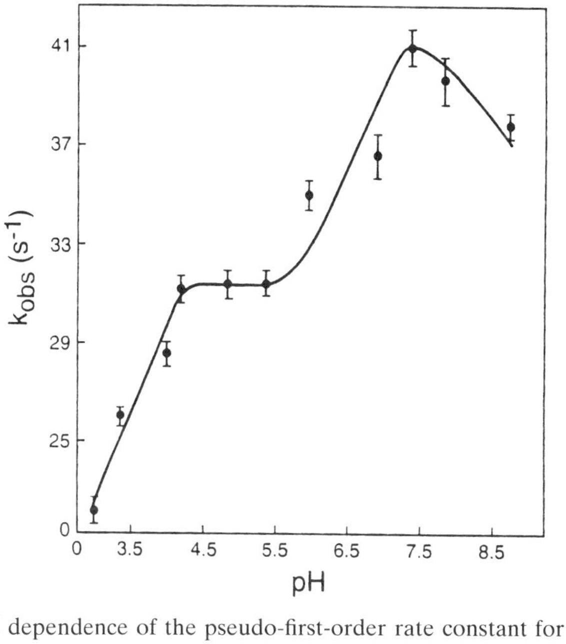

FIG. 15.16. pH dependence of the pseudo-first-order rate constant for the oxidation of L-tyrosine by myeloperoxidase compound I. [Reproduced with permission from L. A. Marquez and H. B. Dunford, *J. Biol. Chem.* **270,** 30434–30440 (1995).]

oxidized myeloperoxidase, an assay based upon dityrosine fluorescence should be used with caution.

15.7.5. Reactions of Compound II with Ascorbate, Acetaminophen, and Iodide

Ascorbic acid, ferrocyanide,[90] 5-aminosalicylic acid,[91] and acetaminophen[92] were found to stimulate the chlorinating activity of myeloperoxidase. This was attributed to reduction of compound II, which is inactive as a chloride ion oxidant, back to native enzyme.

By choosing an appropriate excess of hydrogen peroxide, it is possible to attain a high yield of compound II free of compound III, stable for many minutes, and hence suitable for stopped flow measurements. As an example, this technique was used to study the one-electron redox reactions between the compounds II of three different myeloperoxidases and ascorbate.[36] All behaved in similar fashion, exhibiting saturation kinetics. Thus, for beef spleen myeloperoxidase compound II, the reaction is characterized by a dissociation constant, K, of $(6.3 \pm 0.9) \times 10^{-4}$ M, and a rate constant, k, of $19 \pm 1\ s^{-1}$ at pH 7.1. The one-electron oxidation of iodide by the beef spleen compound II also exhibited saturation kinetics with $K = (1.1 \pm 0.1) \times 10^{-4}$ M, $k = 13.6 \pm 0.8\ s^{-1}$ at pH 7.1.

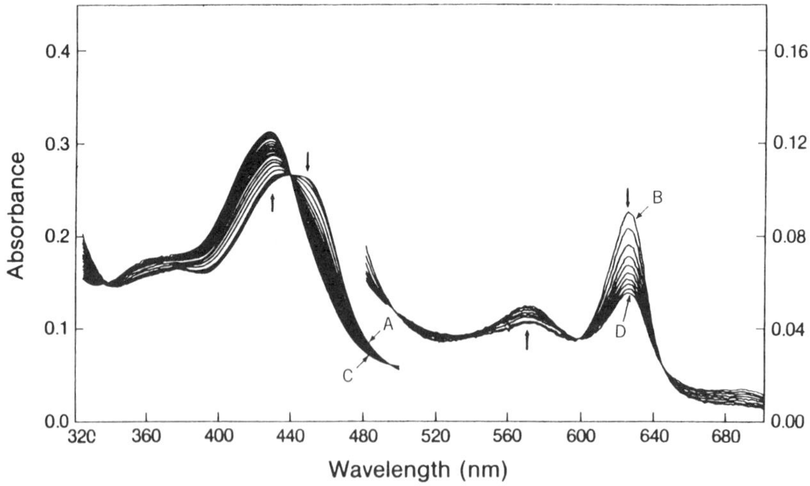

FIG. 15.17. Spontaneous decay of myeloperoxidase compound III to native enzyme. Compound III, stable for 15 min, was prepared from native enzyme using xanthine–xanthine oxidase. Times in minutes of initial and final scans after initial formation of compound III: (A) 15, (B) 18, (C) 60, and (D) 66. Arrows indicate direction of absorption changes. [Reproduced with permission from D. Metodiewa and H. B. Dunford, *Arch. Biochem. Biophys.* **272,** 245–253 (1989).]

15.7.6. Preparation and Reactions of Compound III

Three methods of preparation of myeloperoxidase compound III are discussed here. In the first, compound II is reacted with excess hydrogen peroxide followed by catalase addition.[34] In the second, xanthine–xanthine oxidase is used as a source of superoxide, which reacts with native enzyme to form compound III[93,94]; and in the third, aerobic oxidation of dihydroxyfumarate is used to generate superoxide,[19] a method first used with horseradish peroxidase.[95] Using the beef spleen enzyme, well-defined peaks of compound III at 450 and 625 nm were obtained, with an isosbestic point between native enzyme and compound III at 440 nm. Compound III, prepared using dihydroxyfumarate, is stable for 15 min at pH 4.7. At higher pH it decays more rapidly to native enzyme, a process accelerated by addition of superoxide dismutase. The spontaneous decay of compound III is shown in Fig. 15.17. Ascorbic acid reduces compound III to native enzyme,[94] with a rate constant of $(4.0 \pm 0.1) \times 10^2\ M^{-1}s^{-1}$.

15.8. WHAT IS THE CHLORINATING REAGENT IN MYELOPEROXIDASE REACTIONS?

There is widespread acceptance in the literature that hypochlorous acid, HOCl, is the chlorinating reagent. The chlorination of taurine, described in Section 15.5.3,[111] provides the first concrete evidence for chlorination by an enzyme-bound form of hypochlorous acid. Taurine, the most abundant free amino acid in the neutrophil,[96] may moderate indisciminant oxidative damage by HOCl in the neutrophil. One can hypothesize that if the membrane, of a vacuole in which microbial killing is occurring, is ruptured, then reaction of myeloperoxidase with taurine commences. Taurine thus appears to be a special substrate for myeloperoxidase.

There are reports that molecular chlorine, Cl_2, may be the chlorinating reagent.[97] Certainly Cl_2 can be produced from the reaction of HOCl with Cl^- [reaction (15.4)]. However, there is an equilibrium between HOCl and Cl_2:

$$Cl_2 + H_2O \rightleftarrows HOCl + H^+ + Cl^- \tag{15.6}$$

Even though Cl_2 may be in much higher concentration than HOCl, HOCl could be so much more reactive that it is the sole chlorinating reagent inside a vacuole. In the nonenzymatic iodination of tyrosine, the comparable equilibrium exists between hypoiodous acid, HOI, and molecular iodine, I_2; hence it was not possible to deduce whether HOI or I_2 was the iodinating reagent of tyrosine.[98]

In conclusion, the conventional wisdom that HOCl, free in solution, is the chlorinating agent cannot be disproven. A case can be made for Cl_2 being the agent in certain cases. For microbial killing, it makes sense that a reagent is generated that diffuses from the enzyme, so that it can attack at many locations. Taurine appears to be a special substrate for myeloperoxidase, and in this case the reaction involves enzyme-bound hypochlorous acid. One wonders whether there is another substrate that behaves like taurine.

15.9. REACTIONS WITH HYPOCHLOROUS ACID

It was claimed that the myeloperoxidase chlorinating system produced singlet oxygen.[99] However, it was pointed out that the reagents used to detect singlet oxygen also reacted with hypochlorous acid.[100]

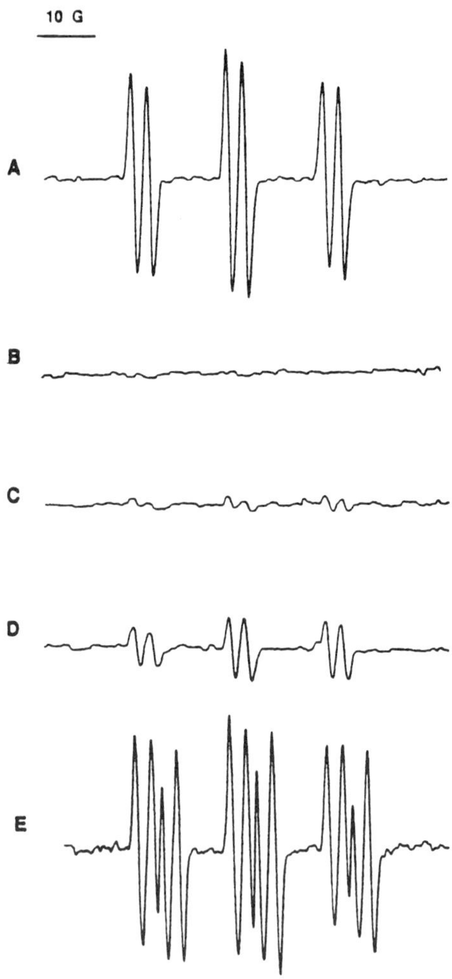

FIG. 15.18. EPR evidence for generation of hydroxyl radicals from stimulated neutrophils using 10 mM of the spin trap 4-pyridyl-1-oxide *N-tert*-butylnitrone in the presence of 170 mM ethanol. (A) Signal after 10 min. (B–D) effect of 15 units/mL of superoxide dismutase, 330 units/mL of catalase, and 1 mM azide, respectively. (E) Same as (A) except ^{13}C ethanol was used. [Reproduced with permission from C. L. Ramos, S. Pou, B. E. Britigan, M. S. Cohen, and G. M. Rosen, *J. Biol. Chem.* **267,** 8307–8312 (1992).]

Hydroxyl radical formation in neutrophils and monocytes was detected by EPR using a spin trap for the α-hydroxyethyl radical, formed from hydroxyl radicals and ethanol (Fig. 15.18).[101] No hydroxyl radical formation occurred in myeloperoxidase-deficient cells. The mechanism of formation was indicated to be:

$$HOCl + O_2\bullet^- \rightarrow \bullet OH + O_2 + Cl^- \qquad (15.7)$$

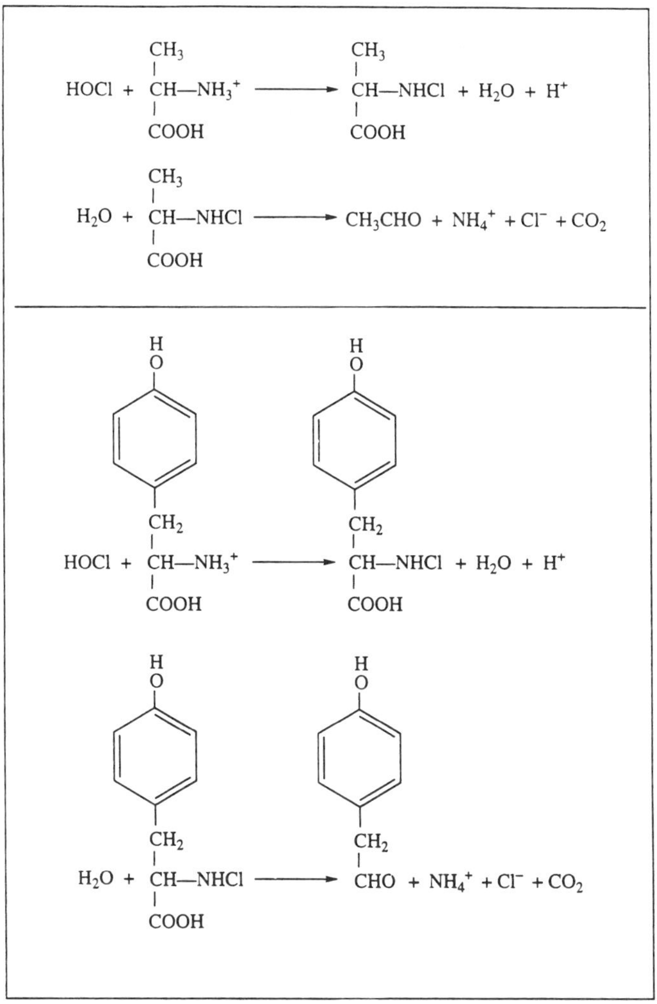

FIG. 15.19. Conversion by hypochlorous acid of alanine into acetaldehyde and tyrosine into *p*-hydroxyphenylacetaldehyde through formation of unstable chloramines.

Hypochlorous acid attacks free amino acids to form chloramines, which are usually unstable.[102,103] Thus, the reactions with analine and tyrosine to form aldehydes are shown in Fig. 15.19. In contrast the chloramine of taurine is stable.

15.10. REDUCED MYELOPEROXIDASE AND ITS NITROSYL COMPLEX

As with other peroxidases, the iron of myeloperoxidase can be reduced to its ferrous state with dithionite.[104] The spectrum of ferrous myeloperoxidase is red-shifted compared to other peroxidases, with its Soret band at 472 nm and the α-band at 636 nm.[105] The EPR spectrum of the nitrosyl compounds of the reduced mammalian peroxidases—myelo-, eosinophil, and lactoperoxidases—indicated that a nitrogen is in the fifth coordination position of these enzymes, now known to be an imidazole side chain.[106]

15.11. STEADY-STATE REACTIONS OF MYELOPEROXIDASE

15.11.1. Peroxidation Reactions

Peroxidation of 4-Chloroaniline

When 4-chloroaniline is incubated with hydrogen peroxide and myeloperoxidase, at least 10 products are formed,[107] none of which is 4-chloronitrosobenzene, the sole product when myeloperoxidase is replaced by chloroperoxidase.

Peroxidation of Thiocyanate

Thiocyanate, SCN^-, is oxidized to hypothiocyanite, $OSCN^-$, by both myelo- and lactoperoxidase.[108] The reactions have sharp maxima between pH 5.0 and 7.5. Thiocyanate is present in saliva and its oxidation product has bactericidal activity. It is not known whether the oxidation proceeds by a two-electron or successive one-electron reactions. Cyanate and thiocyanate are called pseudohalides, since their chemistry often mimics that of halide ions. On this basis one would expect a two-electron mechanism to be operative.

Hydroxylation of Benzoic Acid in the Presence of Hydrazines

Peroxidases normally do not catalyze cytochrome P450 type of reactions. However, in the presence of isoniazid (isonicotinic acid hydrazide), benzoic acid is hydroxylated.[109,110] Phytic acid, which inhibits the ferric ion-catalyzed hydroxylation, stimulates the corresponding myeloperoxidase-catalyzed reaction. Phenylhydrazine exhibits lower reactivity than isoniazid. Autoxidation of the hydrazines produces hydrogen peroxide, which initiates the reaction.

The hydroxylation does not have the characteristics of a hydroxyl radical reaction, but otherwise the mechanism of this interesting reaction is not understood. The combination of isoniazid and phytic acid causes the reduction of the iron(III) of myeloperoxidase to iron(II) with subsequent formation of compound III.

15.11.2. Chlorination Reactions: Formation of Taurine Monochloramine

An alternative to reaction (15.5) is one in which enzyme-bound hypochlorous acid is produced. Evidence for such an intermediate was found in a steady-state study of the chlorination of taurine by myeloperoxidase.[111] A mechanism in which free hypochlorous acid is the chlorinating reagent clearly would not fit the experimental data. The successful mechanism was the following:

$$\text{MPO} + \text{H}_2\text{O}_2 \rightleftarrows \text{MPO-I} + \text{H}_2\text{O} \tag{15.1}$$

$$\text{MPO-I} + \text{Cl}^- \rightarrow \text{MPO-I-Cl}^- \tag{15.8}$$

$$\text{MPO-I-Cl}^- + \text{Taurine} \rightarrow \text{TauNHCl} + \text{H}_2\text{O} \tag{15.9}$$

$$\text{MPO} + \text{Cl}^- \overset{K}{\rightleftarrows} \text{MPO-Cl}^- \tag{15.10}$$

MPO-I-Cl^- is enzyme-bound hypochlorous acid, of the form $\text{Fe}^{\text{IV}}\text{—O—Cl}^-$ and TauNHCl is taurine monochloramine. Chlorination of taurine was shown to be the rate-controlling step. Taurine was an excellent substrate for a chlorination study since it was not an effective reducing substrate for compounds I and II. The study demonstrated that there is no single optimum pH for a chlorination reaction. Rather optimum activity is obtained at a pH that is dependent upon the concentrations of the three substrates, hydrogen peroxide, chloride ion, and taurine.

Myeloperoxidase can catalyze iodination reactions, as do lacto- and thyroid peroxidases. A comparison of the three iodinating systems has been made.[112]

15.12. ROLE OF SUPEROXIDE

The formation of compound III from ferric myeloperoxidase and superoxide was first demonstrated in 1972[113] and was shown to occur in stimulated neutrophils in 1985.[114] This was followed by intensive investigations of the effect of superoxide on the chlorinating activity of myeloperoxidase. Either hypoxanthine or acetaldehyde was reacted with xanthine oxidase to generate superoxide, and the modes of regenerating active enzyme from compounds II and III was examined.[115] The ability of hydrogen peroxide to reduce compound I to compound II, and of superoxide to reduce compound II to native enzyme was noted.[116] Semiquinone radicals generate superoxide, which inhibits the oxidation of hydroquinone.[117] A superoxide generating system plus myeloperoxidase was shown to hydroxylate salicylate.[118] The so-called peroxidase–oxidase reaction, the spontaneous oxidation of nicotinamide adenine dinucleotide (NADH), usually attributed to horseradish peroxidase, also occurs using myeloperoxidase.[119]

The inhibitory effect was examined of a series of drugs on the production of hypochlorous acid by myeloperoxidase.[120,121] It was concluded that the inhibition was caused by compound II production. Superoxide was an antagonist of the anti-inflammatory drugs and was believed to act by reducing compound II back to the native enzyme. In the case of diclofenac, its oxidation by both compounds I and II was demonstrated, with rate constants of 6.0×10^5 $M^{-1}s^{-1}$ and 97.5 $M^{-1}s^{-1}$. Thus, there is a net accumulation of compound II.[122]

15.13. MYELOPEROXIDASE INHIBITORS AND PROTECTORS

A natural inhibitor of myeloperoxidase has been found in normal human serum, a 150-kilodalton protein.[123] Salicylhydroxamic acid acts as a potent inhibitor of the myeloperoxidase-catalyzed luminescence by luminol, but does not stop hypochlorous acid production.[124] Acetaminophen has inhibitory action both by binding to the enzyme and as a competitor of chloride as substrate.[125] Dimethylurea blocks hypochlorous acid formation.[126] Propylthiouracil irreversibly inhibits myeloperoxidase in the presence of hydrogen peroxide.[127] Extracellular hydrogen peroxide can diffuse into intracellular compartments and, when combined with azide, can inactivate myeloperoxidase.[128]

Chlorpromazine, a drug for psychiatric disorders, is a prototype of the phenothiazines, which are widely distributed in bodily tissue. Chlorpromazine acts as a one-electron, not a hydrogen atom, donor to oxidized myeloperoxidase compounds. The reaction is shown in Fig. 15.20. The cation radical can be further oxidized by the enzyme and also by hypochlorous acid. Thus, chlorpromazine offers protection to myeloperoxidase from oxidative damage by hypochlorous acid if the enzyme is released into the extracellular environment.[129]

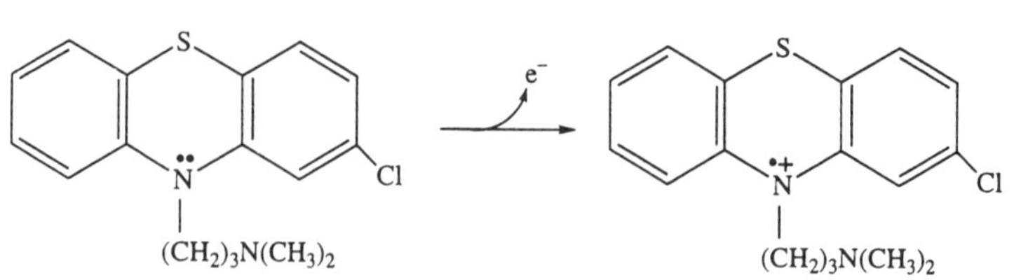

FIG. 15.20. Conversion of chlorpromazine into its cation radical by one-electron oxidation.

15.14. PHYSIOLOGICAL EFFECTS OF MYELOPEROXIDASE

Myeloperoxidase mediates cell adhesion, and an adhesion cell receptor for myeloperoxidase has been identified.[130] Cell adhesion is essential for normal function of multicellular organisms but also plays a role in many pathological conditions.

The conversion of L-tyrosine to the water-soluble *p*-hydroxyphenylacetaldehyde through an unstable chloramine was demonstrated in stimulated human neutrophils (Fig. 15.19). The reaction may exert potent effects in inflammation sites.[131] *p*-Hydroxyphenylacetaldehyde covalently modifies amino groups of lysine residues.[132] Cholesterol is converted into a family of chlorinated sterols (Fig. 15.21).[97,133] In contrast to the behavior of free tyrosine, tyrosyl residues in peptides are converted to chlorotyrosine.[134]

The effects of the oxidation of tyrosine to the tyrosyl radical has been examined in a series of papers. Tyrosyl radicals may attack both high- and low-density lipoprotein.[135–137] They combine to form dityrosine,[138] trityrosine, pulcherosine, and isodityrosine (Fig. 15.22).[139] Recently, it has been shown that a tyrosine peroxide is generated both in a myeloperoxidase-superoxide generating system and by stimulated neutrophils.[140] Tyrosine peroxide production causes a decrease in the amount of dityrosine formed.

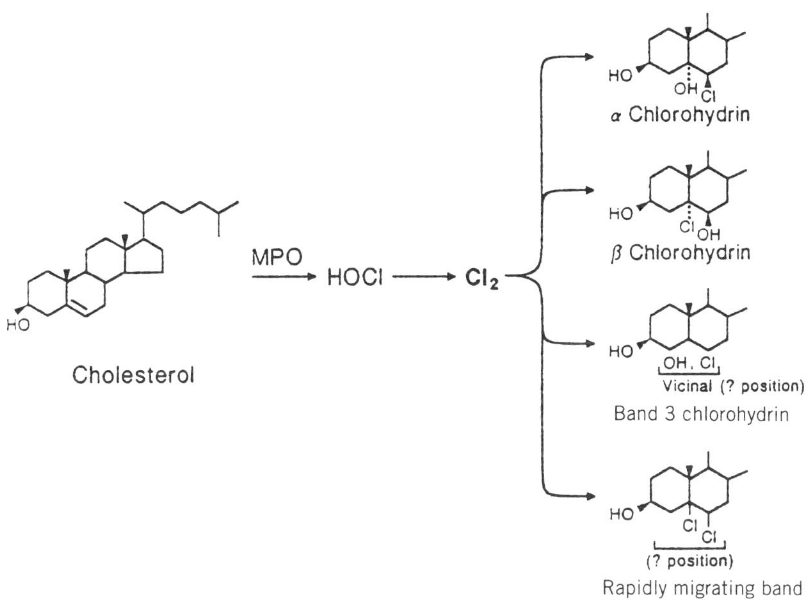

FIG. 15.21. Proposed reaction pathway for the synthesis of cholesterol chlorohydrins and dichlorinated cholesterol by the myeloperoxidase system. [Reproduced with permission from S. R. Hazen, F. F. Hsu, K. Duffin, and J. W. Heinecke, *J. Biol. Chem.* **271,** 23080–23088 (1996).]

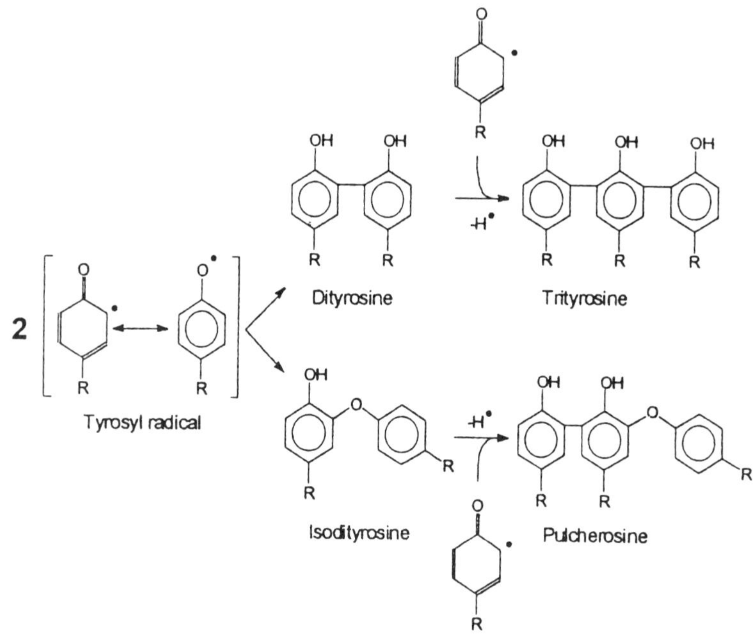

FIG. 15.22. Proposed pathways for formation of products from tyrosyl radicals. [Reproduced with permission from J. S. Jacob, D. P. Cistola, F. F. Hsu, S. Muzaffar, D. M. Mueller, S. L. Hazen, and J. W. Heinecke, *J. Biol. Chem.* **271,** 19950–19956 (1996).]

The myeloperoxidase system is capable of inactivating other enzymes found in granules, such as lysozyme, β-glucuronidase, and collagenase.[141] Inactivation of transferrin[142] and release of its iron,[143] inactivation of iron–sulfur clusters,[144] penicillin-binding protein,[145] human immunodeficiency virus-1,[146] and heparin sulfate proteoglycans[147] have been described. Tissue destruction by neutrophils, in which myeloperoxidase plays a major role, has been reviewed.[148]

The major end product of nitric oxide metabolism is nitrite ion. Nitrite can be oxidizied by myeloperoxidase to form peroxynitrite, a potential additional source of nitric oxide toxicity.[149,150]

The thiobarbituric acid (TBA) assay, dependent largely on production of malondialdehyde, is commonly used to measure the extent of lipid peroxidation. Hypochlorous acid generated from myeloperoxidase attacks the malondialdehyde, so that TBA assays may be inaccurate when neutrophils or myeloperoxidase are involved.[151]

15.15. OXYGEN BURST, PHAGOCYTOSIS, AND MICROBIAL KILLING

The process of ingestion of a foreign organism into a newly formed vacuole of a neutrophil is represented schematically in Fig. 15.23. The neutrophil is

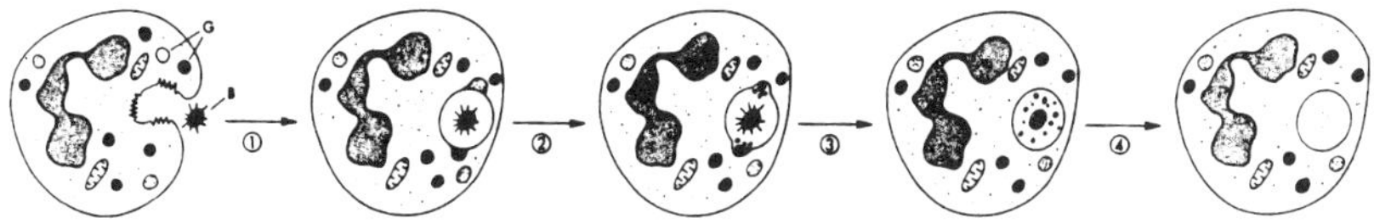

FIG. 15.23. Schematic representation of the phagocytic process in which a microbe is ingested into a vacuole. Granules containing myeloperoxidase are released into the vacuole and microbial killing occurs from hypochlorous acid generation. [Reproduced with permission from M. J. Cline, *The White Cell,* Harvard University Press, Cambridge, MA. Copyright © 1975 by the President and Fellows of Harvard College.]

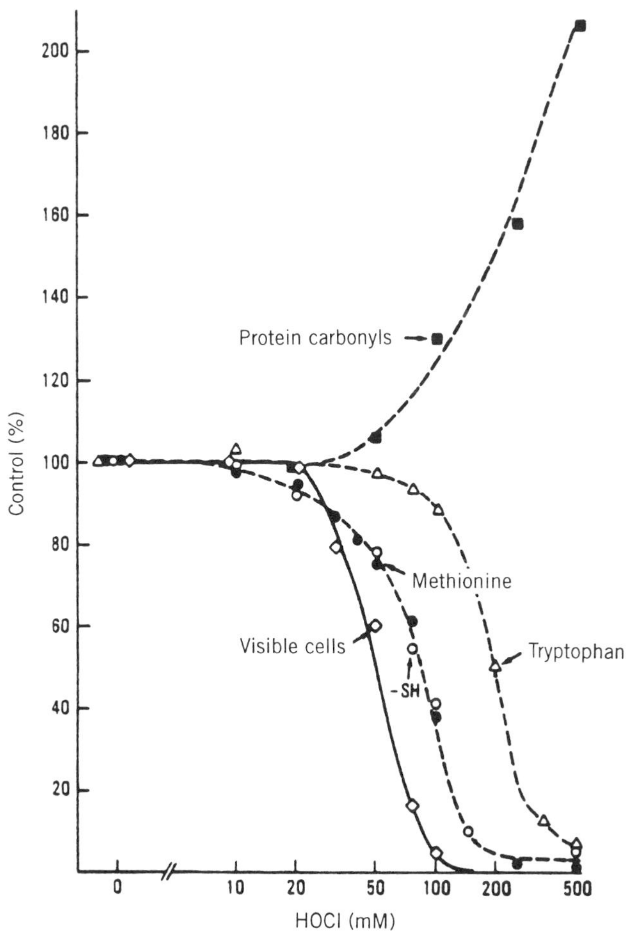

FIG. 15.24. Oxidation of various cellular protein targets after 15 min exposure to hypochlorous acid; 60% of the viable cells remain at a 50-μM concentration of hypochlorous acid. [Reproduced with permission from I. U. Schraufstätter, K. Browne, A. Harris, P. A. Hyslop, J. H. Jackson, O. Quehenberger, and C. G. Cochrane, *J. Clin. Invest.* **85,** 554–562 (1990) by copyright permission of the American Society for Clinical Investigation.]

stimulated, and a burst of oxygen consumption occurs. Granules become attached to the vacuole and empty their contents, including myeloperoxidase, into the vacuole. Hypochlorous acid is generated and the killing process begins. Thus, the primary defence against infection is activated.

The requirements and mechanism for stimulation have been studied, of which a few examples are cited.[152–154] Oxygen is converted into superoxide by the enzyme NADPH oxidase,[155,156] and the superoxide dismutes spontaneously into oxygen and hydrogen peroxide. Hydrogen peroxide forms compound I of myeloperoxidase, which reacts with chloride ion to generate hypochlorous acid. Mechanisms of hypochlorite injury of target cells have been investigated (Fig. 15.24). It was found that a 75-μM dose of hypochlorous acid would lyse all cells.[157]

15.16. HEREDITARY MYELOPEROXIDASE DEFICIENCY

About one in 3000 persons has no neutrophil or monocyte myeloperoxidase. This absence was first noted in 1963.[158] The synthesis and processing of mature myeloperoxidase is a complicated process,[159] and it was suspected that a misstep in the synthesis was blocking formation of the final product. One such misstep has been identified: the missense mutation of Arg569 to tryptophan.[160,161]

15.17. EOSINOPHIL PEROXIDASE

Little did the goddess Eos realize that she would have a dye and a peroxidase named in her honor. Her name was used for the dye eosin, which is a particularly good stain for the type of leukocyte known as eosinophil. The eosinophils, much less abundant than the neutrophils, can readily be detected and quantified using flow cytometry.[162] Persons who suffer from hereditary myeloperoxidase deficiency have normal eosinophils with their normal eosinophil peroxidase content. There is a distinct hereditary eosinophil peroxidase deficiency.[163] Eosinophils are involved in the detoxification of foreign proteins and other substances that appear to have tumoricidal activity.

Much of the pioneering work on isolating, purifying, and testing for toxicity of eosinophil peroxidase was conducted on horse eosinophils.[164–166] An excellent review of the physiological role of eosinophil peroxidase has been published.[167]

Human eosinophil peroxidase was isolated and characterized.[168] The molecular weight of this highly basic protein was found to be 71,000 ± 1000. The enzyme is composed of two subunits of molecular weights 58,000 and 14,000. The molar absorptivity at the Soret maximum of 412 nm was found to be $\varepsilon = 110 \pm 4\ mM^{-1}cm^{-1}$. Electron paramagnetic resonance spectra indicated that eosinophil and lactoperoxidases have similar heme environments, but differences were found in the resonance Raman spectra.[169]

Eosinophil peroxidase does not appear to be efficient at oxidizing chloride to hypochlorous acid,[170] whereas it is highly efficient at oxidizing bromide.[171,172] However, the physiological substrate for eosinophil peroxidase may be the pseudo-halide thiocyanate.[173] Eosinophil peroxidase-dependent hydroxyl radical formation has been detected.[174]

A specific assay for eosinophil peroxidase, capable of distinguishing it from myeloperoxidase, is homovanillic acid.[175] Dapsone is an inhibitor of both myelo- and eosinophil peroxidases.[176]

The gene for eosinophil peroxidase has been cloned[177,178] and shown to have great homology with human myeloperoxidase. The amino acid sequence is listed in Appendix A.

REFERENCES

1. K. Agner, *Acta Physiol. Scand.* **II** (Suppl. VIII), 1–62 (1941).
2. H. Theorell, *Adv. Enzymol.* **7,** 265–303 (1947).
3. K. Agner, *Adv. Enzymol.* **3,** 137–148 (1943).
4. S. J. Klebanoff and R. A. Clark. *The Neutrophil: Function and Clinical Disorders,* North-Holland, Amsterdam, 1978.
5. S. J. Klebanoff in *Inflammation: Basic Principles and Clinical Correlates,* J. I. Gallin, I. M. Goldstein, and R. Snyderman, eds., Raven, New York, 1988, pp. 391–444.
6. S. J. Klebanoff in *Peroxidases in Chemistry and Biology,* J. Everse, K. E. Everse, and M. B. Grisham, eds., CRC Press, Boca Raton, 1991, Vol. I, pp. 1–35.
7. J. C. Davis and B. A. Averill, *J. Biol. Chem.* **256,** 5992–5996 (1981).
8. J. C. Davis and B. A. Averill, *Inorg. Chim. Acta* **93,** L49–L51 (1984).
9. M. Ikeda-Saito, *J. Biol. Chem.* **260,** 11688–11696 (1985).
10. M. Ikeda-Saito, H. C. Lee, K. Adachi, H. S. Eck, R. C. Prince, K. S. Booth, W. S. Caughey, and S. Kimura, *J. Biol. Chem.* **264,** 4559–4563 (1989).
11. K. Agner, *Acta Chem. Scand.* **12,** 89–94 (1958).
12. J. E. Harrison, S. Pabalan, and J. Schultz, *Biochim. Biophys. Acta* **493,** 247–259 (1977).
13. T. Odajima and I. Yamazaki, *Biochim. Biophys. Acta* **206,** 71–77 (1970).
14. A. R. J. Bakkenist, R. Wever, T. Vulsma, H. Plat, and B. F. Van Gelder, *Biochim. Biophys. Acta* **524,** 45–54 (1978).
15. N. R. Matheson, P. S. Wong, and J. Travis, *Biochemistry* **20,** 325–330 (1981).
16. S. O. Pember, R. Shapira, and J. M. Kinkade Jr., *Arch. Biochem. Biophys.* **221,** 391–403 (1983).
17. Y. Morita, H. Iwamoto, S. Aibara, T. Kobayashi, and E. Hasegawa, *J. Biochem.* **99,** 761–770 (1986).

18. G. T. Babcock, R. T. Ingle, W. A. Oertling, J. C. Davis, B. A. Averill, C. L. Hulse, D. J. Stufkens, B. G. J. M. Bolscher, and R. Wever, *Biochim. Biophys. Acta* **828,** 58–66 (1985).
19. L. A. Marquez, Ph.D. Thesis, University of Alberta, 1990.
20. J. Wright, S. Yoshimoto, G. D. Offner, R. A. Blanchard, R. Troxler, and A. I. Tauber, *Biochim. Biophys. Acta* **915,** 68–76 (1987).
21. K. T. Miyasaki, J.-P. Song, and A. R. K. Murthy, *Anal. Biochem.* **193,** 38–44 (1991).
22. J. Wright, N. Bastian, T. A. Davis, C. Zuo, S. Yoshimoto, W. H. Orme-Johnson, and A. I. Tauber, *Blood* **75,** 238–241 (1990).
23. K. Hashinaka, C. Nishio, S.-J. Hur, F. Sakiyama, S. Tsunasawa, and M. Yamada, *Biochemistry* **27,** 5906–5914 (1988).
24. A. Ehrenberg and K. Agner, *Acta Chem. Scand.* **12,** 95–100 (1958).
25. K. S. Booth, S. Kimura, H. C. Lee, M. Ikeda-Saito, and W. S. Caughey, *Biochem. Biophys. Res. Commun.* **160,** 897–902 (1989).
26. W. M. Nauseef, S. McCormick, and H. Yi, *Blood* **80,** 2622–2633 (1992).
27. P. C. Andrews and N. I. Krinsky, *J. Biol. Chem.* **256,** 4211–4218 (1981).
28. R. L. Olsen and C. Little, *Biochem. J.* **222,** 701–709 (1984).
29. R. Wever, W. A. Oertling, H. Hoogland, B. G. J. M. Bolscher, Y. Kim, and G. T. Babcock, *J. Biol. Chem.* **266,** 24308–24313 (1991).
30. P. C. Andrews, C. Parnes, and N. I. Krinsky, *Arch. Biochem. Biophys.* **228,** 439–442 (1984).
31. L. A. Marquez and H. B. Dunford, *Biochem. Cell Biol.* **67,** 187–191 (1989).
32. T. Odajima and I. Yamazaki, *Biochim. Biophys. Acta* **284,** 360–367 (1972).
33. J. E. Harrison, T. Araiso, M. M. Palcic, and H. B. Dunford, *Biochem. Biophys. Res. Commun.* **94,** 34–40 (1980).
34. H. Hoogland, A. van Kuilenburg, C. van Riel, A. O. Muijsers, and R. Wever, *Biochim. Biophys. Acta* **916,** 76–82 (1987).
35. L. A. Marquez, J. T. Huang, and H. B. Dunford, *Biochemistry* **33,** 1447–1454 (1994).
36. L. A. Marquez, H. B. Dunford, and H. Van Wart, *J. Biol. Chem.* **265,** 5666–5670 (1990).
37. K. S. Chang, J. M. Trujillo, R. G. Cook, and S. A. Stass, *Blood* **68,** 1411–1414 (1986).
38. K. Morishita, N. Kubota, S. Asano, Y. Kaziro, and S. Nakata, *J. Biol. Chem.* **262,** 3844–3851 (1987).
39. M. Yamada, S. J. Hur, K. Hashinaka, K. Tsuneoka, T. Saeki, C. Nishio, F. Sakiyama, and S. Tsunasawa, *Arch. Biochem. Biophys.* **255,** 147–155 (1987).
40. K. Hashinaka, C. Nishio, S.-J. Hur, F. Sakiyama, S. Tsunasawa, and M. Yamada, *Biochemistry* **27,** 5906–5914 (1988).
41. S. C. Weil, G. L. Rossner, M. S. Reid, R. L. Chisholm, N. M. Farber, J. K. Spitznagel, and M. S. Swanson, *Proc. Natl. Acad. Sci. U.S.A.* **84,** 2057–2061 (1987).
42. K. L. Taylor, D. J. Uhlinger, and J. M. Kincade, Jr., *Biochem. Biophys. Res. Commun.* **187,** 1572–1578 (1992).
43. A. Jacquet, C. Deby, M. Mathy, N. Moguilevsky, G. Deby-Dupont, A. Thirion, E. Goormaghtigh, L. Garcia-Quintana, A. Bollen, and J. Pincemail, *Arch. Biochem. Biophys.* **291,** 132–138 (1991).

44. N. Moguilevsky, L. Garcia-Quintana, A. Jacquet, C. Tournay, L. Fabry, L. Piérard, and A. Bollen, *Eur. J. Biochem.* **197,** 605–614 (1991).
45. G. E. Austin, W.-G. Zhao, W. Zhang, E. D. Austin, H. W. Findley, and J. J. Murtagh Jr. *Leukemia* **9,** 848–857 (1995).
46. A. M. Ford, C. A. Bennett, L. E. Healy, M. Towatari, M. F. Greaves, and T. Enver, *Proc. Natl. Acad. Sci. U.S.A.* **93,** 10838–10843 (1996).
47. W. M. Nauseef, S. J. McCormick, and R. A. Clark, *J. Biol. Chem.* **270,** 4741–4747 (1995).
48. A. Jacquet, L. Garcia-Quintana, V. Deleersnyder, R. Fenna, A. Bollen, and N. Moguilevsky, *Biochem. Biophys. Res. Commun.* **202,** 73–81 (1994).
49. R. Floris, N. Moguilevsky, G. Puppels, A. Jacquet, R. Renirie, A. Bollen, and R. Wever, *J. Am. Chem. Soc.* **117,** 3907–3912 (1995).
50. J. Zeng and R. E. Fenna, *J. Mol. Biol.* **226,** 185–207 (1992).
51. A. Ehrenberg, *Ark. Kemi.* **19,** 119–127 (1962).
52. S. S. Sibbett and J. K. Hurst, *Biochemistry* **23,** 3007–3013 (1984).
53. L. A. Andersson, T. M. Loehr, C. K. Chang, and A. G. Mauk, *J. Am. Chem. Soc.* **107,** 182–191 (1985).
54. G. T. Babcock, R. T. Ingle, W. A. Oertling, J. C. Davis, B. A. Averill, C. L. Hulse, D. J. Stufkens, B. G. J. M. Bolscher, and R. Wever, *Biochim. Biophys. Acta* **828,** 58–66 (1985).
55. M. Sono, J. H. Dawson, and M. Ikeda-Saito, *Biochim. Biophys. Acta* **873,** 62–72 (1986).
56. L. B. Dugad, G. N. La Mar, H. C. Lee, M. Ikeda-Saito, K. S. Booth, and W. S. Caughey, *J. Biol. Chem.* **265,** 7173–7179 (1990).
57. H. Hori and M. Ikeda-Saito, *Biochemistry* **29,** 7106–7112 (1990).
58. K. L. Taylor, J. Pohl, and J. M. Kinkade, Jr., *J. Biol. Chem.* **267,** 25282–25288 (1992).
59. R. Fenna, J. Zeng, and C. Davey, *Arch. Biochem. Biophys.* **316,** 635–642 (1995).
60. K. L. Taylor, F. Strobel, K. T. Yue, P. Ram, J. Pohl, A. S. Woods, and J. M. Kincade, Jr., *Arch. Biochem. Biophys.* **316,** 635–642 (1995).
61. I. M. Kooter, A. J. Pierik, M. Merkx, B. A. Averill, N. Moguilevsky, A. Bollen, and R. Wever, *J. Am. Chem. Soc.* **119,** 11542–11543 (1997).
62. M. Ikeda-Saito, P. V. Argade, and D. L. Rousseau, *FEBS Lett.* **184,** 52–55 (1985).
63. C. A. Davey and R. E. Fenna, *Biochemistry* **35,** 10967–10973 (1996).
64. P. C. Andrews and N. I. Krinsky, *Methods Enzymol.* **132,** 369–378 (1986).
65. L. A. Marquez and H. B. Dunford, *Biochemistry* **36,** 9349–9355 (1997).
66. A. C. Maehly, *Methods Biochem. Anal.* **1,** 358–408 (1954).
67. K. Uehara, K. Hori, M. Nakano, and S. Koga, *Anal. Biochem.* **199,** 191–196 (1991).
68. M. Nakano and S. Koga, *Methods Enzymol.* **233,** 495–501 (1994).
69. A. J. Kettle and C. C. Winterbourn, *Methods Enzymol.* **233,** 502–512 (1994).
70. B. G. J. M. Bolscher and R. Wever, *Biochim. Biophys. Acta* **788,** 1–10 (1984).
71. M. Ikeda-Saito, *Biochemistry* **26,** 4344–4349 (1987).
72. J. J. Lopez-Garriga, W. A. Oertling, R. T. Kean, H. Hoogland, R. Wever, and G. T. Babcock, *Biochemistry* **29,** 9387–9395 (1990).

73. D. G. Eglington, D. Barber, A. J. Thomson, C. Greenwood, and A. W. Segal, *Biochim. Biophys. Acta* **703,** 187–195 (1982).
74. A. R. J. Bakkenist, J. E. G. De Boer, H. Plat, and R. Wever, *Biochim. Biophys. Acta* **613,** 337–348 (1980).
75. M. Ikeda-Saito and R. C. Prince, *J. Biol. Chem.* **260,** 8301–8305 (1985).
76. C. E. Cooper and E. Odell, *FEBS Lett.* **314,** 58–60 (1992).
77. H. Hori, R. E. Fenna, S. Kimura, and M. Ikeda-Saito, *J. Biol. Chem.* **269,** 8388–8392 (1994).
78. J. K. Hurst in *Peroxidases in Chemistry and Biology,* J. Everse, K. E. Everse, and M. B. Grisham, eds., CRC Press, Boca Raton, FL, 1991, Vol. I, pp. 37–62.
79. H. Iwamoto, T. Kobayashi, E. Hasegawa, and Y. Morita, *J. Biochem.* **101,** 1407–1412 (1987).
80. H. Hoogland, H. L. Dekker, C. van Riel, A. van Kuilenburg, A. O. Muijsers, and R. Wever, *Biochim. Biophys. Acta* **955,** 337–345 (1988).
81. W. Chmulitrat, M. S. Cohen, and R. P. Mason, *Free Rad. Biol. Med.* **11,** 439–445 (1991).
82. R. Floris and R. Wever, *Eur. J. Biochem.* **207,** 697–702 (1992).
83. R. M. J. Palmer, A. G. Ferrige, and S. Moncada, *Nature* **327,** 524–526 (1987).
84. L. G. Ignarro, G. M. Buga, K. S. Wood, R. E. Byrns, and G. Chaudhuri, *Proc. Natl. Acad. Sci. U.S.A.* **84,** 9265–9269 (1987).
85. B. Mayer, K. Schmidt, P. Humbert, and E. Böhme, *Biochem. Biophys. Res. Commun.* **164,** 678–685 (1989).
86. M. A. Tayeh and M. A. Marletta, *J. Biol. Chem.* **264,** 19654–19658 (1989).
87. R. Floris, S. R. Piersma, G. Yang, P. Jones, and R. Wever, *Eur. J. Biochem.* **215,** 767–775 (1993).
88. J. E. Harrison and J. Schultz, *J. Biol. Chem.* **251,** 1371–1374 (1976).
89. L. A. Marquez and H. B. Dunford, *J. Biol. Chem.* **270,** 30434–30440 (1995).
90. B. G. J. M. Bolscher, G. R. Zoutberg, R. A. Cuperus, and R. Wever, *Biochim. Biophys. Acta* **784,** 189–191 (1984).
91. K. W. M. Zuurbier, A. R. J. Bakkenist, R. Wever, and A. O. Muijsers, *Biochim. Biophys. Acta* **1037,** 140–146 (1990).
92. L. A. Marquez and H. B. Dunford, *Arch. Biochem. Biophys.* **305,** 414–420 (1993).
93. D. Metodiewa and H. B. Dunford, *Arch. Biochem. Biophys.* **272,** 245–253 (1989).
94. L. A. Marquez and H. B. Dunford, *J. Biol. Chem.* **265,** 6074–6078 (1990).
95. I. Yamazaki and L. H. Piette, *Biochim. Biophys. Acta* **77,** 47–64 (1963).
96. P. Soupart in *Amino Acid Pools,* J. T. Holden, ed., Elsevier-North Holland, Amsterdam, 1962, pp. 220–262.
97. S. L. Hazen, F. F. Hsu, K. Duffin, and J. W. Heinecke, *J. Biol. Chem.* **271,** 23080–23088 (1996).
98. H. B. Dunford and A. J. Adeniran, *Biochem. Cell. Biol.* **66,** 967–978 (1988).
99. H. Rosen and S. J. Klebanoff, *J. Biol. Chem.* **252,** 4803–4810 (1977).
100. J. E. Harrison, B. D. Watson, and J. Schultz, *FEBS Lett.* **92,** 327–332 (1978).
101. C. L. Ramos, S. Pou, B. E. Britigan, M. S. Cohen, and G. M. Rosen, *J. Biol. Chem.* **267,** 8307–8312 (1992).

102. J. M. Zgliczynski, T. Stelmaszynska, W. Ostrowski, J. Naskalski, and J. Sznajd, *Eur. J. Biochem.* **4,** 540–547 (1968).

103. J. M. Zgliczynski, T. Stelmaszynska, J. Domanski, and W. Ostrowski, *Biochim. Biophys. Acta* **235,** 419–424 (1971).

104. R. Wever and H. Plat, *Biochim. Biophys. Acta* **661,** 235–239 (1981).

105. R. Floris, Y. Kim, G. T. Babcock, and R. Wever, *Eur. J. Biochem.* **222,** 677–685 (1994).

106. B. G. J. M. Bolscher and R. Wever, *Biochim. Biophys. Acta* **791,** 75–81 (1984).

107. A. R. J. Bakkenist, H. Plat, and R. Wever, *Bioorg. Chem.* **10,** 324–328 (1981).

108. R. Wever, W. M. Kast, J. H. Kasinoedin, and R. Boelens, *Biochim. Biophys. Acta* **709,** 212–219 (1982).

109. J. M. van Zyl, K. Basson, R. A. Uebel, and B. J. van der Walt, *Biochem. Pharmacol.* **38,** 2363–2373 (1989).

110. B. J. van der Walt, J. M. van Zyl, and A. Kriegler, *Biochem. Pharmacol.* **47,** 1039–1046 (1994).

111. L. A. Marquez and H. B. Dunford, *J. Biol. Chem.* **269,** 7950–7956 (1994).

112. A. Taurog and M. L. Dorris, *Arch. Biochem. Biophys.* **296,** 239–246 (1992).

113. T. Odajima and I. Yamazaki, *Biochim. Biophys. Acta* **284,** 355–359 (1972).

114. C. C. Winterbourn, R. C. Garcia, and A. W. Segal, *Biochem. J.* **228,** 583–592 (1985).

115. A. J. Kettle and C. C. Winterbourn, *Biochem. J.* **252,** 529–536 (1988).

116. A. J. Kettle and C. C. Winterbourn, *Biochem. J.* **263,** 823–828 (1989).

117. A. J. Kettle and C. C. Winterbourn, *J. Biol. Chem.* **267,** 8319–8324 (1992).

118. A. J. Kettle and C. C. Winterbourn, *J. Biol. Chem.* **269,** 17146–17151 (1994).

119. T. Odajima, *Biochim. Biophys. Acta* **235,** 52–60 (1971).

120. A. J. Kettle and C. C. Winterbourn, *Biochem. Pharmacol.* **41,** 1485–1492 (1991).

121. A. J. Kettle, C. A. Gedye, and C. C. Winterbourn, *Biochem. Pharmacol.* **45,** 2003–2010 (1993).

122. K. W. M. Zuurbier, A. R. J. Bakkenist, R. H. Fokkens, N. M. M. Nibbering, R. Wever, and A. O. Muijsers, *Biochem. Pharmacol.* **40,** 1801–1808 (1990).

123. C. M. Yea, B. Dularay, and C. J. Elson, *Clin. Exp. Immunol.* **84,** 347–352 (1991).

124. B. Davies and S. W. Edwards, *Biochem. J.* **258,** 801–806 (1989).

125. J. M. van Zyl, K. Basson, and B. J. van der Walt, *Biochem. Pharmacol.* **38,** 161–165 (1989).

126. A. L. Sagone, Jr., R. M. Husney, M. D. Wewers, D. J. Herzyk, and W. B. Davis, *J. Appl. Physiol.* **67,** 1056–1062 (1989).

127. E. Lee, Y. Miki, H. Katsura, and K. Kariya, *Biochem. Pharmacol.* **39,** 1467–1471 (1990).

128. Y. Ohno and J. I. Gallin, *J. Biol. Chem.* **260,** 8438–8446 (1985).

129. J. M. van Zyl, K. Basson, A. Kriegler, and B. J. van der Walt, *Biochem. Pharmacol.* **40,** 947–954 (1990).

130. M. W. Johansson, M. Patarroyo, F. Öberg, A. Siegbahn, and K. Nilsson, *J. Cell Sci.* **110,** 1133–1139 (1997).

131. S. L. Hazen, F. F. Hsu, and J. W. Heinecke, *J. Biol. Chem.* **271,** 1861–1867 (1996).
132. S. L. Hazen, J. P. Gaut, F. F. Hsu, J. R. Crowley, A. d'Avignon, and J. W. Heinecke, *J. Biol. Chem.* **272,** 16990–16998 (1997).
133. J. W. Heinecke, W. Li, D. M. Mueller, A. Bohrer, and J. Turk, *Biochemistry* **33,** 10127–10136 (1994).
134. N. M. Domigan, T. S. Charlton, M. W. Duncan, C. C. Winterbourn, and A. J. Kettle, *J. Biol. Chem.* **270,** 16542–16548 (1995).
135. G. A. Francis, A. J. Mendez, E. L. Bierman, and J. W. Heinecke, *Proc. Natl. Acad. Sci. U.S.A.* **90,** 6631–6635 (1993).
136. M. I. Savenkova, D. M. Mueller, and J. W. Heinecke, *J. Biol. Chem.* **269,** 20394–20400 (1994).
137. C. Leeuwenburgh, J. E. Rasmussen, F. F. Hsu, D. M. Mueller, S. Pennathur, and J. W. Heinecke, *J. Biol. Chem.* **272,** 3520–3526 (1997).
138. J. W. Heinecke, W. Li, H. L. Daehnke III, and J. A. Goldstein, *J. Biol. Chem.* **268,** 4069–4077 (1993).
139. J. S. Jacob, D. P. Cistola, F. F. Hsu, S. Muzaffar, D. M. Mueller, S. L. Hazen, and J. W. Heinecke, *J. Biol. Chem.* **271,** 19950–19956 (1996).
140. C. C. Winterbourn, H. Pichorner, and A. J. Kettle, *Arch. Biochem. Biophys.* **338,** 15–21 (1997).
141. M. C. M. Vissers and C. C. Winterbourn, *Biochem. J.* **245,** 277–280 (1987).
142. R. A. Clark and D. W. Pearson, *J. Biol. Chem.* **264,** 9420–9427 (1989).
143. C. C. Winterbourn, H. P. Monteiro, and C. F. Galilee, *Biochim. Biophys. Acta* **1055,** 179–185 (1990).
144. J. K. Hurst, W. C. Barrette Jr., B. R. Michel, and H. Rosen, *Eur. J. Biochem.* **202,** 1275–1282 (1991).
145. R. M. Rakita and H. Rosen, *J. Clin. Invest.* **88,** 750–754 (1991).
146. S. J. Klebanoff and R. W. Coombs, *J. Clin. Invest.* **89,** 2014–2017 (1992).
147. S. J. Klebanoff, M. G. Kinsella, and T. N. Wight, *Am. J. Pathol.* **143,** 907–917 (1993).
148. S. J. Weiss, *New England J. Med.* **320,** 365–376 (1989).
149. C. Leeuwenburgh, M. M. Hardy, S. L. Hazen, P. Wagner, S. Oh-ishi, U. P. Steinbrecher, and J. W. Heinecke, *J. Biol. Chem.* **272,** 1433–1436 (1997).
150. A. van der Vliet, J. P. Eiserich, B. Halliwell, and C. E. Cross, *J. Biol. Chem.* **272,** 7617–7625 (1997).
151. C. C. Winterbourn and A. C. Carr, *Arch. Biochem. Biophys.* **302,** 461–467 (1993).
152. R. Docampo, A. M. Casellas, E. D. Madeira, R. L. Cardoni, S. N. J. Moreno, and R. P. Mason, *FEBS Lett.* **155,** 25–30 (1983).
153. M. P. Wymann, V. von Tscharner, D. A. Deranleau, and M. Baggiolini, *J. Biol. Chem.* **262,** 12048–12053 (1987).
154. I. Eckle, G. Kolb, C. Heiser, and K. Havemann, *Clin. Exp. Immunol.* **81,** 352–356 (1990).
155. B. M. Babior and R. S. Kipnes, *Blood* **50,** 517–524 (1977).
156. G. L. Babior, R. E. Rosin, B. J. McMurrich, W. A. Peters, and B. M. Babior, *J. Clin. Invest.* **67,** 1724–1728 (1981).

157. I. U. Schraufstätter, K. Browne, A. Harris, P. A. Hyslop, J. H. Jackson, O. Quehenberger, and C. G. Cochrane, *J. Clin. Invest.* **85,** 554–562 (1990).

158. Ref. 3, pp. 711–733.

159. H. P. Koeffler, J. Ranyard, and M. Pertcheck, *Blood* **65,** 484–491 (1985).

160. M. Kizaki, C. W. Miller, M. E. Selsted, and H. P. Koeffler, *Blood* **83,** 1935–1940 (1994).

161. W. M. Nauseef, S. Brigham, and M. Cogley, *J. Biol. Chem.* **269,** 1212–1216 (1994).

162. J. W. M. Visser in *Flow Cytometry and Sorting,* 2nd ed., M. R. Melamed, T. Lindmo, and M. L. Mendelsohn, eds., Wiley-Liss, New York, 1990, pp. 669–683.

163. M. Romano, P. Patriarca, C. Melo, F. E. Baralle, and P. Dri, *Proc. Natl. Acad. Sci. U.S.A.* **91,** 12496–12500 (1994).

164. R. K. Archer, *The Eosinophil Leukocytes,* Blackwell, Oxford, 1963.

165. A. Jörg, J.-M. Pasquier, and S. J. Klebanoff, *Biochim. Biophys. Acta* **701,** 185–191 (1982).

166. S. J. Klebanoff, J. M. Agosti, A. Jörg, and A. M. Waltersdorph, *J. Immunol.* **143,** 239–244 (1989).

167. W. R. Henderson, Jr., in *Peroxidases in Chemistry and Biology,* J. Everse, K. E. Everse, and M. B. Grisham, eds., CRC Press, Boca Raton, FL, 1991, Vol. I, pp. 105–121.

168. B. G. J. M. Bolscher, H. Plat, and R. Wever, *Biochim. Biophys. Acta* **784,** 177–186 (1984).

169. S. S. Sibbett, S. J. Klebanoff, and J. K. Hurst, *FEBS Lett.* **189,** 271–275 (1985).

170. S. J. Klebanoff in Ref. 4, 1988, pp. 420–421.

171. A. Slungaard and J. R. Mahoney, Jr., *J. Exp. Med.* **173,** 117–126 (1991).

172. E. L. Thomas, P. M. Bozeman, M. M. Jefferson, and C. C. King, *J. Biol. Chem.* **270,** 2906–2913 (1995).

173. A. Slungaard and J. R. Mahoney, Jr., *J. Biol. Chem.* **266,** 4903–4910 (1991).

174. M. L. McCormick, T. L. Roeder, M. A. Railsback, and B. E. Britigan, *J. Biol. Chem.* **269,** 27914–27919 (1994).

175. R. Menegazzi, G. Zabucchi, P. Zuccato, R. Cramer, C. Piccinini, and P. Patriarca, *J. Immunol. Methods* **137,** 55–63 (1991).

176. P. L. Bozeman, D. B. Learn, and E. L. Thomas, *Biochem. Pharmacol.* **44,** 553–563 (1992).

177. K. Sakamaki, M. Tomonaga, K. Tsukui, and S. Nagata, *J. Biol. Chem.* **264,** 16828–16836 (1989).

178. R. M. Ten, L. R. Pease, D. J. McKean, M. P. Bell, and G. J. Gleich, *J. Exp. Med.* **169,** 1757–1769 (1989).

CHAPTER 16

Prostaglandin Endoperoxide H Synthases

16.1. INTRODUCTION

The biochemistry and physiological roles of the prostaglandins, leukotrienes, and thromboxanes have been intensively studied.[1–4] In this chapter, the enzymology of the initial step of the arachidonic acid cascade, the conversion of arachidonic acid to prostaglandin H_2 by the prostaglandin endoperoxide H synthases (prostaglandin H synthases), is reviewed.

Table 16.1 includes a list of common, abbreviated, and full names for compounds commonly used in studying the prostaglandin synthases.

Some of the earliest studies on enzyme purification were concerned with the catalyst for the conversion of 8,11,14-eicosatrienoic acid (a 20:3 fatty acid: 20 carbon atoms, 3 double bonds) to the corresponding prostaglandins G_1 and H_1.[5,6] In 1976 an enzyme of molecular weight 70,000 was isolated that catalyzed the conversion of arachidonic acid (a 20:4 acid) to prostaglandin G_2 and other products. The structures of arachidonic acid and prostaglandins G_2 and H_2 are represented in Fig. 16.1. Some heme was present, and the enzyme was activated by addition of more heme. Some nonheme iron was believed to be present.[7] The enzyme was shown to be a dimeric membrane-bound glycoprotein with identical subunits and a total molecular weight of approximately 129,000.[8] The role of the iron was debated.[9,10] It was concluded that nonheme iron was not part of the enzyme. This work led to the present-day understanding that the heme, ferriprotoporphyrin IX, is readily lost from the enzyme during the isolation procedure. The heme is readily restored, and a break in the heme titration curve occurs when there is one heme per subunit. Thus, there is one heme bound in the active site of each subunit of the holoenzyme, despite a report to the contrary.[11] The coordination number of the heme iron in both the ferric and ferrous states for a variety of heme concentrations has been examined by resonance Raman spectroscopy.[12] Prostaglandin H synthase has been isolated from both bovine and ovine seminal vesicles, with the latter perhaps the preferred source. The enzymatic preparation and separation of prostaglandins G_2 and H_2 has been described.[13]

TABLE 16.1. Names of Some Substrates of Prostaglandin H Synthase

Common Name	Abbreviation	Full Name
Arachidonic acid	AA	5,8,11,14-Eicosatetraenoic acid
Prostaglandin G_2	PGG_2	15-Hydroperoxy-9,11-peroxidoprosta-5,13-dienoic acid
Prostaglandin H_2	PGH_2	15-Hydroxy-9,11-peroxidoprosta-5,13-dienoic acid
	15-HPETE	15-Hydroperoxy-5,8,11,14-eicosatetraenoic acid
	15-HETE	15-Hydroxy-5,8,11,14-eicosatetraenoic acid
	PPHP	5-Phenyl-4-pentenyl-1-hydroperoxide

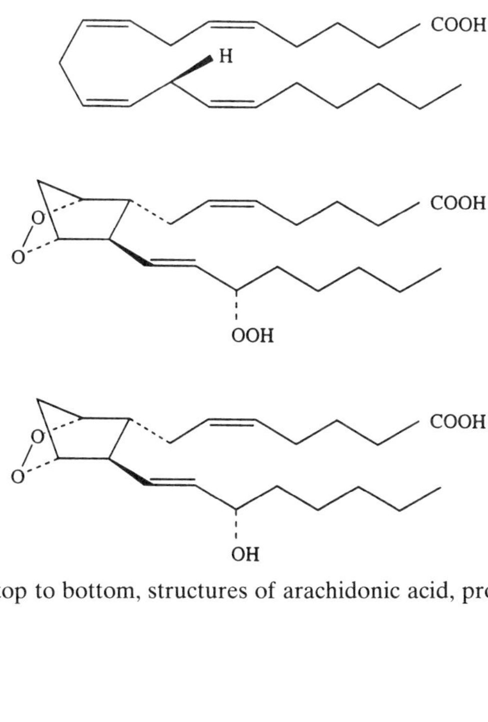

FIG. 16.1. From top to bottom, structures of arachidonic acid, prostaglandin G_2, and prostaglandin H_2.

16.2. INTRODUCTION TO MECHANISM OF ACTION OF PROSTAGLANDIN H SYNTHASE

With suitable enzyme preparations available, attention turned to details of the enzyme action. In a paper with valuable experimental results, the effects of cyanide, carbon monoxide, and hydroperoxide additions were determined.[14] Cyanide was found to bind loosely (K_d = 1.3 mM at pH 9), so that high levels of cyanide were required to inhibit the cyclooxygenase activity. Cyanide inhibition was overcome by adding exogenous hydroperoxide. Competition from added glutathione peroxidase showed that hydroperoxide was required not only to initiate the reaction, but throughout the course of the reaction. On the other hand carbon monoxide failed to inhibit catalysis. These experiments provided evidence that the resting enzyme is in the ferric state, which cannot bind carbon monoxide; also that there is no ferrous intermediate in the reaction cycle. They also demonstrated that the so-called peroxidase reaction, the initial reaction of ferric enzyme with hydroperoxide, must be repeated for the cyclooxygenase reaction to continue, an indication of tight coupling between the peroxidase and cyclooxygenase reactions. However, the results were interpreted in terms of a reduction of the ferric enzyme to a ferrous state for the cycloxygenase reaction.

In a later study of cyanide binding,[15] the binding rate constant was determined to be $(2.8 \pm 0.2) \times 10^3\ M^{-1}s^{-1}$ with the rate affected by an active site acid group of pK_a of 4.15 ± 0.10. Hydrocyanic acid binds to the basic form of the enzyme. This pK_a determination was the first measure of an active site acid dissociation constant for prostaglandin H synthase. The dissociation constant of cyanide from its complex with the enzyme was found to be 65 ± 10 μM at pH 8, indicating much tighter binding, more typical of a peroxidase, than the value cited in the previous paragraph.

Because of the dependence of prostaglandin H synthase reactions on hydroperoxide, search began for peroxidase-like oxidized intermediate species. A species that was identified as compound I,[16] was later identified as compound II,[17] a sequence of events reminiscent of the discoveries of compounds II and I of horseradish peroxidase.

Optical spectral results showed that the enzyme is a peroxidase, which appears to undergo the usual cycle of native enzyme → compound I → compound II → native enzyme.[17] The conversion of native prostaglandin H synthase into compound I by PPHP is shown in Fig. 16.2, and the subsequent apparent conversion into compound II in Fig. 16.3. The conversion of compound I into the apparent compound II appeared to be dependent upon hydroperoxide concentration. A compound-II-like spectrum, developed during the reaction of prostaglandin H synthase and arachidonic acid is shown in Fig. 16.4. The compound-II-like spectra are remarkably similar to those of horseradish peroxidase compound II.

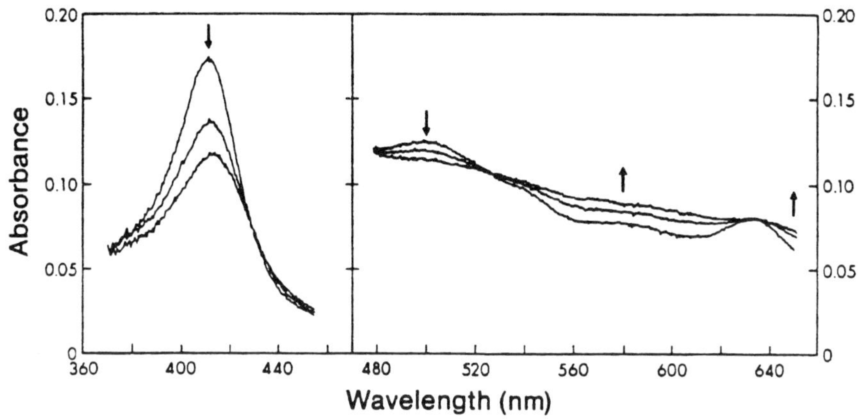

FIG. 16.2. Formation of compound I, as detected by its optical spectrum, from prostaglandin H synthase and the peroxide PPHP. Top spectrum in the Soret region is that of the native enzyme. Subsequent spectra were taken 1 and 3 ms after mixing. The arrows indicate the direction of spectral change with time. [Reproduced with permission from A.-M. Lambeir, C. M. Markey, H. B. Dunford, and L. J. Marnett, *J. Biol. Chem.* **260,** 14894–14896 (1985).]

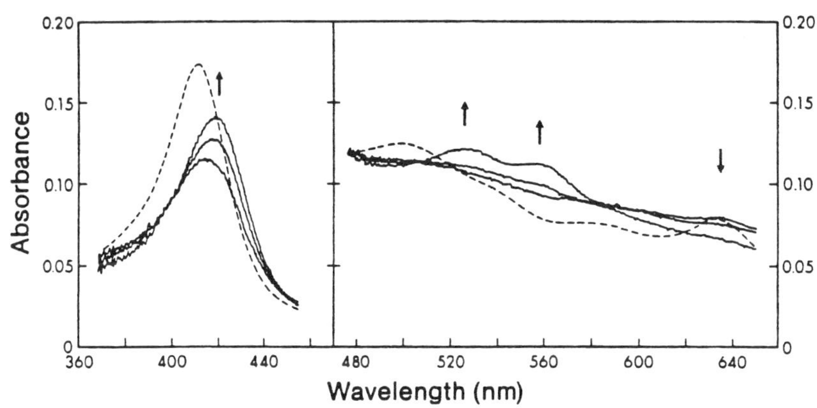

FIG. 16.3. Spontaneous conversion of prostaglandin H synthase compound I to II. The dashed line is the native enzyme spectrum. Soret region spectra recorded 11, 22, and 132 ms after mixing. [Reproduced with permission from A.-M. Lambeir, C. M. Markey, H. B. Dunford, and L. J. Marnett, *J. Biol. Chem.* **260,** 14894–14896 (1985).]

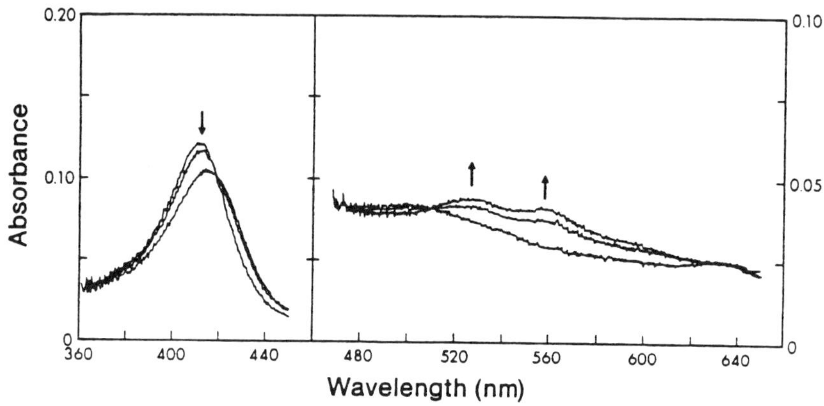

FIG. 16.4. Formation of a compound-II-like species, in the reaction of prostaglandin H synthase with arachidonic acid. The spectra were recorded 1.2, 7.8, and 16.6 s after mixing. [Reproduced with permission from A.-M. Lambeir, C. M. Markey, H. B. Dunford, and L. J. Marnett, *J. Biol. Chem.* **260,** 14894–14896 (1985).]

16.2.1. Normal Peroxidase Cycle for Small Reducing Substrates

As discussed in detail below, it is now known that the arachidonic acid binding site is distinct from the sites for hydroperoxide and small reducing substrate reactions. Therefore, the reactions of small reducing substrates, in the absence of arachidonic acid or other reactive large fatty acids, can be discussed in terms of prostaglandin H synthase as a typical peroxidase. The classic electron paramagnetic resonance (EPR) studies of Ronald Mason on horseradish peroxidase were extended to prostaglandin H synthase in collaboration with Thomas Eling. Oxidized prostaglandin H synthase was shown to catalyze free radical formation from aminopyrine,[18] 3,5,3′,5′-tetramethylbenzidine,[19] bisulfite,[20] benzidine,[21] tetramethylhydrazine,[22] glutathione,[23,24] phenylbutazone,[25] and the calcium ion chelator Quin2 and some analogs.[26] These studies were extended to co-oxidation of tyrosyl residues in proteins,[27] and to cellular[28,29] and *in vivo* systems.[30]

In the presence of PPHP only the rate constant for compound I formation can be measured and has been reported to be $1.3 \times 10^7\ M^{-1}s^{-1}$ and $(1.7 \pm 0.2) \times 10^7\ M^{-1}s^{-1}$ at $4 \pm 1°C$.[17,31] In the absence of added reducing substrate, the enzyme was converted spontaneously into compound II or a species that is spectrally equivalent, and then heme bleaching occurred. Because of ambiguity about the nature of the compound-II-like species in the presence of hydroperoxide only, it will be referred to as intermediate II. In the presence of added phenol, a steady state was attained with a mixture of native enzyme and compound II present. An isosbestic point in their spectra was observed at 420 ± 1 nm. This was followed by disappearance of the compound II spectrum and complete recovery of the spectrum of the native enzyme, which occurred when all of the peroxide was consumed. Therefore, a normal peroxidase cycle was observed with the phenol preventing heme bleaching.[18]

An enzyme-derived free radical is formed during inactivation of prostaglandin H synthase.[32] The compound diethyldithiocarbamate (DDC) is commonly used as a stabilizing agent, sometimes in concentrations in the millimolar range, for the storage of prostaglandin H synthase preparations. Glycerol is also used, sometimes in combination with DDC. A quantitative study was made of the stabilizing effects of DDC and glycerol and their influence on the kinetics of prostaglandin H synthase reactions.[33] It was concluded that DDC is a better stabilizing agent than glycerol. In the presence of a 12-fold excess of DDC over the concentration of enzyme and a 7-fold excess of PPHP, the spectrum of compound I developed over a 16-ms period, with an isosbestic point with the native enzyme spectrum at 426 nm. With other conditions the same but with a 100-fold excess of DDC, a mixture of native enzyme and compound II was observed with the 420 $\pm$ 1 nm isosbestic point. The spectrum of native enzyme returned upon depletion of substrate. Thus, DDC is an excellent reducing substrate for oxidized forms of prostaglandin H synthase, which would appear to be the basis for its stabilizing action. Its rate constant for the reduction of compound I is of the order of 10^7 $M^{-1}s^{-1}$. Its rate of reduction of compound II was estimated by the procedure of reacting enzyme with an equimolar amount of PPHP in the presence of excess DDC, and then following the exponential increase in absorbance at 410 nm. The conversion of compound I to II occurred during the dead time of the experiment so that the compound II reduction rate could be measured. A plot of the pseudo-first-order rate constant for compound II reduction versus DDC concentration was curved. From the initial slope the rate constant was estimated to be 6×10^6 $M^{-1}s^{-1}$. DDC does not interfere with compound I formation.

These studies were extended to the reaction of ascorbic acid in the presence of a constant amount of DDC using PPHP as oxidant.[34] Similarly to the conditions in the previous paragraph, equimolar amounts of enzyme and PPHP were used to avoid recycling of the enzyme and made it possible to measure the reduction of compound II by excess reducing substrate. Upon plotting the pseudo-first-order rate constant for compound II versus ascorbic acid concentration, curved plots were obtained, indicating saturation kinetics with respect to ascorbate. An equation including reductions of compound II by both ascorbate and DDC, and spontaneous decay, was derived that yielded a rate constant of $(1.4 \pm 0.5) \times 10^5$ $M^{-1}s^{-1}$ for compound II reduction by ascorbate. Thus, it was found that ascorbate does not react quite as fast as DDC.

Rate constants of $(5.3 \pm 0.3) \times 10^5$ $M^{-1}s^{-1}$ and $(2.1 \pm 0.4) \times 10^6$ $M^{-1}s^{-1}$ were determined for the reduction of compound II by both phenol and hydroquinone, in similar fashion to the experiments with ascorbic acid. Indomethacin, an inhibitor of the reaction of arachidonic acid, was also shown to inhibit the phenol oxidation.[35] The experiments showed that it is possible to conduct meaningful experiments in the presence of DDC, the removal of which would initiate some enzyme denaturation.

With the lessons learned from the above sometimes complicated experiments, studies proceeded with confidence on some straightforward steady-state experiments. A steady-state kinetic study was made of the reactions of hydrogen peroxide and *p*-coumaric acid catalyzed by prostaglandin H synthase at 25°C in 0.1 M phosphate buffer of pH 8.0.[36] A stopped flow apparatus was used, which decreased the dead time 1000-fold compared to experiments conducted with a conventional spectrophotometer. The experimental data were fit accurately to the equations derived from the conventional peroxidase mechanism involving irreversible steps, which predicts that there is no upper limit in rate (Fig. 16.5).

$$\text{PGH Synthase} + H_2O_2 \rightarrow \text{Compound I} + H_2O$$

$$\text{Compound I} + \text{HOCoumaric} \rightarrow \text{Compound II} + \bullet\text{OCoumaric}$$

$$\text{Compound II} + \text{HOCoumaric} \rightarrow \text{PGH Synthase} + \bullet\text{OCoumaric}$$

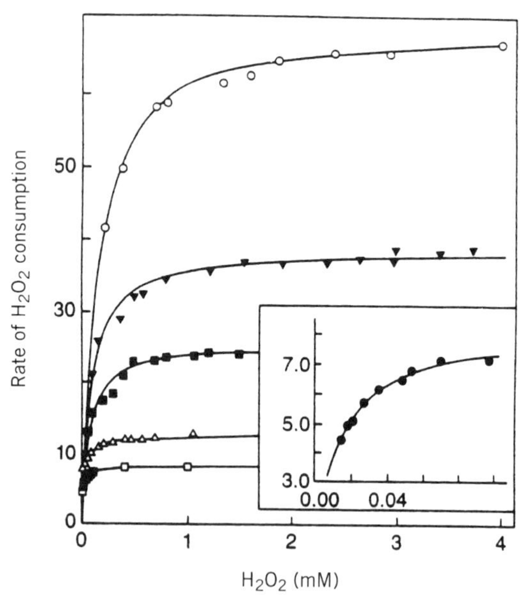

FIG. 16.5. Steady-state results on the oxidation of *p*-coumaric acid by hydrogen peroxide catalyzed by prostaglandin H synthase. Rate of hydrogen peroxide consumption is expressed per second per equivalent of enzyme. Concentrations of *p*-coumaric acid, from lower to upper curve, 5.44, 10.2, 20.4, 30.0, and 55.1 μM. Inset shows the initial portion of the curve for 5.44 μM *p*-coumaric acid. [Reproduced with permission from M. Bakovic and H. B. Dunford, *Biochemistry* **32,** 833–840 (1993).]

in which the phenolic group of *p*-coumaric acid is displayed as OH. The rate constant for compound I formation using hydrogen peroxide was found to be $(9.2 \pm 0.1) \times 10^5\ M^{-1}s^{-1}$, and the rate constant for compound II reduction $(2.5 \pm 0.1) \times 10^6\ M^{-1}s^{-1}$. It was assumed that the rate constant for compound I reduction was significantly faster than for compound II reduction, which is the usual case for peroxidase reactions. Therefore, the rate for compound I reduction cannot be determined from the steady-state results. As a result, there is a lack of data on compound I reduction rates.

Another steady-state study of prostaglandin H synthase reactions was conducted using caffeic acid as reducing substrate.[37] The mechanism was similar to that for *p*-coumaric acid, except that binding of caffeic acid to the native enzyme had an inhibitory effect on the reaction. The dissociation constant of caffeic acid from its complex with the native enzyme was $35 \pm 10\ \mu M$, and the rate constant for compound II reduction $(1.2 \pm 0.1) \times 10^6\ M^{-1}s^{-1}$.

16.2.2. Arachidonic Acid Radical and Scavenging by Dioxygen

Arachidonic acid free radicals were detected by EPR spectroscopy in reactions both inside microsomes[38] and *in vitro* experiments with prostaglandin H synthase.[39] The carbon-centered radicals were proven to belong to arachidonic acid by deuterium substitution and were believed to be centered on C11 or C15. Phenol was used to protect the enzyme against self-destruction. Although a spin trap was used, the dominant spectrum was that of the untrapped arachidonic acid radical. The EPR spectra are shown in Fig. 16.6.

Once a hydrogen atom is abstracted from the C13 pro-*S* position of arachidonic acid, the prostaglandin G_2 (PGG_2) radical is readily formed by the addition of two equivalents of molecular oxygen, as shown in Fig. 16.7. The scavenging of carbon-centered radicals by oxygen is extremely fast. The conversion of arachidonic acid into PGG_2 is commonly called the cyclooxygenase reaction. For this reaction there is no need to invoke, nor any evidence for, an enzyme-assisted addition of molecular oxygen.

The conversion of prostaglandin G_2 into prostaglandin H_2 is commonly called the peroxidase reaction and is widely accepted, although not proven in vivo, to occur via the formation of compound I from the resting enzyme.

$$\text{Prostaglandin H synthase} + \text{ROOH} \rightarrow \text{Compound I} + \text{ROH} \quad (16.1)$$

where ROOH is prostaglandin G_2 and ROH prostaglandin H_2. When the cyclooxygenase reaction is intiated with arachidonic acid, there is a lag period in which an autoxidation product of arachidonic acid must accumulate to initiate the reaction. Hydroperoxides can be removed from a wide variety of compounds, including arachidonic acid, by immobilized borohydride.[40] A wide variety of other hydroperoxides, such as hydrogen peroxide[41] and 13-hydroperoxyoctadeca-9,11-dienoic acid,[42] can also be used to form compound I.

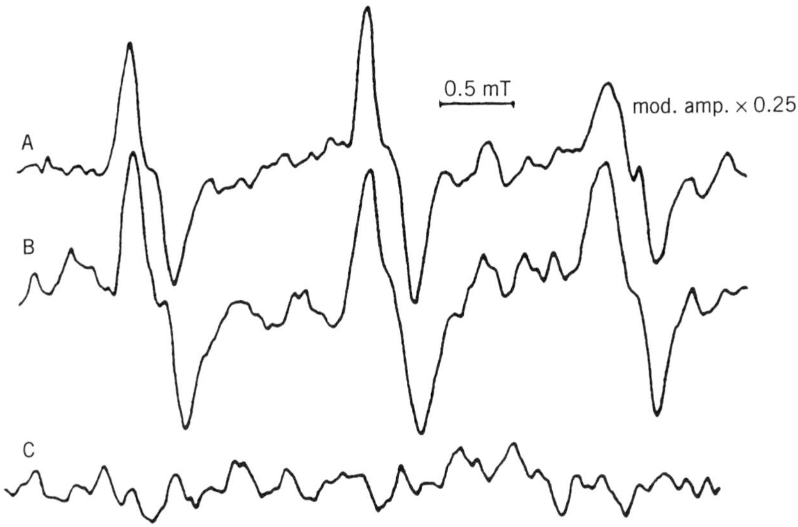

FIG. 16.6. EPR spectra of a carbon-centered radical in the reaction of prostaglandin H synthase with arachidonic acid. (A) holoenzyme, (B) apoenzyme reconstituted with hematin, and (C) effect of the cyclooxygenase inhibitor indomethacin. [Reproduced with permission from J. Schreiber, T. E. Eling, and R. P. Mason, *Arch. Biochem. Biophys.* **249,** 126–136 (1986).]

16.2.3. Detection of a Tyrosyl Radical

A radical on the enzyme was detected during turnover. A $g = 2$ signal was obtained in the first published report, which was speculated to originate from a hydroperoxide radical associated with a ferryl heme.[43] A major step forward was accomplished when the radical was shown to arise from a tyrosyl residue (Fig. 16.8).[44] Two intermediates were detected using stopped flow spectrophotometry. The first was assigned to compound I, as in the earlier optical spectral study.[17] The second was attributed to a species with a compound-II-like spectrum, of the same oxidation state as compound I,[45] which can be represented as $PorFe^{IV}{=}O$, $\bullet^{+}TyrOH$ or $PorFe^{IV}{=}O$, $TyrO\bullet$. In these representations the abbreviations are Por for porphyrin, $\bullet^{+}TyrOH$ for a tyrosyl cation radical, and $TyrO\bullet$ for a neutral tyrosyl radical. If the tyrosyl radical, in either form, is loosely coupled electronically to the chromophoric heme group, then it would have little or no influence on the observed optical spectrum, which would be essentially that of a normal compound II. The compound-II-like species was labeled intermediate II. The role of the tyrosyl radical was questioned[46,47] and discussed in a series of papers in which EPR results were the primary focus.[48–50] The latest EPR papers strongly support a key role for the Tyr385 radical.[51,52] The determination of the crystal structure, discussed below, has provided support for a role for Tyr385. Site-directed mutagenesis in which Tyr385 was replaced by phenylalanine showed that Tyr385 was essential for

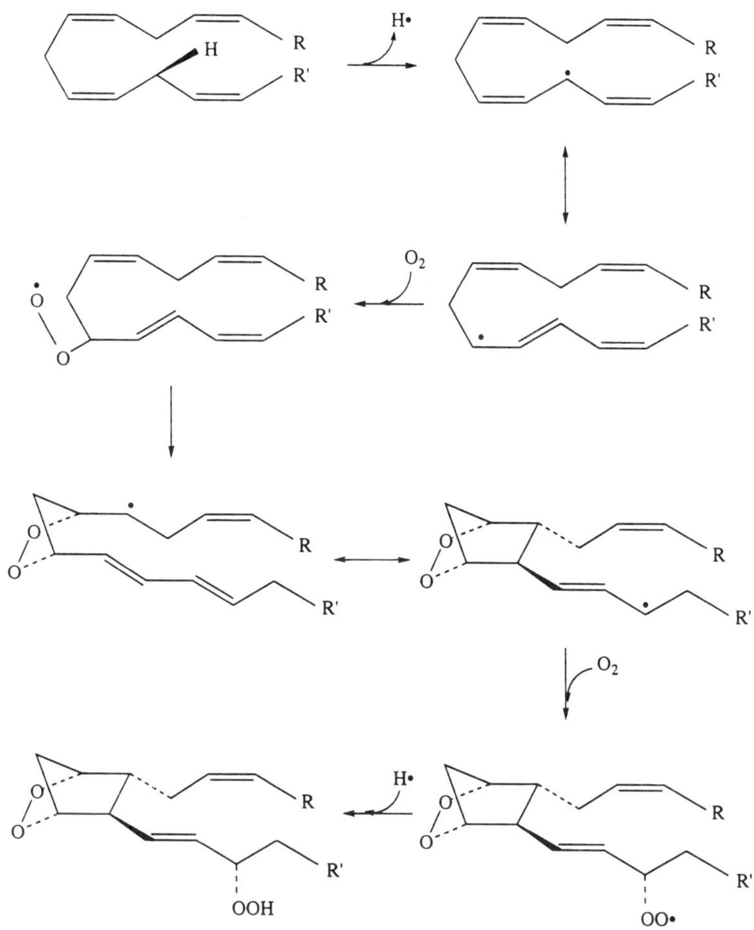

FIG. 16.7. Reaction scheme for the conversion of arachidonic acid into prostaglandin G_2. [Based, with permission, upon the scheme published by J. Schreiber, T. E. Eling, and R. P. Mason, *Arch. Biochem. Biophys.* **249,** 126–136 (1986).]

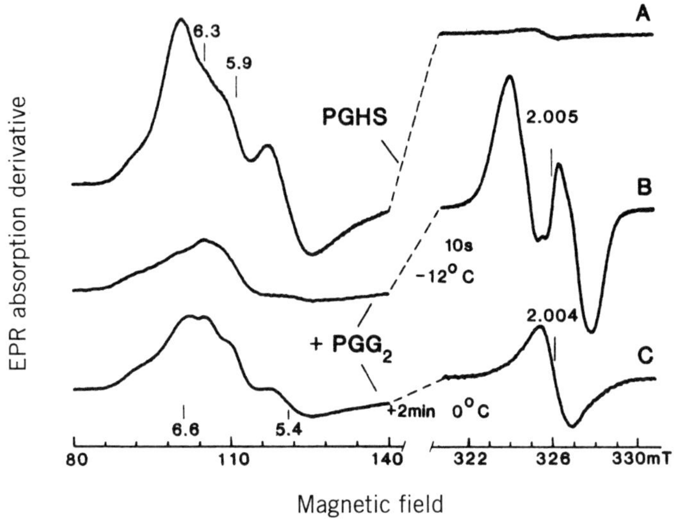

FIG. 16.8. EPR spectra of the reaction of PGH synthase with PGG_2. (*a*) Apoprotein reconstituted with hemin. (*b*) Addition of threefold excess of PGG_2 with sample frozen within 10 s. (*c*) Sample B thawed and incubated for 2 min at 0°C. *g*-Values are shown on the spectra. [Reproduced with permission from R. Karthein, R. Dietz, W. Nastainczyk, and H. H. Ruf, *Eur. J. Biochem.* **171,** 313–320 (1988).]

cyclooxygenase activity. Suitable controls with other tyrosyl residues validated the results.[53]

16.3. CRYSTAL STRUCTURE

The biggest single breakthrough in the prostaglandin H synthase story has been the determination of the crystal structure of the ram seminal vesicle isoenzyme-1 at 3.5 Å resolution.[54] This required first the determination of the primary sequence, which was published in 1988[55,56] and is reproduced in Appendix A. The structure was obtained from crystals grown in the presence of the nonsteroidal anti-inflammatory drug (NSAID) flurbiprofen (Section 3.3.2, Fig. 3.5).

The enzyme is a dimer consisting of two identical monomeric structures consisting largely of helical structures. Each monomer consists of three distinct folding units, all three of which have distinct features that so far have not been found in other peroxidases.

Using the published numbering for the enzyme, which includes the 24-residue signal peptide, residues 34–72 form the first unit, which is very similar in structure to the epidermal growth factor. It contains two small double-stranded β-sheets containing three disulfide linkages. This unit is covalently linked to the main body of the enzyme by a Cys13-Cys135 bridge.

The second distinct unit, residues 73–116, consists of a group of four helices, labeled A–D, which lie along one side of the monomer. These helices have no counterpart in myeloperoxidase. Helices A–D are largely hydrophobic, which facilitates their insertion into the membrane bilayer. The enzyme is described as monotopic since it does not bridge the membrane but reaches from the cytoplasm halfway across the membrane. The interior of these hydrophobic helices provides the arachidonic acid binding site, which extends from the exterior of the protein into the center near the heme. The enzyme has been described as similar to a Portugese man-of-war. It skims the surface of the membrane looking for its prey, which is arachidonic acid.

The third domain, the largest, includes helices numbered 1–19 and residues 117–587. It is a large globular region containing the heme binding site. Distal and proximal sides of the heme are defined as for other peroxidases. The hydrophobic channel, which is the arachidonic acid binding site, is on the proximal side of the heme. This channel intrudes into the third domain, causing the structure of the proximal side of the heme to differ significantly from that of other peroxidases. The proximal-side location of the arachidonic acid binding site came as a huge surprise to many researchers, causing them to rethink their rationales of how the enzyme operates.

Attention is now turned to details of the large domain region, helices 1–19, which are intertwined. The nomenclature for these helices is based on structural homologies with myeloperoxidase. This large domain consists of two distinct lobes. The smaller lobe provides the proximal side of the heme. It consists of helices 1, 8, 12, and 14–16. The heme packs against helices 8 and 12. The larger lobe, the distal side, contains helices 2, 3, 5, 6, 10, 18, and 19. Helix 17 provides part of the connection between the two lobes. (Helices 4, 7, 9, 11, and 13 are unaccounted for. They are either very small or not clearly defined helices). Helices 2, 5, 6, 8, 12, 16, and 17 all have counterparts in myeloperoxidase, lignin peroxidase, and yeast cytochrome *c* peroxidase. His207 and Gln203 of helix 2 are within 5 Å of the heme iron on the distal side. A large part of the heme on the distal side is exposed to solvent, indicating perhaps that prostaglandin G_2 has ready access to the heme iron. His388, a residue that follows residues 379–385 of helix 8, is the proximal ligand.

16.4. MECHANISM OF CONVERSION OF ARACHIDONIC ACID TO PROSTAGLANDIN H_2

As in all complicated scientific problems, the path to a complete solution is long and tortuous, and in the present case the end is not yet in sight. Two of the major stumbling blocks in beginning to understand how prostaglandin H synthase catalyzes the conversion of arachidonic acid into prostaglandin H_2 were: (i) recognition that the enzyme is a peroxidase and (ii) determination of the arachidonic acid binding site. Both of these problems were solved beyond dispute with the determination of the crystal structure. The similarities of the peroxidase part of the prostaglandin H synthase structure to the structure of myeloperoxidase were so great that the numbering of prostaglandin H synthase helices was based on the numbering used for myeloperoxidase.[54] Because the arachidonic acid binding site is on the proximal side of the heme, all mechanisms based on distal-side reactions of arachidonic acid would at first glance appear to be irrelevant. Nevertheless, the reaction of prostaglandin G_2 on the distal side of the heme to form compound I and prostaglandin H_2 is almost universally accepted as the initiation of the cyclooxygenase reaction. A possible alternative, presented here for the first time, is that hydrogen peroxide or another hydroperoxide reacts on the distal side of the heme in vivo to initiate PGG_2 formation on the membrane side. The highly reactive PGG_2 or its radical might play a role at inflammation sites and in platelet aggregation.

Whatever the source of the peroxide, a peroxidase reaction on the distal side is coupled to arachidonic acid oxidation on the proximal side. This implies that electron transfer from the proximal tyrosyl residue to the heme is a key part of the coupling, and the extent of coupling goes to the heart of the mechanism of action of prostaglandin H synthase. The two extreme representations of the extent of coupling will be discussed. The presentation below shows that there are several features in common for the two mechanisms.

16.4.1. Branched-Chain Mechanism

The first published mechanism, which showed the potential role of tyrosyl radicals, was that of Hans Ruf and co-workers, published in advance of the crystal structure.[45] This mechanism is reproduced in Fig. 16.9, with the addition of proton balances that are known for peroxidase cycles and that can be deduced from the stoichiometries in the steps for the oxidation of arachidonic acid. The mechanism accounts nicely for the role of a tyrosyl radical in converting arachidonic acid into its free radical. The authors deduced from modeling calculations that the electron transfer from the tyrosyl residue to the porphyrin π-cation radical is a unimolecular process, not catalyzed by prostaglandin G_2. The model they favored to account for the fate of the enzyme species is the following:

$$\text{E} + \text{ROOH} \xrightarrow{k_1} \text{Compound I} + \text{ROH}$$

$$\text{Compound I} \xrightarrow{k_2} \text{Intermediate II (Tyr}\bullet)$$

$$\text{Intermediate II (Tyr}\bullet) \xrightarrow{k_3} \text{E}$$

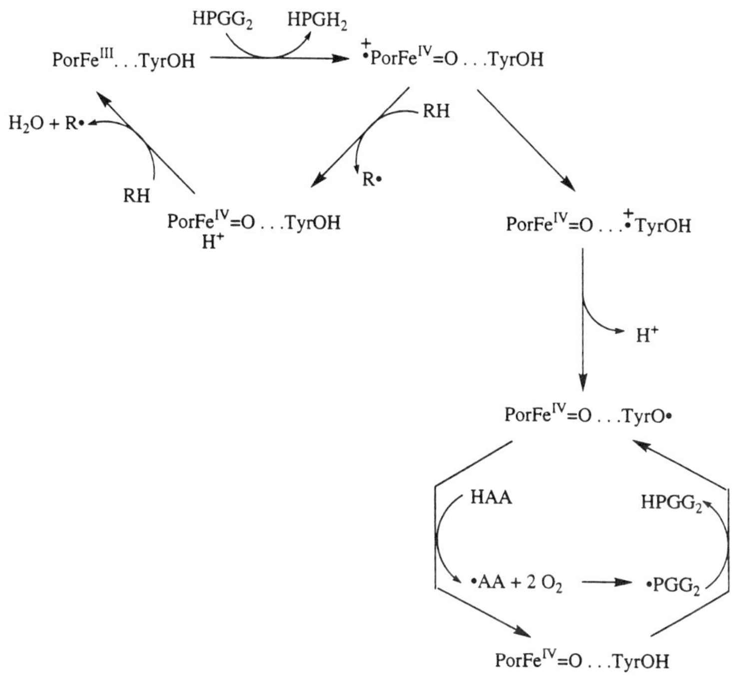

FIG. 16.9. Schematic representation of the role of the tyrosyl radical in the branched chain mechanism of action of prostaglandin H synthase. Labile hydrogen atoms are shown: $HPGG_2$, prostaglandin G_2; $HPGH_2$, prostaglandin H_2; HAA, arachidonic acid; RH, small peroxidase reducing substrate. $\bullet PGG_2$, •AA, •R, corresponding free radicals. $\bullet^{+}$Por, porphyrin π-cation radical of compound I; $\bullet^{+}$TyrOH, π-cation radical of proximal tyrosyl 385; TyrO•, free radical formed by loss of a proton from $\bullet^{+}$TyrOH. The species attacking arachidonic acid, PorFeIV=O . . . TyrO•, called intermediate II, has the same number of oxidizing equivalents as compound I. [Adapted, with permission, from R. Dietz, W. Nastainczyk and H. H. Ruf, *Eur. J. Biochem.* **171,** 321–328 (1988).]

in which E is native enzyme, and intermediate II is the tyrosyl radical form of the enzyme containing the same number of oxidizing equivalents as compound I. The rate constant k_2 is the rate of the electron-transfer reaction from the tyrosyl residue to the heme. Since the rate constant for return of E-II to E, k_3, is small, the sequence of reactions for pseudo-first-order conditions is simply A $\rightarrow$ B $\rightarrow$ C. Thus, compound I is a transient intermediate B, the concentration of which rises to a maximum and then approaches zero. Intermediate II, species C, is the dominant enzyme species, akin to a final product, until the enzyme runs out of substrate. The extent of coupling between the cyclooxygenase reaction and the peroxidase cycle is minimal, and the mechanism by which intermediate II returns to a normal peroxidase intermediate or to native enzyme is unspecified. Thus, there is an unspecified role, if any, for a normal peroxidase compound II.

There are several predictions that can be made from the mechanism shown in Fig. 16.9. Formation of compound I, and subsequently of intermediate II, the chain initiation steps, would result in a large number of molecules of arachidonic acid being converted to prostaglandin G_2. Prostaglandin G_2 could accumulate indefinitely and accelerate the rate of compound I formation. This has been called a branched-chain reaction, a chain reaction initiated by every enzymatic cycle. However, the mechanism is incomplete since it lacks a defined chain termination step. The mechanism predicts a large ratio of prostaglandin G_2 to H_2. The excess of prostaglandin G_2, a hydroperoxide, could attack the enzyme, which would provide a plausible explanation for the enzyme inactivation that occurs in the absence of a small reducing substrate, such as a phenol or ascorbic acid.

Addition of oxygen to form the cyclooxygenase is stereospecific and so would appear to occur in the arachidonic acid binding channel, where the radical could be held in a proper orientation. However, a chain reaction might be expected to deplete the oxygen faster than it could diffuse into the channel. Support for the chain mechanism has been published.[57–59] A detailed tightly coupled mechanism, which can be regarded as an extension of the Ruf model, and which appears to overcome some of the shortcomings of the branched chain mechanism, follows.

16.4.2. Tightly Coupled Mechanism

A mechanism in which peroxidase and cyclooxygenase cycles are tightly coupled is shown in Fig. 16.10. The model is particularly applicable to the likely *in vivo* scenario, where small reductant and hydrogen atom donor molecules are available. The mechanism has features in common with one published on the effect of added indomethacin on intact microsomes.[60] Changes from the branched-chain mechanism are the following.

Hydrogen atom addition, to convert the prostaglandin G_2 radical into prostaglandin G_2, is postulated to occur in a reaction with a hydrogen atom donor. It would be expected to occur more readily outside the binding channel. Diffusion of newly generated free radicals out of the active site location is a common occurrence in enzyme reactions.

An important difference between the two mechanisms depicted in Figs. 16.9 and 16.10 is the conversion of the enzymatic product of the oxidation of arachidonic acid, the species $PorFe^{IV}{=}O$ . . . TyrOH, into peroxidase compound II by simple proton addition. This proton is added to the imidazole ring of the distal histidine residue of peroxidases when compound II is formed.[61] Thus, by proton addition the cyclooxygenase reaction is converted back to a peroxidase reaction. There are several predictions and implications that result from this tightly coupled mechanism.

First, the enzyme is acting as a proton pump. Conversion of the tyrosyl cation radical into a neutral (phenoxyl type) radical means that a proton is transferred to the solution in the binding channel. The tyrosyl radical with no

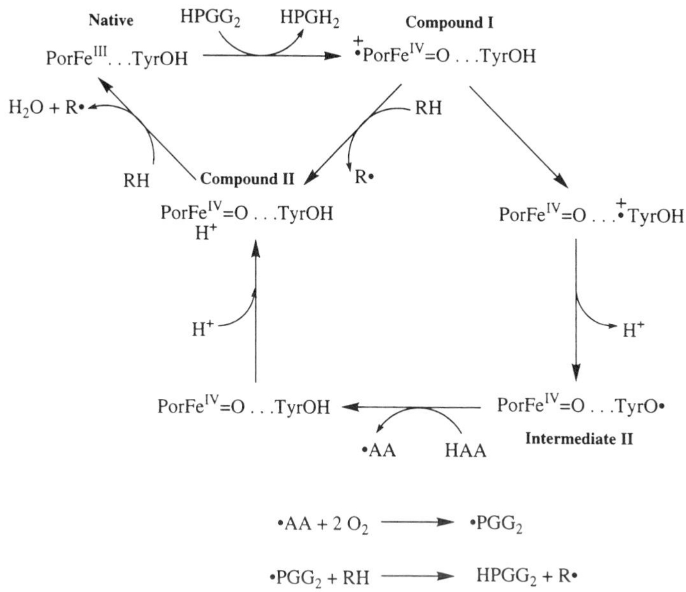

FIG. 16.10. Schematic representation of the tightly coupled mechanism for the action of prostaglandin H synthase. Symbols are the same as for Fig. 16.9.

electronic charge is converted back to an intact tyrosyl residue by hydrogen atom donation from arachidonic acid. The subsequent reactions of the arachidonic acid radical are proton neutral: There is no net consumption or production of protons as the arachidonic acid radical is converted into prostaglandin G_2. The net result is addition of one proton to the bulk medium on the proximal side of the heme. The conversion of the species $PorFe^{IV}=O$ . . . TyrOH into a normal compound II involves proton addition from the bulk medium to the distal imidazole group, so the net result is loss of one proton from the bulk medium to the distal side of the heme. Therefore, protons are being pumped from the bulk solution on the distal side of the heme into the binding channel on the proximal side of the heme, next to the membrane surface. Perhaps a proton gradient provides a control mechanism for the cyclooxygenase reaction.

If conversion of compound I into intermediate II, its tyrosyl radical form, operated with 100% efficiency, then reduction of compound I would be eliminated from the mechanism. Deletion of this step would mean that there is only one cycle. Starting with native enzyme, the cycle would proceed through compound I formation, electron transfer to form the tyrosyl cation radical, loss of a proton to form the phenoxyl radical, oxidation of arachidonic acid and conversion of the resultant enzymatic product to compound II by proton addition, and finally compound II reduction back to the native enzyme. Therefore, the old nomenclature of peroxidase and cyclooxygenase cycles would become obsolete. Instead of talking about two cycles and how tightly they are coupled, one should talk about prostaglandin H_2 synthesis. The ratio of small reductant/arachidonic acid reacted would be 2:1, a prediction that was observed experimentally.[62] In one cycle one small molecule reductant is consumed in converting the prostaglandin G_2 radical into prostaglandin G_2 and another is used to reduce compound II to native enzyme; in the same cycle one molecule of arachidonic acid is oxidized.

What if the conversion of compound I to intermediate II were not 100% efficient? In other words what is the effect of partial reduction of compound I to compound II by small molecule reductants? How would this affect the stoichiometry of the overall reaction? The answer is that there would be a change in the ratio of small reducing substrate/arachidonic acid reacted, but there would be no change in the total amount of small reducing substrate reacted. The reason is that small reducing substrate would now react to reduce compound I to compound II to the exact extent that it is no longer reacting with the prostaglandin G_2 radical to convert it into prostaglandin G_2. A detailed explanation is given in an appendix (Section 16.10).

An efficient tightly coupled mechanism predicts the accumulation of prostaglandin H_2. Starting with native enzyme, one molecule of prostaglandin G_2 is consumed in compound I formation to yield one molecule of prostaglandin H_2, and it is replenished by completion of the cycle. Thus, prostaglandin H_2 steadily accumulates. The extent that prostaglandin G_2 accumulates would be a reflection of its inefficiency in diffusing to the distal side active site. The more efficient the diffusion process, the higher the total yield of prostaglandin products and the higher the percentage of prostaglandin H_2. Further details of some of the evidence for the tightly coupled mechanism are available.[63–65]

The reactions of prostaglandin H synthase, with various combinations of PPHP as oxidant and *p*-coumaric acid as reductant, were examined in the presence and absence of arachidonic acid.[66] The sum of intermediate II and compound II can be monitored spectrophotometrically. Although the amounts of each species cannot be determined, the reaction patterns offer clues as to which species is being observed. In the presence of equimolar PPHP only, the dominant species being formed from the enzyme is intermediate II. At 25°C 1.5 equivalents of *p*-coumaric acid did not interfere with the spontaneous formation of intermediate II but accelerated its reduction after it was formed. At 3°C, for the same reaction conditions as at 25°C, *p*-coumaric acid reduction of compound I to compound II competed successfully with intermediate II formation, so the sum of compound II and intermediate II formation was increased. A 30-fold excess of arachidonic acid had very little effect on the latter results, indicating that at lower temperatures at least there is not a rapid switching mechanism where increased arachidonic acid concentration accelerates intermediate II formation. The results, summarized in Fig. 16.11, show that higher temperatures favor the cyclooxygenase reaction. The cycloxygenase and peroxidase cycles run simultaneously.

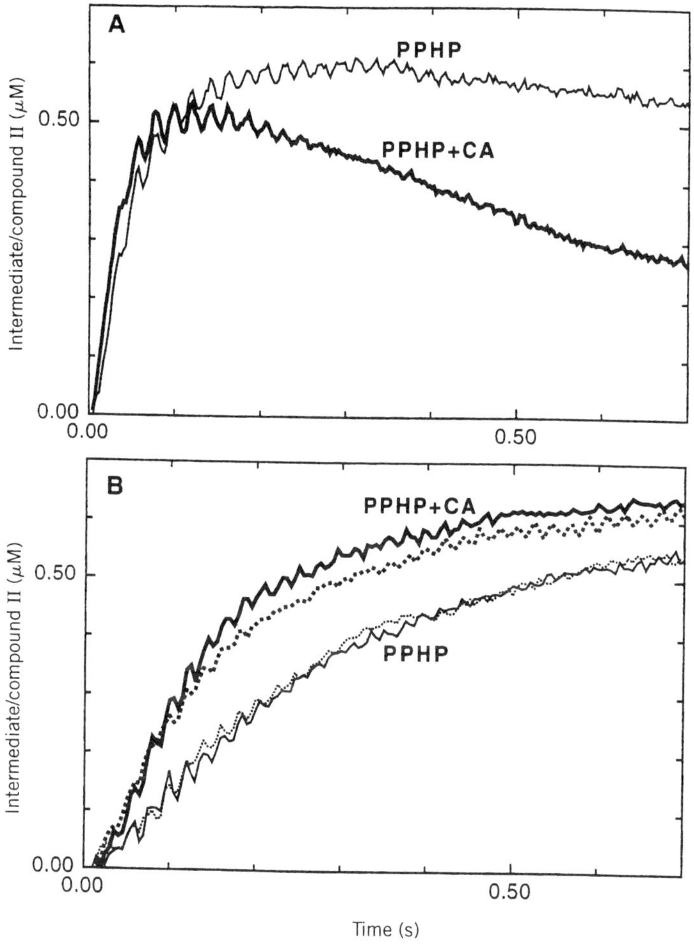

FIG. 16.11. Formation of intermediate II/compound II of prostaglandin H synthase at (A) 25°C and (B) 3°C. A 0.64-μM enzyme was reacted with 0.67 μM of PPHP or with 0.67 μM of PPHP and 0.93 μM of *p*-coumaric acid (CA). The very small effect of 20 μM arachidonic acid is shown by the dotted lines, panel (B). (Reproduced with permission from V. Koshkin and H. B. Dunford, results submitted for publication).

16.5. OTHER PROSTAGLANDIN H SYNTHASE REACTIONS

16.5.1. Manganese-Substituted Enzymes

In experiments in which manganese porphyrins were substituted for ferriprotoporphyrin IX on the enzyme, peroxidase and cyclooxygenase reactions appeared to be decoupled when prostaglandin G_1 was used as the oxidant.[9] The aminopyrine cation radical was detected in reactions with the native enzyme but not with the manganese-substituted enzyme.[67] There have been detailed studies on the properties of manganese-substituted peroxidases. The porphyrins were reacted with manganese(II), which is readily oxidized to manganese(III), and apohorseradish peroxidase and apocytochrome *c* peroxidase were combined with the manganese(III) porphyrins. These Mn-substituted peroxidases were fully reactive with hydrogen peroxide as the oxidant but did not react as rapidly as the native enzymes.[68] The Mn-substituted horseradish and cytochrome *c* peroxidases were further examined by EPR and optical spectroscopy.[69] It appeared that a genuine Mn^{V} species was formed from the Mn^{III} form of yeast enzyme and hydrogen peroxide; no free radical species was detected. When hydrogen peroxide was reacted with Mn^{III} horseradish peroxidase, only a Mn^{IV} species was detected, with no apparent radical on the porphyrin or protein.

There are naturally occurring manganese superoxide dismutases, manganese-containing catalases (with binuclear bridged manganese cores), but no naturally occurring manganese-containing peroxidases. It is clear that although manganese-substituted peroxidases are functional, aspects of their chemistry are dissimilar to that of the native iron-containing species.

Further studies were performed on Mn-substituted prostaglandin H synthases. It was shown that 5-phenyl-4-pentenylhydroperoxide formed a new enzyme species immediately, as detected by optical spectra, and that arachidonic acid formed the same species after a 10-s lag period.[70] Formation of this new enzyme species either coincides with or precedes oxidation of arachidonic acid.[71] A free radical was detected within 1 s of reaction with arachidonic acid, and the Mn enzyme was shown to produce both prostaglandins G_2 and H_2.[72] The latter study was extended to show that two oxidized forms of manganese were produced in a functional enzyme, the first assigned to manganese(V) and the second to manganese(IV). It is now clear that Mn-substituted prostaglandin H synthases are able to perform both peroxidase and cyclooxygenase functions.

16.5.2. Reactions with *cis,cis*-Eicosa-11,14-Dienoic Acid

cis,cis-Eicosa-11,14-dienoic acid offers several advantages over arachidonic acid as a substrate for prostaglandin H synthase. It forms a conjugated diene product that can be observed spectrophotometrically at 235 nm, whereas the arachidonic acid products are spectrally silent. It also reacts more slowly than arachidonic acid, facilitating observation of the reaction in progress.[73,74] The question arises, however, whether the mechanism of reaction is the same as that of the natural substrate. The initiation hydrogen atom abstraction step, originally attributed to compound I,[75] was shown not to occur until after compound II/intermediate II formation.[76] Trolox C was found to act as a typical small molecule reductant/hydrogen atom donor, stimulating the reaction of the dienoic acid at small concentrations and inhibiting at large concentrations. An isosbestic point between Trolox C and its oxidation product at 235 nm facilitates observation of the dienoic acid oxidation.[77] There are conflicting reports about the stoichiometry of the reaction of the dienoic acid.[59,75]

By use of high enzyme concentrations, simultaneous measurements were made of enzyme and eicosadienoic acid transformations in the pre-steady state.[76] The results provided clear evidence that intermediate II formation, the tyrosyl radical form of compound I, preceded substrate oxidation, that the conversion of compound I to intermediate II was a spontaneous unimolecular process at low dienoic acid concentrations, but was accelerated by high dienoic acid concentrations. The acceleration was believed to be an indirect effect caused by increased occupancy of the channel leading into the proximal side of the enzyme. It was concluded that the eicosadienoic acid is oxidized in this channel, as is arachidonic acid.

16.5.3. Reaction with Nitric Oxide

The effects of nitric oxide and nitric-oxide-generating compounds on both the generation of prostaglandin H synthase in murine macrophages and on the reactions of prostaglandin H synthase-1 and -2 were investigated.[78] It was found that there was no influence on enzyme formation, that nitric oxide acts as a peroxidase reductant, and that it does not influence cyclooxygenase activity.

16.6. DISCOVERY OF PROSTAGLANDIN H SYNTHASE-2

An inducible form of prostaglandin H synthase was discovered in 1991. In one study, the cDNA was cloned as a result of rapid induction of the gene using a phorbol acetate stimulus for Swiss 3T3 cells. The deduced amino acid sequence of 604 residues was published,[79] and the enzyme was shown to be distinct from the constitutive form.[80] Dexamethasone was found to inhibit the phorbol acetate stimulus.[81] In another study, mRNA encoding the enzyme was induced using Rous sarcoma virus to stimulate chicken embryo fibroblasts. A 603-amino-acid sequence was determined.[82,83] The mRNA existed in low levels in the unstimulated cell but was inactive because of an intron separating the signal peptide from the remainder of the protein. Further demonstration of the existence of the inducible form of prostaglandin H synthase was obtained using mouse fibroblasts,[84] mouse osteoblasts,[85] granulosa cells of rat preovulatory follicles,[86] and rabbit alveolar macrophages.[87] Both isoenzymes were expressed in a baculovirus system and the recombinant enzymes were isolated in pure form.[88] Details of gene transcription sites have been obtained.[89,90]

Prostaglandin H synthase-2 has the same structural features as isoenzyme-1 but differs considerably in amino acid sequence. The key amino acid sequences in the vicinity of the heme are conserved. Thus, isoenzyme-2 appears to operate by essentially the same mechanism as isoenzyme-1, but the genetic control is different as may be the cellular targets.[91]

The two isoenzymes have been compared with respect to fatty acid substrate specificities,[92] hydroperoxide initiator requirements,[93] and allosteric regulation by arachidonic acid.[94] A tyrosyl radical of prostaglandin H synthase-2 was trapped using nitric oxide.[95] The effects of inhibitors on trytophan fluorescence of isoenzyme-2 has been examined.[96] The cyclooxygenase activity of both isoenzymes can be started using peroxynitrite.[97]

16.7. ASPIRIN AND OTHER NSAID BINDING SITES

Aspirin inhibits prostaglandin formation.[98–101] The cause is acetylation of serine 530 of prostaglandin H synthase-1 by aspirin, which irreversibly inhibits cyclooxygenase activity.[102] Replacement of Ser530 by alanine had little effect on the cyclooxygenase activity, indicating that Ser530 is not an essential catalytic group, but the acetylation introduces a group that is bulky enough to interfere with arachidonic acid binding.[103] These results were obtained with iosenzyme-1. With isoenzyme-2, the effect of aspirin was to change the metabolism of arachidonic acid. Instead of prostaglandin H_2, the primary product was 15-hydroxyeicosatetraenoic acid (15-HETE).[104] The properties of the two isoenzymes treated with aspirin were examined further.[105] In the case of aspirin-treated isoenzyme-2 only, an iron(II) enzyme species was observed upon reaction with ethyl hydroperoxide, indicating oxidation of the peroxide to the corresponding peroxyl radical as the iron is reduced to its ferrous state. It would appear unlikely that this mechanism is applicable to all prostaglandin H synthase reactions, whether aspirin-treated or not, as was suggested. There is no detectable iron(II) species in the normal reaction of isoenzyme-1 with arachidonic acid.[17] Tests on other NSAIDs indicated that the two arachidonic acid binding sites are not identical, which has led to the development of NSAIDs specific for isoenzyme-2.[91,106]

The crystal structure of prostaglandin H synthase-1 was obtained with flurbiprofen bound to the enzyme.[54] The carboxyl group of the flurbiprofen was directed toward the mouth of the arachidonic acid binding channel where it could interact with Arg120. The latter residue appears essential for interaction of prostaglandin H synthase-1 with NSAIDs containing a carboxyl group.[107] Rate constants for the slow binding of flurbiprofen to both isoenzymes-1 and -2 of prostaglandin H synthase have been determined.[108] The binding of flurbiprofen and sulfur-containing isoenzyme-selective inhibitors has been examined by fluorescence spectroscopy. The fluorescence of the inhibitor is quenched by the heme, so that distance from the inhibitor to the heme can be calculated.[109] Affinity labels, bromoacetamido analogs of indomethacin and mefenamic acid, were shown to bind covalently to isoenzyme-1, providing an alternative method of testing for NSAID binding sites.[110]

16.8. NOMENCLATURE

The isoenzymes of prostaglandin H synthase, numbered 1 and 2, are often referred to as COX-1 and COX-2, short for cyclooxygenase-1 and cyclooxygenase-2. These are misnomers, since the enzyme is not an oxygenase. The names prostaglandin H synthase or prostaglandin endoperoxide synthase, with suffixes -1 and -2 for the two isoenzymes, do not contain misleading information. The latter names might be improved, once the details of the reaction cycle are firmly established and widely recognized. However, use of COX-1 and COX-2 is a step in the wrong direction. If, as now appears to be a distinct possibility, the peroxidase and cyclooxygenase cycles are so tightly coupled that they are indistinguishable, then reference to the cyclooxygenase cycle is also erroneous. The peroxidase cycle does exist when only small reducing substrates are present. The terms prostaglandin G_2 synthesis and prostaglandin H_2 synthesis do not convey misleading information.

16.9. SUMMARY

Experiments with absolutely no small molecule reductant present, irrelevant to *in vivo* conditions, lead to rapid enzyme inactivation by products of the arachidonic acid reaction. The center of the controversy about prostaglandin H synthase is the relative merits of the branched-chain and tightly coupled mechanisms when both arachidonic acid and phenol or ascorbate are present. Early reports that the cyclooxygenase and peroxidase cycles operate independently need to be reinterpreted because efficient small molecule reductants are also excellent free radical scavengers. Therefore, phenol molecules, for example, can be diverted from the peroxidase cycle to scavenge prostaglandin G_2 radicals, and the total amount of phenol reacted remains constant. The latter result disproves the branched-chain mechanism that predicts that there would be no relation between the ratio of phenol reacted to arachidonic acid reacted. A major question in both the tightly coupled and branched-chain mechanisms is the efficiency of prostaglandin G_2 diffusion from one side of the enzyme to the other. Another possibility that has not been examined is the possibilty of an equilbrium between two forms of compound II, $Fe^{IV}=O$, Tyr, and Fe^{III}, Tyr•, analogous to what is found in yeast cytochrome *c* peroxidase.

16.10. APPENDIX

In order to understand the constant amount of small reducing substrate consumed, regardless of the efficiency of conversion of compound I to intermediate II, consider Fig. 16.12, a simplified version of Fig. 16.10. The two fast proton transfer steps, explicitly shown in Fig. 16.10, have been combined in

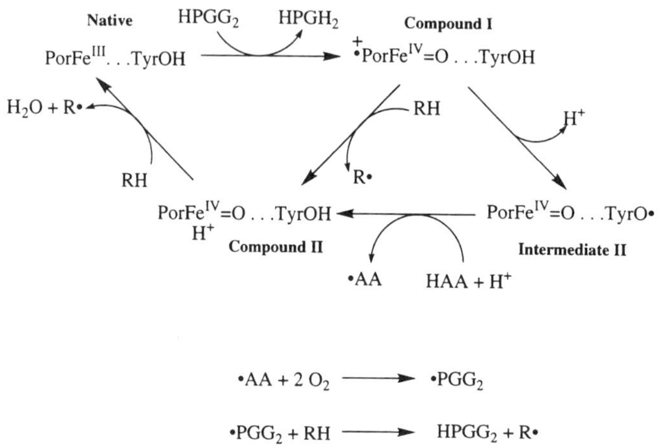

FIG. 16.12. Simplified version of the tightly coupled mechanism shown in Fig. 16.10. Each of the fast proton loss and proton addition reactions has been combined with the previous step.

each case with the previous step. Suppose first that the conversion of compound I to intermediate II is 100% efficient, a process that would eliminate the reduction of compound I by small molecule reductant.

$$\mathrm{E} + \mathrm{PGG_2} \rightarrow \mathrm{I} + \mathrm{PGH_2}$$

$$\mathrm{I} \rightarrow \mathrm{I}\bullet + \mathrm{H^+}$$

$$\mathrm{I}\bullet + \mathrm{AA} + \mathrm{H^+} \rightarrow \mathrm{II} + \mathrm{AA}\bullet$$

$$\mathrm{AA}\bullet + 2\mathrm{O_2} + \mathrm{ROH} \rightarrow \mathrm{PGG_2} + \mathrm{RO}\bullet$$

$$\mathrm{II} + \mathrm{ROH} \rightarrow \mathrm{E} + \mathrm{RO}\bullet$$

$$\mathrm{AA} + 2\mathrm{O_2} + 2\mathrm{ROH} = \mathrm{PGH_2} + 2\mathrm{RO}\bullet$$

where E is native enzyme, I is compound I, I• is intermediate II, II is compound II, AA is arachidonic acid, ROH is small molecule reductant such as a phenol or ascorbate, and AA• and RO• are free radicals. The ratio of small molecule reductant to arachidonic acid reacted is 2:1 and 2 molecules of ROH are reacted per completed reaction cycle starting with native enzyme and returning to native enzyme.

Now consider a reaction where the conversion of compound I is only 25% efficient. In other words small molecule reductant converts 75% of compound I to compound II.

$$E + PGG_2 \rightarrow I + PGH_2$$

$$\tfrac{3}{4}I + \tfrac{3}{4}ROH \rightarrow \tfrac{3}{4}II + \tfrac{3}{4}RO\bullet$$

$$\tfrac{1}{4}I \rightarrow \tfrac{1}{4}I\bullet + \tfrac{1}{4}H^+$$

$$\tfrac{1}{4}I\bullet + \tfrac{1}{4}AA + \tfrac{1}{4}H^+ \rightarrow \tfrac{1}{4}II + \tfrac{1}{4}AA\bullet$$

$$\tfrac{1}{4}AA\bullet + \tfrac{1}{2}O_2 + \tfrac{1}{4}ROH \rightarrow \tfrac{1}{4}PGG_2 + \tfrac{1}{4}RO\bullet$$

$$II + ROH \rightarrow E + RO\bullet$$

$$\tfrac{3}{4}PGG_2 + \tfrac{1}{4}AA + \tfrac{1}{2}O_2 + 2ROH = PGH_2 + 2RO\bullet$$

The ratio of reactants changes, but the total consumption of ROH remains unchanged. An extraneous source of peroxide is required to keep the cycle going. This inefficient mechanism does not make physiological sense if the purpose of the enzyme is solely to produce prostaglandin H_2.

Finally, when no arachidonic acid or arachidonic acid mimic is present, the peroxidase cycle appears to operate with 100% efficiency, with a 2:1 ratio of small molecule reductant consumed to peroxide consumed, and two molecules of small molecule reductant consumed per reaction cycle.

REFERENCES

1. D. A. van Dorp, R. K. Beerthuis, D. H. Nugteren, and H. Vonkeman, *Biochim. Biophys. Acta* **90,** 204–207 (1964).
2. S. Bergström, H. Danielsson, and B. Samuelsson, *Biochim. Biophys. Acta* **90,** 207–210 (1964).
3. B. Samuelsson, *J. Am. Chem. Soc.* **87,** 3011–3013 (1965).
4. B. Samuelsson, *Fed. Proc.* **31,** 1442–1450 (1972).
5. T. Miyamoto, N. Ogino, S. Yamamoto, and O. Hayaishi, *J. Biol. Chem.* **251,** 2629–2636 (1976).
6. S. Okhi, N. Ogino, S. Yamamoto, and O. Hayaishi, *J. Biol. Chem.* **254,** 829–836 (1979).
7. M. Hemler, W. E. M. Lands, and W. L. Smith, *J. Biol. Chem.* **251,** 5575–5579 (1976).
8. F. J. van der Ouderaa, M. Buytenhek, D. H. Nugteren, and D. A. van Dorp, *Biochim. Biophys. Acta* **487,** 315–331 (1977).
9. N. Ogino, S. Ohki, S. Yamamoto, and O. Hayaishi, *J. Biol. Chem.* **253,** 5061–5068 (1978).
10. F. J. van der Ouderaa, M. Buytenhek, F. J. Slikkerveer, and D. A. van Dorp, *Biochim. Biophys. Acta* **572,** 29–42 (1979).
11. R. J. Kulmacz and W. E. M. Lands, *J. Biol. Chem.* **259,** 6358–6363 (1984).
12. S. Gaspard, G. Chottard, J.-P. Mahy, and D. Mansuy, *Eur. J. Biochem.* **238,** 529–537 (1996).

13. G. Graff, *Methods Enzymol.* **86,** 376–385 (1982).
14. M. E. Hemler and W. E. M. Lands, *J. Biol. Chem.* **255,** 6253–6261 (1980).
15. I. D. MacDonald and H. B. Dunford, *J. Inorg. Biochem.* **37,** 35–44 (1989).
16. W. Nastainczyk, D. Schuhn, and V. Ullrich, *Eur. J. Biochem.* **144,** 381–385 (1984).
17. A.-M. Lambeir, C. M. Markey, H. B. Dunford, and L. J. Marnett, *J. Biol. Chem.* **260,** 14894–14896 (1985).
18. J. M. Lasker, K. Sivarajah, R. P. Mason, B. Kalyanaraman, M. B. Abou-Donia, and T. E. Eling, *J. Biol. Chem.* **256,** 7764–7767 (1981).
19. P. D. Josephy, R. P. Mason, and T. E. Eling, *Cancer Res.* **42,** 2567–2570 (1982).
20. C. Mottley, R. P. Mason, C. F. Chignell, K. Sivarajah, and T. E. Eling, *J. Biol. Chem.* **257,** 5050–5055 (1982).
21. P. D. Josephy, R. P. Mason, and T. Eling, *Carcinogenesis* **3,** 1227–1230 (1982).
22. B. Kalyanaraman, K. Sivarajah, T. E. Eling, and R. P. Mason, *Carcinogenesis* **4,** 1341–1343 (1983).
23. T. E. Eling, J. F. Curtis, L. S. Harman, and R. P. Mason, *J. Biol. Chem.* **261,** 5023–5028 (1986).
24. B. H. Stock, J. Schreiber, C. Guenat, R. P. Mason, J. R. Bend, and T. E. Eling, *J. Biol. Chem.* **261,** 15915–15922 (1986).
25. M. F. Hughes, R. P. Mason, and T. E. Eling, *Mol. Pharmacol.* **34,** 186–193 (1988).
26. J. Van der Zee, R. P. Mason, and T. E. Eling, *Arch. Biochem. Biophys.* **271,** 64–71 (1989).
27. G. Lassmann, J. Curtis, B. Liermann, R. P. Mason, and T. E. Eling, *Arch. Biochem. Biophys.* **300,** 132–136 (1993).
28. J. Screiber, G. L. Foureman, M. F. Hughes, R. P. Mason, and T. E. Eling, *J. Biol. Chem.* **264,** 7936–7943 (1989).
29. W. Chamulitrat, R. P. Mason, and M. S. Cohen in *Oxygen Radicals,* K. Yagi, M. Kondon, E. Niki, and T. Yoshikawa, eds., Elsevier, Amsterdam, 1992, pp. 55–57.
30. W. Chamulitrat, S. J. Jordan, and R. P. Mason, *Arch. Biochem. Biophys.* **299,** 361–367 (1992).
31. I. D. MacDonald and H. B. Dunford, *Biochem. Cell Biol.* **67,** 301–305 (1989).
32. B. Kalyanaraman, R. P. Mason, B. Tainer, and T. E. Eling, *J. Biol. Chem.* **257,** 4764–4768 (1982).
33. Y. Hsuanyu and H. B. Dunford, *Biochem. Cell Biol.* **68,** 965–972 (1990).
34. Y. Hsuanyu and H. B. Dunford, *Arch. Biochem. Biophys.* **281,** 282–286 (1990).
35. Y. Hsuanyu and H. B. Dunford, *Arch. Biochem. Biophys.* **292,** 213–220 (1992).
36. M. Bakovic and H. B. Dunford, *Biochemistry* **32,** 833–840 (1993).
37. M. Bakovic and H. B. Dunford, *Prostagland. Leukotri. Essent. Fatty Acids* **51,** 337–345 (1994).
38. R. P. Mason, B. Kalyanaraman, B. E. Tainer, and T. E. Eling, *J. Biol. Chem.* **255,** 5019–5022 (1980).
39. J. Schreiber, T. E. Eling, and R. P. Mason, *Arch. Biochem. Biophys.* **249,** 126–136 (1986).
40. Y. Hsuanyu and H. B. Dunford, *Anal. Biochem.* **198,** 174–178 (1991).
41. M. Bakovic and H. B. Dunford, *Biochem. Cell Biol.* **74,** 117–124 (1996).

42. I. D. MacDonald, G. Graff, L. A. Anderson, and H. B. Dunford, *Arch. Biochem. Biophys.* **272,** 194–202 (1989).
43. R. J. Kulmacz, A.-L. Tsai, and G. Palmer, *J. Biol. Chem.* **262,** 10524–10531 (1987).
44. R. Karthein, R. Dietz, W. Nastainczyk, and H. H. Ruf, *Eur. J. Biochem.* **171,** 313–320 (1988).
45. R. Dietz, W. Nastainczyk, and H. H. Ruf, *Eur. J. Biochem.* **171,** 321–328 (1988).
46. G. Lassmann, R. Odenwaller, J. F. Curtis, J. A. DeGray, R. P. Mason, L. J. Marnett, and T. E. Eling, *J. Biol. Chem.* **266,** 20045–20055 (1991).
47. W. L. Smith, T. E. Eling, R. J. Kulmacz, L. J. Marnett, and A.-L. Tsai, *Biochemistry* **31,** 3–7 (1992).
48. A.-L. Tsai, G. Palmer, and R. J. Kulmacz, *J. Biol. Chem.* **267,** 17753–17759 (1992).
49. J. A. DeGray, G. Lassmann, J. F. Curtis, T. A. Kennedy, L. J. Marnett, T. E. Eling, and R. P. Mason, *J. Biol. Chem.* **267,** 23583–23588 (1992).
50. A.-L. Tsai, R. J. Kulmacz, J.-S. Wang, Y. Wang, H. E. Van Wart, and G. Palmer, *J. Biol. Chem.* **268,** 8554–8563 (1993).
51. A.-L. Tsai, R. J. Kulmacz, and G. Palmer, *J. Biol. Chem.* **270,** 10503–10508 (1995).
52. G. Xiao, A.-L. Tsai, G. Palmer, W. C. Boyar, P. J. Marshall, and R. J. Kulmacz, *Biochemistry* **36,** 1836–1845 (1997).
53. T. Shimokawa, R. J. Kulmacz, D. L. DeWitt, and W. L. Smith, *J. Biol. Chem.* **265,** 20073–20076 (1990).
54. D. Picot, P. J. Loll, and M. Garavito, *Nature* **367,** 243–249 (1994).
55. D. L. DeWitt and W. L. Smith, *Proc. Natl. Acad. Sci. U.S.A.* **85,** 1412–1416 (1988).
56. J. P. Merlie, D. Fagan, J. Mudd, and P. Needleman, *J. Biol. Chem.* **263,** 3550–3553 (1988).
57. R. J. Kulmacz, R. B. Pendleton, and W. E. M. Lands, *J. Biol. Chem.* **269,** 5527–5536 (1994).
58. C. Wei, R. J. Kulmacz, and A.-L. Tsai, *Biochemistry* **34,** 8499–8512 (1995).
59. A.-L. Tsai, G. Wu, and R. J. Kulmacz, *Biochemistry* **36,** 13085–13094 (1997).
60. P. J. Harvison, R. W. Egan, P. H. Gale, G. D. Christian, B. S. Hill, and S. D. Nelson, *Chem.-Biol. Interactions* **64,** 251–266 (1988).
61. S. Hashimoto, Y. Tatsuno, and T. Kitagawa, *Proc. Natl. Acad. Sci. U.S.A.* **83,** 2417–2421 (1986).
62. M. Bakovic and H. B. Dunford, *Biochemistry* **33,** 6475–6482 (1994).
63. Y. Hsuanyu and H. B. Dunford, *J. Biol. Chem.* **267,** 17649–17657 (1992).
64. M. Bakovic and H. B. Dunford, *Biophys. Chem.* **54,** 237–251 (1995); correction *Biophys. Chem.* **61,** 59 (1996).
65. M. Bakovic and H. B. Dunford, *J. Biol. Chem.* **271,** 2048–2056 (1996).
66. V. Koshkin and H. B. Dunford, submitted.
67. T. E. Eling, R. P. Mason, and K. Sivarajah, *J. Biol. Chem.* **260,** 1601–1607 (1985).
68. T. Yonetani and T. Asakura, *J. Biol. Chem.* **244,** 4580–4589 (1969).
69. H. Hori, M. Ikeda-Saito, and T. Yonetani, *Biochim. Biophys. Acta* **912,** 74–81 (1987).
70. R. Odenwaller, K. R. Maddipati, and L. J. Marnett, *J. Biol. Chem.* **267,** 13863–13869 (1992).

71. R. J. Kulmacz, G. Palmer, C. Wei, and A.-L. Tsai, *Biochemistry* **33,** 5428–5439 (1994).
72. A.-L. Tsai, C. Wei, H. K. Baek, R. J. Kulmacz, and H. E. Van Wart, *J. Biol. Chem.* **272,** 8885–8894 (1997).
73. D. H. Nugteren, *Biochim. Biophys. Acta* **380,** 299–307 (1975).
74. M. E. Hemler, C. G. Crawford, and W. E. M. Lands, *Biochemistry* **17,** 1772–1779 (1978).
75. M. Bakovic and H. B. Dunford, *Prostagland. Leukotri. Essent. Fatty Acids* **53,** 423–431 (1995).
76. V. Koshkin and H. B. Dunford, *J. Biol. Chem.* **273,** 6046–6049 (1998).
77. M. Bakovic and H. B. Dunford, *Prostagland. Leukotri. Essent. Fatty Acids* **54,** 341–349 (1996).
78. J. F. Curtis, N. G. Reddy, R. P. Mason, B. Kalyanaraman, and T. E. Eling, *Arch. Biochem. Biophys.* **335,** 369–376 (1996).
79. D. A. Kujubu, B. S. Fletcher, B. C. Varnum, R. W. Lim, and H. R. Herschman, *J. Biol. Chem.* **266,** 12866–12872 (1991).
80. B. S. Fletcher, D. A. Kujubu, D. M. Perrin, and H. R. Herschman, *J. Biol. Chem.* **267,** 4338–4344 (1992).
81. D. A. Kujubu and H. R. Herschman, *J. Biol. Chem.* **267,** 7991–7994 (1992).
82. W. Xie, J. G. Chipman, D. L. Robertson, R. L. Erikson, and D. L. Simmons, *Proc. Natl. Acad. Sci. U.S.A.* **88,** 2692–2696 (1991).
83. W. Xie, D. L. Robertson, and D. L. Simmons, *Drug Develop. Res.* **25,** 249–265 (1992).
84. M. K. O'Banion, H. B. Sadowski, V. Winn, and D. A. Young, *J. Biol. Chem.* **266,** 23261–23267 (1991).
85. K. Yamamoto, T. Arakawa, N. Ueda, and S. Yamamoto, *J. Biol. Chem.* **270,** 313315–31320 (1995).
86. J. Sirois and J. S. Richards, *J. Biol. Chem.* **267,** 6382–6388 (1992).
87. M. G. O'Sullivan, F. H. Chilton, E. M. Huggins, Jr., and C. E. McCall, *J. Biol. Chem.* **267,** 14547–14550 (1992).
88. J. Barnett, J. Chow, D. Ives, M. Chiou, R. Mackenzie, E. Osen, B. Nguyen, S. Tsing, C. Bach, J. Freire, H. Chan, E. Sigal, and C. Ramesha, *Biochim. Biophys. Acta* **1209,** 130–139 (1994).
89. S. A. Kraemer, E. A. Meade, and D. L. DeWitt, *Arch. Biochem. Biophys.* **293,** 391–400 (1992).
90. J. K. Morris and J. S. Richards, *J. Biol. Chem.* **271,** 16633–16643 (1996).
91. W. L. Smith, R. M. Garavito, and D. L. DeWitt, *J. Biol. Chem.* **271,** 33157–33160 (1996).
92. O. Laneuville, D. K. Breuer, N. Xu, Z. H. Huang, D. A. Gage, J. T. Watson, M. Lagarde, D. L. DeWitt, and W. L. Smith, *J. Biol. Chem.* **270,** 19330–19336 (1995).
93. R. J. Kulmacz and L.-H. Wang, *J. Biol. Chem.* **270,** 24019–24023 (1995).
94. D. C. Swinney, A. Y. Mak, J. Barnett, and C. K. Ramesha, *J. Biol. Chem.* **272,** 12393–12398 (1997).
95. M. R. Gunther, L. C. Hsi, J. F. Curtis, J. K. Gierse, L. J. Marnett, T. E. Eling, and R. P. Mason, *J. Biol. Chem.* **272,** 17086–17090 (1997).

96. V. Houtzager, M. Ouellet, J.-P. Falgueyret, L. A. Passmore, C. Bayly, and M. D. Percival, *Biochemistry* **35,** 10974–10984 (1996).
97. L. M. Landino, B. C. Crews, M. D. Timmons, J. D. Morrow, and L. J. Marnett, *Proc. Natl. Acad. Sci. U.S.A.* **93,** 15069–15074 (1996).
98. J. R. Vane, *Nature New Biol.* **231,** 232–234 (1971).
99. J. B. Smith and A. L. Willis, *Nature New Biol.* **231,** 235–237 (1971).
100. S. H. Ferreira, S. Moncada, and J. R. Vane, *Nature New Biol.* **231,** 237–239 (1971).
101. J. Vane, *Nature* **367,** 215–216 (1994).
102. D. L. DeWitt, E. A. El-Harith, S. A. Kraemer, M. J. Andrews, E. F. Yao, R. L. Armstrong, and W. L. Smith, *J. Biol. Chem.* **265,** 5192–5198 (1990).
103. T. Shimokawa and W. L. Smith, *J. Biol. Chem.* **267,** 12387–12392 (1992).
104. E. A. Meade, W. L. Smith, and D. L. DeWitt, *J. Biol. Chem.* **268,** 6610–6614 (1993).
105. M. S. Tang, R. A. Copeland, and T. M. Penning, *Biochemistry* **36,** 7527–7534 (1997).
106. J. K. Gierse, J. J. McDonald, S. D. Hauser, S. H. Rangwala, C. M. Koboldt, and K. Seibert, *J. Biol. Chem.* **271,** 15810–15814 (1996).
107. J. A. Mancini, D. Riendeau, J.-P. Falgueyret, P. J. Vickers, and G. P. O'Neill, *J. Biol. Chem.* **270,** 29372–29377 (1995).
108. O. H. Callan, O.-Y. So, and D. C. Swinney, *J. Biol. Chem.* **271,** 3548–3554 (1996).
109. C. A. Lanzo, J. M. Beechem, J. Talley, and L. J. Marnett, *Biochemistry* **37,** 217–226 (1998).
110. M. S. Tang, L. J. Askonas, and T. M. Penning, *Biochemistry* **34,** 808–815 (1995).

CHAPTER 17

Lactoperoxidase, Thyroid Peroxidase, and Other Animal Peroxidases

17.1. INTRODUCTION

The mucosal surfaces of the body provide protection against potential invading germs and bacteria. If the surfaces become dry, then organisms can more readily enter into the body. First lines of defense are washing and peristaltic movement. Thus, our eyes blink continuously, secretions flow into our noses either spontaneously or are induced by sneezing, and are replaced after we blow our noses. Although saliva moistens, our mouths are particularly susceptible to bacterial influx (except perhaps in tight-lipped individuals).[1] Almost continuous peristaltic movement restricts bacterial growth in the small intestine, but the slow periodic movement in the large intestine allows bacteria to flourish. The exocrine glands and their secretions provide both specific immunological and less specific nonimmunological protection to the mucosal surfaces of eye, nose, mouth, trachiobronchial tree, and intestinal tract. The antibacterial system includes lysozyme, lactoferrin, and peroxidase. Peroxidase also provides an antibacterial system in bovine and human milk, but in different time-dependent ways.

In contrast to lacto-, salivary, intestinal, and uterine peroxidases, which play a defensive role against infection, thyroid peroxidase is essential for the synthesis of thyroxine and the even more potent hormone, triiodothyronine. This chapter is focused on a description of these mammalian peroxidases. The properties of the well-studied myeloperoxidase were described in Chapter 15 and prostaglandin H synthase in Chapter 16.

17.2. BOVINE LACTOPEROXIDASE

17.2.1. Preparation and Properties

Crystalline lactoperoxidase from bovine milk was obtained in 1943 and shown to be green in color as is myeloperoxidase.[2,3] The isolation was greatly facilitated by precipitation of casein with rennet[4] and use of an ion exchanger to isolate lactoperoxidase from the whey.[5] The isolation procedure has been further simplified.[6,7]

Bovine lactoperoxidase is a heme-containing, calcium-containing[8] glycoprotein with one heme group per molecule. The enzyme has a single polypeptide chain.[9] The maximum molar absorptivity is 112.3 $mM^{-1}cm^{-1}$ at 412 nm (Fig. 17.1). There has not been agreement on the nature of the heme group of lactoperoxidase in the literature, which goes back over 40 years, but the number of possibilities has been greatly reduced (Section 17.2.2).

Two major components, A and B, of lactoperoxidase were reported.[4] It was shown that there were six subcomponents of A and four subcomponents of B.[10–12] Eventually it was shown that the components were all modifications of a parent molecule, B1, and did not differ in their chemical properties.[13–15] The microheterogeneity is apparently caused by deamidation of asparagine or glutamine residues, a process that occurs over two days in alkaline ammonium sulfate. The conversion of B into A involves both deamidation and loss of carbohydrate groups. The resting ferric enzyme has a five-coordinate iron,[16] although an EXAFS study claims six-coordination.[17] The imidazole side chain of a histidine residue provides the fifth ligand.[18] The molecular weight of the parent lactoperoxidase molecule B1 is 78,431 based on amino acid composition.[19] Bovine lactoperoxidase has a strong tendency to adhere to surfaces, which facilitates its role as an antimicrobial agent.

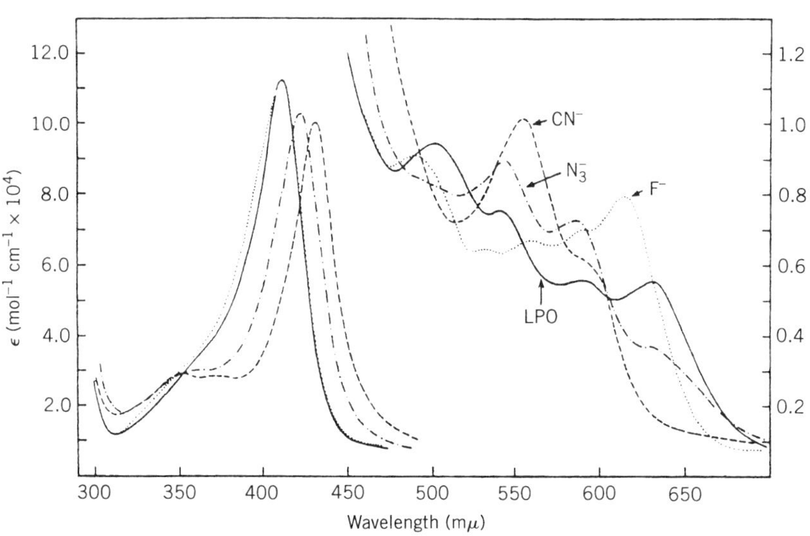

FIG. 17.1. Optical absorption spectra of bovine lactoperoxidase (—) and its cyanide (-----), azide (-·-·), and fluoride (······) complexes obtained at pH 6.0 in 0.2 M phosphate buffer. [Reproduced with permission from A. Carlström, *Acta Chem. Scand.* **23,** 203–213 (1969).]

The amino acid sequence of bovine lactoperoxidase has been determined both by sequencing of peptide fragments,[20] and from the cloned cDNA.[21] The peptide sequence was used to distinguish the N-terminal residue of the mature protein from signal and propeptide residues. Two residues in the peptide sequences were corrected using the DNA sequence, and the combined results are listed in Appendix A. There is 55, 54, and 45% sequence identity with the human enzymes myeloperoxidase, eosinophil peroxidase, and thyroid peroxidase. Sequence alignments showed that critical regions are conserved, confirmation that the mammalian peroxidases belong in one superfamily.

17.2.2. Nature of the Heme Group

Several groups concluded that the heme of lactoperoxidase is ferriprotoporphyrin IX and that the difficulty in abstracting it was caused by tight binding in a restricted active site.[15,19,22] In the latter study the heme was isolated using Pronase digestion. However, others have maintained that the heme is not ferriprotoporphyrin IX and that it is covalently bound.[23] Evidence was presented for a covalent sulfur linkage that involved a CH_2S group, in place of the methyl group normally found in position 18 of the heme ring.[24] The heme was extracted using mercaptoethanol in 8 M urea. A disulfide bridge was believed to occur involving a cysteine residue and the CH_2S group. Support is found in the comparison of mammalian peroxidase sequences,[20,21] which showed 15 half-cystines on all the peroxidases, and analysis that showed no free cysteine.[13]

Several different procedures to isolate the heme were developed in a careful study.[25] It was concluded that one or more hydroxyl groups were attached to side chains, and the heme was linked to the protein through an ester bond. A repeat of Pronase extraction of the heme indicated that the heme contains modifications to ferriprotoporphyrin IX and is indeed covalently bound. However, Pronase extraction was believed to destroy the covalently attached sulfur.[24] In a recent important paper, the heme of native bovine lactoperoxidase was isolated by the Pronase method and compared to the end product of ferriprotoporphyrin IX, which was processed by recombinant lactoperoxidase. The hemes obtained by the two methods were found to be identical and to contain two hydroxylated methyl groups. Therefore, covalent attachment appears to occur through ester linkages. Two methyl and two methylene groups on the heme of heat-denatured lactoperoxidase were detected by nuclear magnetic resonance (NMR).[26] Lactoperoxidase appears capable of processing and covalently attaching its own heme, unlike cytochrome *c*, which requires another enzyme, cytochrome *c* lyase, to complete the heme attachment.[27] Magnetic circular dichroism studies showed that the heme of lactoperoxidase is different from that of myeloperoxidase.[28] On the other hand, Fourier transform infrared difference spectroscopy shows great similarities.[29]

The current state of knowledge of the heme prosthetic group of bovine lactoperoxidase can be summarized as follows. The heme is covalently attached and is a modified form of ferriprotoporphyrin IX. There is conflicting evidence as to whether it differs from the covalently attached heme of myeloperoxidase (see Section 15.2). The hemes of both enzymes are responsible for their green color. The question of whether a covalent attachment of heme to lactoperoxidase involving a sulfur atom is not yet resolved beyond dispute. A legitimate question is whether the mercaptoethanol isolation procedure, described as a mild treatment, somehow introduces a sulfur atom, or whether Pronase or other methods of heme isolation destroy a heme-bound sulfur. It would be interesting to determine how recombinant apolactoperoxidase processes hemes other than ferriprotoporhyrin IX.

17.2.3. Ligand and Aromatic Donor Binding to Lactoperoxidase

Spectra of lactoperoxidase and its cyanide, azide, and fluoride complexes are shown in Fig. 17.1. Fluoride binding to lactoperoxidase is similar to that with other peroxidases, but it occurs at a much slower rate.[30] From pH 3.8 to 5.4, a plot of the log of the apparent rate constant versus pH has a slope of –1. The rate data fit the equation

$$k_{\mathrm{app}} = \frac{k}{1 + K_{\mathrm{L}}/[\mathrm{H}^+]}$$

at ionic strength 0.11 and 25°C. The acid dissociation constant, K_{L}, of the ligand HF is 2.91 and the value of the pH-independent rate constant, k, is $(9.7 \pm 0.4) \times 10^2\ \mathrm{M}^{-1}\mathrm{s}^{-1}$.

Whereas cyanide binding to other peroxidases exhibits essentially the same lack of pH dependence as that of compound I formation, cyanide binding to lactoperoxidase is unique.[31] A skewed bell-shaped curve is obtained when the rate constant for HCN binding is plotted versus pH (Fig. 17.2). The mechanism is adequately described by

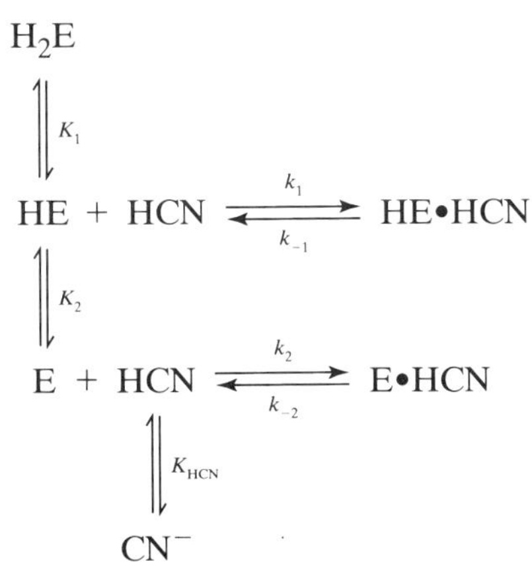

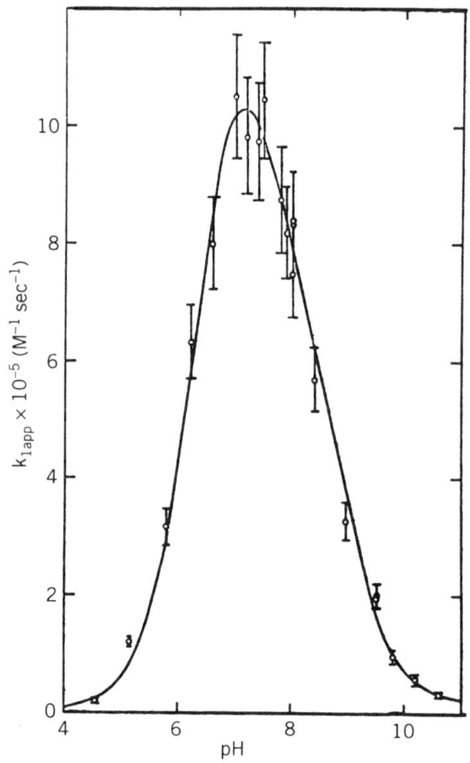

FIG. 17.2. pH dependence of the rate constant for cyanide binding to lactoperoxidase. [Reproduced with permission from D. Dolman, H. B. Dunford, D. M. Chowdhury, and M. Morrison, *Biochemistry* **7,** 3991–3996 (1968).]

where pK_1 is 6.3, pK_2 is 7.6, pK_{HCN} is 9.0, $k_1 = (1.3 \pm 0.2) \times 10^6\ M^{-1}s^{-1}$, and $k_2 = (6.9 \pm 1.4) \times 10^2\ M^{-1}s^{-1}$. The value of k_{-1} averages $15 \pm 7\ s^{-1}$ (below pH 6.4) and k_{-2} $31 \pm 3\ s^{-1}$ (above pH 6.4). The strong pH dependence of the cyanide binding rate, combined with the somewhat weaker binding of cyanide to lactoperoxidase ($K_{Diss} \sim 3 \times 10^{-5}$ M) compared to other peroxidases, and the slow rate of fluoride binding is indicative of a restricted active site, a result deduced from circular dichroism studies,[32] and resonance Raman measurements indicating an unusual bent structure for the cyanide complex.[33]

From an NMR investigation of aromatic donor molecules, it was concluded that the donor protons were 9.4–11.1 Å from the ferric ion, 2–3 Å farther away than in horseradish peroxidase. Binding was facilitated by protonation of an active site group of pK_a 6.1, assumed to be the distal histidine. Iodide and thiocyanate weakened aromatic molecule binding and appeared to compete for the same binding site.[34] From optical difference spectra, circular dichroism and NMR, it was shown that L-tyrosine binds preferentially to lactoperoxidase, compared to D-tyrosine.[35]

17.2.4. Formation of Lactoperoxidase Compounds

Despite the indications of a restricted active site discussed in Section 17.2.3, compound I formation using hydrogen peroxide as oxidant occurs at close to the rate observed for other peroxidases. Because of the instability of lactoperoxidase compound I, the rate constant for compound I formation was measured by a steady-state method, where the concentration of reducing substrate (hydroquinone or guaiacol) was sufficiently large, and hydrogen peroxide concentration was sufficiently small, that compound I formation was rate controlling.[36] The rate constant for compound I formation was pH-independent over the range from pH 3.1 to 10.8 with a value of $(9.2 \pm 0.9) \times 10^6\ M^{-1}s^{-1}$ at 25°C and ionic strength 0.05 M. At low pH formate binds strongly to the enzyme, inhibiting compound I formation. An earlier transient-state kinetic result indicated a rate constant of $2 \times 10^7\ M^{-1}s^{-1}$ at pH 7 for lactoperoxidase compound I formation using hydrogen peroxide,[37] and a rate constant of $1.2 \times 10^7\ M^{-1}s^{-1}$ over the pH range 5.5–10.8 was obtained using transient-state kinetics.[38] In the latter two studies compound I formation was directly observed (Fig. 17.3), an indication of highly purified enzyme. Acetate buffer probably was the cause of a lowered rate of compound I formation below pH 5.5.[38]

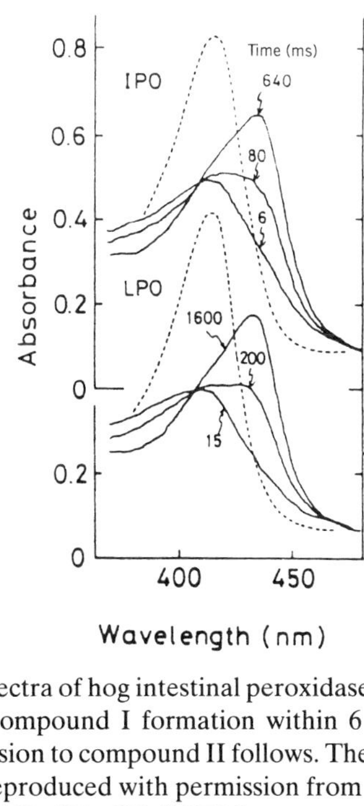

FIG. 17.3. Soret region spectra of hog intestinal peroxidase (IPO) and bovine lactoperoxidase (LPO) showing compound I formation within 6 ms for IPO and 15 ms for LPO. Spontaneous conversion to compound II follows. The dashed lines are the spectra of the native enzymes. [Reproduced with permission from S. Kimura and I. Yamazaki, *Arch. Biochem. Biophys.* **198,** 580–588 (1979).]

In recent studies it was implied that both compounds I and II of lactoperoxidase may exist in forms where an oxidizing equivalent is stored on the protein.[39] However, there appears to be no other evidence in the literature for protein radical forms of lactoperoxidase compounds.

Compound II is formed spontaneously upon addition of one equivalent of hydrogen peroxide to native lactoperoxidase, and is stable for about 20 min. A threefold excess of hydrogen peroxide can also be used to obtain pure compound II, and addition of a 117-fold excess of hydrogen peroxide to the same solution produces pure compound III.[40] Eventually excess hydrogen peroxide inactivates the enzyme.[41–43] Compound II has an absorption maximum at 430 nm and an isosbestic point with native lactoperoxidase at 422 nm. Spectra of compounds II and III are shown in Fig. 17.4. The twin peaks in the visible region are characteristic of peroxidase compounds II and III (535 and 567 nm for lactoperoxidase compound II, and 550 and 588 nm for lactoperoxidase compound III). Resonance Raman studies on lactoperoxidase compound III, prepared from both the ferrous enzyme and by addition of

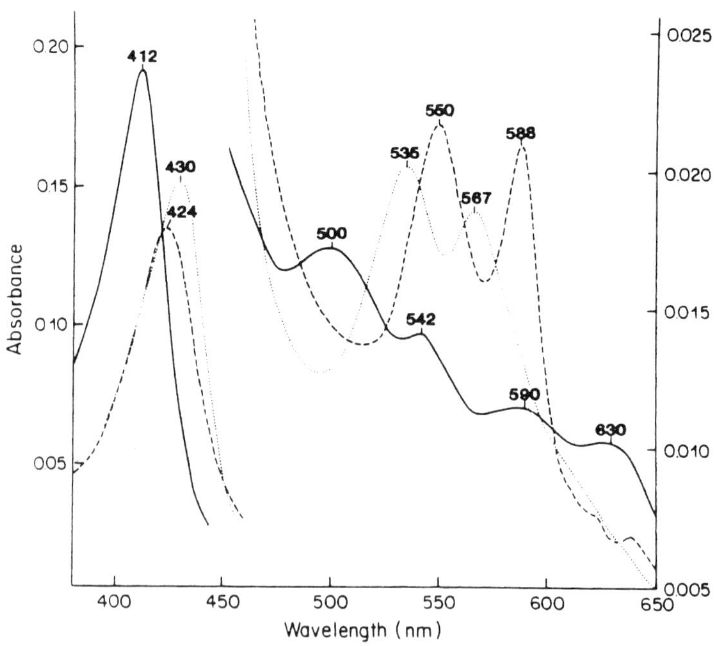

FIG. 17.4. Spectra of native bovine lactoperoxidase (—), compound II (······), and compound III (-----). The lactoperoxidase (1.7 μM in 67 mM phosphate buffer, pH 7.0) spectrum was recorded first. Compound II was formed by adding 6 μM H_2O_2, and compound III by further addition of 200 μM H_2O_2. [Reproduced with permission from H. Kohler, A. Taurog, and H. B. Dunford, *Arch. Biochem. Biophys.* **264,** 438–449 (1988).]

excess hydrogen peroxide to the ferric enzyme, show that it is a low-spin dioxygen adduct. Isotopic labeling shows unambiguously that when compound II is converted to compound III, the ferryl oxygen atom is displaced and both oxygen atoms of compound III come from hydrogen peroxide.[44]

17.2.5. Reactions of Oxidized Forms of Lactoperoxidase

There appear to be no reports of accurate rate measurements of lactoperoxidase compound I reactions. Reduction of compound II to native enzyme by both iodide and *p*-cresol were studied as a function of pH.[45,46] Lactoperoxidase compound II reacts with iodide an order-of-magnitude faster than horseradish peroxidase compound II. In addition, the reaction is accelerated by excess iodide.

$$\text{LPO-II} + \text{I}^- \xrightarrow[\text{H}^+]{k_1} \text{LPO} + \bullet\text{I} + \text{H}_2\text{O}$$

$$\text{LPO-II} \underset{}{\overset{K_1,\ \text{I}^-}{\rightleftharpoons}} \text{LPO-II-I}^-$$

$$\text{LPO-II-I}^- + \text{I}^- \xrightarrow[\text{H}^+]{k_2'} \text{LPO} + \bullet\text{I} + \text{H}_2\text{O} + \text{I}^-$$

The log of the rate constant k_1 ($M^{-1}s^{-1}$) exhibits a perfect inverse dependence on pH, within a small experimental error. The product of K_1 and k_2' provides a third-order rate constant k_2 ($M^{-2}s^{-1}$), and it exhibits the same pH dependence as k_1. Values of k_1 range from about 5×10^6 $M^{-1}s^{-1}$ at pH 3 to near 0.5 $M^{-1}s^{-1}$ at pH 10. Values of k_2 range from about 5×10^{10} $M^{-2}s^{-1}$ at pH 3 to near 5×10^3 $M^{-2}s^{-1}$ at pH 10. Hydriodic acid, HI, cannot be the reactant since the diffusion-controlled limit would be exceeded by several orders of magnitude. Therefore, iodide ion, I^-, is the reactant, and the source of the proton required for compound II reduction is unknown. If the proton originates on the enzyme, then the pK_a of the acid group must be well below a value of 3 (protonation and dissociation of the imidazole fifth ligand?). Perhaps the simplest explanation is that the proton originates from a hydronium ion, H_3O^+.

p-Cresol reacts in similar fashion with lactoperoxidase compound II as it does with horseradish peroxidase compound II. The reaction exhibits a broad region where it has little pH dependence. The ionization of *p*-cresol (pK_a 10.3) and of an enzyme group (pK_a 9.7) both inhibit the reaction at high pH (Fig. 17.5).

Lactoperoxidase is capable of both bromination[47] and iodination reactions.[48] Thiocyanate, SCN^-, is found in body fluids and is present in milk. Addition of hydrogen peroxide and thiocyanate is used to help preserve milk in tropical climates. The main oxidation product of thiocyanate in the presence of hydrogen peroxide and peroxidase is hypothiocyanite anion, $OSCN^-$.[49–51] The conjugate acid, hypothiocyanous acid, HOSCN, has a pK_a value of 5.3.

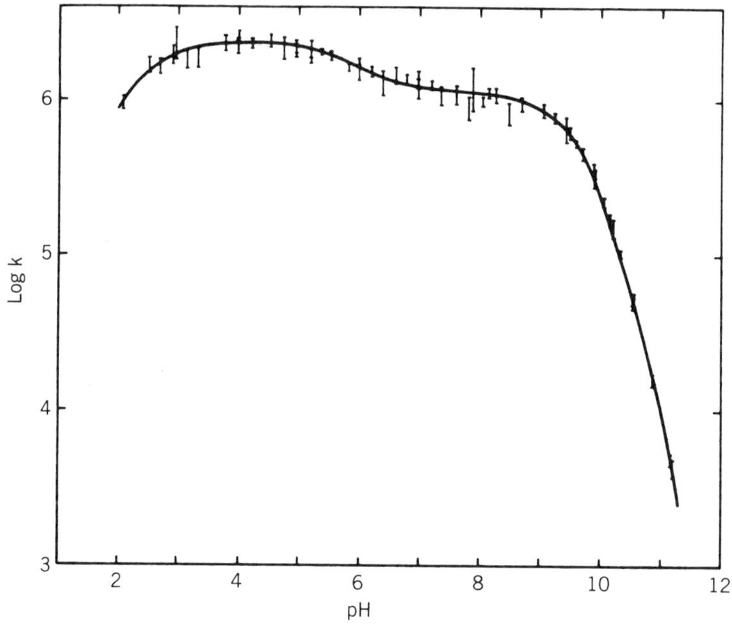

FIG. 17.5. pH dependence of the rate constant for the one-electron oxidation of *p*-cresol by lactoperoxidase compound II. [Reproduced with permission from R. J. Maguire and H. B. Dunford, *Can. J. Chem.* **51,** 1721–1723 (1973).]

Formation of $OSCN^-$ or its conjugate acid may occur by the one-electron oxidation of SCN^- to •SCN as follows.

$$2SCN^- \rightarrow 2e^- + 2\ \bullet SCN$$

$$2\ \bullet SCN \rightarrow (SCN)_2$$

$$(SCN)_2 + H_2O \rightarrow HOSCN + SCN^- + H^+$$

Alternatively, it could occur by direct oxygen atom addition, a two-electron reaction.[52]

$$SCN^- + H_2O_2 \rightarrow OSCN^- + H_2O$$

The complete lactoperoxidase system exhibits better antimicrobial action than exogenous hypothiocyanite.[53,54] Therefore, another agent may be generated in the enzyme system, which has been suggested to be the thiooxime isomer, $SCNO^-$.[52]

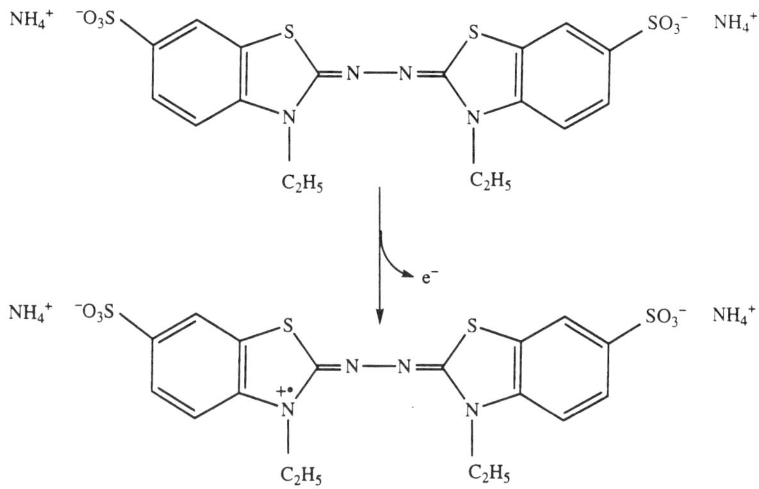

FIG. 17.6. 2,2′-Azino-di-(3-ethylbenzthiazoline-6-sulfonic acid), (ABTS), top, and its cation radical product formed by one-electron oxidation.

17.2.6. Lactoperoxidase Assays

The reagent 2,2′-azino-di-(3-ethylbenzthiazoline-6-sulfonic acid), commonly supplied as the ammonium salt of the acid, and known as ABTS, was introduced as an assay reagent for lactoperoxidase in 1975.[55] ABTS has its maximum absorbance at 340 nm. It acts as a one-electron donor to form a cation radical that absorbs strongly at 414 nm ($\varepsilon = 3.6 \times 10^4\ M^{-1}cm^{-1}$). The reaction is shown in Fig. 17.6. The cation radical is sufficiently stable to provide the basis for an accurate assay. It slowly disproportionates to ABTS and a dication. The advantages and potential pitfalls in use of ABTS have been well documented.[56] Comparisons were made of pyrogallol, guaiacol, ABTS, and thiocyanate assays for animal peroxidases.[57]

17.3. THYROID PEROXIDASE

17.3.1. Thyroxine and Triiodothyronine

There is an admirable summary of knowledge about the thyroid gland, traced back to early Greek civilization.[58] The thyroid hormone thyroxine, which helps to control the level of basal metabolism, was isolated in 1926 and its structure determined the following year (Fig. 17.7).[59,60] Later triiodothyronine, which lacks one of the iodine atoms of thyroxine, was discovered and found to be a more potent hormone than thyroxine.[61] The majority of triiodothyronine formation occurs via removal of one iodine atom from thyroxine, with the

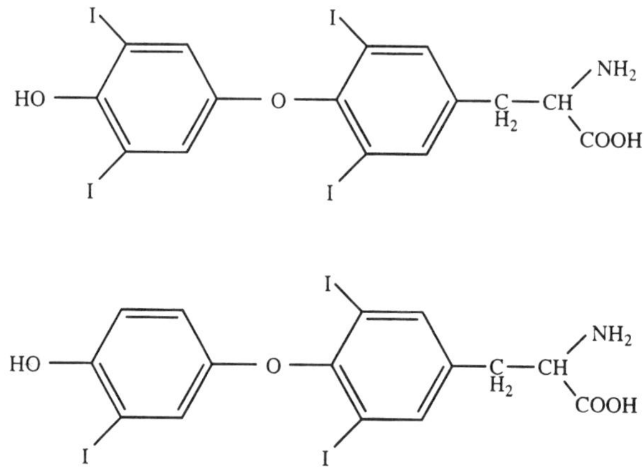

FIG. 17.7. 3′,5′,3,5-Tetraiodothyronine (thyroxine), top; and 3′,3,5-triiodothyronine, bottom.

remainder formed by coupling of mono- and diiodotyrosine.[62,63] Because hydrogen peroxide is one of the few biological agents capable of oxidizing iodide ion, peroxidases were suspected to be responsible for at least a part of thyroxine synthesis.[64]

Thyroxine and triiodothyronine synthesis involves the following steps[65]: active iodide transport, iodination of tyrosyl residues contained in the protein thyroglobulin, coupling of iodinated tyrosyl residues to form thyroxine and triiodothyronine, hydrolysis of thyroglobulin to release the hormones, and secretion of hormones into the bloodstream. This review is concentrated on properties of thyroid peroxidase and the mechanisms of the iodination and coupling reactions. The steady-state iodination of tyrosyl residues in serum albumin by the thyroid peroxidase system, as well as the oxidation of iodide, guaiacol, and ABTS, has been studied.[66]

17.3.2. Purification, Cloning, and Properties of Thyroid Peroxidase

Early studies on thyroid peroxidase were hampered by difficulty in obtaining sufficient enzyme in highly purified form. Highly purified thyroid peroxidase was obtained in 1979 after treatment of the membrane fraction of hog thyroid glands with trypsin and detergent.[67] The heme-containing enzyme has a reinheitzahl (RZ) value (A_{410}/A_{280}) of 0.54. It contains about 10% carbohydrate by weight, has a molecular weight of approximately 90,000, and an isoelectric point of 5.75. Glycosylation sites were shown to be located in the N-terminal third of the polypeptide chain, believed to be extracellular.[68] Purified human thyroid peroxidase, obtained from surgery on patients with Graves disease, using a similar procedure to that for the hog enzyme, was reported in 1990.[69]

Studies were conducted on human enzyme reactions with microsomal autoantibodies.[70]

The amino acid sequences of human thyroid peroxidases were published in 1987,[71] and of hog thyroid peroxidase in 1989,[72] both deduced from cDNAs. There are two different human enzymes, designated human thyroid peroxidase-1 (hTPO-1) and human thyroid peroxidase-2 (hTPO-2).[73] The first human isoenzyme consists of 933 amino acid residues, whereas the second differs mainly in lacking 57 residues in the middle of the sequence. Both isoenzymes were found in all tissues examined, but the amounts of each varied. By sequence analogy with myeloperoxidase, it was shown that the catalytically active distal residues in hTPO-1 are His239 and Arg396.[74] The proximal ligand is His494,[75] which is hydrogen bonded to Asn579. The Asn579 residue is missing in hTPO-2, which may account for its lack of reactivity.[74] The amino acid sequences of human thyroid peroxidase-1 and porcine thyroid peroxidase are reproduced in Appendix A.

The human thyroid peroxidase cDNA was successfully expressed in nonthyroidal eukaryotic cells. In one study human liver Hep G2 cells using a vaccinia virus cDNA expression system were utilized.[76] In another Chinese hamster ovary cells were used.[77] The cloning, expression, and role of thyroid peroxidase as an autoantigen in autoimmune thyroid disease has been reviewed.[78] Work is progressing in defining the immunodominant region recognized by thyroid peroxidase antibodies.[79] The baculovirus expression system in insect cells, so successful for other heme proteins, was used to overexpress thyroid peroxidase. It was shown that large quantities of thyroid peroxidase could be produced.[80] Both normal and mutated thyroid peroxidases from patients of total iodide organification defects (TIOD) disease have been expressed.[75,81] Fluorescence *in situ* hybridization (FISH) was used to localize the human thyroid peroxidase gene to human chromosone 2 and the mouse enzyme to mouse chromosone 12.[82] Similarities in the genes for myelo- and thyroid peroxidase have been emphasized.[83]

It would appear that all known mammalian peroxidases, except prostaglandin H synthase, contain covalently linked heme prosthetic groups. Thyroid and lactoperoxidase exhibit similar spectral properties and are both capable of catalyzing iodination and coupling reactions.[84] Therefore, the two enzymes would be expected to have very similar, if not identical, prosthetic groups (Section 17.2.2).

17.3.3. Mechanism of Thyroxine Synthesis

Using horseradish peroxidase as a model, it was shown that the rate constants for iodination of tyrosine are larger than for phenolic oxidation and coupling.[85–87] If iodination did not happen first, tyrosyl radicals would couple at the ortho positions. With the ortho positions blocked by iodine atoms, then coupling to form thyroxine is the favored reaction.

Reactions of iodide ion with thyroid and lactoperoxidases were examined. It was shown that iodide can catalyze catalatic activity, the decomposition of hydrogen peroxide to oxygen and water.[88,89] The iodothyronines can also catalyze the catalatic reaction by iodide released from them.[90] The results indicated that reaction occurred via enzyme-bound hypoiodous acid, EOI, in the case of thyroid peroxidase, but that free hypoiodous acid, HOI, was generated from lactoperoxidase. In his 1986 review, Alvin Taurog lists three mechanisms for the iodination reactions[91]: via free radicals, via I^+ formation, and via enzyme-bound hypoiodous acid, which can be represented as FeOI. The free radical mechanism can be deleted since studies on a variety of peroxidases show that halide ion oxidation proceeds by a two-electron mechanism to produce an ion in a +1 oxidation state. The iodine in free hypoiodous acid, HOI, or in FeOI, has a formal +1 oxidation state, and the hypothetical I^+ species can be deleted from the list of possibilities. The question of whether HOI is liberated or remains enzyme-bound was addressed in a comparative study using horseradish and lactoperoxidases.[92] The conclusion was that free HOI was liberated by horseradish peroxidase, but iodination of tyrosine was accomplished by enzyme-bound hypoiodous acid, in the form Fe—O—I. Lactoperoxidase is about 10 times as efficient as the horseradish enzyme in accomplishing the iodination. In a reaction of hypoiodous acid with excess iodide, molecular iodine is liberated and an equilibrium with I_3^- is established. In an enzyme-free system, it is not possible to distinguish between I_2 and HOI as iodinating agents.[93] However, in the thyroid gland, despite its amazing ability to trap iodide, excess iodide is not present, and the efficient route to iodination is through the Fe—O—I species. Therefore, the conversion of tyrosine to monoiodotyrosine occurs as follows:

$$Fe + H_2O_2 \longrightarrow Fe{=}O + H_2O$$

$$Fe{=}O + I^- \longrightarrow Fe{-}O{-}I$$

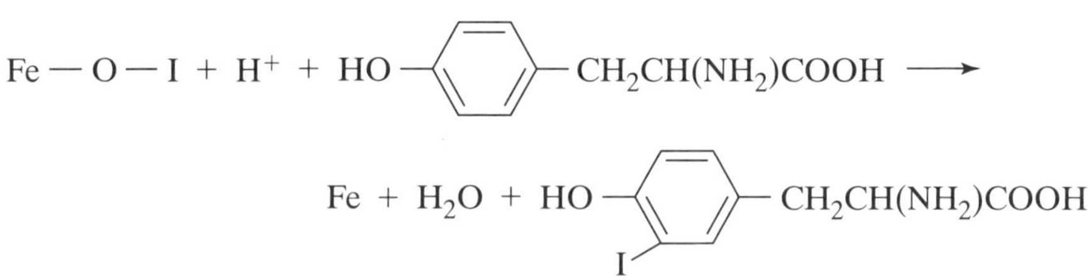

For simplicity, formal charges and oxidation states are ignored. Fe=O represents compound I. A repeat of the above reaction using monoiodotyrosine as substrate would produce diiodotyrosine.

The one- and two-electron oxidations of tyrosine, monoiodotyrosine and diiodotyrosine by thyroid peroxidase have been intensively investigated.[94–97] A free radical chain mechanism was deduced for coupling of iodinated tyrosyl radicals in thyroglobulin.[98] A reaction scheme showing thyroxine synthesis within the thyroglobulin molecule is shown in Fig. 17.8. The occurrence of iodination and coupling on the thyroglobulin molecule is an amazing sequence of reactions.[99] Subsequent protein hydrolysis to liberate the hormones requires a great deal of energy but appears to be a highly conserved mechanism.[58]

In a recent paper, a possible role of a protein radical form of compound I was examined and found not to exist.[100] Although protein radical forms of mammalian peroxidase compounds I are a possibility that has not been conclusively disproven, I find the evidence presented for their existence to be less than compelling.[101,102]

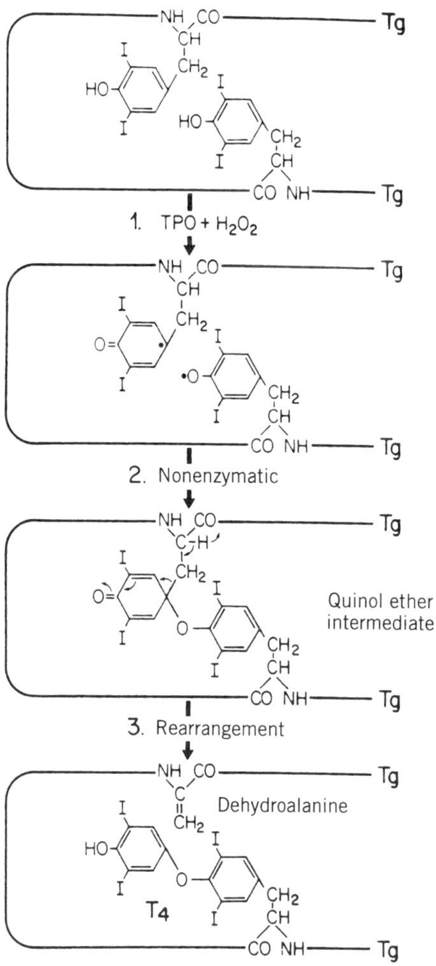

FIG. 17.8. Hypothetical coupling scheme for formation of thyroxine (T_4) within the thyroglobulin (Tg) molecule, catalyzed by thyroid peroxidase (TPO). [Reproduced with permission from A. Taurog, *Recent Prog. Horm. Res.* **26,** 189–247 (1970).]

17.3.4. Compounds of Thyroid Peroxidase

The formation of compounds I, II, and III of hog thyroid peroxidase was demonstrated.[103] Compound I formation was incomplete. The spectra of compounds II and III, compared to that of the native enzyme, are shown in Fig. 17.9.[40]

17.3.5. Inhibition of Thyroid Peroxidase

The antithyroid drugs 6-propylthiouracil and 1-methyl-2-mercaptoimidazole inhibit the thyroid peroxidase-catalyzed iodination of thyroglobulin. Metabolites of both drugs were examined using ^{35}S- and ^{14}C-labels and product identification by high-pressure liquid chromatography (HPLC).[104,105] It was shown that both reversible and irreversible inhibition occurred. Suicide inhibition of both thyroid and lactoperoxidases by resorcinol derivatives was demonstrated.[106]

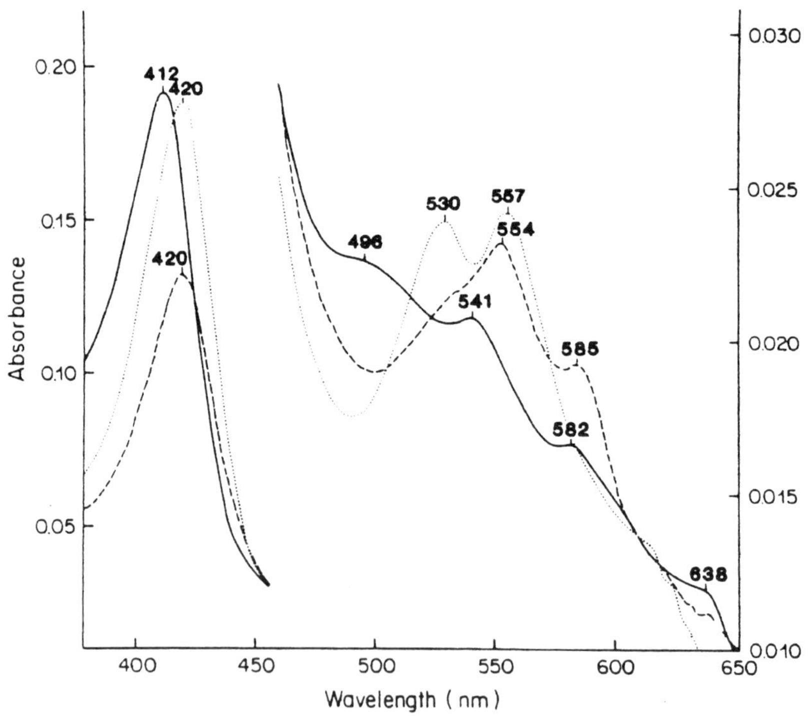

FIG. 17.9. Spectra of native hog thyroid peroxidase (—), compound II (······), and compound III (------). The spectra were recorded under exactly the same conditions as described for lactoperoxidase in Fig. 17.1. [Reproduced with permission from H. Kohler, A. Taurog, and H. B. Dunford, *Arch. Biochem. Biophys.* **264,** 438–449 (1988).]

17.4. OTHER ANIMAL PEROXIDASES

There have so far not been the intensive studies on other animal peroxidases as on bovine lactoperoxidase and thyroid peroxidase described in this chapter. Probably the most intensively studied of the other animal peroxidases is salivary peroxidase. It is closely related to lactoperoxidase. Human salivary peroxidase and bovine lactoperoxidase are immunologically cross-reactive.[107] The properties of salivary peroxidase have been reviewed,[108] as has its genetic variation.[109] It has been reported that there is no peroxidase secreted in human milk.[110] The peroxidase activity appears to be derived from human milk leukocytes.

Intestinal peroxidase, found in the mucus of the small intestine, was purified in 1971.[111] Optical spectra were measured (Fig. 17.3), and an electron paramagnetic resonance (EPR) investigation was made of the enzyme and its complexes.[112] Coupling reactions of tyrosine and some derivatives, catalyzed by intestinal peroxidase, were studied,[113] as was inhibition by nonsteroidal antiinflammatory drugs.[114]

Several papers describe properties and behavior of mouse uterine peroxidase. However, the enzyme may be identical to eosinophil peroxidase brought to the uterus by eosinophils after estrogen stimulation.[115]

A new area of peroxidase research started in 1988, with the identification of a cell adhesion factor from crayfish (*Pacifastacus leniusculus*) blood cells.[116] This 76-kilodalton protein was purified. Its cDNA was isolated, cloned, and sequenced. The protein sequence was found to be significantly similar to human myeloperoxidase.[117] Thus, the cell adhesion properties of human myeloperoxidase were uncovered by research on the crayfish.[118] (Tell that to your granting agency!) Peroxidases have also been found in the following invertabrates: squid (*Euprymna scolopes*),[119] *Drosophila,*[120] the nematode *Caenorhabditis elegans,*[121] and cuttlefish (*Sepia officinalis*).[122] Sequence alignments of the five invertabrate peroxidases listed here, with five mammalian peroxidases, show that key peroxidase residues are conserved.[74]

17.5. PEROXIDASE NOMENCLATURE

The *Drosophila* peroxidase and the crayfish peroxidase, discussed in the previous section, have been named peroxidasin and peroxonectin by their discoverers, which are not conventional enzyme names. There is bound to be a profusion of new names in the years to come. Perhaps it is time for peroxidase researchers to invent a new nomenclature, as the cytochrome P450 researchers did several years ago. The cumbersome, formal International Union of Biochemistry (IUB) system does not distinguish families from specific members of families, and is, in my view, inadequate.

REFERENCES

1. I. D. Mandel and S. A. Ellison in *The Lactoperoxidase System. Chemistry and Biological Significance,* K. M. Pruitt and J. O. Tenovuo, eds., Marcel Dekker, New York, 1985, pp. 1–14.
2. H. Theorell and Å. Åkeson, *Arkiv Kemi Mineral. Geol.,* **17B**(7), 1–6 (1943).
3. H. Theorell and K. G. Paul, *Arkiv Kemi Mineral. Geol.,* **18A**(12), 1–23 (1944).
4. B. D. Polis and H. W. Schmukler, *J. Biol. Chem.* **201,** 475–500 (1953).
5. M. Morrison, H. B. Hamilton, and E. Stotz, *J. Biol. Chem.* **228,** 767–776 (1957).
6. M. Morrison and D. E. Hultquist, *J. Biol. Chem.* **238,** 2847–2849 (1963).
7. K. G. Paul, P. I. Ohlsson, and A. Henriksson, *FEBS Lett.* **110,** 200–204 (1980).
8. K. S. Booth, S. Kimura, H. C. Lee, M. Ikeda-Saito, and W. S. Caughey, *Biochem. Biophys. Res. Commun.* **160,** 897–902 (1989).
9. G. Sievers, *FEBS Lett.* **127,** 253–256 (1981).
10. A. Carlström, *Acta Chem. Scand.* **19,** 2387–2394 (1965).
11. A. Carlström, *Acta Chem. Scand.* **20,** 1426–1427 (1966).
12. A. Carlström and O. Vesterberg, *Acta Chem. Scand.* **21,** 271–278 (1967).
13. A. Carlström, *Acta Chem. Scand.* **23,** 171–184 (1969).
14. A. Carlström, *Acta Chem. Scand.* **23,** 185–202 (1969).
15. A. Carlström, *Acta Chem. Scand.* **23,** 203–213 (1969).
16. S. Modi, D. V. Behere, and S. Mitra, *J. Inorg. Biochem.* **38,** 17–25 (1990).
17. C.-S. Chang, R. Sinclair, S. Khalid, I. Yamazaki, S. Nakamura, and L. Powers, *Biochemistry* **32,** 2780–2786 (1993).
18. G. S. Lukat, K. R. Rodgers, and H. M. Goff, *Biochemistry* **26,** 6927–6932 (1987).
19. K. G. Paul and P. I. Ohlsson in Ref. 1, pp. 15–29.
20. M.-M. Cals, P. Mailliart, G. Brignon, P. Anglade, and B. R. Dumas, *Eur. J. Biochem.* **198,** 733–739 (1991).
21. T. J. Dull, C. Uyeda, A. D. Strosberg, G. Nedwin, and J. J. Seilhamer, *DNA Cell Biol.* **9,** 499–509 (1990).
22. G. Sievers, *Biochim. Biophys. Acta* **579,** 181–190 (1979).
23. D. B. Morell and P. S. Clezy, *Biochim. Biophys. Acta* **71,** 157–164 (1963).
24. A. W. Nichol, L. A. Angel, T. Moon, and P. S. Clezy, *Biochem. J.* **247,** 147–150 (1987).
25. D. E. Hultquist and M. Morrison, *J. Biol. Chem.* **238,** 2843–2846 (1963).
26. T. D. Rae and H. M. Goff, *J. Am. Chem. Soc.* **118,** 2103–2104 (1996).
27. Z. Xie and S. Merchant, *J. Biol. Chem.* **271,** 4632–4639 (1996).
28. L. A. Andersson, S. A. Bylkas, and A. E. Wilson, *J. Biol. Chem.* **271,** 3406–3412 (1996).
29. I. M. Kooter, A. J. Pierik, M. Merkx, B. A. Averill, N. Moguilevsky, A. Bollen, and R. Wever, *J. Am. Chem. Soc.* **119,** 11542–11543 (1997).
30. R. Segal, H. B. Dunford, and M. Morrison, *Can. J. Biochem.* **46,** 1471–1474 (1968).

31. D. Dolman, H. B. Dunford, D. M. Chowdhury, and M. Morrison, *Biochemistry* **7,** 3991–3996 (1968).
32. G. Sievers, *Biochim. Biophys. Acta* **624,** 249–259 (1980).
33. S. Hu, R. W. Treat, and J. R. Kincaid, *Biochemistry* **32,** 10125–10130 (1993).
34. S. Modi, D. V. Behere, and S. Mitra, *Biochim. Biophys. Acta* **996,** 214–225 (1989).
35. L. Casella, M. Gullotti, S. Poli, M. Bonfà, R. P. Ferrari, and A. Marchesini, *Biochem. J.* **279,** 245–250 (1991).
36. R. J. Maguire, H. B. Dunford, and M. Morrison, *Can. J. Biochem.* **49,** 1165–1171 (1971).
37. B. Chance, *Science* **109,** 204–208 (1949).
38. S. Kimura and I. Yamazaki, *Arch. Biochem. Biophys.* **198,** 580–588 (1979).
39. E. Monzani, A. L. Gatti, A. Profumo, L. Casella, and M. Gullotti, *Biochemistry* **36,** 1918–1926 (1997).
40. H. Kohler, A. Taurog, and H. B. Dunford, *Arch. Biochem. Biophys.* **264,** 438–449 (1988).
41. H. Jenzer, W. Jones, and H. Kohler, *J. Biol. Chem.* **261,** 15550–15556 (1986).
42. H. Jenzer, H. Kohler, and C. Broger, *Arch. Biochem. Biophys.* **258,** 381–390 (1987).
43. H. Kohler and H. Jenzer, *Free Rad. Biol. Med.* **6,** 323–339 (1989).
44. S. Hu and J. R. Kincaid, *J. Am. Chem. Soc.* **113,** 7189–7194 (1991).
45. R. J. Maguire and H. B. Dunford, *Biochemistry* **11,** 937–941 (1972).
46. R. J. Maguire and H. B. Dunford, *Can. J. Chem.* **51,** 1721–1723 (1973).
47. A. Taurog and M. L. Dorris, *Arch. Biochem. Biophys.* **287,** 288–296 (1991).
48. C. Illy, N. M. Thielens, J. Gagnon, and G. J. Arlaud, *Biochemistry* **30,** 7135–7141 (1991).
49. T. M. Aune and E. L. Thomas, *Eur. J. Biochem.* **80,** 209–214 (1977).
50. J. O. Tenovuo in Ref. 1, pp. 101–122.
51. B. Reiter in Ref. 1, pp. 123–141.
52. J. R. Pollock and H. M. Goff, *Biochim. Biophys. Acta* **1159,** 279–285 (1992).
53. L. Bjoerck and O. Claesson, *J. Dairy Sci.* **63,** 919–922 (1980).
54. J. Tenovuo, B. Mansson-Rahemtulla, K. M. Pruitt, and R. Arnold, *Infect. Immunol.* **34,** 208–214 (1981).
55. R. E. Childs and W. E. Bardsley, *Biochem. J.* **145,** 93–103 (1975).
56. W. E. Bardsley in Ref. 1, pp. 55–87.
57. K. M. Pruitt, D. N. Kamau, K. Miller, B. Månsson-Rahemtulla, and F. Rahemtulla, *Anal. Biochem.* **191,** 278–286 (1990).
58. S. C. Werner in *Werner's The Thyroid. A Fundamental and Clinical Text,* 5th ed., S. H. Ingbar, and L. E. Braverman, eds., J. B. Lippincott, Philadelphia, 1986, pp. 3–6.
59. C. R. Harington, *Biochem. J.* **20,** 293–299 (1926).
60. C. R. Harington and G. Barger, *Biochem. J.* **21,** 169–181 (1927).
61. J. Gross and R. Pitt-Rivers, *Recent Prog. Horm. Res.* **10,** 109–128 (1954).
62. L. F. Braverman, S. H. Ingbar, and K. Sterling, *J. Clin. Invest.* **49,** 855–864 (1970).

63. J. Brown, I. J. Chopra, and J. M. Cornell, *Ann. Int. Med.* **81,** 68–81 (1974).
64. A. Taurog, *Recent Prog. Horm. Res.* **26,** 189–247 (1970).
65. A. Taurog in Ref. 58, pp. 53–97.
66. P. R. Kootstra, R. Wever, and J. J. M. de Vijlder, *Acta Endocrinol.* **129,** 328–331 (1993).
67. A. B. Rawitch, A. Taurog, S. B. Chernoff, and M. L. Dorris, *Arch. Biochem. Biophys.* **194,** 244–257 (1979).
68. A. B. Rawitch, G. Pollock, S.-X. Yang, and A. Taurog, *Arch. Biochem. Biophys.* **297,** 321–327 (1992).
69. A. Taurog, M. L. Dorris, N. Yokoyama, and C. Slaughter, *Arch. Biochem. Biophys.* **278,** 333–341 (1990).
70. N. Yokoyama, A. Taurog, M. L. Dorris, and G. G. Klee, *J. Clin. Endocrinol. Metab.* **70,** 758–765 (1990).
71. S. Kimura, T. Kotani, O. W. McBride, K. Umeki, K. Hirai, T. Nakayama, and S. Ohtaki, *Proc. Natl. Acad. Sci. U.S.A.* **84,** 5555–5559 (1987).
72. R. P. Magnusson, J. Gestautas, A. Taurog, and B. Rapoport, *J. Biol. Chem.* **262,** 13885–13888 (1987).
73. S. Kimura, Y.-S. Hong, T. Kotani, S. Ohtaki, and F. Kikkawa, *Biochemistry* **28,** 4481–4489 (1989).
74. A. Taurog and M. Wall, *Thyroid* **8,** 185–191 (1998).
75. H. Bikker, F. Baas, and J. J. M. de Vijlder, *J. Clin. Endocrinol. Metab.* **82,** 649–653 (1997).
76. S. Kimura, T. Kotani, S. Ohtaki, and T. Aoyama, *FEBS Lett.* **250,** 377–380 (1989).
77. K. D. Kaufman, B. Rapoport, P. Seto, G. D. Chazenbalk, and R. P. Magnusson, *J. Clin. Invest.* **84,** 394–403 (1989).
78. S. M. McLachlan and B. Rapoport, *Endocrine Rev.* **13,** 192–206 (1992).
79. T. Nishikawa, B. Rapoport, and S. M. McLachlan, *J. Clin. Endocrinol. Metab.* **79,** 1648–1654 (1994).
80. D. L. Kendler, V. Brennan, T. F. Davies, and R. P. Magnusson, *Mol. Cell. Endocrinol.* **93,** 199–206 (1993).
81. H. Bikker, J. J. J. Waelkens, B. Bravenboer, and J. J. M. de Vijlder, *J. Clin. Endcrinol. Metab.* **81,** 2076–2079 (1996).
82. Y. Endo, S. Onogi, K. Umeki, I. Yamamoto, T. Kotani, S. Ohtaki, and T. Fujita, *Genomics* **25,** 760–761 (1995).
83. S. Kimura and M. Ikeda-Saito, *Proteins: Structure, Function Genetics* **3,** 113–120 (1988).
84. A. Taurog, M. L. Dorris, and L. Lamas, *Endocrinol.* **94,** 1286–1294 (1974).
85. I. Ralston and H. B. Dunford, *Can. J. Biochem.* **56,** 1115–1119 (1978).
86. I. M. Ralston and H. B. Dunford, *Can. J. Biochem.* **58,** 1270–1276 (1980).
87. H. B. Dunford and I. M. Ralston, *Biochem. Biophys. Res. Commun.* **116,** 639–643 (1983).
88. R. P. Magnusson, A. Taurog, and M. L. Dorris, *J. Biol. Chem.* **259,** 197–205 (1984).
89. R. P. Magnusson, A. Taurog, and M. L. Dorris, *J. Biol. Chem.* **259,** 13783–13790 (1984).

90. M. Nakamura, I. Yamazaki, and S. Ohtaki, *J. Biochem.* **108,** 804–810 (1990).
91. A. Taurog in Ref. 58, pp. 53–97.
92. W. Sun and H. B. Dunford, *Biochemistry* **32,** 1324–1331 (1993).
93. H. B. Dunford and A. J. Adeniran, *Biochem. Cell Biol.* **66,** 967–978 (1988).
94. S. Ohtaki, H. Nakagawa, S. Kimura, and I. Yamazaki, *J. Biol. Chem.* **257,** 13398–13403 (1982).
95. M. Nakamura, I. Yamazaki, T. Kotani, and S. Ohtaki, *J. Biol. Chem.* **260,** 13546–13552 (1985).
96. M. Nakamura, I. Yamazaki, S. Ohtaki, and S. Nakamura, *J. Biol. Chem.* **261,** 13923–13927 (1986).
97. M. Nakamura, I. Yamazaki, T. Kotani, and S. Ohtaki, *J. Biol. Chem.* **264,** 12909–12913 (1989).
98. A. Taurog, M. Dorris, and D. R. Doerge, *Arch. Biochem. Biophys.* **315,** 82–89 (1994).
99. L. Lamas and A. Taurog, *Endocrinology* **100,** 1129–1136 (1977).
100. A. Taurog, M. L. Dorris, and D. R. Doerge, *Arch. Biochem. Biophys.* **330,** 24–32 (1996).
101. D. Dème, A. Virion, J. Michot, and J. Pommier, *Arch. Biochem. Biophys.* **236,** 559–566 (1985).
102. F. Courtin, D. Dème, A. Virion, J. L. Michot, J. Pommier, and J. Nunez, *Eur. J. Biochem.* **124,** 603–609 (1982).
103. S. Ohtaki, H. Nakagawa, S. Kimura, and I. Yamazaki, *J. Biol. Chem.* **256,** 805–810 (1981).
104. A. Taurog, M. L. Dorris, and F. S. Guziec, Jr., *Endocrinology* **124,** 30–39 (1989).
105. A. Taurog, M. L. Dorris, F. S. Guziec, Jr., and J. P. Uetrecht, *Endocrinology* **124,** 3030–3037 (1989).
106. R. L. Divi and D. R. Doerge, *Biochemistry* **33,** 9668–9674 (1994).
107. B. Månson-Rahemtulla, F. Rahemtulla, and M. G. Humphreys-Beher, *J. Dent. Res.* **69,** 1839–1846 (1990).
108. J. O. Tenovuo, in Ref. 1, pp. 101–122.
109. E. A. Azen in Ref. 1, pp. 89–99.
110. Z. Moldoveanu, J. Tenovuo, J. Mestecky, and K. M. Pruitt, *Biochim. Biophys. Acta* **718,** 103–108 (1982).
111. T. Stelmaszynska and J. M. Zglinczynski, *Eur. J. Biochem.* **19,** 56–63 (1971).
112. M. Ikeda-Saito and S. Kimura, *Arch. Biochem. Biophys.* **283,** 351–355 (1990).
113. M. Valoti, K. F. Tipton, and G. P. Sgaragli, *Biochem. Pharmacol.* **43,** 945–951 (1992).
114. R. Chatterjee, U. Bandyopadhyay, D. Bhattacharyya, and R. K. Banerjee, *Biochim. Biophys. Acta* **1161,** 168–176 (1993).
115. P. H. Jellinck, R. R. Newbold, and J. A. McLachlan, *Steroids* **56,** 162–167 (1991).
116. M. W. Johnansson and K. Söderhäll, *J. Cell Biol.* **106,** 1795–1803 (1988).
117. M. W. Johnansson, M. I. Lind, T. Holmblad, P.-O. Thörnquist, and K. Söderhäll, *Biochem. Biophys. Res. Commun.* **216,** 1079–1087 (1995).
118. M. W. Johansson, M. Patarroyo, F. Öberg, A. Siegbahn, and K. Nilsson, *J. Cell Sci.* **110,** 1133–1139 (1997).

119. S. I. Tomarev, R. D. Zinovieva, V. M. Weis, A. B. Chepelinsky, J. Piatigorsky, and M. J. McFall-Ngai, *Gene* **132,** 219–226 (1993).
120. R. E. Nelson, L. I. Fessler, Y. Takagi, B. Blumberg, D. R. Keene, P. F. Olson, C. G. Parker, and J. H. Fessler, *EMBO J.* **13,** 3438–3447 (1994).
121. R. Wilson, R. Ainscough, K. Anderson, et al., *Nature* **368,** 32–38 (1994).
122. A. Palumbo and I. J. Jackson, *Biochim. Biophys. Acta* **1247,** 173–178 (1995).

CHAPTER 18

Catalases

18.1. INTRODUCTION

The early work on catalases has been well reviewed.[1,2] Modern research on the catalases started in earnest in the 1940s.[3–7] Catalases are found in most aerobic organisms. They are tetrameric heme enzymes, and the tetramer structure is required for their activity. Their principal function is catalysis of the disproportionation of hydrogen peroxide, the *catalatic* reaction:

$$2H_2O_2 = O_2 + 2H_2O$$

The reaction is very exothermic (-23.5 kcal/mole of H_2O_2) and occurs so rapidly in the presence of a high concentration of catalase that foaming and boiling of the solution can be observed. The exothermicity of the catalatic reaction is utilized by the bombardier beetle to steam-drive its stream of obnoxious repellant aimed at potential enemies.[8]

Work on the mechanism of the catalatic reaction, to 1968,[9–11] was reviewed and extended.[12,13] Peter Jones provided the best combination of careful, comprehensive experimental investigations and theoretical interpretation, up to the time of the X-ray crystallographic stuctural determination.[14–16]

18.2. REACTIONS WITH LIGANDS

Catalases form complexes with typical heme ligands.[17–19] Spectra of cyanide, azide, and fluoride complexes, as well of native human erythrocyte catalase, are shown in Fig. 18.1, and molar absorptivities of peaks are listed in Table 18.1. The high-spin electron paramagnetic resonance (EPR) spectrum of the native enzyme and low-spin spectrum of its cyanide complex are shown in Fig. 18.2.

18.3. COMPOUND I FORMATION AND THE CATALATIC REACTION

Catalase, when premixed with certain buffers, is inhibited. In the absence of buffer effects the catalatic reaction is essentially pH-independent from pH

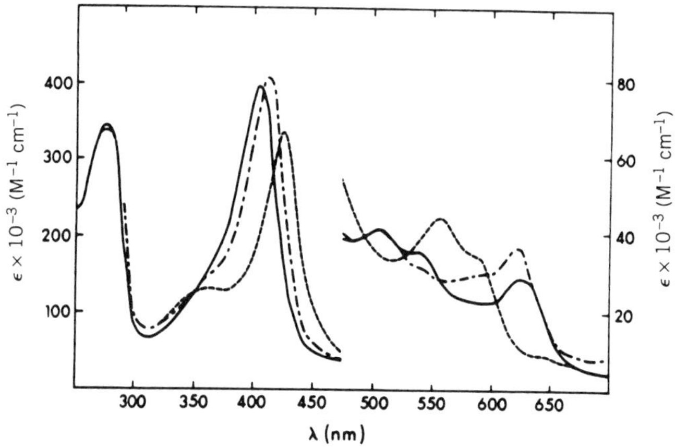

FIG. 18.1. Optical absorption spectra of human erythrocyte catalase (solid line), its cyanide (-----), and its azide complex (—-—) at pH 7.0 and 25°C. [Reproduced with permission from M. Palcic and H. B. Dunford, *Can. J. Biochem.* **57,** 321–329 (1979).]

4.7 to 10.5. The catalatic reaction is accompanied by some compound II formation.[12] Detection of pure compound I, formed from the reaction of hydrogen peroxide with catalase, is not possible. It has widely been assumed that the cause is the rapid decomposition of compound I in the second half of the catalatic reaction:

$$\text{Catalase} + H_2O_2 \xrightarrow{k_1} \text{Compound I} + H_2O \tag{18.1}$$

$$\text{Compound I} + H_2O_2 \xrightarrow{k_2} \text{Catalase} + O_2 + H_2O \tag{18.2}$$

(See, however, Section 18.4.) Reaction (18.1) can be subdivided into two parts: first formation of a catalase–hydrogen peroxide complex, a precursor of compound I, followed by completion of compound I formation. At the risk of oversimplification, the complex formation is shown as a process in which equilibrium is maintained:

$$\text{Catalase} + H_2O_2 \overset{K_{1a}}{\rightleftarrows} \text{Catalase}\bullet H_2O_2 \tag{18.1a}$$

$$\text{Catalase}\bullet H_2O_2 \xrightarrow{k_{1b}} \text{Compound I} + H_2O \tag{18.1b}$$

If a large concentration of hydrogen peroxide is used, then the temperature rises considerably, which was sometimes overlooked.[20] If a very small concentration of catalase is used, then dissociation of the enzyme into subunits becomes important.[21,22] If the reaction is followed for an extended period of

TABLE 18.1. Light Absorption and Circular Dichroic Spectra of Human Erthrocyte Catalase and Its Complexes

		Light Absorption		Circular Dichroism	
Ligand	pH	λ_{max}	ε	λ_{max}	$[\theta]_M \times 10^{-4}$
None	7.0[b]	405	397.0	294	16.1
		505	40.5	388	−21.7[c]
		540	35.2	420	10.0
		624	28.0	550	1.2
				640	−1.6[d]
Cyanide, 0.1 mM	7.0	360	119.0	294	19.2
		424	325.0	340	5.6
		555	44.9	390	−15.3
		580–590 (s)[e]	34.4	410	−16.1
				455	−8.4
				530	4.4
				590	3.8
				640	−1.2
Azide, 0.1 M	7.0	413	411.0	294	11.6
		505	40.5	330	−2.0
		540	31.2	380	−8.0
		580–590 (s)[e]	30.0	420	26.9
		620	36.1	530	1.2
				640	−1.6
Fluoride, 0.1 M	5.0	405	365.0	298	8.1
		485	47.4	385	−10.5
		530	39.1	416	10.0
		600	43.8	550	1.0

[a] Units: wavelength (λ_{max}) in nm, molar absorptivity (ε) in $mM^{-1}cm^{-1}$, molar ellipticity ($[\theta]_M$) in $deg \cdot cm^2 \cdot dmol^{-1}$.

[b] Optical spectrum the same at pH 5.0. All results obtained in phosphate buffer.

[c] Changes to −17.3 at pH 5.0.

[d] Disappears at pH 5.0.

[e] Shoulder.

(Reproduced with permission from M. M. Palcic and H. B. Dunford, *Can. J. Biochem.* **57,** 321–329 (1979).

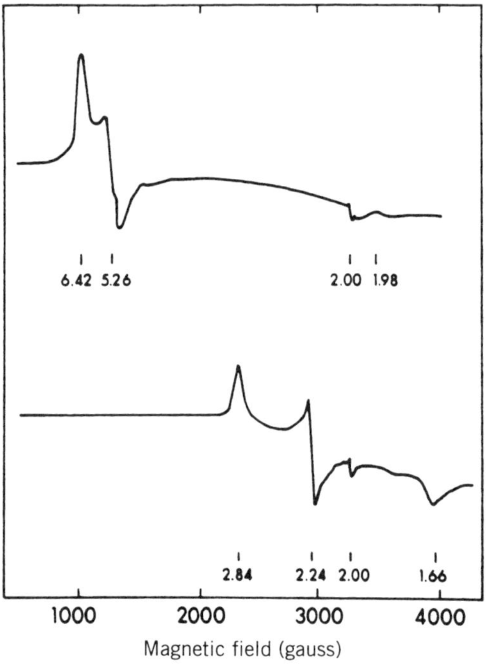

FIG. 18.2. First-derivative EPR spectra of human erythrocyte catalase (upper) and its cyanide complex (lower), with *g* values of peaks, at 77 K and pH 7.0. [Reproduced with permission from M. Palcic and H. B. Dunford, *Can. J. Biochem.* **57,** 321–329 (1979).]

time, then some compound II formation occurs.[23] Hydrogen peroxide generating systems have been used, such as notatin (glucose–glucose oxidase) in combination with catalase. In such systems determination of the steady-state concentration of hydrogen peroxide becomes difficult.[24] Provided all of these potential pitfalls are avoided, or are accurately corrected, and assuming that reactions (18.1a,b) and (18.2) do represent the reaction pathway, then determination of the equilibrium constant K_{1a} and the two rate constants k_{1a} and k_2 defines the rate of the catalatic reaction. If K_{1a} is expressed in units of M^{-1}, then the overall rate of compound I formation, k_0, in units of $M^{-1}s^{-1}$ is given by $K_{1a}k_{1b}$. Rate constants, for the formation of compound I from different catalases and oxidants,[25–28] are summarized in Table 18.2.

Chloroperoxidase, which has about 20% of the catalatic activity toward hydrogen peroxide as a catalase subunit, can also undergo a catalatic reaction with a variety of ROOH oxidants.[29] This is not the case for catalase, where only hydrogen peroxide is efficiently disproportionated. Thus, formation of relatively pure compound I is possible using reagents such as methyl hydroperoxide, ethyl hydroperoxide, and peroxoacetic (peracetic) acid. The optical

TABLE 18.2. Second-Order Rate Constants, k ($M^{-1}s^{-1}$), for the Formation of Catalase Compounds I Using Different Oxidants

Catalase	H_2O_2	CH_3OOH	C_2H_5OOH	$CH_3\overset{\overset{O}{\|}}{C}OOH$	References
Human erythrocyte		$(1.40 \pm 0.04) \times 10^6$	$(2.7 \pm 0.1) \times 10^4$	$(3.8 \pm 0.2) \times 10^4$	24, 25
	3×10^7	8×10^5	2×10^4		
Horse liver	3.0×10^7, 1.2×10^7				6, 7
Beef liver	$(5.6 \pm 0.5) \times 10^6$		2×10^4	$(1.44 \pm 0.03) \times 10^4$	15, 26, 28
Micrococcus lysodeikticus	6×10^7	2.2×10^5	2.8×10^3	$(5.72 \pm 0.08) \times 10^2$	15, 27

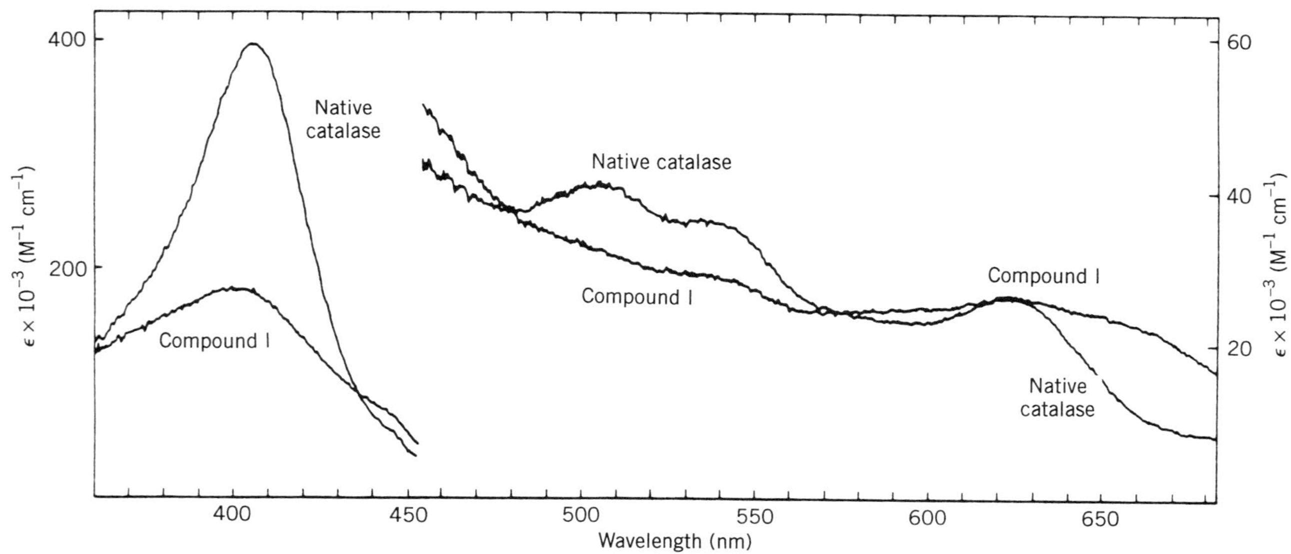

FIG. 18.3. Optical absorption spectra of human erythrocyte catalase and its compound I, formed using methyl hydroperoxide, at pH 7.1 and 25°C. [Reproduced with permission from M. Palcic and H. B. Dunford, *J. Biol. Chem.* **255,** 6128–6132 (1980).]

spectrum of human erythrocyte catalase compound I, generated using methyl hydroperoxide, is shown in Fig. 18.3.

18.4. MECHANISM OF THE CATALATIC REACTION

It is the unionized form of all ROOH oxidants that is utilized in compound I formation, and of hydrogen peroxide in the oxygen evolution reaction. In a prescient paper, a detailed mechanism for the catalatic reaction was proposed in 1968.[13] The mechanism is reproduced in Fig. 18.4, with the modification that the sixth coordination position of the native heme enzyme is not occupied by a water molecule. On the basis of reaction dependencies on pH, the distal groups BH^+ and A^- were postulated to be side chains of arginine and aspartate or glutamate residues. From the X-ray crystallographic results on bovine liver catalase (Section 18.8), these are now known to be side chains of Asn147 and His74. A feature of the mechanism is that the components of the hydrogen peroxide molecule remain bound to the enzyme until the catalatic reaction is complete. In today's terminology, compound I formation is not completed before the second molecule of hydrogen peroxide is reacted.

The mechanism accounts nicely for the lack of catalatic activity of substrates such as CH_3OOH, since the alkyl group does not possess the lability of a proton. An important feature is that it is in accord with retention of both ^{18}O atoms from doubly labeled hydrogen peroxide in the evolved dioxygen.[30] It is also in accord with the most negative enthalpies and entropies of activation known for heme protein reactions with hydrogen peroxide.[31] The mechanism might be described as side-by-side, since it has a transition state in which the oxygen atoms of two hydrogen peroxide molecules occupy two opposite sides of a rectangle.

An alternative mechanism, in which a linear array of three oxygen atoms exists in the transition state, is shown in Fig. 18.5.[32] In this mechanism, compound I formation is completed before reaction of the second molecule of hydrogen peroxide. The mechanism of compound I formation is believed to be similar to that in yeast and plant peroxidases, although the role of a distal arginine residue in the plant peroxidases is replaced by a distal asparagine.

The currently accepted structure for compound I is a ferryl ($Fe^{IV}{=}O$) porphyrin π-cation radical. Therefore, if compound I is formed from a 50–50 mixture of hydrogen peroxide doubly labeled with ^{18}O, and normal hydrogen peroxide with both oxygen atoms ^{16}O, then the oxygen bound to iron in compound I (α) will be a 50–50 mixture of ^{18}O and ^{16}O. Addition of hydrogen peroxide to form the trilinear complex means that both the β and γ oxygen atoms will consist of the same isotope. Therefore, for the evolved dioxygen in the catalatic reaction to be either $^{16}O^{16}O$ or $^{18}O^{18}O$ with no isotopic scrambling, the cleavage of the triatom complex leading to O_2 formation must occur

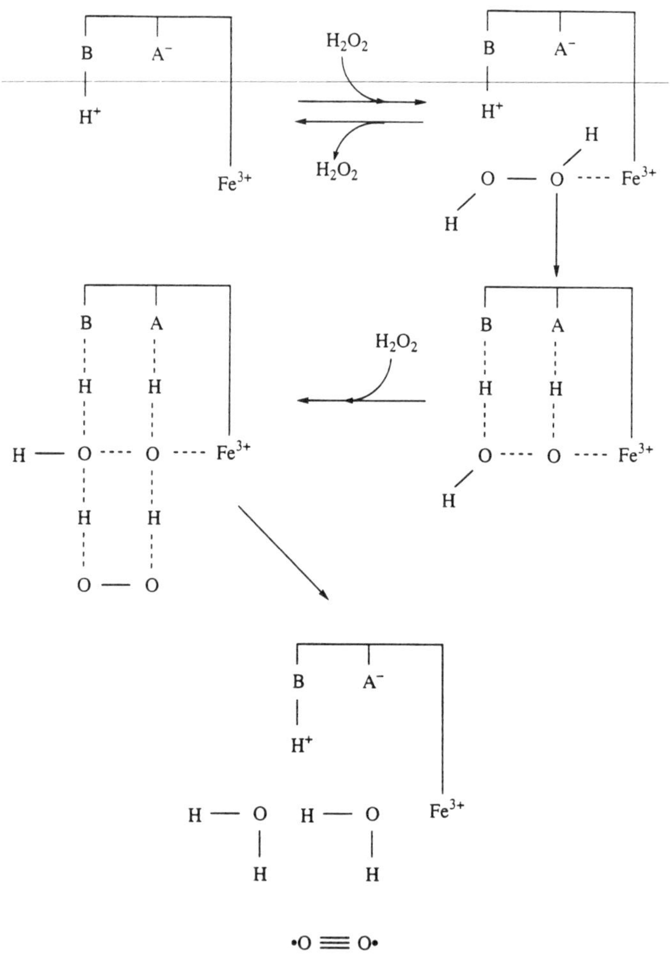

FIG. 18.4. Schematic representation of the catalatic reaction. [Based upon, with permission, P. Jones and A. Suggett, *Biochem. J.* **110,** 621–629 (1968).]

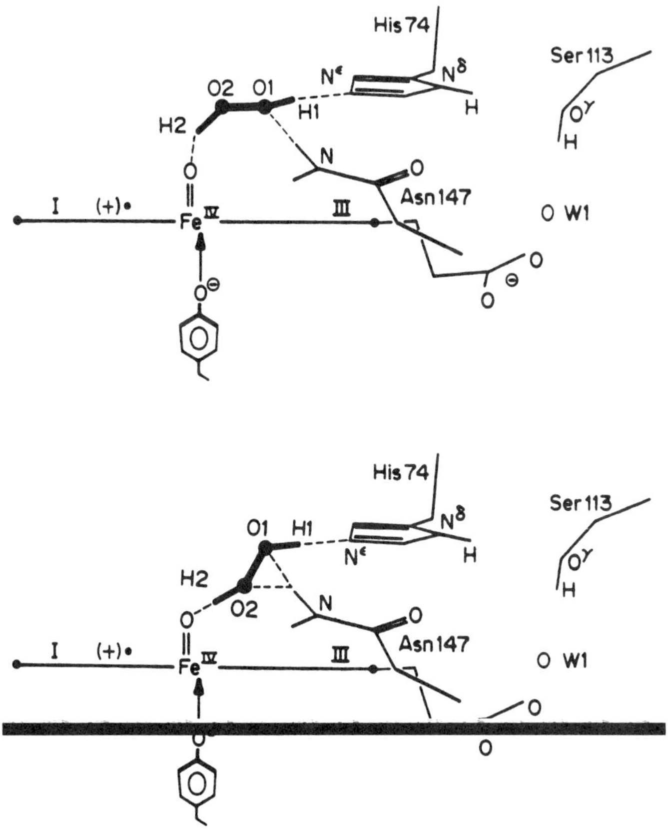

FIG. 18.5. Two representations of the possible linear complexes of hydrogen peroxide with compound I of catalase. [Reproduced with permission from I. Fita and M. G. Rossmann, *J. Mol. Biol.* **185,** 21–37 (1985) by permission of the publisher Academic Press Ltd., London.]

between the α and β oxygen atoms. In other words, both atoms in one dioxygen molecule are derived from a single hydrogen peroxide molecule.

$$\underset{\gamma}{\mathrm{O}} - \underset{\beta}{\mathrm{O}} \cdot\!\S\!\cdot \underset{\alpha}{\mathrm{O}} - \mathrm{Fe}$$

In the chloroperoxidase catalatic reaction with *m*-chloroperbenzoic acid, the double oxygen label is scrambled, since O_2 is formed from the α and β atoms.

Both the side-by-side and linear mechanisms account for the salient features of the catalatic reaction.[12] It is esthetically pleasing to contemplate that the most sophisticated heme enzyme may be capable of utilizing a compound I precursor, containing a total hydrogen peroxide molecule, to convert a second molecule of hydrogen peroxide to dioxygen. However, unless the active site is sufficiently flexible, there may not be enough room for the side-by-side reaction to occur. Existing mechanisms do not explain the necessity for the tetrameric structure for catalatic activity.

18.5. COMPOUND II REACTIONS OF CATALASE

Reactions of catalase compound II have been studied.[33] Both ferrocyanide and ascorbate can effectively compete with hydrogen peroxide for available compound I, reducing it to compound II.[9] Phenols react with catalase compound II but much more slowly than with peroxidase compounds II. Rate constants for catalase compounds II reacting with pyrogallol and *p*-cresol are 70 and 6 $M^{-1}s^{-1}$. Nitrite reacts very rapidly with compound II: $k = 1.3 \times 10^6$ $M^{-1}s^{-1}$. Salicylic acid acts as an electron donor to catalase and so inhibits its catalatic reaction. A signaling role of salicylic acid to activate defenses has been investigated.[34]

18.6. CATALASE REACTION WITH ETHANOL

Catalase can convert ethanol to acetaldehyde.[35] By adding both hydrogen peroxide and ethanol to beef liver catalase, competition for compound I was established.[36] The mechanism was assumed to be reactions (18.1) and (18.2) followed by

$$\text{Compound I} + CH_3CH_2OH \xrightarrow{k_3} \text{Catalase} + CH_3CH=O + H_2O \qquad (18.3)$$

The values found for the rate constants were: $k_1 = (5.6 \pm 0.5) \times 10^6\ M^{-1}s^{-1}$, $k_2 = (1.2 \pm 0.1) \times 10^7\ M^{-1}s^{-1}$, and $k_3 = (1.8 \pm 0.4) \times 10^2\ M^{-1}s^{-1}$. Therefore, a 10^5- to 10^6-fold excess of ethanol over hydrogen peroxide is required to observe competition for compound I. It would appear that catalase does not play an important role in the metabolism of ethanol.

If the sole substrate is ethyl hydroperoxide, a different situation arises.

$$\text{Catalase} + \text{CH}_3\text{CH}_2\text{OOH} \xrightarrow{k_4} \text{Compound I} + \text{CH}_3\text{CH}_2\text{OH} + \text{H}_2\text{O} \qquad (18.4)$$

Since the substrate product of compound I formation is ethanol, then it is already in place for reaction (18.3) to occur. Reaction (18.4) followed by reaction (18.3) has been modeled based on the X-ray structure using a computer graphics system.[37] The results are shown in Figs. 18.6 and 18.7.

18.7. PHYSIOLOGICAL ROLE OF CATALASE

As described in the previous section, ethanol metabolism by catalase is an inefficient process. This leaves the known catalatic reaction as the primary physiological function of catalase. Erythrocyte catalase and glutathione peroxidase are equally active in the detoxification of hydrogen peroxide in human erthrocytes.[38] Overexpression of catalase retards age-related oxidative damage in *Drosophilia melanogaster*[39] and suppresses doxorubicin cardiotoxicity in the hearts of transgenic mice.[40] In the presence of magnesium ion at alkaline pH, β-adrenergic agonists such as epinephrine appear to act as electron donors to catalase, and to interfere with O_2 production from H_2O_2.[41] Thus, a harmful effect on the catalatic reaction occurs, and the potential of a new role for catalase is suggested.

18.8. CRYSTAL STRUCTURES OF CATALASES

An introductory description of the four domains of a catalase subunit structure was given in Section 3.3.3. The complete amino acid sequence for beef liver catalase, listed in Appendix A, was fitted to the diffraction data of 2.5 Å resolution.[42] The beef liver enzyme contains ferriprotoporphyrin IX (heme *b*) as its prosthetic group. Starting from the N-terminal end, residues 1–75 form a domain consisting of two helices in an arm that is involved in intersubunit contacts. Residues 76–320 form the second domain, an eight-stranded anti-parallel β-barrel sheet, which contains most of the residues on the distal side of the heme, including the key catalytic residues His74 and Asn 147. This structure is strikingly different from the distal-side helical structure of the peroxidases. The third domain, consisting of residues 321–436 has no obvious similarities to previously observed structures. It is a "wrapping domain," forming an outer layer to each subunit. It also has a helix that contains the proximal tyrosinate ligand to the heme iron, Tyr357. The final domain, residues

A

B

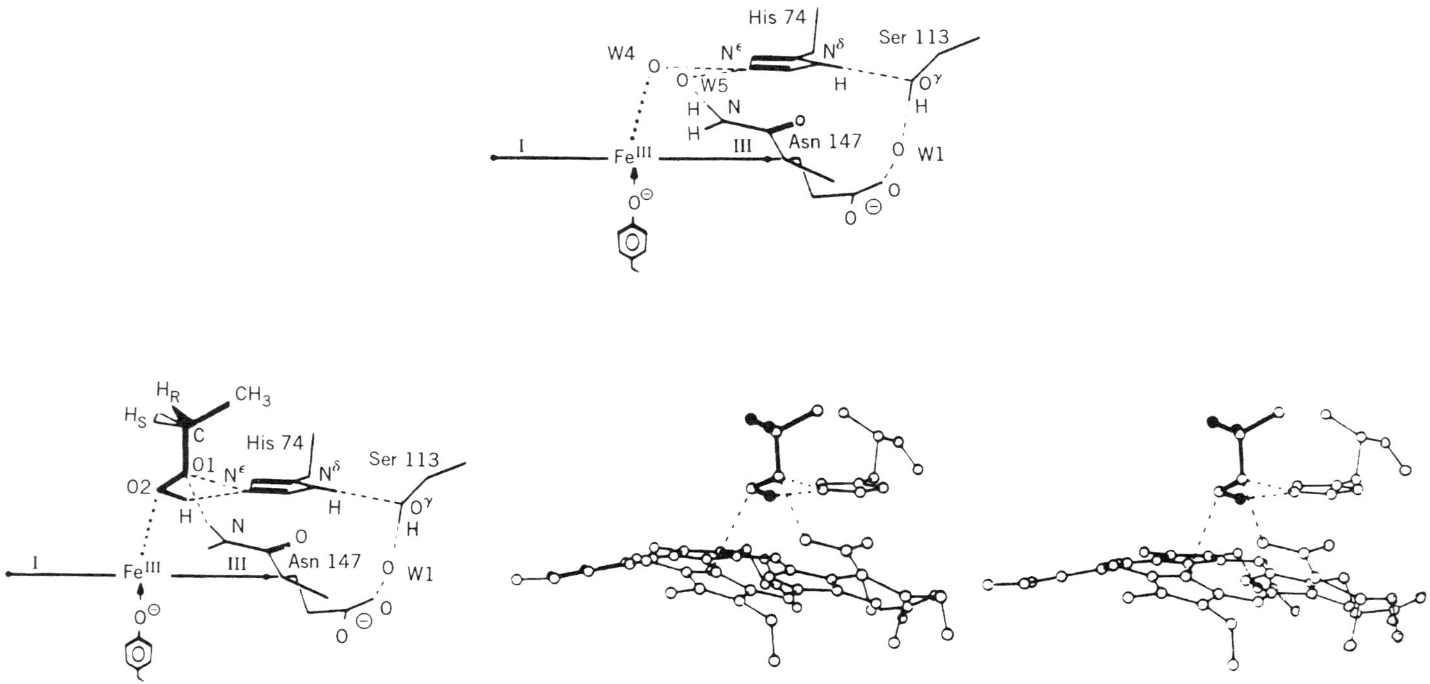

FIG. 18.6. Diagrammatic representation (left) and corresponding stereoscopic views (right) of successive steps in the formation of catalase compound I using ethyl hydroperoxide as oxidant. [Reproduced from I. Fita and M. G. Rossmann, *J. Mol. Biol.* **185,** 21–37 (1985) by permission of the publisher Academic Press Ltd., London.] Continued on next page.

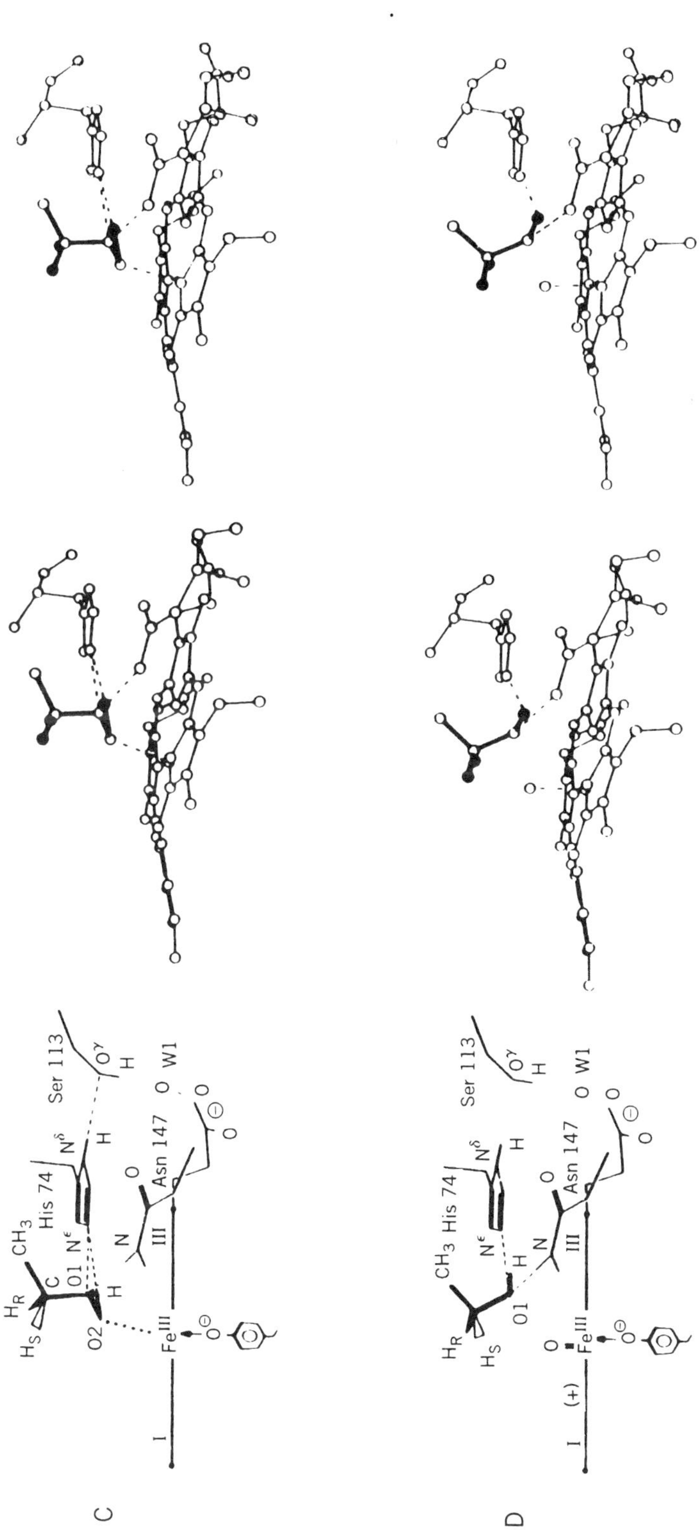

FIG. 18.6. *(Concluded)*.

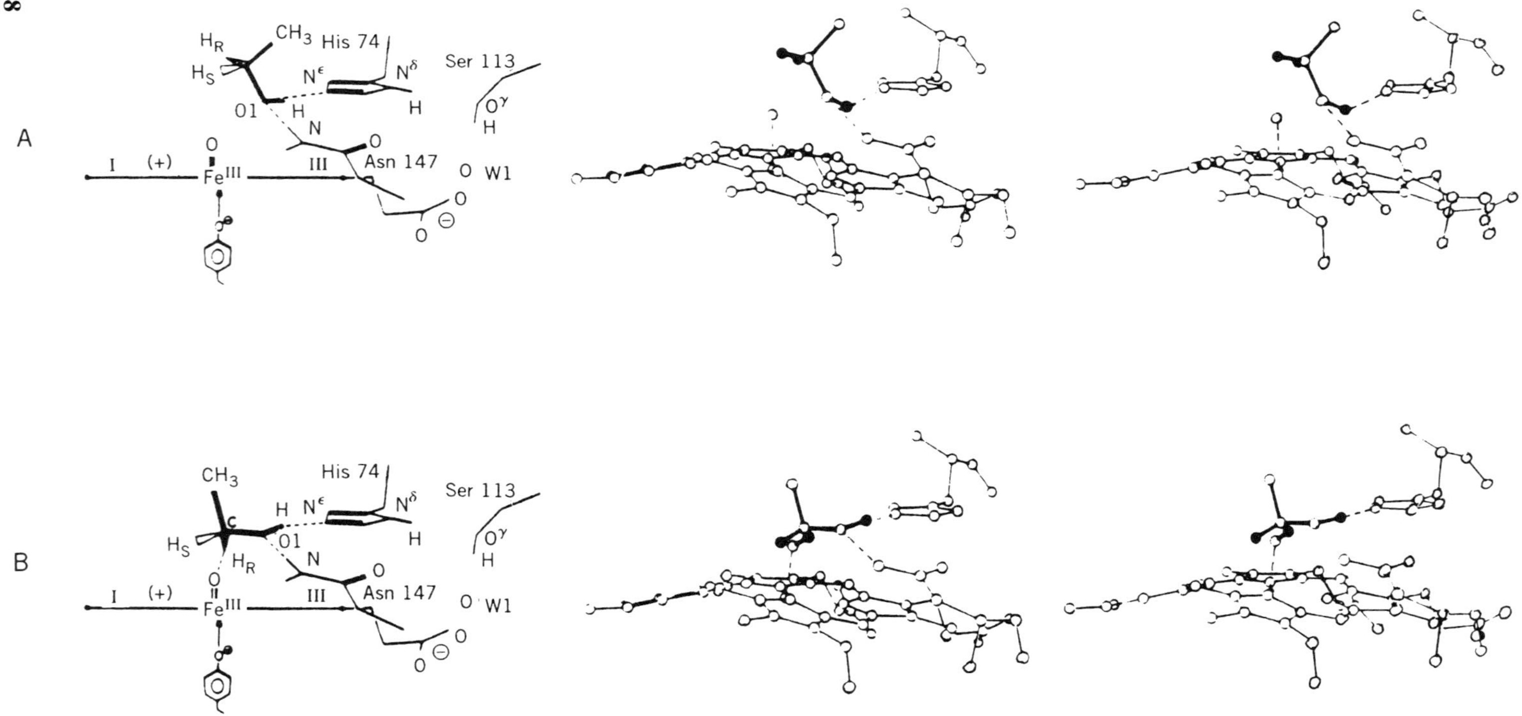

FIG. 18.7. Diagrammatic representation and corresponding stereoscopic views (right) of successive steps in the reduction of catalase compound I by ethanol. [Reproduced from I. Fita and M. G. Rossmann, *J. Mol. Biol.* **185,** 21–37 (1985) by permission of the publisher Academic Press Ltd., London.] Continued on next page.

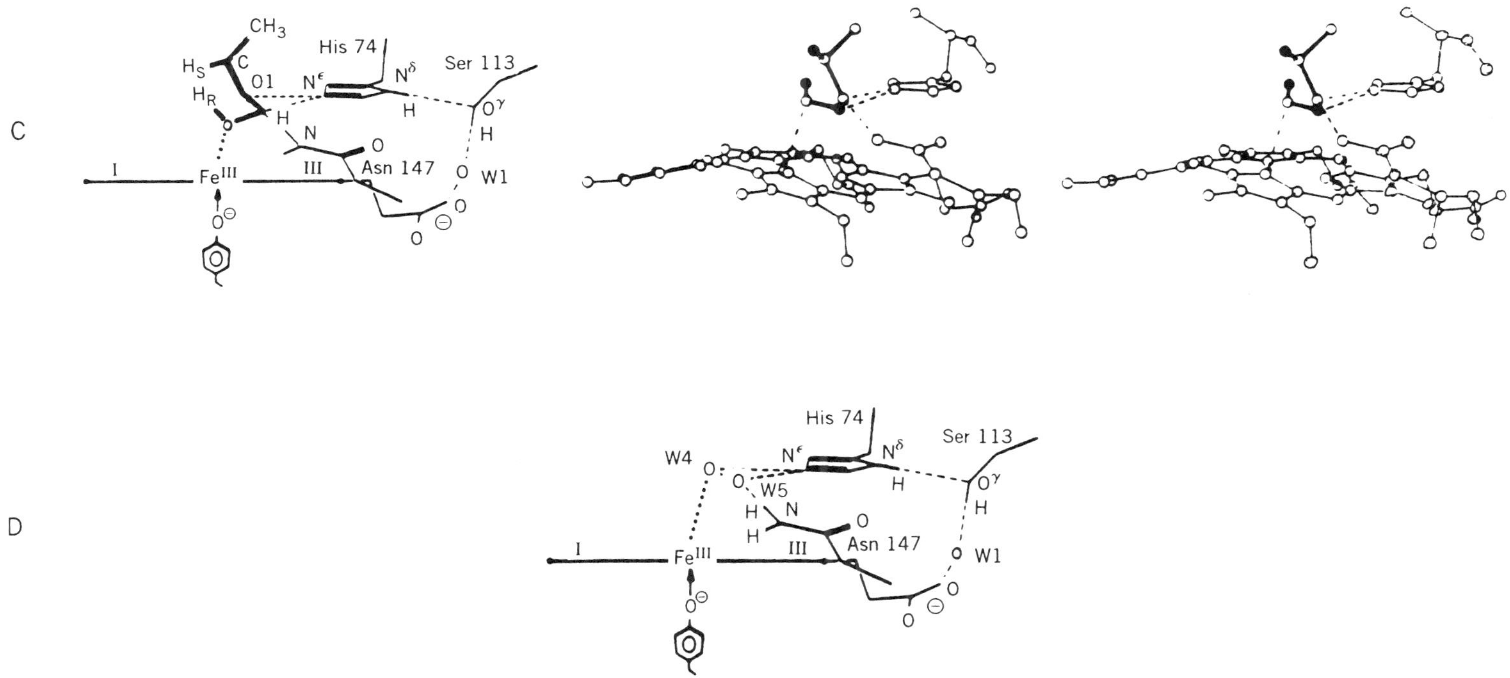

FIG. 18.7. (*Concluded*).

437–506, consists of four α-helices, which contibute to a funnel-shaped hydrophobic channel leading to the heme. The originally proposed structure was refined and corrected,[43] to accommodate the finding that nicotinamide adenine dinucleotide phosphate (NADP) binds tightly to catalase.[37] The NADP molecule is bound near the carboxyl ends of helices α5 and α10 (Fig. 18.8). Points of contact for the NADP molecule range from Phe197 to His304 of the distal side sheet structure of the second domain, and from Gln441 to Val449 of the fourth domain, forming part of the opening to a possible secondary channel leading to the heme. This secondary channel has a narrow opening but a fairly wide hydrophilic interior. The NADP and heme groups are within 13.7 Å of each other with no direct interaction. However, interaction mediated by a substrate molecule traveling along the secondary channel is possible, although there is no known function for the NADP molecule. A problem with beef liver catalase is that some of the hemes are degraded to bile pigments, so that

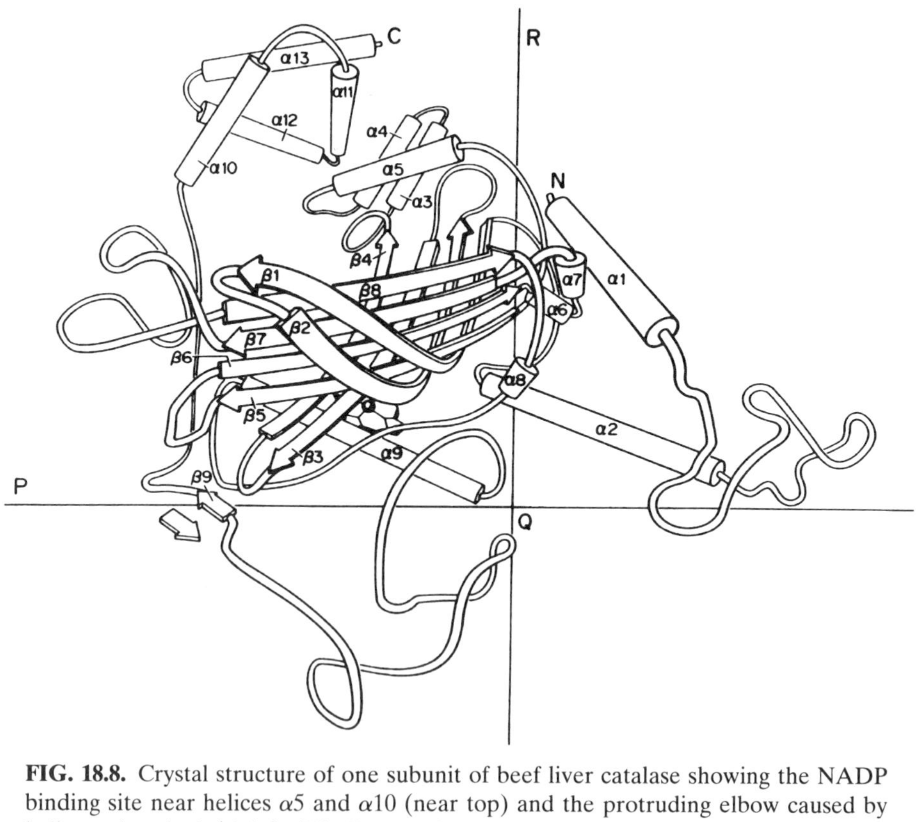

FIG. 18.8. Crystal structure of one subunit of beef liver catalase showing the NADP binding site near helices α5 and α10 (near top) and the protruding elbow caused by helices α1 and α2 (right). (Kindly supplied by Dr. Michael Rossmann, Purdue University).

the recorded prosthetic group structure is an average of intact and degraded heme.

The stucture of the *Penicillium vitale* catalase is similar to that of the beef liver catalase, with two important exceptions. There is an extended arm at the C-terminal end, a flavodoxin-like domain,[44] and the prosthetic group is a chlorin, designated as heme *d*.[45] Crystal structures have also been determined for tetrameric catalases from *Escherichia coli*,[45] *Micrococcus lysodeikticus*,[46] and *Proteus mirabilis*.[47] The *P. mirabilis* enzyme shares with beef liver catalase the ability to tightly bind one NADPH molecule per subunit. Electron paramagnetic resonance spectra of compound I of the latter two enzymes, formed using peroxoacetic acid, indicate that the porphyrin π-cation radical is present but can be converted to a proximal tyrosyl π-cation radical.[48] The possibility that the latter structure may be important in electron transfer from the NADPH molecule is discussed.

The first domain of a catalase subunit is like a bent elbow that sticks into a cavity in its neighbor (Fig. 18.8). This places restrictions on how the tetrameric structure of a viable catalase molecule can be formed. The tetramer is assembled from holosubunits with the following three essential steps:

Monomer + Heme $\rightarrow$ Holo monomer
2 Holo monomer $\rightarrow$ Holo dimer
2 Holo dimer $\rightarrow$ Holo tetramer

18.9. SUMMARY

Aside from valuable structural determinations on other catalases, there has not been much recent work on catalases. The reason is that, although there are still more questions than answers about the catalatic mechanism, there appears to be no obvious way to obtain further answers. Here is a quote from Peter Jones[49]: "A very basic problem is that activity measurements cannot be made at concentrations of enzyme anywhere near those necessary for e.g. stopped flow studies of intermediates. So although one can see compound I formed with micromolar enzyme under non-steady state conditions (using a peroxy acid rather than hydrogen peroxide), this does not mean that compound I is formed under turnover with nanomolar concentrations of enzyme. What handles are there on the enzyme under turnover conditions? Precious few other than activity measurements." Another valuable comment from the same source: "Our model heme studies show that heme monomer catalatic activity at pH 13 is closely similar to catalase at pH 7, but declines proportionately to H^+. I suggest that the catalase tetrameric structure is involved in controlling the pH effect on the intrinsic activity of the heme. The peculiarity of catalase vis a vis the peroxidases is in the reaction with the second molecule of hydrogen peroxide, and I suggest the tetrameric structure relates to this." Who will accomplish the next major breakthrough and when will it occur?

REFERENCES

1. A. Deisseroth and A. L. Dounce, *Physiol. Rev.* **50,** 319–375 (1970).
2. P. Nicholls and G. R. Schonbaum *The Enzymes,* 2nd ed., in P. D. Boyer, H. Lardy and K. Myrbäck, eds., Academic, New York, 1963, Vol. 8, Chapter 6, pp. 147–225.
3. R. K. Bonnichsen, *Acta Chem. Scand.* **1,** 114–117 (1947).
4. P. George, *Nature* **160,** 41–43 (1947).
5. B. Chance, *Acta Chem. Scand.* **1,** 236–267 (1947).
6. R. K. Bonnichsen, B. Chance, and H. Theorell, *Acta Chem. Scand.* **1,** 685–709 (1947).
7. P. George, *Biochem. J.* **44,** 197–205 (1949).
8. D. J. Aneshanley and T. Eisner, *Science* **165,** 61–63 (1969).
9. D. Keilin and P. Nicholls, *Biochim. Biophys. Acta* **29,** 302–307 (1958).
10. P. Nicholls, *Trans. Faraday Soc.* **60,** 137–140 (1964).
11. P. Jones and W. F. K. Wynne-Jones, *Trans. Faraday Soc.* **58,** 1148–1158 (1962).
12. P. Jones and A. Suggett, *Biochem. J.* **110,** 617–620 (1968).
13. P. Jones and A. Suggett, *Biochem. J.* **110,** 621–629 (1968).
14. P. Jones and P. G. Perkins, *Nature* **215,** 129–132 (1967).
15. P. Jones and D. N. Middlemiss, *Biochem. J.* **130,** 411–415 (1972).
16. P. Jones and H. B. Dunford, *J. Theor. Biol.* **69,** 457–470 (1977).
17. D. Keilin and E. F. Hartree, *Biochem. J.* **49,** 88–104 (1951).
18. A. S. Brill and R. J. P. Williams, *Biochem. J.* **78,** 246–253 (1961).
19. M. M. Palcic and H. B. Dunford, *Can. J. Biochem.* **57,** 321–329 (1979).
20. Y. Ogura, *Arch. Biochem. Biophys.* **57,** 288–300 (1955).
21. P. Jones and A. Suggett, *Biochem. J.* **108,** 833–838 (1968).
22. P. Jones, R. H. Pain, and A. Suggett, *Biochem. J.* **118,** 319–323 (1970).
23. P. Nicholls, *Trans. Faraday Soc.* **60,** 137–140 (1964).
24. B. Chance, *Biochem. J.* **46,** 387–402 (1950).
25. M. M. Palcic and H. B. Dunford, *J. Biol. Chem.* **255,** 6128–6132 (1980).
26. B. Chance, *J. Biol. Chem.* **180,** 947–959 (1949).
27. B. Chance and D. Herbert, *Biochem. J.* **46,** 402–414 (1950).
28. M. L. Kremer, *Biochim. Biophys. Acta* **198,** 199–209 (1970).
29. W. D. Hewson and L. P. Hager in *The Porphyrins,* Vol. VII, Part B, D. Dolphin, ed., Academic, New York, 1979, pp. 295–332.
30. R. C. Jarnagin and J. H. Wang, *J. Am. Chem. Soc.* **80,** 786–787 (1958).
31. D. Job, P. Jones, and H. B. Dunford, *J. Phys. Chem.* **97,** 9259–9262 (1993).
32. I. Fita and M. G. Rossmann, *J. Mol. Biol.* **185,** 21–37 (1985).
33. P. George, *Biochem. J.* **52,** xix (1952).

34. J. Durner and D. F. Klessig, *J. Biol. Chem.* **271,** 28492–28501 (1996).
35. D. Keilin and E. F. Hartree, *Biochem. J.* **39,** 293–301 (1945).
36. M. L. Kremer, *Biochim. Biophys. Acta* **198,** 199–209 (1970).
37. I. Fita and M. G. Rossmann, *Proc. Natl. Acad. Sci. U.S.A.* **82,** 1604–1608 (1985).
38. G. F. Gaetani, S. Galiano, L. Canepa, A. M. Ferraris, and H. N. Kirkman, *Blood* **73,** 334–339 (1989).
39. R. S. Sohal, A. Agarwal, S. Agarwal, and W. C. Orr, *J. Biol. Chem.* **270,** 15671–15674 (1995).
40. Y. J. Kang, Y. Chen, and P. N. Epstein, *J. Biol. Chem.* **271,** 12610–12616 (1996).
41. M. E. Percy, *Can. J. Biochem. Cell Biol.* **62,** 1006–1014 (1984).
42. M. R. N. Murthy, T. J. Reid III, A. Sicignano, N. Tanaka, and M. G. Rossmann, *J. Mol. Biol.* **152,** 465–499 (1981).
43. H. N. Kirkman and G. F. Gaetani, *Proc. Natl. Acad. Sci. U.S.A.* **81,** 4343–4348 (1984).
44. B. K. Vainshtein, W. R. Melik-Adamyan, V. V. Barynin, A. A. Vagin, A. I. Grebenko, V. V. Borisov, K. S. Bartels, I. Fita, and M. G. Rossmann, *J. Mol. Biol.* **188,** 49–61 (1986).
45. G. N. Murshudov, A. I. Grebenko, V. Barynin, Z. Dauter, K. S. Wilson, B. K. Vainshtein, W. Melik-Adamyan, J. Bravo, J. M. Ferrán, J. C. Ferrer, J. Switala, P. C. Loewen, and I. Fita, *J. Biol. Chem.* **271,** 8863–8868 (1996).
46. G. N. Murshudov, W. R. Melik-Adamyan, A. I. Grebenko, V. V. Barynin, A. A. Vagin, B. K. Vainshtein, Z. Dauter, and K. S. Wislon, *FEBS Lett.* **312,** 127–131 (1992).
47. P. Gouet, H.-M. Jouve, and O. Dideberg, *J. Mol. Biol.* **249,** 933–954 (1995).
48. A. Ivancich, H. M. Jouve, B. Sartor, and J. Gaillard, *Biochemistry* **36,** 9356–9364 (1997).
49. Personal communication from P. Jones, Department of Chemistry, University of Newcastle Upon Tyne, United Kingdom, March 2, 1998.

APPENDIX

PEROXIDASE AMINO ACID SEQUENCES

Conversions from Three-Letter to One-Letter Code

ALIPHATIC			S-CONTAINING			DICARBOXYLIC		
Alanine	Ala	A	Cysteine	Cys	C	Aspartic acid	Asp	D
Glycine	Gly	G	Methionine	Met	M	Asparagine	Asn	N
Isoleucine	Ile	I				Asp or Asn	Asx	B
Leucine	Leu	L	IMINO			Glutamic acid	Glu	E
Valine	Val	V	Proline	Pro	P	Glutamine	Gln	Q
						Glu or Gln	Glx	Z

HYDROXY			BASIC			AROMATIC		
Serine	Ser	S	Arginine	Arg	R	Phenylalanine	Phe	F
Threonine	Thr	T	Histidine	His	H	Tryptophan	Trp	W
			Lysine	Lys	K	Tyrosine	Tyr	Y

List of Peroxidase Amino Acid Sequences in This File

Plant Class I

Yeast cytochrome *c* peroxidase
Pea cytosolic peroxidase
E. coli catalase-peroxidase

Plant Class II

Lignin peroxidase-2
Lignin peroxidase-8
Manganese peroxidase-1
Coprinus cinereus peroxidase

Plant Class III

Horseradish peroxidase C
Horseradish peroxidase E5
Turnip peroxidase-7
Tobacco peroxidase-A
Barley peroxidase-1
Wheat peroxidase
Tomato peroxidase
Peanut peroxidases 1 and 2

Animal

Human myeloperoxidase
Ram prostaglandin H synthase
Human thyroid peroxidase
Porcine thyroid peroxidase
Bovine lactoperoxidase
Human eosinophil peroxidase

Catalases	*Other Peroxidases*
Bovine liver catalase	*Caldariomyces fumago* peroxidase
	Pseudomonas aeruginosa peroxidase

PLANT PEROXIDASE SUPERFAMILY

Class I. Prokaryotic Origin

Yeast Cytochrome c Peroxidase

K. Takio, K. Titani, L. H. Ericsson, and T. Yonetani, *Arch. Biochem. Biophys.* **203,** 615–629 (1980).

H. F. Bosshard, H. Anni, and T. Yonetani in *Peroxidases in Chemistry and Biology,* J. Everse, K. E. Everse and M. B. Grisham, eds., CRC Press, Boca Raton, 1991, Vol. II, pp. 51–84.

1 10	20	30
TTPLVHVASV	EKGRSYEDFQ	KVYNAIALKL
40	50	60
REDDEYDNYI	GYGPVLVRLA	WHISGTWDKH
70	80	90
DNTGGSYGGT	YRFKKEFNDP	SNAGLQNGFK
100	110	120
FLEPIHKEFP	WISSGDLFSL	GGVTAVQEMQ
130	140	150
GPKIPWRCGR	VDTPEDTTPD	NGRLPDADKD
160	170	180
AGYVRTFFQR	LNMNDREVVA	LMGAHALGKT
190	200	210
HLKNSGYEGP	WGAANNVFTN	EFYLNLLNED
220	230	240
WKLEKNDANN	EQWDSKSGYM	MLPTDYSLIQ
250	260	270
DPKYLSIVKE	YANDQDKFFK	DFSKAFEKLL
280	290	294
ENGITFPKDA	PSPFIFKTLE	EQGL

Pea Cytosolic Ascorbate Peroxidase

R. Mittler and B. A. Zilinskas, *J. Biol. Chem.* **267,** 21802–21807 (1992).

1 10	20	30
MGKSYPTVSP	DYQKAIEKAK	RKLRGFIAEK
40	50	60
KCAPLILRLA	WHSAGTFDSK	TKTGGPFGTI

70	80	90
KHQAELAHGA	NNGLDIAVRL	LEPIKEQFPI
100	110	120
VSYADFYQLA	GVVAVEITGG	PEVPFHPGRE
130	140	150
DKPEPPPEGR	LPDATKGSDH	LRDVFGKAMG
160	170	180
LSDQDDIVALS	GGHTIGAAHK	ERSGFEGPWT
190	200	210
SNPLIFDNSY	FTELLTGEKD	GLLQLPSDKA
220	230	240
LLTDSVFRPL	VEKYAADEDV	FFADYAEAHL
250		
KLSELGFAEA		

Escherichia coli Catalase-Peroxidase

B. L. Triggs-Raine, B. W. Doble, M. R. Mulvey, P. A. Sorby, and P. C. Loewen, *J. Bacteriol.* **170,** 4415–4419 (1988).

1 10	20	30
MSTSDDIHNT	TATGKCPFHQ	GGHDQSAGAG
40	50	60
TTTRDWWPNQ	LRVDLLNQHS	NRSNPLGEDF
70	80	90
DYRKEFSKLD	YYGLKKDLKA	LLTESQPWWP
100	110	120
ADWGSYAGLF	IRMAWHGAGT	YRSIDGRGGA
130	140	150
GRGQQRFAPL	NSWPDNVSLD	KARRLLWPIK
160	170	180
QKYGQKISWA	DLFILAGNVA	LENSGFRTFG
190	200	210
FGAGREDVWE	PDLDVNWGDE	KAWLTHRHPE
220	230	240
ALAKAPLGAT	EMGLIYVNPE	GPDHSGEPLS
250	260	270
AAAAIRATFG	NMGMNDEETV	ALIAGGHTLG
280	290	300
KTHGAGPTSN	VGPDPEAAPI	EEQGLGWAST
310	320	330
YGSGVGADAI	TSGLEVVWTQ	TPTQWSNYFF
340	350	360
ENLFKYEWVQ	TRSPAGAIQF	EAVDAPEIIP
370	380	390
DPFDPSKKRK	PTMLVTDLTL	RFDPEFEKIS

400	410	420
RRFLNDPQAF	NEAFARAWFK	LTHRDMGPKS
430	440	450
RYIGPEVPKE	DLIWQDPLPQ	PIYNPTEQDI
460	470	480
IDLKFAIADS	GLSVSELVSV	AWASASTFRG
490	500	510
GDKRGGANGA	RLALMPQRDW	DVNAAAVRAL
520	530	540
PVLEKIQKES	GKASLADIIV	LAGVVGVEKA
550	560	570
ASAAGLSIHV	PFAPGRVDAR	QDQTDIEMFE
580	590	600
LLEPIADGFR	NYRARLDVST	TESLLIDKAQ
610	620	630
QLTLTAPEMT	ALVGGMRVLG	GNFDGSKNGV
640	650	660
FTDRVGVLSN	DFFVNLLDMR	YEWKATDESK
670	680	690
ELFEGRDRET	GEVKFTASRA	DLVFGSNSVL
700	710	720
RAVAEVYASS	DAHEKFVKDF	VAAWVKVMNL
726		
DRFDLL		

Class II. Extracellular Fungal Peroxidases

Lignin Peroxidase 2

T. G. Ritch, Jr., V. J. Nipper, L. Akileswaran, A. J. Smith, D. G. Pribnow, and M. H. Gold, *Gene* **107,** 119–126 (1991).

1 10	20	30
ATCANGLTVG	DASCCAWFDV	LDDIQANMFH
40	50	60
GGQCGAEAHE	SIRLVFHDSI	AISPAMEAKG
70	80	90
KFGGGGADGS	IMIFDTIETA	FHPNIGLDEV
100	110	120
VAMGLPFVQK	HGVTPGDFIA	FAGAVALSNC
130	140	150
PGAPQMNFFT	GRKPATQPAP	DGLVPEPFHT
160	170	180
VDQIIARVND	AGEFDELELV	WMLSAHSVAA
190	200	210
VNDVDPTVQG	LPFDSTPGIF	DSQFFVETQF

220	230	240
RGTLFPGSGG	NQGEVESGMA	GEIRIQTDHT
250	260	270
LARDSRTACE	WQSFVGNQSK	LVDDFQFIFL
280	290	300
ALTQLGQDPN	AMTDCSDVIP	LSKPIPGNGP
310	320	330
FSFFPPGKSH	SDIEQACAET	PFPSLVTLPG
340	343	
PATSVARIPP	HKA	

Lignin Peroxidase 8

M. Tien and C.-P. D. Tu, *Nature* **326,** 520–523 (1987).

1 10	20	30
ATCSNGKTVG	DASSCAWFDV	LDDIQQNLFH
40	50	60
GGQCGAEAHE	SIRLVFHDSI	AISPAMEAQG
70	80	90
KFGGGGADGS	IMIFDDIETA	FHPNIGLDEI
100	110	120
VKLQKPFVQK	HGVTPGDFIA	FAGRVALSNC
130	140	150
PGAPQMNFFT	GRAPATQPAP	DGLVPEPFHT
160	170	180
VDQIINRVND	AGEFDELELV	WMLSAHSVAA
190	200	210
VNDVDPTVQG	LPFDSTPGIF	DSQFFVETQL
220	230	240
RGTAFPGSGG	NQGEVESPLP	GEIRIQSDHT
250	260	270
IARDSRTACE	WQSFVNNQSK	LVDDFQFIFL
280	290	300
ALTQLGQDPN	AMTDCSDVIP	QSKPIPGNLP
310	320	330
FSFFPAGKTI	KDVEQACAES	PFRLSPLSRA
340	345	
PRRPSSASLR	LRVLK	

Manganese Peroxidase 1

B. J. Godfrey, M. B. Mayfield, J. A. Brown, and M. H. Gold, *Gene* **93,** 119–124 (1990)

1 10	20	30
AVCPDGTRVS	HAACCAFIPL	AQDLQETIFQ
40	50	60
NECGEDAHEV	IRLTFHDAIA	ISRSQGPKAG

70	80	90
GGADGSMLLF	PTVEPNFSAN	NGIDDSVNNL
100	110	120
IPFMQKHNTI	SAADLVQFAG	AVALSNCPGA
130	140	150
PRLEFLAGRP	NKTIAAVDGL	IPEPQDSVTK
160	170	180
ILQRFEDAGG	FTPFEVVSLL	ASHSVARADK
190	200	210
VDQTIDAAPF	DSTPFTFDTQ	VFLEVLLKGV
220	230	240
GFPGSANNTG	EVASPLPLGS	GSDTGEMRLQ
250	260	270
SDFALAHDPR	TACIWQGFVN	EQAFMAASFR
280	290	300
AAMSKLAVLG	HNRNSLIDCS	DVVPVPKPAT
310	316	
GQPAMFPAST	GPQDLE	

Coprinus cinereus Peroxidase

L. Baunsgaard, H. Dalbøge, G. Houen, E. M. Rasmussen, and K. G. Welinder, *Eur. J. Biochem.* **213,** 605–611 (1993).

H. Sawai-Hatanaka, T. Ashikari, Y. Tanaka, Y. Asada, T. Nakayama, H. Minakata, N. Kunishima, K. Fukuyama, H. Yamada, Y. Shibano, and T. Amachi, *Biosci. Biotech. Biochem.* **59,** 1221–1228 (1995).

1 10	20	30
QGPGGGGSVT	CPGGQSTSNS	QCCVWFDVLD
40	50	60
DLQTQFYQGS	KCESPVRLIL	RIVFHDAIGF
70	80	90
SPALTAAGQF	GGGGADGSII	AHSNIDLAFP
100	110	120
ANGGLTDTID	ALRAVGINHG	VSFGDLIQFA
130	140	150
TAVGMSNCPG	SPRLDFLTGR	SNSSQPSPPS
160	170	180
LIPGPGNTVT	AILDRMGDAG	FSPDEVVDLL
190	200	210
AAHSLASQEG	LNSAIFRSPL	DSTPQVFDTQ
220	230	240
FYIETLLKGT	TQPGPSLGFA	EELSPFPGEF
250	260	270
RMRSDALLAR	DSRTACRWQS	MTSSNEVMGQ
280	290	300
RYRAAMAKMS	VLGFDRNALT	DCSDVIPSAV

310	320	330
SNNAAPVIPG	GLTVDDIEVS	CPSEPFPEIA
340	343	
TASGPLPSLA	PAP	

Class III. Classical Plant Secretory Peroxidases

Horseradish Peroxidase C

K. G. Welinder, *FEBS Lett.* **72,** 19–23 (1976).

1 10	20	30
ELTPTFYDNS	CPNVS**N**IVRD	TIVNELRSDP
40	50	60
RIAASILRLH	FHDCFVNGCD	ASILLD**N**TTS
70	80	90
FRTEKDAFGN	ANSARGFPVI	DRMKAAVESA
100	110	120
CPRTVSCADL	LTIAAQQSVT	LAGGPSWRVP
130	140	150
LGRRDSLQAF	LDLANANLPA	PFFTLPQLKD
160	170	180
SFRNVGL**N**RS	SDLVALSGGH	TFGKNQCRFI
190	200	210
MDRLY**N**FSNT	GLPDPTL**N**TT	YLQTLRGLCP
220	230	240
LNG**N**LSALVD	FDLRTPTIFD	NKYYVNLEEQ
250	260	270
KGLIQSDQEL	FSSP**N**ATDTI	PLVRSFA**N**ST
280	290	300
QTFFNAFVEA	MDRMGNITPL	TGTQGQIRLN
308		
CRVVNSNS		

Horseradish Peroxidase E5

Y. Morita, B. Mikami, H. Yamashita, J. Y. Lee, S. Aibara, M. Sato, Y. Katsube, and N. Tanaka in *Biochemical, Molecular, and Physiological Aspects of Plant Peroxidases,* J. Lobarzewski, H. Greppin, C. Penel, and Th. Gaspar, eds., University of Geneva Press, Geneva, 1991, pp. 81–88.

1 10	20	30
ZLRPDFYSRT	CPSVFNIIKN	VIVDELQTDP
40	50	60
RIAASILRLH	FHDCFVRGCD	ASILLDTSKS
70	80	90
FRTEKDAAPN	VNSARGFNVI	DRMKTALERA

100	110	120
CPRTVSCADI	LTIASQISVL	LSGGPSWAVP
130	140	150
LGRRDSVEAF	FDLANTALPS	PFFTLAQLKK
160	170	180
AFADVGLNRP	SDLVALSGGH	TFGRARCLFV
190	200	210
TARLYNFNGT	NRPDPTLNPS	YLADLRRLCP
220	230	240
RNGNGTVLVN	FDVMTPNTFD	NQFYTNLRNG
250	260	270
KGLIQSDQEL	FSTPGADTIP	LVNLYSSNTL
280	290	300
SFFGAFADAM	IRMGNLRPLT	GTQGEIRQNC
306		
RVVNSR		

Turnip Peroxidase 7

K. G. Welinder and L. Norskov-Lauritsen in *Molecular and Physiological Aspects of Plant Peroxidases,* H. Greppin, C. Penel, and Th. Gaspar, eds., University of Geneva Press, Geneva, 1986, pp. 61–70.

K. G. Welinder, *Eur. J. Biochem.* **151,** 497–504 (1985).

1 10	20	30
ZLTTNFYSTS	CPNLLSTVKS	GVKSAVSSQP
40	50	60
RMGASILRLF	FHDCFVNGCD	GSILLDDTSS
70	80	90
FTGEQNAGPN	RNSARGFTVI	NDIKSAVEKA
100	110	120
CPGVVSCADI	LAIAARDSVV	QLGGPNWVK
130	140	150
VGRRDAKTAS	QAAANSNIPA	PSMSLSQLIS
160	170	180
SFSAVGLSTR	DMVALSGAHT	IGQSRCVNFR
190	200	210
ARVYNETNIN	AAFATLRQRS	CPRAAGSGDA
220	230	240
NLAPLDINSA	TSFDNSYFKN	LMAQRGLLHS
250	260	270
DQVLFNGGST	DSIVRGYSNS	PSSFNSDFAA
280	290	296
AMIKMGDISP	LTGSSGEIRK	VCGKTN

Tobacco Peroxidase

L. M. Lagrimini, W. Burkhart, M. Moyer, and S. Rothstein, *Proc. Natl. Acad. Sci. U.S.A.* **84,** 7542–7546 (1987).

1 10	20	30
QLSATFYDTT	CPNVTSIVRG	VMDQRQRTDA
40	50	60
RAGAKIIRLH	FHDCFVNGCD	GSILLDTDGT
70	80	90
QTEKDAPANV	GAGGFDIVDD	IKTALENVCP
100	110	120
GVVSCADILA	LASEIGVVLA	KGPSWQVLFG
130	140	150
RKDSLTANRS	GANSDIPSPF	ETLAVMIPQF
160	170	180
TNKGMDLTDL	VALSGAHTFG	RARCGTFEQR
190	200	210
LFNFNGSGNP	DLTVDATFLQ	TLQGICFQGG
220	230	240
NNGATFTNLD	ISTPNDFDND	YFTNLQSNQG
250	260	270
LLQTDQELFS	TSGSATIAIV	NRYAGSQTQF
280	290	300
FDDFVSSMIK	LGNISPLTGT	NGQIRTDCKR
302		
VN		

Barley Peroxidase 1

A. Johansson, S. K. Rasmussen, J. E. Hartill, and K. G. Welinder, *Plant Mol. Biol.* **18,** 1151–1161 (1992).

1 10	20	30
AEPPVAPGLS	FDFYWQTCPR	AESIVREFVQ
40	50	60
EAVRKDIGLA	AGLLRLHFHD	CFVQGSDASV
70	80	90
LLDGSATGPG	EQQAPPNLTL	RPSAFKAVND
100	110	120
IRDRLERECR	GAVVSCSDIL	ALAARDSVVV
130	140	150
SGGPDYRVPL	GRRDSRSFAS	TQDVLSDLPG
160	170	180
PSSNVQSLLA	LLGRLGLDAT	DLVTISGGHT
190	200	210
IGLAHCSSFE	DRLFPRPDPT	ISPTFLSRLK

220	230	240
RTCPAKGTDR	RTVLDVRTPN	VFDNKYYIDL
250	260	270
VNREGLFVSD	QDLFTNAITR	PIVERFAQSQ
280	290	300
QDFFEQFGVS	IGKMGQMRVR	TSDQGEVRRN
309		
CSVRNPGPG		

Wheat Peroxidase 1

G. Rebmann, C. Hertig, J. Bull, F. Mauch, and R. Dudler, *Plant Mol. Biol.* **16,** 329–331 (1991).

1 10	20	30
ZLSPTFYDTS	CPRALAIIKS	GVMAAVSSDP
40	50	60
RMGASLLRLH	FHDCFVQGCD	ASVLLSGMEQ
70	80	90
NAIPNAGSLR	GFGVIDSIKT	QIEAICHQTV
100	110	120
SCADILTVAA	RDSVVALGGP	SWTVPLGRRD
130	140	150
STDANEAAAN	SDLPGFTSSR	SDLELAFRNK
160	170	180
GLLTIDMVAL	SGAHTIGQAQ	CGTFKDRIYN
190	200	210
ETNIDTAFAT	SLRANCPRSN	GDGSLANLDT
220	230	240
TTANTFDNAY	YTNLMSQKGL	LHSDQVLFNN
250	260	270
DTTDNTVRNF	ANPSAAFSSA	FTTAMIKMGN
280	290	
IAPKTGTQGQ	IRLSCSRVNS	

Tomato Peroxidase 1

E. Roberts and P. E. Kolattukudy, *Mol. Gen. Genetics* **217,** 223–232 (1989).

1 10	20	30
ZLTPEACVFS	AVRAVVDSAI	DAETRMGASL
40	50	60
IRLHFHDCFV	DGCDGGILLD	DINGTFTGEQ
70	80	90
NSPPNANSAR	GYEVIAQAKQ	SVINTCPNVS
100	110	120
VSCADILAIA	ARDSVAKLGG	QTYSVALGRS

130 140 150
DARTANFSGA INQLPAPFDN LTVQIQKFSD
160 170 180
KNFTLREMVA LAGAHTVGFA RCSTVCTSGN
190 200 210
VNPAAQLQCN CSATLTDSDL QQLDTTPTMF
220 230 240
DKVYYDNLNS NQGIMFSDQV LTGDATTAGF
250 260 270
VTDYSNDVNV FLGDFAAAMI KMGDLPPSAG
280 290
AQLEIRDVCS RVNPTSVASM

Peanut Peroxidases

D. Buffard, C. Breda, R. B. van Huystee, O. Asemota, M. Pierre, D. B. D. Ha, and R. Esnault, *Proc. Natl. Acad. Sci. U.S.A.* **87,** 8874–8878 (1990). The numbering of residues in this publication is wrong for peanut peroxidase 1, beginning with residue 160, and has been corrected below.

Cationic Peanut Peroxidase 1

D. J. Schuller, N. Ban, R. B. van Huystee, A. McPherson, and T. L. Poulos, *Structure* **4,** 311–321 (1996).

This is the dominant cationic peanut peroxidase. Residue 25 has been changed from Cys to Ala in accordance with the X-ray results. Residue 1 was modeled as a pyrrolidone carboxylic acid or as pyroglutamate in the crystal structure.

1 10 20 30
QLSSNFYATK CPNALSTIKS AVNSAVAKEA
40 50 60
RMGASLLNLH FHDCFVQGCD ASVLLDDTSN
70 80 90
FTGEKTAGPN ANSIRGFEVI DTIKSQVESL
100 110 120
CPGVVSCADI LAVAARDSVV ALGGASWNVL
130 140 150
LGRRDSTTAS LSSANSDLPA PFFNLSGLIS
160 170 180
AFSNKGFTTK ELVTLSGAHT IGQAQCTAFR
190 200 210
TRIYNDSNID PTYAKSLQAN CPSVGGDTNL
220 230 240
SPFDVTTPNK FDNAYYINLR NKKGLLHSDQ
250 260 270
QLFNGVSTDS GVTAYSNNAA TFNTDFGNAM
280 290 294
IKMGNLSPLT GTSGQIRTNC RKTN

Cationic Peanut Peroxidase 2

1 10	20	30
ITTVHGQGTR	VGFYSRTCPR	AESIVRSTVR
40	50	60
SHVNSDPTLA	AKILRMHFHD	CFVQGCDGSI
70	80	90
LISGPATEKT	AFANLGLRGY	EIIDDAKTQL
100	110	120
EAACPGVVSC	ADILALAARD	SVVLSGGLSY
130	140	150
QVPTGRRDGR	VSQASDVSNL	PAPSDSVDVQ
160	170	180
KQKFAAKGLN	TGDLVTLVGG	HTIGTSECQF
190	200	210
FSNRLFNFNG	TAAADPAIDP	SFVSNLQALC
220	230	240
PQNTGAANRV	ALDTGSQFKF	DTSYFSNLRN
250	260	270
RRGVLQSDQA	LWNDPSTKSF	VQRYLGLRGF
280	290	300
LGLTFNVEFG	KSMVKMSNIG	VKTGTDGEIR
307		
KICSAFN		

THE MAMMALIAN PEROXIDASE SUPERFAMILY

Human Myeloperoxidase

J. Zeng and R. E. Fenna, *J. Mol. Biol.* **226,** 185–207 (1992).

K. Morishita, N. Kubota, S. Asano, Y. Kaziro, and S. Nagata, *J. Biol. Chem.* **262,** 3844–3851 (1987).

1 10	20	30
CPEQDKYRTI	TGMCNNRRSP	TLGASNRAFV
40	50	60
RWLPAEYEDG	FSLPYGWTPG	VKRNGFPVAL
70	80	90
ARAVSNEIVR	FPTDQLTPDQ	ERSLMFMQWG
100	110	120
QLLDHDLDFT	PEPAARASFV	TGVNCETSCV
130	140	150
QQPPCFPLKI	PPNDPRIKNQ	ADCIPFFRSC

160	170	180
PACPGSNITI	RNQINALTSF	VDASMVYGSE
190	200	210
EPLARNLRNM	SNQLGLLAVN	QRFQDNGRAL
220	230	240
LPFDNLHDDP	CLLTNRSARI	PCFLAGDTRS
250	260	270
SEMPELTSMH	TLLLREHNRL	ATELKSLNPR
280	290	300
WDGERLWQEA	RKIVGAMVQI	ITYRDYLPLV
310	320	330
LGPTAMRKYL	PTYRSYNDSV	DPRIANVFTN
340	350	360
AFRYGHTLIQ	PFMFRLDNRY	QPMEPNPRVP
370	380	390
LSRVFFASWR	VVLEGGIDPI	LRGLMATPAK
400	410	420
LNRQNQIAVD	EIRERLFEQV	MRIGLDLPAL
430	440	450
NMQRSRDHGL	PGYNAWRRFC	GLPQPETVGQ
460	470	480
LGTVLRNLKL	ARKLMEQYGT	PNNIDIWMGG
490	500	510
VSEPLKRKGR	VGPLLACTIG	TQFRKLRDGD
520	530	540
RFWWENEGVF	SMQQRQALAQ	ISLPRIICDN
550	560	570
TGITTVSKNN	IFMSNSYPRD	FVNCSTLPAL
578		
NLASWREA		

Ram Prostaglandin H Synthase

D. L. DeWitt and W. L. Smith, *Proc. Natl. Acad. Sci. U.S.A.* **85,** 1412–1416 (1988).

1 10	20	30
ADPGAPAPVN	PCCYYPCQHQ	GICVRFGLDR
40	50	60
YQCDCTRTAI	PAPTAPSRRY	GPGSGRLCGP
70	80	90
APLSSTFMLT	HGRWLWDFVN	ATFIRDTLMR
100	110	120
LVLTVRSNLI	PSPPTYNIAH	DYISWESFSN
130	140	150
VSYYTRILPS	VPRDCPTPMG	TKGKKQLPDA

160	170	180
EFLSRRFLLR	RKFIPDPQGT	NLMFAFFAQH
190	200	210
FTHQFFKTSG	KMGPGFTKAL	GHGVDLGHIY
220	230	240
GDNLERQYQL	RLFKDGKLKY	QMLNGEVYPP
250	260	270
SVEEAPVLMH	YPRGIPPQSQ	MAVGQEVFGL
280	290	300
LPGLMLYATI	WLREHNRVCD	LLKAEHPTWG
310	320	330
DEQLFQTARL	ILIGETIKIV	IEEYVQQLSG
340	350	360
YFLQLKFDPE	LLFGAQFQYR	NRIAMEFNQL
370	380	390
YHWHPLMPDS	FRVGPQDYSY	EQFLFNTSML
400	410	420
VDYGVEALVD	AFSRQPAGRI	GGGRNIDHHI
430	440	450
LHVAVDVIKE	SRVLRLQPFN	EYRKRFGMKP
460	470	480
YTSFQELTGE	KEMAAELEEL	YGDIDALEFY
490	500	510
PGLLLEKCHP	NSIFGESMIE	MGAPFSLKGL
520	530	540
LGNPICSPEY	WKASTFGGEV	GFNLVKTATL
550	560	570
KKLVCLNTKT	CPYVSFHVPD	PRQEDRPGVE
576		
RPPTEL		

Bovine Lactoperoxidase

T. J. Dull, C. Uyeda, A. D. Strosberg, G. Nedwin and J. J. Seilhamer, *DNA Cell Biol.* **9,** 499–509 (1990).

M.-M. Cals, P. Mailliart, G. Brignon, P. Anglade, and B. R. Dumas, *Eur. J. Biochem.* **198,** 733–739 (1991).

1 10	20	30
DTTLTNVTDP	SLDLTALSWE	VGCGAPVPLV
40	50	60
KCDENSPYRT	ITGDCNNRRS	PALGAANRAL
70	80	90
ARWLPAEYED	GLALPFGWTQ	RKTRNGFRVP

100	110	120
LAREVSNKIV	GYLDEEGVLD	QNRSLLFMQW
130	140	150
GQIVDHDLDF	APETELGSNE	HSKTQCEEYC
160	170	180
IQGDNCFPIM	FPKNDPKLKT	QGKCMPFFA
190	200	210
GFVCPTPPYQ	SLAREQINAV	TSFLDASLVY
220	230	240
GSEPSLASRL	RNLSSPLGLM	AVNQEAWDHG
250	260	270
LAYLPFNNKK	PSPCEFINTT	ARVPCFLAGD
280	290	300
FRASEQILLA	TAHTLLLREH	NRLARELKKL
310	320	330
NPHWNGEKLY	QEARKILGAF	IQIITFRDYL
340	350	360
PIVLGSEMQK	WIPPYQGYNN	SVDPRISNVF
370	380	390
TFAFRFGHME	VPSTVSRLDE	NYQPWGPEAE
400	410	420
LPLHTLFFNT	WRIIKDGGID	PLVRGLLAKK
430	440	450
SKLMNQDKMV	TSELRNKLFQ	PTHKIHGFDL
460	470	480
AAINLQRCRD	HGMPGYNSWR	GFCGLSQPKT
490	500	510
LKGLQTVLKN	KILAKKLMDL	YKTPDNIDIW
520	530	540
IGGNAEPMVE	RGRVGPLLAC	LLGRQFQQIR
550	560	570
GDGRFWWENP	GVFTEKQRDS	LQKVSFSRLI
580	590	600
CDNTHITKVP	LHAFQANNYP	HDFVDCSTVD
610	612	
KLDLSPWASR	EN	

Human Thyroid Peroxidase

S. Kimura, T. Kotani, O. W. McBride, K. Umeki, K. Hirai, T. Nakayama, and S. Ohtaki, *Proc. Natl. Acad. Sci. U.S.A.* **84,** 5555–5559 (1987).

1 10	20	30
MRALAVLSVT	LVMACTEAFF	PFISRGKELL
40	50	60
WGKPEESRVS	SVLEESKRLV	DTAMYATMQR

70	80	90
NLKKRGILSP	AQLLSFSKLP	EPTSGVIARA
100	110	120
AEIMETSIQA	MKRLVNLKTQ	QSQHPTDALS
130	140	150
EDLLSIIANM	SGCLPYMLPP	KCPNTCLANK
160	170	180
YRPITGACNN	RDHPRWGASN	TALARWLPPV
190	200	210
YEDGFSQPRG	WNPGFLYNGF	PLPPVREVTR
220	230	240
HVIQVSNEVV	TDDDRYSDLL	MAWGQYIDHD
250	260	270
IAFTPQSTSK	AAFGGGADCQ	MTCENQNPCF
280	290	300
PIQLPEEARP	AAGTACLPFY	RSSAACGTGD
310	320	330
QGALFGQLST	ANPRQQMNGL	TSFLDASTVY
340	350	360
GSSPALDRQL	RNWTSADGLL	RVHARLRDSG
370	380	390
RAYLPFVPPR	RPAACAPQPG	IPGETRGPCF
400	410	420
LAGDGRASEV	PSLTALHTLW	LREHNRLAAA
430	440	450
LKALNAHWSA	DAVYQEARKV	VGALHQIITL
460	470	480
RDYIPRILGP	EAFQQYVGPY	EGYDSTANPT
490	500	510
VSNVFSTAAF	RFGHATIHPL	VRRLDASFQE
520	530	540
HPDLPGLWLH	QAFFSPWTLL	RGGGLDPLIR
550	560	570
GLLARPAKLQ	VQDQLMNEEL	TERLFVLSNS
580	590	600
STLDLASINL	QRGRDHGLPG	YNEWREFCGL
610	620	630
PRLETPADLS	TAIASRSVAD	KILDLYKHPD
640	650	660
NIDVWLGGLA	ENFLPRARTG	PLFACLIGKQ
670	680	690
MKALRDGDWF	WWENSHVFTD	AQRRELEKHS
700	710	720
LSRVICDNTG	LTRVPMDAFQ	VGKFPEDFES
730	740	750
CDSIPGMNLE	AWRETFPQDD	KCGFPESVEN

760	770	780
GDFVHCEESG	RRVLVYSCRH	GYELQGREQL
790	800	810
TCTQEGWDFQ	PPLCKDVNEC	ADGAHPPCHA
820	830	840
SARCRNTKGG	FQCLCADPYE	LGDDGRTCVD
850	860	870
SGRLPRATWI	SMSLAALLIG	GFAGLTSTVI
880	890	900
CRWTRTGTKS	TLPISETGGG	TPELRCGKHQ
910	920	930
AVGTSPQRAA	AQDSEQESAG	MEGRDTHRLP
933		
RAL		

Porcine Thyroid Peroxidase

R. P. Magnusson, J. Gestautas, A. Taurog, and B. Rapoport, *J. Biol. Chem.* **262,** 13885–13888 (1987).

1 10	20	30
MGARAVLGVT	LAVACAGAFF	ASILRRKDLL
40	50	60
GGDTEASGVA	GLVEASRLLV	DEAIHTTMRR
70	80	90
NLRKRGIFSP	SQLLSFSKLP	EPTSRTASRA
100	110	120
AEIMETAVQE	VKRRVCRRRD	TDQLPTDVLS
130	140	150
EELLSTIAQL	SGCLPHMLPP	SCPHTCLANK
160	170	180
YRLITGACNN	RDHPRWGASN	TALARWLPPA
190	200	210
YEDGVTEPRG	WNPHFLYNGL	PLPPVREVTR
220	230	240
QVIHVSNEAV	TEDGQYSDLL	MAWGQYIDHD
250	260	270
IAFTPQSTSK	AAFAGGADCQ	LTCEQRSPCF
280	290	300
PIQLPTQASG	AAGATCLPFY	RSSAACGSGR
310	320	330
QGALVGQLSW	AAPRQQMNGL	TSFLDASTVY
340	350	360
GSSPAQEQRL	RNWTSAEGLL	RVNTRHRDAG
370	380	390
RAFLPFAPPP	APPACAPEPG	TPAARAPCFL

400	410	420
AGDSRASEVP	GLTALHTLWL	REHNRLAAAF
430	440	450
KALNAHWSAD	TVYQEARKVV	GALHQIVTLR
460	470	480
DYVPKILGAE	AFGQHVGPYQ	GYDPAVDPTV
490	500	510
SNVFSTAAFR	FGHATIHPLV	RRLDARFQEH
520	530	540
PGSHLPLRAA	FFQPWRLLRE	GGVDPVLRGL
550	560	570
LARPAKLQVQ	DQLMNEELTE	RLFVLSNSGT
580	590	600
LDLASINLQR	GRDHGLPGYN	EWREFCGLSR
610	620	630
LETWADLSAA	TANGRVADRI	DVWLGGLAES
640	650	660
FLPGARTGPL	FACIIGKQMR	ALRDGDRFWW
670	680	690
ENPGVFTEAQ	RRELSRHSMS	RVICDNSGLS
700	710	720
HVPLDAFRVG	QWPQEFEPCA	SIQGMDLGAW
730	740	750
REAPPSGDAC	GFPDPVEDGG	FLLCEERGQR
760	770	780
VLVFSCRHGF	RLRGPAQITC	TPRGWDSPPP
790	800	810
LCKDINECED	ETDPPCHASA	RCKNTKGGVL
820	830	840
CECSDPLVLG	EDGRTCVDAG	RLPRASVVSI
850	860	870
ALGAVLVCGL	AGLAWTVVCR	WTHADARPLL
880	890	900
PVGEGEGDGK	SPSLPLPGCG	NRRDVGAAPA
910	916	
LEVEQDLSCG	SRGLCE	

Human Eosinophil Peroxidase

K. Sakamaki, M. Tomonaga, K. Tsukui, and S. Nagata, *J. Biol. Chem.* **264,** 16828–16836 (1989).

1 10	20	30
RCSDKYRTIT	GRCNNKRRPL	LGASNQALAR
40	50	60
WLPAEYEDGL	SLPFGWTPSR	RRNGFLLPLV

70	80	90
RAVSNQIVRF	PNERLTSDRG	RALMFMQWGQ
100	110	120
FIDHDLDFSP	ESPARVAFTA	GVDCERTCAQ
130	140	150
LPPCFPIKIP	PNDPRIKNQR	DCIPFFRSAP
160	170	180
SCPQNKNRVR	NQINALTSFV	DASMVYGSEV
190	200	210
SLSLRLRNRT	NYLGLLAINQ	RFQDNGRALL
220	230	240
PFDNLHDDPC	LLTNRSARIP	CFLAGDTRST
250	260	270
ETPKLAAMHT	LFMREHNRLA	TELRRLNPRW
280	290	300
NGDKLYNEAR	KIMGAMVQII	TYRDFLPLVL
310	320	330
GKARARRTLG	HYRGYCSNVD	PRVANVFTLA
340	350	360
FRFGHTMLQP	FMFRLDSQYR	ASAPNSHVPL
370	380	390
SSAFFASWRI	VYEGGIDPIL	RGLMATPAKL
400	410	420
NRQDAMLVDE	LRDRLPRQVR	RIGLDLAALN
430	440	450
MQRSRDHGLP	GYNAWRRFCG	LSQPRNLAQL
460	470	480
SRVLKNQDLA	RKFLNLYGTP	DNIDIWIGAI
490	500	510
AEPLLPGARV	GPLLACLFEN	QFRRARDGDR
520	530	540
FWWQKRGVFT	KRQRKALSRI	SLSRIICDNT
550	560	570
GITTVSRDIF	RANIYPRGFV	NCSRIPRLNL
576		
SAWRGT		

THE CATALASE SUPERFAMILY

Bovine Liver Catalase

W. A. Schroeder, J. R. Shelton, J. B. Shelton, B. Robberson, G. Apell, R. S. Fang, and J. Bonaventura, *Arch. Biochem. Biophys.* **214,** 397–421 (1982).

1 10	20	30
ADNRDPASDQ	MKHWKEQRAA	QKPDVLTTGG
40	50	60
GNPVGDKLNS	LTVGPRGPLL	VQDVVFTDEM
70	80	90
AHFDRERIPE	RVVHAKGAGA	FGYFEVTHDI
100	110	120
TRYSKAKVFE	HIGKRTPIAV	RFSTVAGESG
130	140	150
SADTVRDPRG	FAVKFYTEDG	NWDLVGNNTP
160	170	180
IFFIRDALLF	PSFIHSQKRN	PQTHLKDPDM
190	200	210
VWDFWSLRPE	SLHQVSFLFS	DRGIPDGHRH
220	230	240
MDGYGSHTFK	LVNADGEAVY	CKFHYKTDQG
250	260	270
IKNLSVEDAA	RLAHEDPDYG	LRDLFNAIAT
280	290	300
GNYPSWTLYI	QVMTFSEAEI	FPFNPFDLTK
310	320	330
VWPHGDYPLI	PVGKLVLNRN	PVNYFAEVEQ
340	350	360
LAFDPSNMPP	GIEPSPDKML	QGRLFAYPDT
370	380	390
HRHRLGPNYL	QIPVNCPYRA	RVANYQRDGP
400	410	420
MCMMDNQGGA	PNYYPNSFSA	PEHQPSALEH
430	440	450
RTHFSGDVQR	FNSANDDNVT	QVRTFYLKVL
460	470	480
NEEQRKRLCE	NIAGHLKDAQ	LFIQKKAVKN
490	500	506
FSDVHPEYGS	RIQALLDKYN	EEKPKN

Caldariomyces fumago Chloroperoxidase

M. J. Nuell, G-H. Fang, M. J. Axley, P. Keningsberg, and L. P. Hager, *J. Bacteriol.* **170,** 1007–1011 (1988).

M. Sundaramoorthy, J. Terner, and T. L. Poulos, *Structure* **3,** 1367–1377 (1995).

The numbering starts from 0, which is a pyroglutamic acid residue. Thus, the total sequence consists of 299 residues, numbered from 0 to 298.

10	20	30
EPGSGIGYPY	DNNTLPYVAP	GPTDSRRPCP
40	50	60
ALNALANHGY	IPHDGRAISR	ETLQNAFLNH
70	80	90
MGIANSVIEL	ALTNAFVVCE	YVTGSDCGDS
100	110	120
LVNLTLLAEP	HAFEHDHSFS	RKDYKQGVAN
130	140	150
SNDFIDNRNF	DAETFQTSLD	VVAGKTHFDY
160	170	180
ADMNEIRLQR	ESLSNELDFP	GWFTESKPIQ
190	200	210
NVESGFIFAL	VSDFNLPDND	ENPLVRIDWW
220	230	240
KYWFTNESFP	YHLGWHPPSP	AREIEFVTSA
250	260	270
SSAVLAASVT	STPSSLPSGA	IGPGAEAVPL
280	290	298
SFASTMTPFL	LATNAPYYAQ	DPTLGPND

Pseudomonas aeruginosa Cytochrome c Peroxidase

C. J. Ridout, R. James, and C. Greenwood, *FEBS Lett.* **365,** 152–154 (1995).

1 10	20	30
MQSSQLLPLG	SLLLSFATPL	AQADALHDQA
40	50	60
SALFKPIPEQ	VTELRGQPIS	EQQRELGKKL
70	80	90
FFDPRLSRSH	VLSCNTCHNV	GTGGADNVPT
100	110	120
SVGHGWQKGP	RNSPTVFNAV	FNAAQFWDGR
130	140	150
AKDLGEQAKG	PIQNSVEMHS	TPQLVEQTLG
160	170	180
SIPEYVDAFR	KAFPKAGKPV	SFDNMALAIE
190	200	210
AYEATLVTPD	SPFDLYLKGD	DKALDAQQKK
220	230	240
GLKAFMDSGC	SACHNGINLG	GQAYFPFGLV
250	260	270
KKPDASVLPS	GDKGRFAVTK	TQSDEYVFRA

280	290	300
APLRNVALTA	PYFHSGQVWE	LKDAVAIMGN
310	320	330
AQLGKQLAPD	DVENIVAFLH	SLSGKQPRVE
340	346	
YPLLPASTET	TPRPAE	

INDEX